Introduction to Bayesian Data Analysis for Cognitive Science

This book introduces Bayesian data analysis and Bayesian cognitive modeling to students and researchers in cognitive science (e.g., linguistics, psycholinguistics, psychology, computer science), with a particular focus on modeling data from planned experiments. The book relies on the probabilistic programming language Stan and the R package brms, which is a front-end to Stan. The book only assumes that the reader is familiar with the statistical programming language R, and has basic high school exposure to pre-calculus mathematics; some of the important mathematical constructs needed for the book are introduced in the first chapter.

Through this book, the reader will be able to develop a practical ability to apply Bayesian modeling within their own field. The book begins with an informal introduction to foundational topics such as probability theory, and univariate and bi-/multivariate discrete and continuous random variables. Then, the application of Bayes' rule for statistical inference is introduced with several simple analytical examples that require no computing software; the main insight here is that the posterior distribution of a parameter is a compromise between the prior and the likelihood functions. The book then gradually builds up the regression framework using the brms package in R, ultimately leading to hierarchical regression modeling (aka the linear mixed model). Along the way, there is detailed discussion about the topic of prior selection, and developing a well-defined workflow. Later chapters introduce the Stan programming language, and cover advanced topics using practical examples: contrast coding, model comparison using Bayes factors and cross-validation, hierarchical models and reparameterization, defining custom distributions, measurement error models and meta-analysis, and finally, some examples of cognitive models: multinomial processing trees, finite mixture models, and accumulator models. Additional chapters, appendices, and exercises are provided as online materials and can be accessed here: https://github.com/bnicenboim/bayescogsci.

Introduction to Bayesian Data Analysis for Cognitive Science

Bruno Nicenboim, Daniel J. Schad, and Shravan Vasishth

CRC Press
Taylor & Francis Group
Boca Raton London New York

CRC Press is an imprint of the
Taylor & Francis Group, an **informa** business
A CHAPMAN & HALL BOOK

First edition published 2026
by CRC Press
2385 NW Executive Center Drive, Suite 320, Boca Raton FL 33431

and by CRC Press
4 Park Square, Milton Park, Abingdon, Oxon, OX14 4RN

CRC Press is an imprint of Taylor & Francis Group, LLC

ISBN: 978-0-367-35851-8 (hbk)
ISBN: 978-0-367-35933-1 (pbk)
ISBN: 978-0-429-34264-6 (ebk)

DOI: 10.1201/9780429342646

Typeset in CMR10 font
by KnowledgeWorks Global Ltd.

Publisher's Note: This book has been prepared from camera-ready copy provided by the author.

This book is dedicated to the cognitive science community.

Contents

Preface

This book is intended to be a relatively gentle introduction to carrying out Bayesian data analysis and cognitive modeling using the probabilistic programming language Stan (Carpenter et al. 2017) and the front-end to Stan called `brms` (Bürkner 2024). Our target audience is cognitive scientists (e.g., linguists, psychologists, and computer scientists) who carry out planned behavioral experiments and are interested in learning the Bayesian data analysis methodology from the ground up and in a principled manner. Our aim is to make Bayesian statistics a standard part of the data analysis toolkit for experimental linguistics, psycholinguistics, psychology, and related disciplines.

Many excellent introductory textbooks already exist for Bayesian data analysis. Why write yet another book? Our text is different from other attempts in two respects. First, our main focus is on showing how to analyze data from planned experiments involving repeated measures; this type of experimental data involves unique complexities. We provide many examples of data sets involving time measurements (e.g., self-paced reading, eye-tracking while reading, voice onset time), event-related potentials, pupil sizes, accuracies (e.g., recall tasks, yes-no questions), categorical answers (e.g., picture naming), choice-reaction time (e.g, the Stroop task, the motion detection task), etc. Second, from the very outset, we emphasize a particular workflow that has as its centerpiece simulating data; we aim to teach a philosophy that involves thinking hard about the assumed underlying generative process, **even before the data are collected**. The data analysis approach that we hope to teach through this book involves a cycle of prior predictive and posterior predictive checks, sensitivity analyses, and model validation using simulated data. We try to inculcate a sense of how inferences can be drawn from the posterior distribution of theoretically interesting parameters without resorting to binary decisions like "significant" or "not-significant." We are hopeful that this will set a new standard for reporting and interpreting results of data analyses in a more nuanced manner, and lead to more measured claims in the published literature.

Please report typos, errors, or suggestions for improvement at `https://github.com/bnicenboim/bayescogsci/issues`.

Why read this book, and what is its target audience?

A commonly held belief in psychology, psycholinguistics, and other areas is that statistical data analysis is secondary to the science and should be quick and easy. For example, a senior mathematical psychologist once told the last author of this book: "if you need to run anything more complicated than a paired t-test, you are asking the wrong question." We take a different perspective here: the science and the statistical modeling are one unitary thing. The statistical model should represent some reasonable approximation of the latent cognitive processes that are assumed to be in play.

The target audience for this book is students and researchers who want to treat statistics as an equal partner in their scientific work. We expect that the reader is willing to take the time to both understand and run the computational analyses.

Any rigorous introduction to Bayesian data analysis requires at least a passive knowledge of probability theory, calculus, and linear algebra. However, we do not require that the reader has this background when they start the book. Instead, the relevant ideas are introduced informally and just in time, as soon as they are needed. The reader is never required to have an active ability to solve probability problems, to solve integrals or compute derivatives, or to carry out matrix computations (such as inverting matrices) by hand. There are a few places where the discussion becomes technical and requires some knowledge of calculus or related topics. However, the reader unfamiliar with the required mathematics can simply skip those sections, as these are not really necessary to follow the main thread of the book.

What we do expect is familiarity with arithmetic, basic set theory and elementary probability theory (e.g., sum and product rules, conditional probability), simple matrix operations like addition and multiplication, and simple algebraic operations. A quick look through chapter 1 of Gill (2006) before starting this book is highly recommended. We also presuppose that when the need arises, the reader is willing to look up concepts that they might have forgotten (e.g., logarithms). We also provide a self-paced course on some basic concepts in mathematics (aimed at non-mathematicians) which the reader can work through: see `https://vasishth.github.io/FoM/`.

We also expect that the reader already knows and/or is willing to learn enough of the programming language R (R Core Team 2023) to reproduce the examples presented and to carry out the exercises. If the reader is completely unfamiliar with R, before starting this book they should first consult books like R for data science (`https://r4ds.had.co.nz/`), and Efficient R programming (`https://csgillespie.github.io/efficientR/`). Readers familiar with python may find the python port of the first five chapters by Jozsef Arato useful (`https://github.com/jozsarato/bayescogdat`).

We also assume that the reader has encountered simple linear modeling and linear mixed models (Bates et al. 2015; Baayen, Davidson, and Bates 2008). What this means in practice is that the reader should have used the `lm()` and `lmer()` functions in R. A passing acquaintance with basic statistical concepts, like the correlation between two variables, is also taken for granted.

This book is not appropriate for complete beginners to data analysis. Newcomers to data analysis should start with a freely available textbook like Kerns (2014), and then read our introduction to frequentist data analysis, which is also available freely online (Vasishth, Schad, et al. 2021). The latter book will prepare the reader well for the material presented here.

Developing the right mindset for this book

One very important characteristic that the reader should bring to this book is a can-do spirit. There will be many places where the going will get tough, and the reader will have to slow down and play around with the material, or refresh their understanding of arithmetic or middle-school algebra. The basic principles of such a can-do spirit are nicely summarized in the book by Burger and Starbird (2012); also see Levy (2021). Although we cannot summarize all the insights from these books in a few words, inspired by the Burger and Starbird (2012) book, here is a short enumeration of the kind of mindset the reader will need to cultivate:

- Spend time on the basic, apparently easy material; make sure you understand it deeply. Look for gaps in your understanding. Reading different presentations of the same material (in different books or articles) can yield new insights.
- Let mistakes and errors be your teacher. We instinctively recoil from our mistakes, but errors are ultimately our friends; they have the potential to teach us more than our correct answers can. In this sense, a correct solution can be less interesting than an incorrect one.
- When you are intimidated by some exercise or problem, give up and admit defeat immediately. This relaxes the mind; you've already given up, there's nothing more to do. Then, after a while, try to solve a simpler version of

the problem. Sometimes, it is useful to break the problem down into smaller parts, each of which may be easier to solve.
- Create your own questions. Don't wait to be asked questions; develop your own problems and then try to solve them.
- Don't expect to understand everything in the first pass. Just mentally note the gaps in your understanding, and return to them later and work on these gaps.
- Step back periodically to try to sketch out a broader picture of what you are learning. Writing down what you know, without looking up anything, is one helpful way to achieve this. Don't wait for the teacher to give you bullet-point summaries of what you should have learned; develop such summaries yourself.
- Develop the art of finding information. When confronted with something you don't know, or with some obscure error message, use Google to find some answers.
- Do not hesitate to re-read a chapter; often, one only understands a topic after one revisits the material.

As instructors, we have noticed over the years that students with such a mindset generally do very well. Some students already have that spirit, but others need to explicitly develop it. We firmly believe that everyone can develop such a mindset, but one may have to work on acquiring it. In any case, such an attitude is necessary for a book of this sort.

How to read this book

The chapters in this book are intended to be read in sequence, but during the first pass through the book, readers should feel free to skip the optional, in-depth materials available online. These resources provide a more formal development (useful to transition to more advanced textbooks like Gelman et al. 2014) or deal with tangential aspects of the topics presented in the chapter.

Here are some suggested paths through this book, depending on the reader's goals:

- For a short course for complete beginners, read chapters 1 to 5. We usually cover these five chapters in a five-day summer school course that we teach annually. Most of the material in these chapters are also covered in a free four-week course available online: `https://open.hpi.de/courses/bayesian-statistics2023`.

- For a course that focuses on regression models with the R package `brms`, read chapters 1 to 7 and, optionally, 13.
- For an advanced course that focuses on complex models involving Stan, read chapters 8 to 18.

Some conventions used in this book

We adopt the following conventions:

- All distribution names are lower-case unless they are also a proper name (e.g., Poisson, Bernoulli).
- The univariate normal distribution is parameterized by the mean and standard deviation (not variance).
- The code for figures is provided only in some cases, where we consider it to be pedagogically useful. In other cases, the code remains hidden, but it can be found in the web version of the book (`https://bruno.nicenboim.me/bayescogsci/`). Notice that all the R code from the book can be extracted from the Rmd source files for each chapter, which are released with the book (`https://github.com/bnicenboim/bayescogsci`).

Online materials

The entire book, including all data and source code, is available online for free at `https://bruno.nicenboim.me/bayescogsci/`. Additional optional materials and exercises for each chapter are also available online. Solutions to exercises can be provided upon request.

Software needed

Before you start, please install

- R (`https://cran.r-project.org/`) and RStudio (`https://www.rstudio.com/`), or any other Integrated Development Environment that you prefer, such as Visual Studio Code (`https://code.visualstudio.com/`) and Emacs Speaks Statistics (`https://ess.r-project.org/`).
- The R package `rstan`. At the time of writing this book, the CRAN version of `rstan` lags behind the latest developments in Stan so it is recommended

to install `rstan` from `https://mc-stan.org/r-packages/` as indicated in `https://github.com/stan-dev/rstan/wiki/RStan-Getting-Started`

- The R packages `dplyr`, `purrr`, `tidyr`, `extraDistr`, `brms`, `hypr`, and `lme4` are used in many chapters of the book and can be installed the usual way: `install.packages(c("dplyr","purrr","tidyr", "extraDistr", "brms","hypr","lme4"))`. We use `ggplot2` for figures; please consult the relevant documentation if you are unfamiliar with `ggplot2` (e.g., Wickham, Chang, et al. 2024).
- The following R packages are optional: `tictoc`, `rootSolve`, `SHELF`, `cmdstanr`, and `SBC`.
- Some packages and their dependencies, such as `intoo`, `barsurf`, `bivariate`, and `SIN` could require manual installation from archived or github versions. For such packages, go to the CRAN archive: `https://cran.r-project.org/src/contrib/Archive/`; download the tar.gz of the relevant package; and then install it from the command line using the command `R CMD INSTALL package.tar.gz`.
- The data and Stan models used in this book can be installed using `remotes::install_github("bnicenboim/bcogsci")`. This command uses the function `install_github()` from the package `remotes`. (Thus this package should be in the system as well.)

In every R session, load these packages, and set the options shown below for Stan.

```
library(MASS)
## be careful to load dplyr after MASS
library(dplyr)
library(tidyr)
library(purrr)
library(extraDistr)
library(ggplot2)
library(loo)
library(bridgesampling)
library(brms)
library(bayesplot)
library(tictoc)
library(hypr)
library(bcogsci)
library(lme4)
library(rstan)
# This package is optional, see https://mc-stan.org/cmdstanr/:
library(cmdstanr)
```

```r
# This package is optional, see https://hyunjimoon.github.io/SBC/:
library(SBC)
library(SHELF)
library(rootSolve)

## Save compiled models:
rstan_options(auto_write = FALSE)
## Parallelize the chains using all the cores:
options(mc.cores = parallel::detectCores())
# To solve some conflicts between packages:
select <- dplyr::select
extract <- rstan::extract
```

Acknowledgments

We are grateful to the many generations of students at the University of Potsdam, various summer schools at ESSLLI, the LOT winter school, the Open HPI MOOC on Bayesian statistics (`https://open.hpi.de/courses/bayesian-statistics2023`), other short courses we have taught at various institutions, and the annual summer school on Statistical Methods for Linguistics and Psychology (SMLP) (`https://vasishth.github.io/smlp/`) held annually at Potsdam, Germany. The participants in these courses helped us considerably in improving the material presented here. A special thanks to Anna Laurinavichyute, Paula Lissón, and Himanshu Yadav for co-teaching the Bayesian courses at SMLP. We are also grateful to members of Vasishth lab, especially Dorothea Pregla, for comments on earlier drafts of this book. We would also like to thank Douglas Bates, Ben Bolker, Christian Robert (otherwise known as Xi'an), Robin Ryder, Nicolas Chopin, Michael Betancourt, Andrew Gelman, the Stan developers (especially Bob Carpenter and Paul-Christian Bürkner), Philip D. Loewen, and Leendert Van Maanen for their comments and advice; to Pavel Logačev for his feedback, and Athanassios Protopapas, Patricia Mirabile, Masataka Ogawa, Alex Swiderski, Andrew Ellis, Jakub Szewczyk, Chi Hou Pau, Alec Shaw, Patrick Wen, Riccardo Fusaroli, Abdulrahman Dallak, Elizabeth Pankratz, João Veríssimo, Jean-Pierre Haeberly, Chris Hammill, Florian Wickelmaier, Ole Seeth, Jules Bouton, Siqi Zheng, Michael Gaunt, Benjamin Senst, Chris Moreh, Richard Hatcher, Noelia Stetie, Robert Lew, Leonardo Cerliani, Stefan Riedel, Raluca Rilla, Arne Schernich, Sven Koch, Joy Sarow, Iñigo Urrestarazu-Porta, Jan Winkowski, Adrian Staub, Brian Dillon, Job Schepens, Katja Politt, Cui

Ding, Marc Tortorello, Michael Vrazitulis, Marisol Murujosa, Carla Bombi Ferrer, and Ander Egurtzegi for catching typos, unclear passages, and errors in the book. A special thanks to Daniel Heck, Alexandre Cremers, Henrik Singmann, and Martin Modrák for reading (parts of) the book and catching many mistakes and typos. Our apologies to anyone that we have forgotten to mention. Thanks also go to Jeremy Oakley and other statisticians at the School of Mathematics and Statistics, University of Sheffield, UK, for helpful discussions, and ideas for exercises that were inspired from the MSc program taught online at Sheffield.

Finally, we extend our heartfelt thanks to Iohanna, Oliver, Luc, Milena, Luisa, Andrea, and Atri for their love and support throughout the long process of writing this book.

This book would have been impossible to write without the following software: R (Version 4.3.2; R Core Team 2023) and the R-packages *afex* (Singmann et al. 2020), *barsurf* (Version 0.7.0; Spurdle 2020), *bayesplot* (Version 1.11.1; Gabry and Mahr 2024), *bcogsci* (Version 0.0.0.9000; Nicenboim, Schad, and Vasishth 2024), *bibtex* (Francois 2017), *bivariate* (Version 0.7.0; Spurdle 2021), *bookdown* (Version 0.39; Xie 2024a), *brms* (Version 2.21.0; Bürkner 2024), *citr* (Aust 2019), *cmdstanr* (Version 0.8.0; Gabry et al. 2024), *cowplot* (Version 1.1.3; Wilke 2024), *digest* (Version 0.6.35; Antoine Lucas et al. 2021), *dplyr* (Version 1.1.4; Wickham, François, et al. 2023), *DT* (Xie, Cheng, and Tan 2019), *extraDistr* (Version 1.10.0; Wolodzko 2023), *forcats* (Wickham 2019a), *gdtools* (Gohel et al. 2019), *ggplot2* (Version 3.5.1; Wickham, Chang, et al. 2024), *gridExtra* (Version 2.3; Auguie 2017), *htmlwidgets* (Version 1.6.4; Vaidyanathan et al. 2023), *intoo* (Version 0.4.0; Spurdle and Bode 2020), *kableExtra* (Version 1.4.0; Zhu 2024), *knitr* (Version 1.47; Xie 2024b), *lme4* (Version 1.1.35.3; Bates et al. 2015), *loo* (Version 2.7.0; Yao et al. 2017), *MASS* (Version 7.3.60; Ripley 2023), *Matrix* (Version 1.6.1.1; Bates, Maechler, and Jagan 2023), *miniUI* (Version 0.1.1.1; Cheng 2018), *papaja* (Version 0.1.2; Aust and Barth 2020), *pdftools* (Version 3.4.0; Ooms 2023), *purrr* (Version 1.0.2; Wickham and Henry 2023), *Rcpp* (Version 1.0.12; Eddelbuettel et al. 2024), *readr* (Wickham, Hester, and Francois 2018), *RefManageR* (McLean 2017), *remotes* (Version 2.5.0; Hester et al. 2021), *rethinking* (Version 2.40; McElreath 2021), *RJ-2021-048* (Bengtsson 2021), *rmarkdown* (Version 2.27; Allaire et al. 2024), *rootSolve* (Version 1.8.2.4; Soetaert and Herman 2009), *rstan* (Version 2.35.0.9000; Guo et al. 2024), *SBC* (Version 0.3.0.9000; Kim et al. 2024), *servr* (Xie 2019), *SHELF* (Version 1.10.0; Oakley 2024), *SIN* (Version 0.6; Drton 2013), *StanHeaders* (Version 2.35.0.9000; Goodrich et al. 2024), *stringr* (Version 1.5.1; Wickham 2019b), *texPreview* (Sidi and Polhamus 2020), *tibble* (Version 3.2.1; Müller and Wickham 2020), *tictoc* (Version 1.2.1; Izrailev 2024), *tidyr* (Version 1.3.1; Wickham, Vaughan, and Girlich 2024), *tidyverse* (Wickham, Averick, et al. 2019), *tinylabels* (Version 0.2.4; Barth 2023), and *webshot* (Chang 2018).

Bruno Nicenboim (Tilburg, the Netherlands), Daniel Schad (Potsdam, Germany), and Shravan Vasishth (Potsdam, Germany)

About the Authors

Bruno Nicenboim (`https://bruno.nicenboim.me`) is an assistant professor in the department of Cognitive Science and AI at Tilburg University, the Netherlands, where he is the principal investigator (PI) of the Computational Psycholinguistics research line. He started studying Electronic Engineering at the National University of Rosario, Argentina, then transitioned to Human Sciences and spent eight years in Israel, where he completed a Bachelor's degree in Sociology and Linguistics and a Master's degree in Linguistics at Tel Aviv University. During this time, he also worked in several IT companies. He then moved to Germany, where he completed a PhD in Cognitive Science at the University of Potsdam and worked for two years as a postdoctoral researcher. His research interests include Bayesian methods, computational cognitive modeling, sentence comprehension, memory processes, decision-making, and predictive processing. He regularly teaches short courses on Bayesian data analysis at workshops and summer schools, and he teaches Bayesian modeling and statistics courses in the master's programs Cognitive Science & AI and Data Science and Society at Tilburg University.

Daniel J. Schad (`https://danielschad.github.io/`) is professor of Quantitative Methods in the Psychology department at the HMU Health and Medical University, at Potsdam, Germany. He studied Psychology at the University of Potsdam, Germany, and at the University of Michigan, Ann Arbor, USA. He did a PhD in Cognitive and Mathematical Psychology at the University of Potsdam, working on computational models of eye-movement control and on mindless reading. He then did a five-year post-doc in the field of Computational Psychiatry at the Charité - Universitätsmedizin Berlin, Germany (partly also at the University of Potsdam), with research visits at the ETH Zürich, Switzerland, and University College London, UK, working on model-free and model-based decision-making and Pavlovian-instrumental transfer in alcohol dependence and on the cognitive and brain mechanisms underlying Pavlovian conditioning. He has worked as a postdoctoral researcher at the University of Potsdam, conducting research on quantitative methods in Cognitive Science, including contrasts, properties of significance tests, Bayesian Workflow, and Bayes factor analyses, and has been assistant professor at the Department of Cognitive Science and AI at Tilburg University.

Shravan Vasishth (`http://vasishth.github.io`) is professor of Psycholinguistics at the University of Potsdam, Germany. He holds the chair for Psycholinguistics and Neurolinguistics (Language Processing). After completing his Bachelor's degree in Japanese from Jawaharlal Nehru University, New Delhi, India, he spent a total of five years in Osaka, Japan. During his time in Japan, he studied Japanese at the Osaka University of Foreign Studies, conducted research at Osaka University, and worked as an in-house translator in a patent law firm in Osaka. He completed an MS in Computer and Information Science (2000–2002) and a PhD in Linguistics (1997–2002) from the Ohio State University, Columbus, Ohio, USA, and an MSc in Statistics (2011–2015) from the School of Mathematics and Statistics, University of Sheffield, UK. He is a chartered statistician (the Royal Statistical Society, UK; id 128307). He is on the editorial board of the open access journal *Glossa Psycholinguistics*. His research focuses on computational modeling of sentence processing in unimpaired and impaired populations, and the application of mathematical, computational, experimental, and statistical methods (particularly Bayesian methods) in linguistics and psychology. He created an annual summer school, Statistical Methods in Linguistics and Psychology (SMLP), which started in 2017: vasishth.github.io/smlp. He regularly teaches short courses on statistical data analysis (Bayesian and frequentist methods).

Part I

Foundational ideas

1

Introduction

The central idea we will explore in this book is how to use Bayes' theorem to quantify uncertainty about our belief regarding a scientific question of interest, given some data. Before we delve into the details of the underlying theory and its application, it is important to have some familiarity with the following topics: basic concepts of probability, the concept of random variables, probability distributions, and the concept of likelihood. Therefore, we will begin with these topics.

Some of these concepts might seem abstract at first, but they are very relevant for conducting a Bayesian analysis. When reading this book for the first time, it might be helpful to do a quick pass through this chapter and return to it as needed while progressing through the rest of the book.

1.1 Probability

Informally, we all understand what the term *probability* means. We routinely talk about things like the probability of it raining today. However, there are two distinct ways to think about probability. One can think of the probability of something happening with reference to the frequency with which it might occur in repeated observations. Such a conception of probability is easy to imagine in cases where something can, at least in principle, occur repeatedly.

An example would be obtaining a 6 when tossing a die again and again. However, this frequentist view of probability is difficult to justify when talking about one-of-a-kind things, such as earthquakes; here, probability is expressing our uncertainty about the earthquake happening.

Both the frequency-based and the uncertain-belief perspective have their place in statistical inference, and depending on the situation, we are going to rely on both ways of thinking.

The statements below are not formal definitions of the axioms of probability theory; for more details (and more precise formulations), see Blitzstein and Hwang (2014), Ross (2002), Kerns (2014), Resnick (2019), or Kolmogorov

DOI: 10.1201/9780429342646-2

(1933) (among many other books). Keep in mind that different textbooks have slightly different ways of presenting the underlying structure of what constitutes a probability space (defined below).

The probability of something happening is defined to be constrained in the way described below. A concrete example of "something happening" is an outcome–call it ω–such as obtaining nine correct (c) answers and one incorrect (i) when we ask a subject 10 yes-no questions (say, about the meaning of a sentence). An example would be a sequence of correct (c) and incorrect (i) answers, such as $iiiiciiiii$. Another possible outcome is $cciicicici$. The outcomes are thus all possible sequences of correct and incorrect answers, and the sample space, Ω, is the set of all possible outcomes.

Another important concept is events, E, next; the event corresponding to obtaining one correct answer when 10 questions are asked, is the subset of outcomes $\{ciiiiiiiii, iciiiiiiii, iiciiiiiii, \dots\}$; in other words, the event corresponding to obtaining one correct answer is a set containing 10 elements. Similarly, the event corresponding to obtaining nine correct answers when 10 questions are asked is the subset of outcomes $\{ccccccccci, cccccccccic, \dots\}$.

When we conduct an experiment, if we get a particular outcome like $ciiiiiiiii$, then we say that the event $\{ciiiiiiiii, iciiiiiiii, iiciiiiiii, \dots\}$ occurred. A probability space involves two more key ingredients: A collection of subsets of Ω denoted by F and called the *event space*;[1] and a real-valued function named P that assigns a number (a probability) to each set in F.

Table 1.1 summarizes the notation.

The probability axioms refer to the sample space Ω, event space F, and probability P as follows:

- For every event E in the event space F, the probability $P(E)$ is a real number between 0 and 1.
- The event $E = \Omega$ belongs to F, and $P(\Omega) = 1$.
- If the events $A_1, A_2, A_3, \dots$ are mutually exclusive (in other words, if no two of these subsets of Ω overlap), then the probability of the event "one of A_1

[1]When the sample space Ω is finite, any subset of Ω can be an event; in this case, the event space F is the collection of all possible subsets of Ω. In set theory, a set of all the subsets of another set is called a power set, and the number of subsets of any set with n elements is 2^n. For example, for a standard six-sided die, the sample space is the set $\Omega = \{1, 2, 3, 4, 5, 6\}$ and the event space F will contain $2^6 = 64$ sets: the empty set, six one-element sets, 15 two-element sets, 20 three-element sets, ..., and one six-element set. Here are three assumptions we will always make about any event space: (a) Both the empty set and universal set (Ω) belong to the event space F; (b) if E is an event, then so is the complement of E; and (c) for any list of events $A_1, A_2, \dots$ (finite or infinite), the phrase "A_1 or A_2 or ..." describes another event.

TABLE 1.1: Basic probability terminology for the 10-question example.

Symbol	Name	Brief explanation	Example
Ω	Sample space	The set of all possible outcomes	All possible sequences of correct (c) or incorrect (i) answers of length 10; that is $2^{10} = 1024$ sequences.
F	Event space	A collection of subsets of Ω	All sequences with exactly 1 correct answer, or exactly 9 correct answers, etc.
P	Prob. function	Assigns probabilities to events in F	If each question is answered correctly with probability θ, $P(\{ciiiiiiiii\}) = (1-\theta)^9\theta$.
ω	Outcome	An element of Ω	The particular sequence $ciiiiiiiii$.
E	Event	A subset of Ω	A collection of sequences $\{ciiiiiiiii, iciiiiiiii, iiciiiiiii, \ldots\}$.

or A_2 or A_3 or ..." is given by the sum of the probability of A_1 occurring, of A_2 occurring, of A_3 occurring, ... (this sum could be finite or infinite).

Together, the triplet (Ω, F, P) is called a *probability space*.

In the context of data analysis, we will talk about probability in the following way. Consider some data that we might have collected. This could be discrete $0, 1$ responses in a question-response accuracy task, or continuous measurements of reading times in milliseconds from an eyetracking study, or multi-category responses like "yes," "no," or "don't know," or electrical potentials on the microvolts scale.

In any such case, we will say that the data are being generated from a *random variable*, which we will designate with a capital letter such as Y.[2]

The actually observed outcome (a $0, 1$ response; reading time, a response like "yes," "no," "don't know," etc.) will be distinguished from the random variable that generated it by using lowercase y for the observed outcome. We can call y an instance of Y; every new observed outcome can be different due to random variability.

We can summarize the above informal concepts relating to random variables very compactly if we re-state them in mathematical form. A mathematical

[2]Here, we use Y, but we could have used any letter, such as $X, Z, \ldots$. Later on, in some situations we will use Greek letters like θ, μ, σ to represent a random variable.

statement has the advantage not only of brevity but also of reducing (and hopefully eliminating) ambiguity.

So, stating the definition of random variables formally (following Blitzstein and Hwang 2014), we define a random variable Y as a function from a sample space Ω of possible outcomes ω to the real number system:[3]

$$Y : \Omega \rightarrow \mathbb{R} \tag{1.1}$$

The random variable associates to each outcome $\omega \in \Omega$ exactly one number $Y(\omega) = y$. The number y is a variable that represents all the possible values that the random variable generates; these values are taken to belong to the support of the random variable Y: $y \in S_Y$.

In our running example of asking 10 questions and obtaining a correct or incorrect response in each of the 10 trials, $Y(ciiiiiiiii) = 1$, $Y(cciiiiiiii) = 2$, etc., we will say that the number of correct responses from a subject is generated from a random variable Y. Because in our running example only discrete responses are possible (the number of correct responses can be $0, 1, 2, \dots, 10$), this is an example of a *discrete random variable*.

This particular random variable Y will be assumed to have a parameter θ that represents the probability of producing a particular number of correct responses (as discussed below, you will see that the random variable in question is the binomial random variable). Given some observed data that is assumed to come from a particular distribution, typically our goal is to obtain an estimate of the (unknown) value of the parameter associated with that distribution. More generally, if there is more than one parameter involved in the distribution, then our goal is to obtain an estimate of these parameters.

Now, suppose that in our above example, the random variable Y gives us exactly one correct answer; this can be (somewhat sloppily) written as $Y = 1$. The outcomes that could produce 1 are any of this set of 10 possible outcomes: $ciiiiiiiii$, $iciiiiiiii$, $iiciiiiiii$, $iiiciiiiii$,.... The set $\{ciiiiiiiii, iciiiiiiii, iiciiiiiii, iiiciiiiii, \dots\}$ is an element of F, so it is an event. The number $y = 1$ thus represents this event, and will have a probability of occurring associated with it; this is defined next.

A discrete random variable Y has associated with it a function called a *probability mass function* or PMF. This function, which is written $p(y)$, gives us the probability of obtaining each of these 11 possible values (from 0 correct responses to 10). We are using lower-case $p(y)$ here to denote a function of y.

[3]The actual formal definition of random variable is more complex, and it is based on measure theory. A more rigorous definition can be found in, for example, Steyer and Nagel (2017) and Resnick (2019).

When we want to talk about the probability of observing y, we will use $P(y)$. In discrete random variables the value $p(y)$ will be the same as $P(y)$; but when we turn to continuous random variables, this equality will not hold.

We will write that this PMF $p(y)$ depends on, or is conditional on, a particular fixed but unknown value for θ; the PMF will be written $p(y|\theta)$.[4]

In frequentist approaches to data analysis, only the observed data y are used to draw inferences about θ. A typical question that we ask in the frequentist paradigm is: does θ have a particular value θ_0? One can obtain estimates of the unknown value of θ from the observed data y, and then draw inferences about how different–or more precisely how far away–this estimate is from the hypothesized θ_0. This is the essence of null hypothesis significance testing. The conclusions from such a procedure are framed in terms of either rejecting the hypothesis that θ has value θ_0 or failing to reject this hypothesis. Here, rejecting the null hypothesis is the primary goal of the statistical hypothesis test.

Bayesian data analysis begins with a different question. What is common to the frequentist paradigm is the assumption that the data are generated from a random variable Y and that there is a function $p(y|\theta)$ that depends on the parameter θ. Where the Bayesian approach diverges from the frequentist one is that an important goal is to express our uncertainty about θ. In other words, we treat the parameter θ itself as a random variable, which means that we assign a probability distribution $p(\theta)$ to this random variable. This distribution $p(\theta)$ is called the *prior distribution* on θ; such a distribution could express our belief about the probability of correct responses before we observe the data y.

In later chapters, we will spend some time trying to understand how such a prior distribution can be defined for a range of different research problems.

Given such a prior distribution and some data y, the end-product of a Bayesian data analysis is what is called the *posterior distribution* of the parameter (or parameters) given the data: $p(\theta|y)$. This posterior distribution is the probability distribution of θ after conditioning on y, i.e., after the data has been observed and is therefore known. All our statistical inference is based on this posterior

[4]A notational aside: In frequentist treatments, the PMF would usually be written $p(y;\theta)$, i.e., with a semi-colon rather than the conditional distribution marked by the vertical bar. The semi-colon is intended to indicate that in the frequentist paradigm, the parameters are fixed point values; by contrast, in the Bayesian paradigm, parameters are random variables. This has the consequence that for the Bayesian, the distribution of y, $p(y)$ is really a conditional distribution, conditional on a random variable, here θ. For the frequentist, $p(y)$ requires some point value for θ, but it cannot be a conditional distribution because θ is not a random variable. We define conditional distributions later in this section.

distribution of θ; we can even carry out hypothesis tests that are analogous (but not identical) to the likelihood ratio-based frequentist hypothesis tests.

We already mentioned conditional probability above when discussing the probability of the data given some parameter θ, which we wrote as the PMF $p(y|\theta)$. Conditional probability is an important concept in Bayesian data analysis, not least because it allows us to derive Bayes' theorem. Let's look at the definition of conditional probability next.

1.2 Conditional probability

Suppose that A stands for some discrete event; an example would be "the streets are wet." Suppose also that B stands for some other discrete event; an example is "it has been raining." We can talk about the probability of the streets being wet given that it has been raining; or more generally, the probability of A given that B has happened.

This kind of statement is written as $Prob(A|B)$ or more simply $P(A|B)$. This is the conditional probability of A given B. Conditional probability is defined as follows.

$$P(A|B) = \frac{P(A,B)}{P(B)} \text{ where } P(B) > 0 \tag{1.2}$$

The conditional probability of A given B is thus defined to be the joint probability of A and B, divided by the probability of B. We can rearrange the above equation so that we can talk about the joint probability of both events A and B happening. This joint probability can be computed by first taking $P(B)$, the probability that event B (it has been raining) happens, and multiplying this by the probability that A happens conditional on B, i.e., the probability that the streets are wet given it has been raining. This multiplication will give us $P(A, B)$, the joint probability of A and B, i.e., that it has been raining and that the streets are wet. We will write the above description as: $P(A, B) = P(A|B)P(B)$.

Now, since the probability A and B happening is the same as the probability of B and A happening, i.e., since $P(B, A) = P(A, B)$, we can equate the expansions of these two terms:

$$P(A,B) = P(A|B)P(B) \text{ and } P(B,A) = P(B|A)P(A) \tag{1.3}$$

Equating the two expansions, we get:

$$P(A|B)P(B) = P(B|A)P(A) \tag{1.4}$$

Dividing both sides by $P(B)$:

$$P(A|B) = \frac{P(B|A)P(A)}{P(B)} \tag{1.5}$$

The above statement is Bayes' rule, and is the basis for all the statistical inference we will do in this book.

1.3 The law of total probability

Related to the above discussion of conditional probability is the law of total probability. Suppose that we have $A_1, \ldots, A_n$ distinct events that are pairwise disjoint which together make up the entire sample space Ω; see Figure 1.1. Then, $P(B)$, the probability of B happening, will be the sum of the probabilities $P(B \cap A_i)$, i.e., the sum of the joint probabilities of B and each A occurring (the symbol $\cap$ is the "and" operator used in set theory). The use of $\cap$ to represent joint probability is just an alternative notation to the one we used earlier ($P(B, A_i)$); you may see both notational variants in textbooks.

Formally:

$$P(B) = \sum_{i=1}^{n} P(B \cap A_i) \tag{1.6}$$

Because of the conditional probability rule, we can rewrite this as:

$$P(B) = \sum_{i=1}^{n} P(B|A_i)P(A_i) \tag{1.7}$$

Thus, the probability of B is the sum of the conditional probabilities $P(B|A_i)$ weighted by the probability $P(A_i)$. We will see the law of total probability in action below when we talk about *marginal likelihood*.

For now, this is all the probability theory we need to know!

The sections that follow expand on the idea of a random variable, the probability distributions associated with the random variable, what it means to specify a

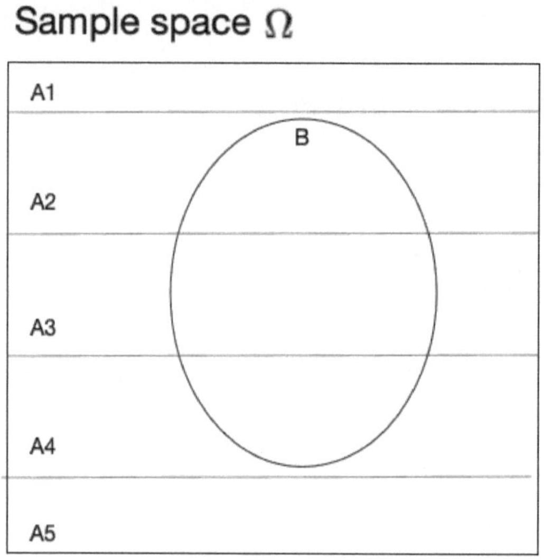

FIGURE 1.1: An illustration of the law of total probability.

prior distribution on a parameter, and how the prior and data can be used to derive the posterior distribution of θ.

To make the discussion concrete, we will use an example of a discrete random variable, the binomial. After discussing this discrete random variable, we present another example, this time involving a continuous random variable, the normal random variable.

The binomial and normal cases serve as the canonical examples that we will need in the initial stages of this book. We will introduce other random variables as needed: in particular, we will need the uniform and beta distributions. In other textbooks, you will encounter distributions like the Poisson, gamma, and exponential. The most commonly used distributions and their properties are discussed in most textbooks on statistics (see Further Reading at the end of this chapter).

1.4 Discrete random variables: An example using the binomial distribution

Consider the following sentence:

"It's raining, I'm going to take the ...."

Suppose that our research goal is to estimate the probability, call it θ, of the word "umbrella" appearing in this sentence, versus any other word. If the sentence is completed with the word "umbrella," we will refer to it as a success; any other completion will be referred to as a failure. This is an example of a binomial random variable: given n trials, there can be only two possible outcomes in each trial, a success or a failure, and there is some true unknown probability θ of success that we want to estimate. When the number of trials n is one, the random variable is called a Bernoulli distribution. So the Bernoulli distribution is the binomial distribution with the number of trials $n = 1$.

One way to empirically estimate this probability of success is to carry out a *cloze task*. In a cloze task, subjects are asked to complete a fragment of the original sentence, such as "It's raining, I'm going to take the ...." The predictability or cloze probability of "umbrella" is then calculated as the proportion of times that the target word "umbrella" was produced as an answer by subjects.

Assume for simplicity that 10 subjects are asked to complete the above sentence; each subject does this task only once. This gives us 10 independent trials that are either coded as a success ("umbrella" was produced) or as a failure (some other word was produced). We can sum up the number of successes to calculate how many of the 10 trials had "umbrella" as a response. For example, if 8 instances of "umbrella" are produced in 10 trials, we would estimate the cloze probability of producing "umbrella" to be 8/10.

We can repeatedly generate simulated sequences of the number of successes in R (later on we will demonstrate how to generate such random sequences of simulated data). Here is a case where we carry out 20 experiments, and each experiment will have 10 trials.

Before we look at the R code for generating such simulated data, some notational conventions are needed to avoid confusion. In the function we use below (`rbinom()`) for generating simulated data, `n` refers to the number of experiments, which in the R function's terminology is the number of observations. By contrast, `size` is the number of trials. This means that in the example below, `n` refers to the number of experiments conducted; because in R-speak these are called observations, we are also going to adopt this terminology. So, if 20 experiments are done, each with 10 trials, we will say that we have 20 observations, each with 10 trials.

```
## n=the number of observations (no. of experiments)
## size=the number of trials in each observation
rbinom(n = 20, size = 10, prob = 0.5)
```

```
##  [1] 5 4 2 5 6 4 1 5 4 4 5 5 8 7 4 8 7 2 2 3
```

The number of successes in each of the 20 simulated observations above is being generated by a discrete random variable Y with a probability distribution $p(y|\theta)$ called the *binomial distribution.*

For discrete random variables such as the binomial, the probability distribution $p(y|\theta)$ is called a probability mass function (PMF). The PMF defines the probability of each possible event. In the above example, with the number of trials size $= 10$, there are in principle 11 possible events that could produce $y = 0, 1, 2, ..., 10$ successes. Which of these is most probable depends on the parameter θ in the binomial distribution that represents the probability of success.

The left-hand side plot in Figure 1.2 shows an example of a binomial PMF with 10 trials, with the parameter θ fixed at 0.5. Setting θ to 0.5 leads to a PMF where the most probable outcome is 5 successes out of 10. If we had set θ to, say 0.1, then the most probable outcome would be 1 success out of 10; and if we had set θ to 0.9, then the most probable outcome would be 9 successes out of 10.

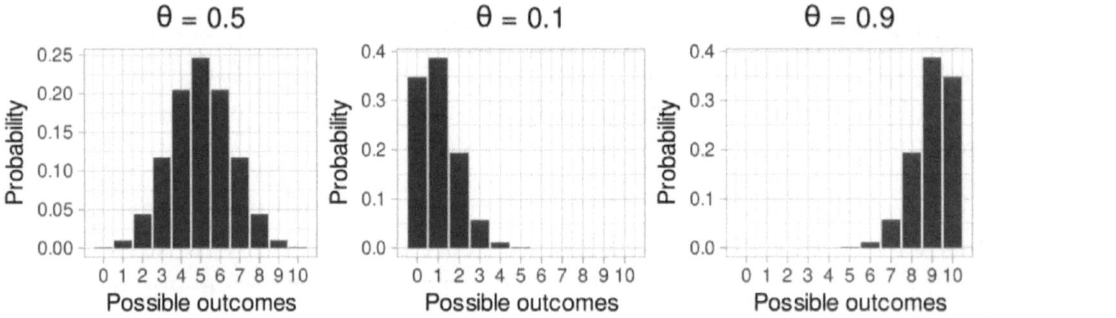

FIGURE 1.2: Probability mass functions of a binomial distribution assuming 10 trials, with 50%, 10%, and 90% probability of success.

The probability mass function for the binomial is written as follows. Here, it is again important to pay attention to the notation, as there is potential for confusion given the conventions in the `rbinom` function we saw earlier.

$$Binomial(k|n, \theta) = \binom{n}{k} \theta^k (1 - \theta)^{n-k} \tag{1.8}$$

Here, n represents the total number of **trials**, and corresponds to `size` in the `rbinom` function; the confusion that can arise here is that in the `rbinom` function `n` is used to represent the number of observations (or the number of experiments). It is important to remember that in the definition of the probability mass function above, n represents the total number of trials in a single observation (or experiment).

The variable k the number of successes (this could range from 0 to 10 in our running example), and θ the probability of success. The term $\binom{n}{k}$, pronounced n-choose-k, represents the number of ways in which one can obtain k successes in an experiment with n trials. For example, 1 success out of 10 can occur in 10 possible ways: the very first observation could be a 1, or the second observation could be a 1, etc. The term $\binom{n}{k}$ expands to $\frac{n!}{k!(n-k)!}$. In R, it is computed using the function `choose(n,k)`, with n and k representing any real number (although of course, the `choose()` function only makes sense for positive integers).

When we want to express the fact that the data is assumed to be generated from a binomial random variable, we will write $Y \sim Binomial(n, \theta)$. If the data is generated from a random variable that has some other probability distribution $f(\theta)$, we will write $Y \sim f(\theta)$. In this book, we use $f(\cdot)$ synonymously with $p(\cdot)$ to represent a probability density/mass function.

1.4.1 The mean and variance of the binomial distribution

It is possible to analytically compute the mean (expectation) and variance of the PMF associated with the binomial random variable Y.

The expectation of a discrete random variable Y with probability mass function f(y), is defined as

$$E[Y] = y_1 \cdot f(y_1) + \cdots + y_n \cdot f(y_n) = \sum_{i=1}^{n} y_i \cdot f(y_i) \tag{1.9}$$

As a simple example, suppose that we toss a fair coin once. The possible events are Tails (represented as $y_1 = 0$) and Heads (represented as $y_2 = 1$), each with equal probability, 0.5. The expectation is:

$$E[Y] = \sum_{i=1}^{2} y_i \cdot f(y_i) = 0 \cdot 0.5 + 1 \cdot 0.5 = 0.5 \tag{1.10}$$

The expectation has the interpretation that if we were to do the experiment a large number of times and calculate the sample mean of the observations, in the long run we would approach the value 0.5. Another way to look at the above definition is that the expectation gives us the weighted mean of the possible outcomes, weighted by the respective probabilities of each outcome.

Without getting into the details of how these are derived mathematically (Kerns 2014), we just state here that the mean of Y (the expectation $E[Y]$) and the variance of Y (written $Var(Y)$) of a binomial distribution with parameter θ and n trials are $E[Y] = n\theta$ and $Var(Y) = n\theta(1-\theta)$, respectively.

In the binomial example above, n is a fixed number because we decide on the total number of trials before running the experiment. In the PMF, $\binom{n}{k}\theta^k(1-\theta)^{n-k}$, θ is also a fixed value; the only variable in a PMF is k. In real experimental situations, we never know the true point value of θ. But θ can be estimated from the data. From the observed data, we can compute the estimate of θ, $\hat{\theta} = k/n$. The quantity $\hat{\theta}$ is the observed proportion of successes and is called the *maximum likelihood estimate* of the true (but unknown) parameter θ.[5]

What does the term "maximum likelihood estimate" mean? In order to understand this term, it is necessary to first understand what a likelihood function is. Recall that in the discussion above, the PMF $p(k|n,\theta)$ assumes that θ and n are fixed, and k will have some value between 0 and 10 when the experiment is repeated multiple times.

The *likelihood function* refers to the PMF $p(k|n,\theta)$, treated as a function of θ. Once we have observed a particular value for k, this value is now fixed, along with the number of trials n. Once k and n are fixed, the function $p(k|n,\theta)$ only depends on θ. Thus, the likelihood function is the same function as the PMF, but assumes that the data from the completed experiment is fixed and only the parameter θ varies (from 0 to 1). The likelihood function is written $\mathcal{L}(\theta|k,n)$, or simply $\mathcal{L}(\theta)$.

For example, suppose that we have $n = 10$ trials, and observe $k = 7$ successes. The likelihood function is:

$$\mathcal{L}(\theta|k=7,n=10) = \binom{10}{7}\theta^7(1-\theta)^{10-7} \tag{1.11}$$

If we now plot the likelihood function for all possible values of θ ranging from 0 to 1, we get the plot shown in Figure 1.3(a).

What is important about this plot is that it shows that, given the data, the maximum point is at the point 0.7, which corresponds to the estimated mean using the formula shown above: $k/n = 7/10$. Thus, the maximum likelihood estimate (MLE) gives us the most likely value that the parameter θ has, given the data. In the binomial, the proportion of successes k/n can be shown to be the maximum likelihood estimate of the parameter θ (e.g., see p. 339-340 of Miller and Miller 2004).

[5]Looking ahead to the rest of the book, in the Bayesian approach, the parameter of interest, θ in the present example, never has just a true unknown point value associated with it; instead, we use a probability distribution to express our belief about plausible values of the parameter. However, throughout this book, when generating simulated data, we will often use point values for parameters. These point values are adopted simply to evaluate the behavior of the model being investigated.

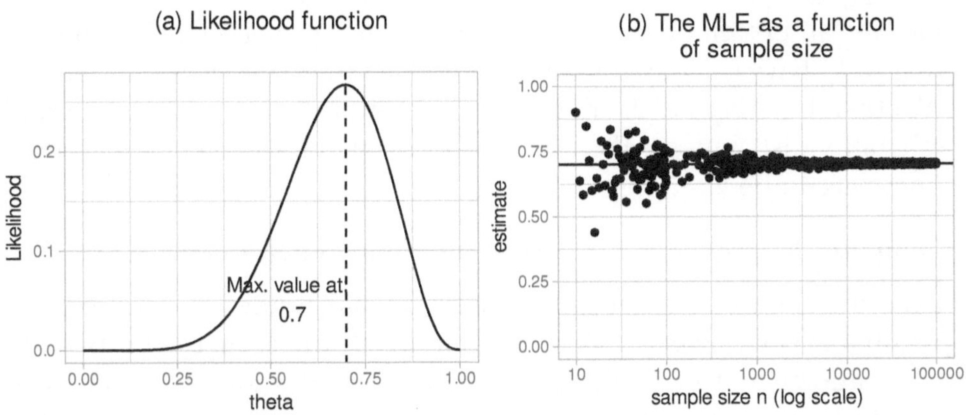

FIGURE 1.3: (a) The likelihood function for 7 successes out of 10. (b) The plot shows the estimate of the mean proportion of successes sampled from a binomial distribution with true probability of success 0.7, with increasing sample sizes. As the sample size increases, the estimate converges to the true value of 0.7.

A crucial point: the "most likely" value of the parameter is with respect to the data at hand. The goal is to estimate an unknown parameter value from the data. This estimated parameter value is chosen such that the probability (discrete case) or probability density (continuous case) of getting the sample values (i.e., the data) is a maximum. This parameter value is the maximum likelihood estimate (MLE).

This MLE from a particular sample of data need not invariably give us an accurate estimate of θ. For example, if we run our experiment with 10 trials and get 1 success out of 10, the MLE is 0.10. We could have just happened to observe only one success out of 10 by chance, even if the true θ were 0.7. If we were to repeatedly run the experiment with increasing sample sizes, as the sample size increases, the MLE would converge to the true value of the parameter. Figure 1.3(b) illustrates this point. The key point here is that with a smaller sample size, the MLE from a particular data set may or may not point to the true value.

1.4.2 What information does a probability distribution provide?

In Bayesian data analysis, we will constantly be asking the question: what information does a probability distribution give us? In particular, we will treat each parameter θ as a random variable; this will raise questions like: "what is the probability that the parameter θ lies between two values a and b"; and "what is the range over which we can be 95% certain that the parameter lies"? In order to be able to answer questions like these, we need to know what information we can obtain once we have decided on a probability distribution that is assumed to have generated the data, and how to extract this information using R. We therefore discuss the different kinds of information we can obtain from a probability distribution. For now we focus only on the binomial random variable introduced above.

1.4.2.1 Compute the probability of a particular outcome (discrete case only)

The binomial distribution shown in Figure 1.2 already shows the probability of each possible event under a different value for θ. In R, there is a built-in function that allows us to calculate the probability of k successes out of n, given a particular value of k (this number constitutes our data), the number of trials n, and given a particular value of θ; this is the `dbinom()` function. For example, the probability of 5 successes out of 10 when θ is 0.5 is (note: we are using k to represent the number of successes, but the `dbinom()` function below uses `x` instead):

```
dbinom(x = 5, size = 10, prob = 0.5)
```

```
## [1] 0.246
```

The probabilities of success when θ is 0.1 or 0.9 can be computed by replacing 0.5 above by each of these probabilities. One can just do this by giving `dbinom()` a vector of probabilities:

```
dbinom(x = 5, size = 10, prob = c(0.1, 0.9))
```

```
## [1] 0.00149 0.00149
```

The probability of a particular outcome like $k = 5$ successes is only computable in the discrete case. In the continuous case, the probability of obtaining a particular point value will always be zero (we discuss this when we turn to continuous probability distributions below).

1.4.2.2 Compute the cumulative probability of k or less (more) than k successes

Using the `dbinom()` function, we can compute the cumulative probability of obtaining 1 or less, 2 or less successes, etc. This is done through a simple summation procedure:

```
## the cumulative probability of obtaining
## 0, 1, or 2 successes out of 10 trials,
## with theta=0.5:
```

```
dbinom(x = 0, size = 10, prob = 0.5) +
  dbinom(x = 1, size = 10, prob = 0.5) +
  dbinom(x = 2, size = 10, prob = 0.5)
```

```
## [1] 0.0547
```

Mathematically, we could write the above summation as:

$$\sum_{k=0}^{2} \binom{n}{k} \theta^k (1-\theta)^{n-k} \tag{1.12}$$

An alternative to the cumbersome addition in the R code above is this more compact statement, which is identical to the above mathematical expression:

```
sum(dbinom(0:2, size = 10, prob = 0.5))
```

```
## [1] 0.0547
```

R has a built-in function called `pbinom()` that does this summation for us. If we want to know the probability of 2 or fewer than 2 successes as in the above example, we can write:

```
pbinom(2, size = 10, prob = 0.5, lower.tail = TRUE)
```

```
## [1] 0.0547
```

The specification `lower.tail = TRUE` (the default value) ensures that the summation goes from 2 to numbers smaller than 2 (which lie in the lower tail of the distribution in Figure 1.2). If we wanted to know what the probability is of obtaining 3 or more successes out of 10, we can set `lower.tail` to `FALSE`:

```
pbinom(2, size = 10, prob = 0.5, lower.tail = FALSE)
```

```
## [1] 0.945
```

```
## equivalently:
## sum(dbinom(3:10,size = 10, prob = 0.5))
```

The cumulative distribution function or CDF for a random variable Y is thus related to the corresponding probability mass/density function, and is written as follows:

$$F_Y(a) = P(Y \leq a) \quad (1.13)$$

The CDF can be plotted by computing the cumulative probabilities for any value k or less than k, where k ranges from 0 to 10 in our running example. The CDF is shown in Figure 1.4(a).

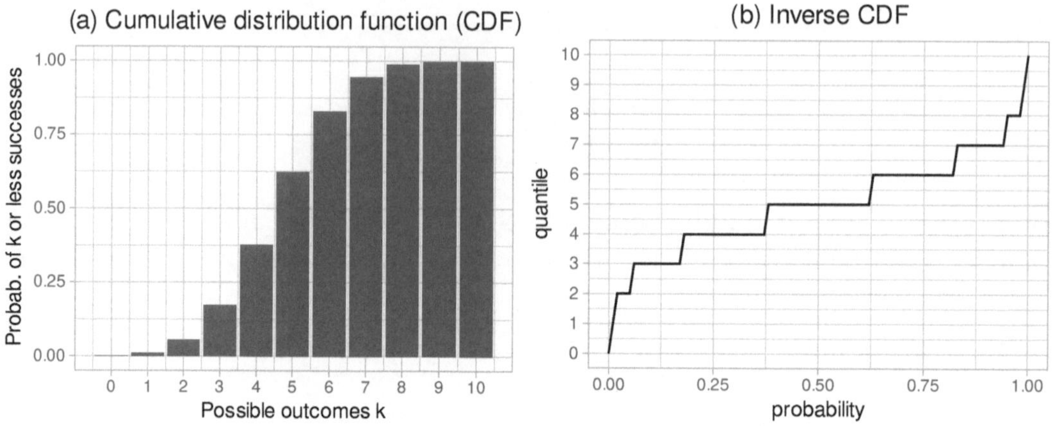

FIGURE 1.4: (a) Cumulative distribution function (CDF) of a binomial distribution with 10 trials and a 50% probability of success. (b) Inverse CDF for the same binomial distribution.

1.4.2.3 Compute the inverse of the cumulative distribution function (the quantile function)

We can also find out the value of the variable k (the quantile) such that the probability of obtaining k or less than k successes is some specific probability value p. If we switch the x and y axes of Figure 1.4(a), we obtain another very useful function, the inverse of the CDF.

The inverse of the CDF (known as the quantile function in R because it returns the quantile, the value k) is available in R as the function `qbinom()`. The usage is as follows: to find out what the value k of the outcome is such that the probability of obtaining k or less successes is 0.37, type:

```
qbinom(p = 0.37, size = 10, prob = 0.5)
```

```
## [1] 4
```

One can visualize the inverse CDF of the binomial as in Figure 1.4(b).

1.4.2.4 Generate simulated data from a *Binomial*(n, θ) distribution

We can generate simulated data from a binomial distribution by specifying the number of observations or experiments (`n`), the number of trials (`size`), and the probability of success θ. In R, we do this as follows:

```
rbinom(n = 1, size = 10, prob = 0.5)
```

```
## [1] 6
```

The above code generates the number of successes in an experiment with 10 trials. Repeatedly run the above code; we will get different numbers of successes each time.

As mentioned earlier, if there is only one trial, then instead of the binomial distribution, we have a Bernoulli distribution. For example, if we have 10 experiments from a Bernoulli distribution, where the probability of success is 0.5, we can simulate data as follows using the function `rbern()` from the package `extraDistr`.

```
rbern(n = 10, prob = 0.5)
```

```
##  [1] 1 1 0 0 1 1 0 1 1 1
```

The above kind of output can also be generated by using the `rbinom()` function: `rbinom(n = 10, size = 1, prob = 0.5)`. When the data are generated using the `rbinom()` function in this way, one can calculate the number of successes by just summing up the vector, or computing its mean and multiplying by the number of trials, here 10:

```
(y <- rbinom(n = 10, size = 1, prob = 0.5))
```

```
##  [1] 1 0 1 0 0 1 1 1 0 0
```

```
mean(y) * 10
```

```
## [1] 5
```

```
sum(y)
```

```
## [1] 5
```

1.5 Continuous random variables: An example using the normal distribution

We will now revisit the idea of the random variable using a continuous distribution. Imagine that we have a vector of reading time data y measured in milliseconds and coming from a normal distribution. The normal distribution is defined in terms of two parameters: the *location*, its mean value μ, which determines its center, and the *scale*, its standard deviation, σ, which determines how much spread there is around this center point.

The probability density function (PDF) of the normal distribution is defined as follows:

$$Normal(y|\mu,\sigma) = f(y) = \frac{1}{\sqrt{2\pi\sigma^2}} \exp\left(-\frac{(y-\mu)^2}{2\sigma^2}\right) \tag{1.14}$$

Here, μ is some true, unknown mean, and σ^2 is some true, unknown variance of the normal distribution that the reading times have been sampled from. There is a built-in function in R that computes the above function once we specify the mean μ and the standard deviation σ (in R, this parameter is specified in terms of the standard deviation rather than the variance).

Figure 1.5 visualizes the normal distribution for particular values of μ and σ, as a PDF (using `dnorm()`), a CDF (using `pnorm()`), and the inverse CDF (using `qnorm()`). It should be clear from the figure that these are three different ways of looking at the same information.

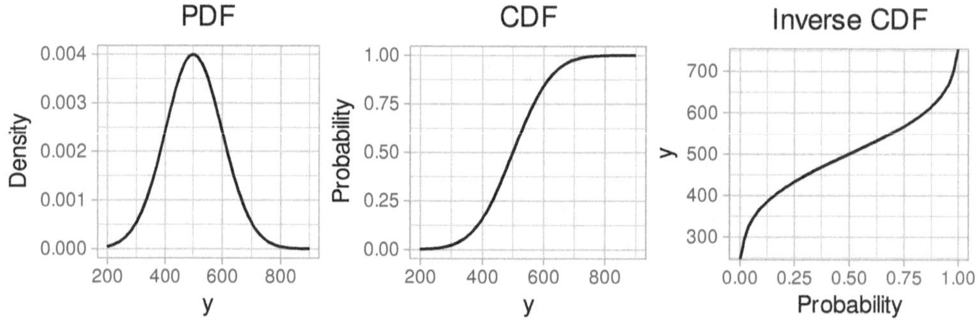

FIGURE 1.5: The PDF, CDF, and inverse CDF for the $Normal(\mu = 500, \sigma = 100)$.

As in the discrete example, the PDF, CDF, and inverse of the CDF allow us to ask questions like:

- What is the probability of observing values between a and b from a normal distribution with mean μ and standard deviation σ? Using the above example, we can ask what the probability is of observing values between 200 and 700 ms:

```
pnorm(700, mean = 500, sd = 100) - pnorm(200, mean = 500, sd = 100)
```

```
## [1] 0.976
```

The probability that the random variable takes any specific point value is zero. This is because the probability in a continuous probability distribution is the area under the curve, and the area at any point on the x-axis is always zero. The implication here is that it is only meaningful to ask about probabilities between two different point values; e.g., the probability that Y lies between a and b, or $P(a < Y < b)$. Notice that $P(a < Y < b)$ is the same statement as $P(a \leq Y \leq b)$. Of course, for any particular point value, the PDF itself does not return the value zero; but the value returned by the PDF is not the probability of that particular value occurring. It is the density of that particular value; and if the PDF is seen as a function of the parameters, it is the likelihood of that particular value.

- What is the quantile q such that the probability is p of observing that value q or a value more extreme than q? For example, we can work out the quantile q such that the probability of observing q or some value less than it is 0.975, in the $Normal(500, 100)$ distribution. Formally, we would write this as $P(Y < a)$.

```
qnorm(0.975, mean = 500, sd = 100)
```

```
## [1] 696
```

The above output says that the quantile value q such that $Prob(X < q) = 0.975$ is $q = 696$.

- Generate simulated data. Given a vector of n independent and identically distributed data y, i.e., given that each data point is being generated independently from $Y \sim Normal(\mu, \sigma)$ for some values of the parameters, the sample mean and standard deviation[6] are:

[6] R will compute the standard deviation by dividing by $n-1$, not n; this is because dividing by n gives a biased estimate (chapter 10 of Miller and Miller 2004). This is not an important detail for our purposes, and in any case for large n it doesn't really matter whether one divides by n or $n-1$.

$$\bar{y} = \frac{\sum_{i=1}^{n} y_i}{n} \tag{1.15}$$

$$sd(y) = \sqrt{\frac{\sum_{i=1}^{n} (y_i - \bar{y})^2}{n}} \tag{1.16}$$

For example, we can generate 10 data points using the `rnorm()` function, and then use the simulated data to compute the mean and standard deviation:

```
y <- rnorm(10, mean = 500, sd = 100)
mean(y)
```

```
## [1] 494
```

```
sd(y)
```

```
## [1] 117
```

Again, the sample mean and sample standard deviation computed from a particular (simulated or real) data set need not necessarily be close to the true values of the respective parameters. Especially when sample size is small, one can end up with mis-estimates of the mean and standard deviation.

Incidentally, simulated data can be used to generate all kinds of statistics. For example, we can compute the lower and upper quantiles such that 95% of the simulated data are contained within these quantiles:

```
quantile(y, probs = c(0.025, 0.975))
```

```
##  2.5% 97.5%
##   302   675
```

Later on, this function will be used to generate summary statistics once we have obtained samples of a parameter after we have fit a model using Stan/brms.

1.5.1 An important distinction: Probability vs. density in a continuous random variable

In continuous distributions like the normal discussed above, it is important to understand that the probability density function or PDF, $p(y|\mu, \sigma)$ defines a mapping from the y values (the possible values that the data can have) to a quantity called the density of each possible value. We can see this function in action when we use `dnorm()` to compute, say, the density value corresponding to $y = 1$ and $y = 2$ in the standard normal distribution, that is, a normal distribution with a mean of zero and a standard deviation of one.

```
## density:
dnorm(1, mean = 0, sd = 1)
```

```
## [1] 0.242
```

```
dnorm(2, mean = 0, sd = 1)
```

```
## [1] 0.054
```

If the density at a particular point value like 1 is high compared to some other value (2 in the above example) then this point value 1 has a higher likelihood than 2 in the standard normal distribution.

The quantity computed for the values 1 and 2 are *not* the probability of observing 1 or 2 in this distribution. As mentioned earlier, probability in a continuous distribution is defined as the area under the curve, and this area will always be zero at any point value (because there are infinitely many different possible values). If we want to know the probability of obtaining values between an upper and lower bound b and a, i.e., $P(a < Y < b)$ where these are two distinct values, we must use the cumulative distribution function or CDF: in R, for the normal distribution, this is the `pnorm()` function. For example, the probability of observing a value between +2 and −2 in a normal distribution with mean 0 and standard deviation 1 is:

```
pnorm(2, mean = 0, sd = 1) - pnorm(-2, mean = 0, sd = 1)
```

```
## [1] 0.954
```

The situation is different in discrete random variables. These have a probability mass function (PMF) associated with them—an example is the binomial distribution that we saw earlier. There, the PMF maps the possible y values to the probabilities of those values occurring. That is why, in the binomial distribution, the probability of observing exactly 2 successes when sampling from a $Binomial(n = 10, \theta = 0.5)$ can be computed using either `dbinom()` or `pbinom()`:

```
dbinom(2, size = 10, prob = 0.5)
```

```
## [1] 0.0439
```

```
pbinom(2, size = 10, prob = 0.5) - pbinom(1, size = 10, prob = 0.5)
```

```
## [1] 0.0439
```

In the second line of code above, we are computing the cumulative probability of observing two or less successes, minus the probability of observing one or less successes. This gives us the probability of observing exactly two successes. The `dbinom()` gives us this same information.

1.5.2 Truncating a normal distribution

In the above discussion, the support for the normal distribution ranges from minus infinity to plus infinity. One can define PDFs with a more limited support; an example would be a normal distribution whose PDF $f(x)$ is such that the lower bound is truncated at 0 to allow only positive values. In such a case, the area under the range minus infinity to zero ($\int_{-\infty}^{0} f(x)\,\mathrm{d}x$) will be 0 because the range lies outside the support of the truncated normal distribution. Also, if one truncates the standard normal ($Normal(0,1)$) at 0, in order to make the area between zero and plus infinity sum up to 1, we would have to multiply it by 2, because we just halved the area under the curve. More formally and more generally, we would have to multiply the truncated distribution $f(x)$ by some factor k such that the following integral sums to 1:

$$k \int_{0}^{\infty} f(x)\,\mathrm{d}x = 1 \tag{1.17}$$

Clearly, this factor is $k = \frac{1}{\int_{0}^{\infty} f(x)\,\mathrm{d}x}$. For the standard normal, this integral is easy to compute using R; we just calculate the complement of the cumulative distribution (CCDF):

```
pnorm(0, mean = 0, sd = 1, lower.tail = FALSE)
```

```
## [1] 0.5
```

```
## alternatively:
1 - pnorm(0, mean = 0, sd = 1, lower.tail = TRUE)
```

```
## [1] 0.5
```

The above calculation implies that k is indeed 2, as we informally argued ($k = \frac{1}{0.5} = 2$).

Also, if we had truncated the distribution at 0 to the right instead of the left (allowing only negative values), we would have to find the factor k in the same way as above, except that we would have to find k such that:

$$k \int_{-\infty}^{0} f(x)\,\mathrm{d}x = 1 \tag{1.18}$$

For the standard normal case, in R, this factor would require us to use the CDF:

```
pnorm(0, mean = 0, sd = 1, lower.tail = TRUE)
```

```
## [1] 0.5
```

Later in this book, we will be using such truncated distributions when doing Bayesian modeling, and when we use them, we will want to multiply the truncated distribution by the factor k to ensure that it is still a proper PDF whose area under the curve sums to 1. Truncated normal distributions are discussed in more detail in the online section A.2.

1.6 Bivariate and multivariate distributions

So far, we have only discussed univariate distributions; these are distributions that involve only one variable. For example, when we talk about data generated from a Binomial distribution, or from a normal distribution, these are univariate distributions.

It is also possible to specify distributions with two or more dimensions. Some examples will make it clear what a bivariate (or more generally, multivariate) distribution is.

1.6.1 Example 1: Discrete bivariate distributions

Starting with the discrete case, consider the discrete bivariate distribution shown below. These are data from an experiment where, inter alia, in each trial a Likert acceptability rating and a question-response accuracy were recorded

TABLE 1.2: The joint PMF for two random variables X and Y.

	$x = 1$	$x = 2$	$x = 3$	$x = 4$	$x = 5$	$x = 6$	$x = 7$
$y = 0$	0.018	0.023	0.04	0.043	0.063	0.049	0.055
$y = 1$	0.031	0.053	0.086	0.096	0.147	0.153	0.142

(the data are from a study by Laurinavichyute 2020, used with permission here). Load the data by loading the R package `bcogsci`.

```
data("df_discreteagrmt")
```

Figure 1.6 shows the *joint probability mass function* of two random variables X and Y. The random variable X consists of 7 possible values (this is the 1–7 Likert response scale), and the random variable Y is question-response accuracy, with 0 representing an incorrect response, and 1 representing a correct response. One can also display Figure 1.6 as a table; see Table 1.2.

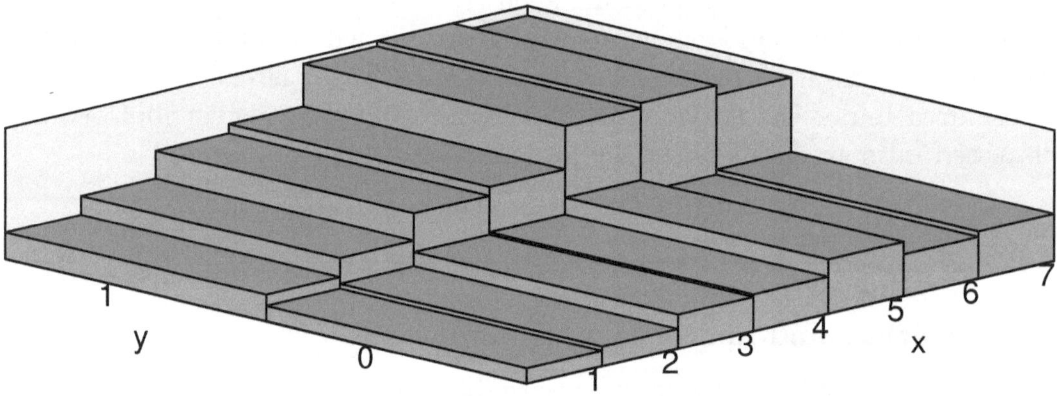

FIGURE 1.6: Example of a discrete bivariate distribution. In these data, in every trial, two pieces of information were collected: Likert responses and yes-no question responses. The random variable X represents Likert scale responses on a scale of 1–7. and the random variable Y represents 0, 1 (incorrect, correct) responses to comprehension questions.

For each possible pair of values of X and Y, we have a joint probability $p_{X,Y}(x, y)$. Given such a bivariate distribution, there are two useful quantities we can compute: the *marginal* distributions (p_X and p_Y), and the *conditional* distributions ($p_{X|Y}$ and $p_{Y|X}$). Table 1.2 shows the joint probability mass function $p_{X,Y}(x, y)$.

1.6.1.1 Marginal distributions

The marginal distribution p_Y is defined as follows. S_X is the support of X, i.e., all the possible values of X.

$$p_Y(y) = \sum_{x \in S_X} p_{X,Y}(x, y). \tag{1.19}$$

Similarly, the marginal distribution p_X is defined as:

$$p_X(x) = \sum_{y \in S_Y} p_{X,Y}(x, y). \tag{1.20}$$

p_Y is computed, by summing up the rows; and p_X by summing up the columns. We can see why this is called the marginal distribution; the result appears in the margins of the table. In the code below, the object `probs` contains bivariate PMF shown in Table 1.2.

```
# P(Y)
(PY <- rowSums(probs))
```

```
##   y=0   y=1
## 0.291 0.709
```

```
sum(PY) ## sums to 1
```

```
## [1] 1
```

```
# P(X)
(PX <- colSums(probs))
```

```
##    x=1    x=2    x=3    x=4    x=5    x=6    x=7
## 0.0491 0.0766 0.1257 0.1394 0.2102 0.2020 0.1969
```

```
sum(PX) ## sums to 1
```

```
## [1] 1
```

The marginal probabilities sum to 1, as they should. Table 1.3 shows the marginal probabilities.

To compute the marginal distribution of X, one is summing over all the Y's; and to compute the marginal distribution of Y, one sums over all the X's. We say

TABLE 1.3: The joint PMF for two random variables X and Y, along with the marginal distributions of X and Y.

	$x=1$	$x=2$	$x=3$	$x=4$	$x=5$	$x=6$	$x=7$	$P(Y)$
$y=0$	0.018	0.023	0.04	0.043	0.063	0.049	0.055	0.291
$y=1$	0.031	0.053	0.086	0.096	0.147	0.153	0.142	0.709
$P(X)$	0.049	0.077	0.126	0.139	0.21	0.202	0.197	

that we are *marginalizing out* the random variable that we are summing over. One can also visualize the two marginal distributions using barplots (Figure 1.7).

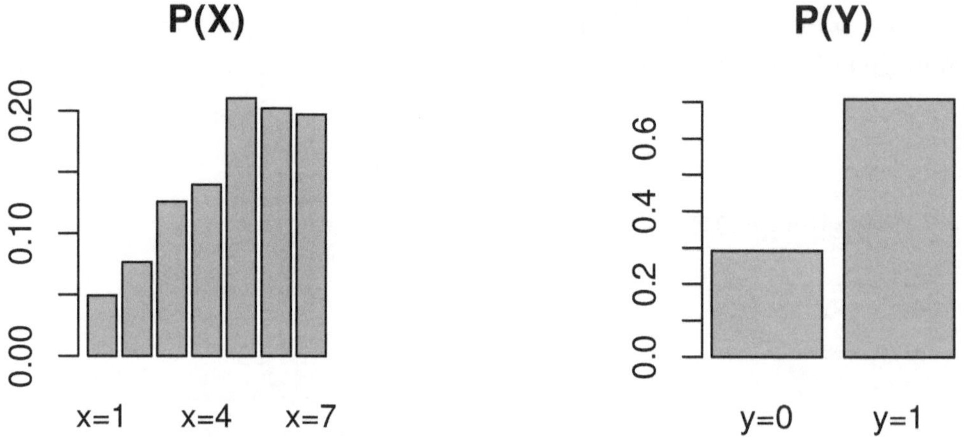

FIGURE 1.7: The marginal distributions of the random variables X and Y, presented as barplots.

1.6.1.2 Conditional distributions

For computing conditional distributions, recall that conditional probability (see section 1.2) is defined as:

$$p_{X|Y}(x \mid y) = \frac{p_{X,Y}(x,y)}{p_Y(y)} \tag{1.21}$$

and

$$p_{Y|X}(y \mid x) = \frac{p_{X,Y}(x,y)}{p_X(x)} \tag{1.22}$$

The conditional distribution of a random variable X given that $Y = y$, where y is some specific (fixed) value, is:

TABLE 1.4: A table for listing conditional distributions of X given Y.

	x = 1	x = 2	x = 3	x = 4	x = 5	x = 6	x = 7
$p_{X\|Y(x\|y=0)}$	0.062						
$p_{X\|Y(x\|y=1)}$							

$$p_{X|Y}(x \mid y) = \frac{p_{X,Y}(x, y)}{p_Y(y)} \quad \text{provided } p_Y(y) = P(Y = y) > 0 \tag{1.23}$$

As an example, let's consider how $p_{X|Y}$ would be computed. The possible values of y are $0, 1$, and so we have to find the conditional distribution (defined above) for each of these values. That is, we have to find $p_{X|Y}(x \mid y = 0)$, and $p_{X|Y}(x \mid y = 1)$.

Let's do the calculation for $p_{X|Y}(x \mid y = 0)$.

$$\begin{aligned} p_{X|Y}(1 \mid 0) &= \frac{p_{X,Y}(1,0)}{p_Y(0)} \\ &= \frac{0.018}{0.291} \\ &= 0.062 \end{aligned} \tag{1.24}$$

This conditional probability value will occupy the cell $X = 1, Y = 0$ in Table 1.4 summarizing the conditional probability distribution $p_{X|Y}$. In this way, one can fill in the entire table, which will then represent the conditional distributions $p_{X|Y=0}$ and $p_{X|Y=1}$. The reader may want to take a few minutes to complete Table 1.4. After the conditional probabilities have been computed, the rows should sum up to 1.

Similarly, one can construct a table that shows $p_{Y|X}$.

1.6.1.3 Covariance and correlation

Here, we briefly define the covariance and correlation of two discrete random variables. For detailed examples and discussion, see the references at the end of the chapter. Informally, if there is a high probability that large values of a random variable X are associated with large values of another random variable Y, we will say that the covariance between the two random variable X and Y, written $Cov(X, Y)$, is positive.

The covariance of two (discrete) random variables X and Y is defined as follows. $E[\cdot]$ refers to the expectation of a random variable.

$$Cov(X,Y) = E[(X - E[X])(Y - E[Y])] \tag{1.25}$$

It is possible to show that this is equivalent to:

$$Cov(X,Y) = E[XY] - E[X]E[Y] \tag{1.26}$$

The expectation E[XY] is defined to be:

$$E[XY] = \sum_x \sum_y xy f_{X,Y}(x,y) \tag{1.27}$$

If the standard deviations of the two random variables is σ_X and σ_Y, the correlation between the two random variables, ρ_{XY}, is defined as:

$$\rho_{XY} = \frac{Cov(X,Y)}{\sigma_X \sigma_Y} \tag{1.28}$$

1.6.2 Example 2: Continuous bivariate distributions

Consider now the continuous bivariate case; this time, we will use simulated data. Consider two normal random variables X and Y, each of which coming from, for example, a $Normal(0,1)$ distribution, with some correlation $\rho_{X,Y}$ between the two random variables.

A bivariate distribution for two random variables X and Y, each of which comes from a normal distribution, is expressed in terms of the means and standard deviations of each of the two distributions, and the correlation ρ_{XY} between them. The standard deviations and correlation in a bivariate distribution are expressed in a special form of a 2×2 matrix called a variance-covariance matrix Σ. If ρ_{XY} is the correlation between the two random variables, and σ_X and σ_Y the respective standard deviations, the variance-covariance matrix is written as:

$$\Sigma = \begin{pmatrix} \sigma_X^2 & \rho_{XY}\sigma_X\sigma_Y \\ \rho_{XY}\sigma_X\sigma_Y & \sigma_Y^2 \end{pmatrix} \tag{1.29}$$

The off-diagonals of this matrix contain the covariance between X and Y.

The joint distribution of X and Y is defined as follows:

$$\begin{pmatrix} X \\ Y \end{pmatrix} \sim \mathcal{N}_2\left(\begin{pmatrix} 0 \\ 0 \end{pmatrix}, \Sigma\right) \tag{1.30}$$

The joint PDF is written with reference to the two variables $f_{X,Y}(x,y)$. It has the property that the volume under the surface that is bounded by the density function sums to 1—this sum-to-1 property is the same idea that we encountered in the univariate cases (the normal and binomial distributions), except that we are talking about a bivariate distribution here, so we talk about the volume under the surface rather than the area under the curve.

Formally, we would write the volume as a double integral: we are summing up the volume under the surface representing the joint density for X and Y (hence the two integrals).

$$\iint_{S_{X,Y}} f_{X,Y}(x, y)\, \mathrm{d}x\mathrm{d}y = 1 \tag{1.31}$$

Here, the terms dx and dy express the fact that we are summing up the volume under the surface.

The joint CDF would be written as follows. The equation below gives us the probability of observing a value like (u, v) or some value smaller than that (i.e., some (u', v'), such that $u' < u$ and $v' < v$).

$$\begin{aligned} F_{X,Y}(u, v) &= P(X < u, Y < v) \\ &= \int_{-\infty}^{u} \int_{-\infty}^{v} f_{X,Y}(x, y)\, \mathrm{d}y\mathrm{d}x \end{aligned} \tag{1.32}$$

Just as in the discrete case, the marginal distributions can be derived by marginalizing out the other random variable:

$$f_X(x) = \int_{S_Y} f_{X,Y}(x, y)\, \mathrm{d}y, \quad f_Y(y) = \int_{S_X} f_{X,Y}(x, y)\, \mathrm{d}x \tag{1.33}$$

Here, S_X and S_Y are the respective supports.

Here, the integral sign $\int$ is the continuous equivalent of the summation sign $\sum$ in the discrete case. Luckily, we will never have to compute such integrals ourselves; but it is important to appreciate how a marginal distribution arises from a bivariate distribution—by integrating out or marginalizing out the other random variable.

A visualization will help. The figures below show a bivariate distribution with zero correlation (Figure 1.8), a negative correlation (Figure 1.9) and a positive correlation (Figure 1.10).

In this book, we will make use of such multivariate distributions a lot, and it will soon become important to know how to generate simulated bivariate or multivariate data that is correlated. So let's look at that next.

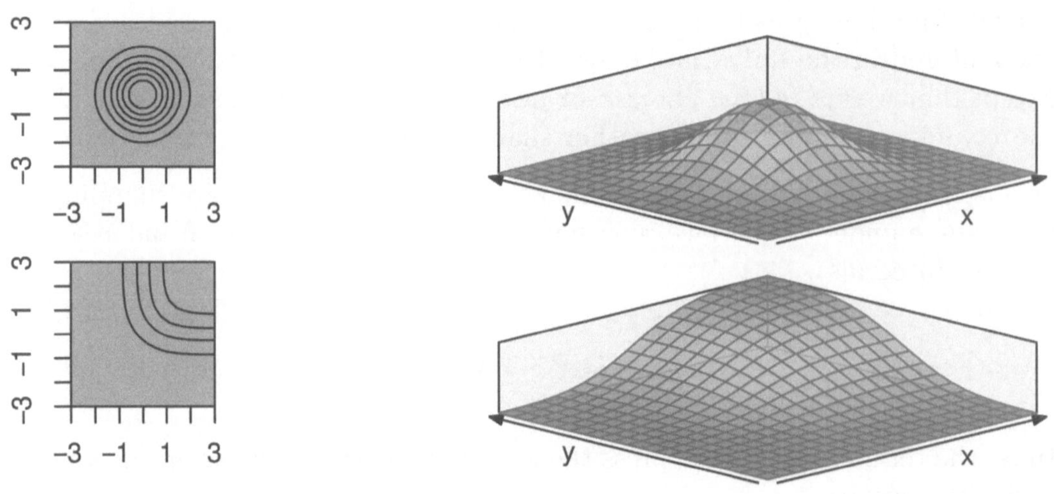

FIGURE 1.8: A bivariate normal distribution with zero correlation. Shown are four plots: the top-right plot shows the three-dimensional bivariate density, the top-left plot the contour plot of the distribution (seen from above). The lower plots show the cumulative distribution function from two views, as a three-dimensional plot and as a contour plot.

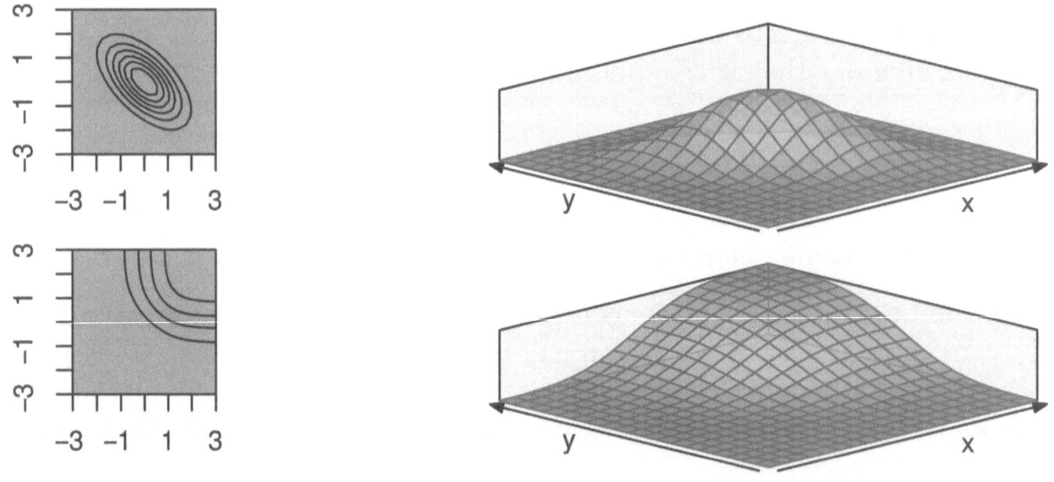

FIGURE 1.9: A bivariate normal distribution with a negative correlation of −0.6. Shown are four plots: the top-right plot shows the three-dimensional bivariate density and the top-left plot the contour plot of the distribution (seen from above). The lower plots show the cumulative distribution function from two views, as a three-dimensional plot and as a contour plot.

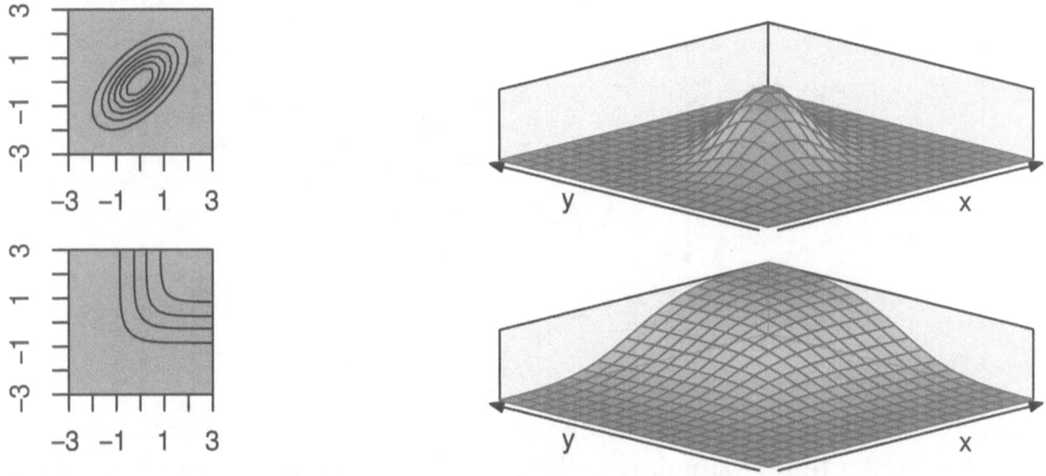

FIGURE 1.10: A bivariate normal distribution with a positive correlation of 0.6. Shown are four plots: the top-right plot shows the three-dimensional bivariate density and the top-left plot the contour plot of the distribution (seen from above). The lower plots show the cumulative distribution function from two views, as a three-dimensional plot and as a contour plot.

1.6.3 Generate simulated bivariate (or multivariate) data

Suppose we want to generate 100 pairs of correlated continuous data, with correlation $\rho = 0.6$. The two random variables both have a normal PDF, and have mean 0, and standard deviations 5 and 10, respectively.

Here is how we would generate such data. First, define a variance-covariance matrix; then, use the multivariate analog of the `rnorm()` function, `mvrnorm()`, from the `MASS` package to generate 100 data points.

```
## define a variance-covariance matrix:
Sigma <- matrix(c(5 ^ 2, 5 * 10 * 0.6, 5 * 10 * 0.6, 10 ^ 2),
                byrow = FALSE, ncol = 2)
## generate data:
u <- mvrnorm(n = 100, mu = c(0, 0), Sigma = Sigma)
head(u, n = 3)
```

```
##        [,1]  [,2]
## [1,]   4.54 12.42
## [2,]   7.13  1.11
## [3,]  -2.85  1.96
```

Figure 1.11 confirms that the simulated data are positively correlated.

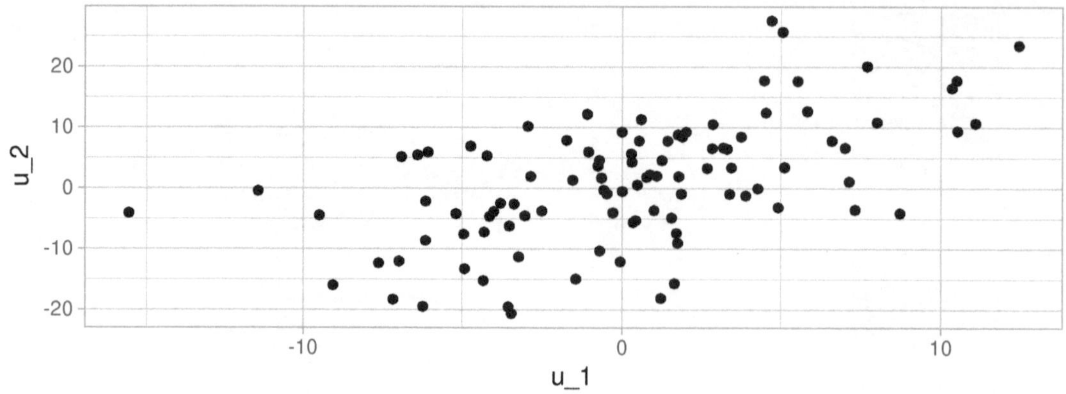

FIGURE 1.11: The relationship between two positively correlated random variables, generated by simulating data using the R function mvrnorm from the MASS library.

1.6.4 Decomposing a variance-covariance matrix

One final useful fact about the variance-covariance matrix—one that we will need later—is that it can be decomposed into the component standard deviations and an underlying correlation matrix. For example, consider the matrix above:

```
Sigma
```

```
##      [,1] [,2]
## [1,]   25   30
## [2,]   30  100
```

One can decompose the matrix as follows. The matrix can be seen as the product of a diagonal matrix of the standard deviations and the correlation matrix:

$$\begin{pmatrix} 5 & 0 \\ 0 & 10 \end{pmatrix} \begin{pmatrix} 1.0 & 0.6 \\ 0.6 & 1.0 \end{pmatrix} \begin{pmatrix} 5 & 0 \\ 0 & 10 \end{pmatrix} \tag{1.34}$$

One can reassemble the variance-covariance matrix by pre-multiplying and post-multiplying the correlation matrix with the diagonal matrix containing the standard deviations:[7]

[7]There is a built-in convenience function, `sdcor2cov` in the `SIN` package that does this calculation, taking the vector of standard deviations (not the diagonal matrix) and the correlation matrix to yield the variance-covariance matrix: `sdcor2cov(stddev = sds, corr = corrmatrix)`.

$$\begin{pmatrix} 5 & 0 \\ 0 & 10 \end{pmatrix} \begin{pmatrix} 1.0 & 0.6 \\ 0.6 & 1.0 \end{pmatrix} \begin{pmatrix} 5 & 0 \\ 0 & 10 \end{pmatrix} = \begin{pmatrix} 25 & 30 \\ 30 & 100 \end{pmatrix} \quad (1.35)$$

Using R (the symbol `%*%` is the matrix multiplication operator in R):

```
## sds:
(sds <- c(5, 10))
```

```
## [1]  5 10
```

```
## diagonal matrix:
(sd_diag <- diag(sds))
```

```
##      [,1] [,2]
## [1,]    5    0
## [2,]    0   10
```

```
## correlation matrix:
(corrmatrix <- matrix(c(1, 0.6, 0.6, 1), ncol = 2))
```

```
##      [,1] [,2]
## [1,]  1.0  0.6
## [2,]  0.6  1.0
```

```
sd_diag %*% corrmatrix %*% sd_diag
```

```
##      [,1] [,2]
## [1,]   25   30
## [2,]   30  100
```

This decomposition and reassembly of the variance-covariance matrix will become important when we start building hierarchical models in Stan.

1.7 An important concept: The marginal likelihood (integrating out a parameter)

Here, we introduce a concept that will turn up many times in this book. The concept we unpack here is called "integrating out a parameter." We will need this when we encounter Bayes' rule in the next chapter, and when we use Bayes factors later in the book (chapter 13).

Integrating out a parameter refers to the following situation. The example used here discusses the binomial distribution, but the approach is generally applicable for any distribution.

Suppose we have a binomial random variable Y with PMF $p(Y)$. Suppose also that this PMF is defined in terms of parameter θ that can have only three possible values, $0.1, 0.5, 0.9$, each with equal probability. In other words, the probability that θ is $0.1, 0.5$, or 0.9 is $1/3$ each.

We stick with our earlier example of $n = 10$ trials and $k = 7$ successes. The likelihood function then is

$$p(k = 7, n = 10|\theta) = \binom{10}{7} \theta^7 (1-\theta)^3 \tag{1.36}$$

There is a related concept of marginal likelihood, which we can write here as $p(k = 7, n = 10)$. Marginal likelihood is the likelihood computed by "marginalizing" out the parameter θ: for each possible value that the parameter θ can have, we compute the likelihood at that value and multiply that likelihood with the probability/density of that θ value occurring. Then we sum up each of the products computed in this way. Mathematically, this means that we carry out the following operation.

In our example, there are three possible values of θ, call them $\theta_1 = 0.1$, $\theta_2 = 0.5$, and $\theta_3 = 0.9$. Each has probability $1/3$; so $p(\theta_1) = p(\theta_2) = p(\theta_3) = 1/3$. Given this information, we can compute the marginal likelihood as follows:

$$\begin{aligned} p(k = 7, n = 10) = & \binom{10}{7} \theta_1^7 (1-\theta_1)^3 \times p(\theta_1) \\ & + \binom{10}{7} \theta_2^7 (1-\theta_2)^3 \times p(\theta_2) \\ & + \binom{10}{7} \theta_3^7 (1-\theta_3)^3 \times p(\theta_3) \end{aligned} \tag{1.37}$$

Writing the θ values and their probabilities, we get:

$$
\begin{aligned}
p(k=7, n=10) = & \binom{10}{7} 0.1^7 (1-0.1)^3 \times \frac{1}{3} \\
& + \binom{10}{7} 0.5^7 (1-0.5)^3 \times \frac{1}{3} \\
& + \binom{10}{7} 0.9^7 (1-0.9)^3 \times \frac{1}{3}
\end{aligned} \tag{1.38}
$$

Taking the common factors $\left(\frac{1}{3} \text{ and } \binom{10}{7}\right)$ out:

$$
\begin{aligned}
p(k=7, n=10) = & \frac{1}{3}\binom{10}{7} [0.1^7 (1-0.1)^3 \\
& + 0.5^7 (1-0.5)^3 \\
& + 0.9^7 (1-0.9)^3] \\
= & 0.058
\end{aligned} \tag{1.39}
$$

Thus, a marginal likelihood is a kind of weighted sum of the likelihood, weighted by the possible values of the parameter.[8]

The above example was contrived because we stated that the parameter θ has only three possible discrete values. Now consider the case where the parameter θ can have all possible values between 0 and 1; every possible value is equally likely. Formally, what this means is that the possible values of θ can be described in terms of the uniform distribution with lower bound 0 and upper bound 1 (for more details on this distribution, see section 5.2 in Blitzstein and Hwang 2014). This is a continuous distribution, and because the area under the distribution has to sum to 1, the density of every possible value of θ is 1.

In this example, the summation now has to be done over a continuous space [0, 1]. The way this summation is expressed in mathematics is through the integral symbol:

$$
p(k=7, n=10) = \int_0^1 \binom{10}{7} \theta^7 (1-\theta)^3 \, p(\theta) d\theta \tag{1.40}
$$

Because $p(\theta) = 1$ for all θ (in this particular case), we can just write:

$$
p(k=7, n=10) = \int_0^1 \binom{10}{7} \theta^7 (1-\theta)^3 \, d\theta \tag{1.41}
$$

[8]Where does the above formula come from? It falls out from the law of total probability discussed above.

This statement is computing something similar to what we computed above with the three discrete parameter values, except that the summation is being done over a continuous space ranging from 0 to 1. We say that the parameter θ has been integrated out, or marginalized. Integrating out a parameter will be a very common operation in this book, but we will never have to do the calculation ourselves. For the above case, we can compute the integral in R:

```
BinLik <- function(theta) {
  choose(10, 7) * theta^7 * (1 - theta)^3
}
integrate(BinLik, lower = 0, upper = 1)$value
```

```
## [1] 0.0909
```

The value that is output by the `integrate()` function above is the marginal likelihood.

This completes our discussion of random variables and probability distributions. Next, we summarize what we have learned so far about univariate distributions.

1.8 Summary of some useful R functions

Table 1.5 summarizes the different functions relating to PMFs and PDFs, using the binomial and normal as examples.

TABLE 1.5: Important R functions relating to random variables.

	Discrete	Continuous
Example:	$Binomial(y\|n, \theta)$	$Normal(y\|\mu, \sigma)$
Likelihood function	`dbinom()`	`dnorm()`
Prob Y=y	`dbinom()`	always 0
Prob $Y \geq y, Y \leq y, y_1 < Y < y_2$	`pbinom()`	`pnorm()`
Inverse CDF	`qbinom()`	`qnorm()`
Generate simulated data	`rbinom()`	`rnorm()`

Later on, we will use other distributions, such as the uniform, beta, etc., and each of these has their own set of `d-p-q-r` functions in R. One can look up these different distributions in, for example, Blitzstein and Hwang (2014).

1.9 Summary

This chapter briefly reviewed some very basic concepts in probability theory, univariate discrete and continuous random variables, and bivariate distributions. An important set of functions we encountered are the d-p-q-r family of functions for different distributions; these are very useful for understanding the properties of commonly used distributions, visualizing distributions, and for simulating data. Distributions will play a central role in this book; for example, knowing how to visualize distributions will be important for deciding on prior distributions for parameters. Other important ideas we learned about were marginal and conditional probability, marginal likelihood, and how to define multivariate distributions; these concepts will play an important role in Bayesian statistics.

1.10 Further reading

An informal but useful discussion about probability appears in Spiegelhalter (2024). A quick review of the mathematical foundations needed for statistics is available in the short book by Fox (2009), as well as Gill (2006). A more comprehensive introduction to the mathematical background needed for advanced Bayesian modeling is Jordan and Smith (2008). Morin (2016) and Blitzstein and Hwang (2014) are accessible introductions to probability theory. Ross (2002) offers a more advanced treatment which discusses random variable theory and illustrates applications of probability theory. Some basic results about random variables are also discussed in the above-mentioned textbooks (for example, the expectation and variance of sums of random variables); some of these results will be needed in later chapters of the present book. A good formal introduction to mathematical statistics (covering classical frequentist theory) is Miller and Miller (2004). The freely available book by Kerns (2014) introduces frequentist and Bayesian statistics from the ground up in a very comprehensive and systematic manner; the source code for the book is available from `https://github.com/gjkerns/IPSUR`. The open-access book, *Probability and Statistics: a simulation-based introduction*, by Bob Carpenter is also worth studying: `https://github.com/bob-carpenter/prob-stats`. A thorough introduction to the matrix algebra needed for statistics, with examples using R, is provided in Fieller (2016). Commonly used probability distributions are presented in detail in Miller and Miller (2004), Blitzstein and Hwang (2014), and Ross (2002). A useful reference for continuous univariate distributions is Johnson, Kotz, and Balakrishnan (1995).

2

Introduction to Bayesian data analysis

Before we can start analyzing realistic data sets using Bayes' rule, it is important to understand the application of Bayes' rule in one of the simplest of cases, data involving the binomial likelihood. This simple case is important to understand because it encapsulates the essence of the Bayesian approach to data analysis, and because it allows us to analytically work out the posterior distribution of the parameter of interest, using just pen and paper. This simple case also helps us to appreciate a crucial point: The posterior distribution of a parameter is a compromise between the prior and the likelihood. This important insight will play a central role in the realistic data analysis situations we will cover in the remainder of this book.

2.1 Bayes' rule

Recall Bayes' rule: When A and B are observable discrete events (such as "it has been raining" or "the streets are wet"), we can state the rule as follows:

$$P(A \mid B) = \frac{P(B \mid A)P(A)}{P(B)} \tag{2.1}$$

Given a vector of data $\boldsymbol{y}$, Bayes' rule allows us to work out the posterior distributions of the parameters of interest, which we can represent as the vector of parameters $\boldsymbol{\Theta}$. This computation is achieved by rewriting Equation (2.1) as (2.2). What is different here is that Bayes' rule is written in terms of probability distributions. Here, $p(\cdot)$ is a probability density function (continuous case) or a probability mass function (discrete case).

$$p(\boldsymbol{\Theta}|\boldsymbol{y}) = \frac{p(\boldsymbol{y}|\boldsymbol{\Theta}) \times p(\boldsymbol{\Theta})}{p(\boldsymbol{y})} \tag{2.2}$$

The above statement can be rewritten in words as follows:

$$\text{Posterior} = \frac{\text{Likelihood} \times \text{Prior}}{\text{Marginal Likelihood}} \tag{2.3}$$

DOI: 10.1201/9780429342646-3

The terms here have the following meanings. We elaborate on each point with an example below.

- The *Posterior*, $p(\boldsymbol{\Theta}|\boldsymbol{y})$, is the probability distribution of the parameters conditional on the data.
- The *Likelihood*, $p(\boldsymbol{y}|\boldsymbol{\Theta})$ is as described in chapter 1: it is the PMF (discrete case) or the PDF (continuous case) expressed as a function of $\boldsymbol{\Theta}$.
- The *Prior*, $p(\boldsymbol{\Theta})$, is the initial probability distribution of the parameter(s), before seeing the data.
- The *Marginal Likelihood*, $p(\boldsymbol{y})$, was introduced in chapter 1 and standardizes the posterior distribution to ensure that the area under the curve of the distribution sums to 1, that is, it ensures that the posterior is a valid probability distribution.

An example will clarify all these terms.

2.2 Deriving the posterior using Bayes' rule: An analytical example

Recall our cloze probability example earlier. Subjects are shown sentences like

"It's raining. I'm going to take the ..."

Suppose that 100 subjects are asked to complete the sentence. If 80 out of 100 subjects complete the sentence with "umbrella," the estimated cloze probability or predictability (given the preceding context) would be $\frac{80}{100} = 0.8$. This is the maximum likelihood estimate of the probability of producing this word; we will designate the estimate with a "hat" on the parameter name: $\hat{\theta} = 0.8$. In the frequentist paradigm, $\hat{\theta} = 0.8$ is an estimate of an unknown point value θ "out there in nature."

A crucial point to notice here is that the proportion 0.8 that we estimated above from the data can vary from one data set to another, and the variability in the estimate will be influenced by the sample size (the number of trials). For example, assuming that the true value of the θ parameter is in fact 0.8, if we repeatedly carry out the above experiment with say 10 trials, we will get some variability in the estimated proportion. Let's check this by carrying out 100 simulated experiments and computing the variability of the estimated means under repeated sampling:

```
estimated_means <- rbinom(100, size = 10, prob = 0.8) / 10
sd(estimated_means)
```

```
## [1] 0.117
```

The repeated runs of the (simulated) experiment are the sole underlying cause for the variability (shown by the output of the `sd(estimated)` command above) in the estimated proportion; the parameter $\theta = 0.80$ itself is invariant here (we are repeatedly estimating this point value).

However, consider now an alternative, radical idea: what if we treat θ as a random variable? That is, suppose now that θ has a PDF associated with it. This PDF would now represent our belief about possible values of θ, even before we have seen any data. For example, if at the outset of the experiment, we believe that all possible values between 0 and 1 are equally likely, we could represent that belief by stating that $\theta \sim Uniform(0, 1)$. The radical new idea here is that we now have a way to represent our prior belief or knowledge about plausible values of the parameter.

Now, if we were to run our simulated experiments again and again, there would be *two* sources of variability in the estimate of the parameter: the data as well as the uncertainty associated with θ.

```
theta <- runif(100, min = 0, max = 1)
estimated_means <- rbinom(100, size = 10, prob = theta) / 10
sd(estimated_means)
```

```
## [1] 0.309
```

The higher standard deviation is now coming from the uncertainty associated with the θ parameter. To see this, assume a "tighter" PDF for θ, say $\theta \sim Uniform(0.3, 0.8)$, then the variability in the estimated means would again be smaller, but not as small as when we assumed that θ was a point value:

```
theta <- runif(100, min = 0.3, max = 0.8)
estimated_means <- rbinom(n=100, size = 10, prob = theta) / 10
sd(estimated_means)
```

```
## [1] 0.199
```

In other words, the greater the uncertainty associated with the parameter θ, the greater the variability in the data.

The Bayesian approach to parameter estimation makes a radical departure from the standard frequentist assumption, which assumes that the true, unknown value of θ is some point value. In the Bayesian approach, θ is a random variable with a probability density/mass function associated with it. This PDF is called a prior distribution, and represents our prior belief or prior knowledge about possible values of this parameter. Once we obtain data, these data serve to modify our prior belief about this distribution; this updated probability density function of the parameter is called the posterior distribution. These ideas are unpacked in the sections below.

2.2.1 Choosing a likelihood

Under the assumptions we have set up above, the responses follow a binomial distribution, and so the PMF can be written as follows.

$$p(k|n,\theta) = \binom{n}{k}\theta^k(1-\theta)^{n-k} \tag{2.4}$$

where k indicates the number of times "umbrella" is given as an answer, and n is the number of trials. In our running example, k can therefore be any whole number going from 0 to 100.

In a particular experiment that we carry out, if we collect 100 data points ($n = 100$) and it turns out that $k = 80$, these data are now a fixed quantity. The only variable now in the PMF above is θ:

$$p(k = 80|n = 100,\theta) = \binom{100}{80}\theta^{80}(1-\theta)^{20} \tag{2.5}$$

The above function is a now a continuous function of the value θ, which has possible values ranging from 0 to 1. Compare this to the PMF of the binomial, which treats θ as a fixed value and defines a discrete distribution over the $n + 1$ possible discrete values k that we can observe (the possible number of successes).

Recall that the PMF and the likelihood are the same function seen from different points of view. The only difference between the two is what is considered to be fixed and what is varying. The PMF treats data as varying from experiment to experiment and θ as fixed, whereas the likelihood function treats the data that have been collected as fixed and the parameter θ as varying.

We now turn our attention back to our main goal, which is to find out, using Bayes' rule, the posterior distribution of θ given our data: $p(\theta|n,k)$. In order to use Bayes' rule to calculate this posterior distribution, we need to define a prior distribution over the parameter θ. In doing so, we are explicitly expressing our prior uncertainty about plausible values of θ.

2.2.2 Choosing a prior for θ

For the choice of prior for θ in the binomial distribution, we need to assume that the parameter θ is a random variable that has a PDF whose range lies within [0,1], the range over which θ can vary (this is because θ represents a probability). The beta distribution, which is a PDF for a continuous random variable, is commonly used as prior for parameters representing probabilities. One reason for this choice is that its PDF ranges over the interval $[0, 1]$. The other reason for this choice is that it makes the Bayes' rule calculation remarkably easy.

The beta distribution has the following PDF.

$$p(\theta|a,b) = \frac{1}{B(a,b)}\theta^{a-1}(1-\theta)^{b-1} \tag{2.6}$$

The term $B(a,b)$ expands to $\int_0^1 \theta^{a-1}(1-\theta)^{b-1}\,\mathrm{d}\theta$, and is a normalizing constant that ensures that the area under the curve sums to one. In some textbooks, you may see the PDF of the beta distribution with the normalizing constant $\frac{\Gamma(a+b)}{\Gamma(a)\Gamma(b)}$ (the expression $\Gamma(n)$ is defined as (n-1)!):

$$p(\theta|a,b) = \frac{\Gamma(a+b)}{\Gamma(a)\Gamma(b)}\theta^{a-1}(1-\theta)^{b-1}$$

These two statements for the beta distribution are identical because $B(a,b)$ can be shown to be equal to $\frac{\Gamma(a)\Gamma(b)}{\Gamma(a+b)}$ (Ross 2002).

The beta distribution's parameters a and b can be interpreted as expressing our prior beliefs about the probability of success; a represents the number of "successes," in our case, answers that are "umbrella" and b the number of failures, the answers that are not "umbrella." Figure 2.1 shows the different beta distribution shapes given different values of a and b.

As in the binomial and normal distributions that we saw in chapter 1, one can analytically derive the formulas for the expectation and variance of the beta distribution. These are:

$$\mathrm{E}[X] = \frac{a}{a+b} \qquad \mathrm{Var}(X) = \frac{a \times b}{(a+b)^2(a+b+1)} \tag{2.7}$$

As an example, choosing $a = 4$ and $b = 4$ would mean that the answer "umbrella" is as likely as a different answer, but we are relatively unsure about this. We could express our uncertainty by computing the region over which we are 95% certain that the value of the parameter lies; this is the *95% credible interval.*

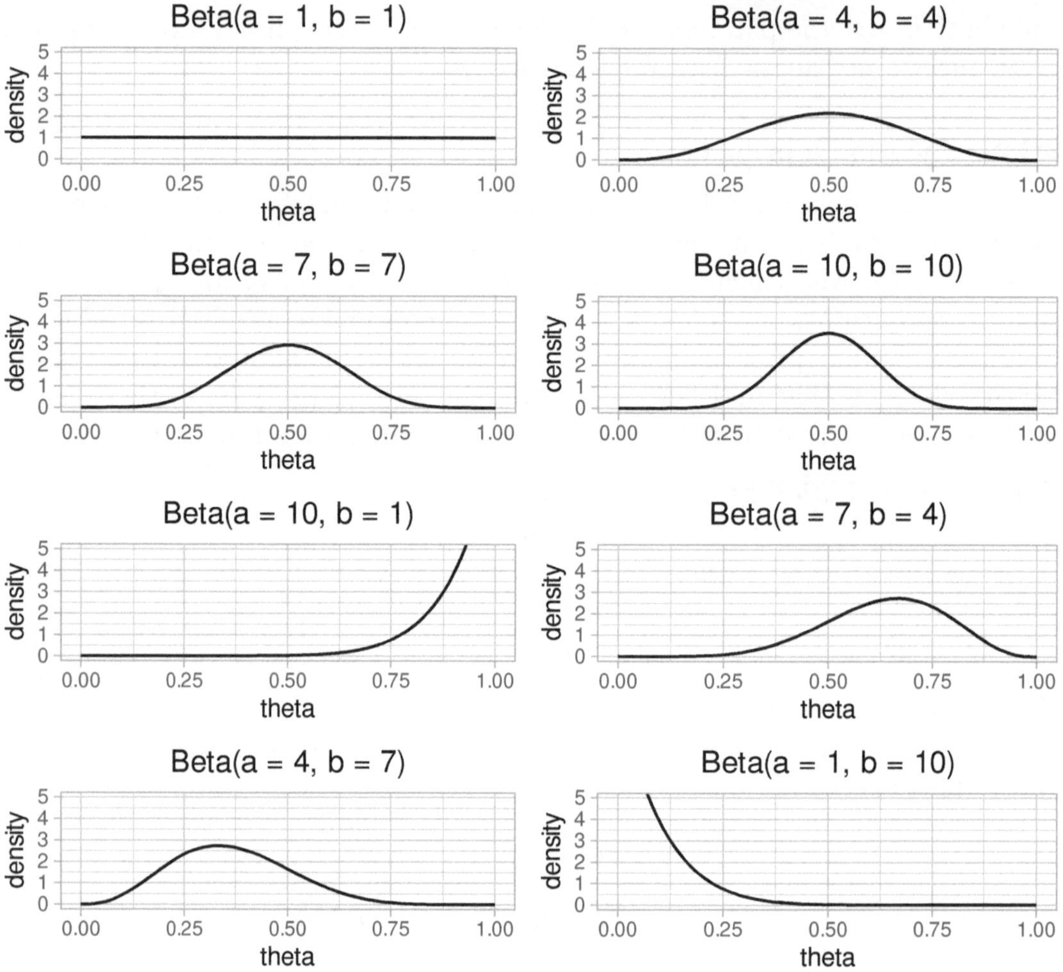

FIGURE 2.1: Examples of beta distributions with different parameters.

For this, we would use the `qbeta()` function in R; the parameters a and b are called `shape1` and `shape2` in R.

```
qbeta(c(0.025, 0.975), shape1 = 4, shape2 = 4)
```

```
## [1] 0.184 0.816
```

The credible interval chosen above is an equal-tailed interval: the area below the lower bound and above the upper bound is the same (0.025 in the above case). One could define alternative intervals; for example, in a distribution with only one mode (one peak; a unimodal distribution), one could choose to use the narrowest interval that contains the mode. This is called the highest posterior density interval (HDI). In skewed posterior distributions, the equal-tailed credible interval and the HDI will not be identical, because the HDI will have unequal tail probabilities. Some authors, such as Kruschke (2014), prefer to report the HDI. We will use the equal-tailed interval in this book, simply because this is the standard output in Stan and `brms`.

If we were to choose $a = 10$ and $b = 10$, we would still be assuming that a priori the answer "umbrella" is just as likely as some other answer, but now our prior uncertainty about this mean is lower, as the 95% credible interval computed below shows.

```
qbeta(c(0.025, 0.975), shape1 = 10, shape2 = 10)
```

```
## [1] 0.289 0.711
```

In Figure 2.1, we can see also the difference in uncertainty in these two examples graphically.

Which prior should we choose? In a real data analysis problem, the choice of prior would depend on what prior knowledge we want to bring into the analysis (see the online chapter E). If we don't have much prior information, we could use $a = b = 1$; this gives us a uniform prior (i.e., $Uniform(0, 1)$). This kind of prior goes by various names, such as *flat, non-informative prior*, or *uninformative prior*. By contrast, if we have a lot of prior knowledge and/or a strong belief (e.g., based on a particular theory's predictions, or prior data) that θ has a particular range of plausible values, we can use a different set of a, b values to reflect our belief about the parameter. Generally speaking, the larger the values of the parameters a and b, the narrower the spread of the distribution; i.e., the lower our uncertainty about the mean value of the parameter.

We will discuss prior specification in detail in the online chapter E. For the moment, just for illustration, we choose the values $a = 4$ and $b = 4$ for the beta prior. Then, our prior for θ is the following beta PDF:

$$p(\theta) = \frac{1}{B(4,4)}\theta^3(1-\theta)^3 \tag{2.8}$$

Having chosen a likelihood, and having defined a prior on θ, we are ready to carry out our first Bayesian analysis to derive a posterior distribution for θ.

2.2.3 Using Bayes' rule to compute the posterior $p(\theta|n, k)$

Having specified the likelihood and the prior, we will now use Bayes' rule to calculate $p(\theta|n, k)$. Using Bayes' rule simply involves replacing the likelihood and the prior we defined above into the equation we saw earlier:

$$\text{Posterior} = \frac{\text{Likelihood} \times \text{Prior}}{\text{Marginal Likelihood}} \tag{2.9}$$

Replace the terms for likelihood and prior into this equation:

$$p(\theta|n = 100, k = 80) = \frac{\left[\binom{100}{80}\theta^{80} \times (1-\theta)^{20}\right] \times \left[\frac{1}{B(4,4)} \times \theta^{3}(1-\theta)^{3}\right]}{p(k = 80)} \tag{2.10}$$

where $p(k = 80)$ is $\int_0^1 p(k = 80|n = 100, \theta)p(\theta)\, d\theta$. This term will be a constant once the number of successes k is known; this is the marginal likelihood we encountered in chapter 1. In fact, once k is known, there are several constant values in the above equation; they are constants because none of them depend on the parameter of interest, θ. We can collect all of these together:

$$p(\theta|n = 100, k = 80) = \left[\frac{\binom{100}{80}}{B(4,4) \times p(k = 80)}\right][\theta^{80}(1-\theta)^{20} \times \theta^{3}(1-\theta)^{3}] \tag{2.11}$$

The first term in square brackets, $\frac{\binom{100}{80}}{B(4,4) \times p(k=80)}$, is all the constants collected together, and is the normalizing constant we have seen before; it makes the posterior distribution $p(\theta|n = 100, k = 80)$ sum to one. Since it is a constant, we can ignore it for now and focus on the two other terms in the equation. Because we are ignoring the constant, we will now say that the posterior is proportional to the right-hand side.

$$p(\theta|n = 100, k = 80) \propto [\theta^{80}(1-\theta)^{20} \times \theta^{3}(1-\theta)^{3}] \tag{2.12}$$

A common way of writing the above equation is:

$$\text{Posterior} \propto \text{Likelihood} \times \text{Prior} \tag{2.13}$$

Resolving the right-hand side now simply involves adding up the exponents! In this example, computing the posterior really does boil down to this simple addition operation on the exponents.

$$p(\theta|n = 100, k = 80) \propto [\theta^{80+3}(1-\theta)^{20+3}] = \theta^{83}(1-\theta)^{23} \tag{2.14}$$

The expression on the right-hand side corresponds to a beta distribution with parameters $a = 84$ and $b = 24$. This becomes evident if we rewrite the right-hand side such that it represents the core part of a beta PDF (see equation

(2.6)). All that is missing is a normalizing constant which would make the area under the curve sum to one.

$$\theta^{83}(1-\theta)^{23} = \theta^{84-1}(1-\theta)^{24-1} \tag{2.15}$$

This core part of any PDF or PMF is called the kernel of that distribution. Without a normalizing constant, the area under the curve will not sum to one. Let's check this:

```
PostFun <- function(theta) {
  theta^84 * (1 - theta)^24
}
(AUC <- integrate(PostFun, lower = 0, upper = 1)$value)
```

```
## [1] 1.42e-26
```

So the area under the curve (AUC) is not 1—the posterior that we computed above is not a proper probability distribution. What we have just done above is to compute the following integral:

$$\int_0^1 \theta^{84}(1-\theta)^{24} \tag{2.16}$$

We can use this integral to figure out what the normalizing constant is. Basically, we want to know what the constant k is such that the area under the curve sums to 1:

$$k\int_0^1 \theta^{84}(1-\theta)^{24} = 1 \tag{2.17}$$

We know what $\int_0^1 \theta^{84}(1-\theta)^{24}$ is; we just computed that value (called `AUC` in the R code above). So, the normalizing constant is:

$$k = \frac{1}{\int_0^1 \theta^{84}(1-\theta)^{24}} = \frac{1}{AUC} \tag{2.18}$$

So, all that is needed to make the kernel $\theta^{84}(1-\theta)^{24}$ into a proper probability distribution is to include a normalizing constant, which, according to the definition of the beta distribution (Equation (2.6)), would be $\boldsymbol{B}(84, 24)$. This term is in fact the integral we computed above.

So, what we have is the distribution of θ given the data, expressed as a PDF:

$$p(\theta|n = 100, k = 80) = \frac{1}{B(84, 24)}\theta^{84-1}(1 - \theta)^{24-1} \tag{2.19}$$

Now, this function will sum to one:

```
PostFun <- function(theta) {
  theta^84 * (1 - theta)^24 / AUC
}
integrate(PostFun, lower = 0, upper = 1)$value
```

```
## [1] 1
```

2.2.4 Summary of the procedure

To summarize, we started with data ($n = 100, k = 80$) and a binomial likelihood, multiplied it with the prior probability density function $\theta \sim Beta(4, 4)$, and obtained the posterior $p(\theta|n, k) \sim Beta(84, 24)$. The constants were ignored when carrying out the multiplication; we say that we computed the posterior *up to proportionality*. Finally, we showed how, in this simple example, the posterior can be rescaled to become a probability distribution, by including a proportionality constant.

The above example is a case of a *conjugate* analysis: the posterior on the parameter has the same form (belongs to the same family of probability distributions) as the prior. The above combination of likelihood and prior is called the beta-binomial conjugate case. There are several other such combinations of Likelihoods and Priors that yield a posterior that has a PDF that belongs to the same family as the PDF on the prior; some examples will appear in the exercises.

Formally, conjugacy is defined as follows: Given the likelihood $p(y|\theta)$, if the prior $p(\theta)$ results in a posterior $p(\theta|y)$ that has the same form as $p(\theta)$, then we call $p(\theta)$ a conjugate prior.

For the beta-binomial conjugate case, we can derive a very general relationship between the likelihood, prior, and posterior. Given the binomial likelihood up to proportionality (ignoring the constant) $\theta^k(1 - \theta)^{n-k}$, and given the prior, also up to proportionality, $\theta^{a-1}(1 - \theta)^{b-1}$, their product will be:

$$\theta^k(1 - \theta)^{n-k}\theta^{a-1}(1 - \theta)^{b-1} = \theta^{a+k-1}(1 - \theta)^{b+n-k-1} \tag{2.20}$$

Thus, given a $Binomial(n, k|\theta)$ likelihood, and a $Beta(a, b)$ prior on θ, the posterior will be $Beta(a + k, b + n - k)$.

2.2.5 Visualizing the prior, likelihood, and posterior

We established in the example above that the posterior is a beta distribution with parameters $a = 84$ and $b = 24$. We visualize the likelihood, prior, and the posterior side by side in Figure 2.2.

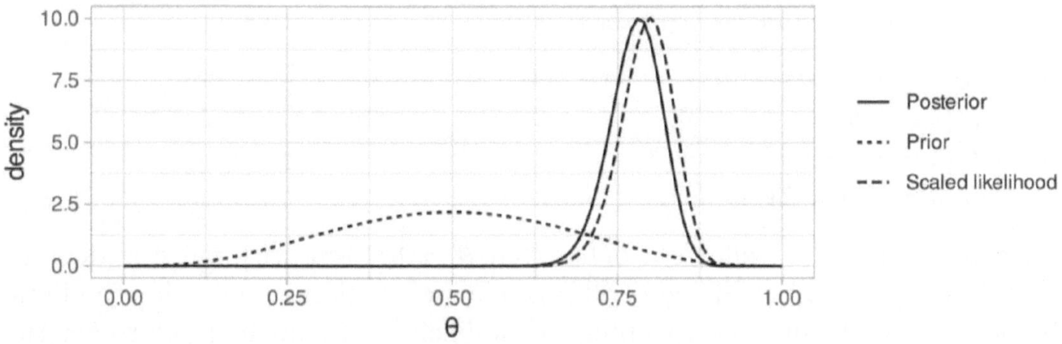

FIGURE 2.2: The (scaled) likelihood, prior, and posterior in the beta-binomial conjugate example. The likelihood is scaled to integrate to 1 to make it easier to compare to the prior and posterior distributions.

We can summarize the posterior distribution either graphically as we did above, or summarize it by computing the mean and the variance. The mean gives us an estimate of the cloze probability of producing "umbrella" in that sentence (given the model, i.e., given the likelihood and prior):

$$\mathrm{E}[\hat{\theta}] = \frac{84}{84 + 24} = 0.78 \tag{2.21}$$

$$\mathrm{var}[\hat{\theta}] = \frac{84 \times 24}{(84 + 24)^2(84 + 24 + 1)} = 0.0016 \tag{2.22}$$

We could also display the 95% credible interval, the range over which we are 95% certain that θ lies, given the data and model.

```
qbeta(c(0.025, 0.975), shape1 = 84, shape2 = 24)
```

```
## [1] 0.695 0.851
```

Typically, we would summarize the results of a Bayesian analysis by displaying the posterior distribution of the parameter (or parameters) graphically, along with the above summary statistics: the mean, the standard deviation or variance, and the 95% credible interval. You will see many examples of such summaries later.

2.2.6 The posterior distribution is a compromise between the prior and the likelihood

Recall from the preceding sections that the a and b parameters in the beta distribution determine the shape of the prior distribution on the θ parameter. Just for the sake of illustration, let's take four different beta priors, which reflect increasing prior certainty about θ.

- $Beta(a = 2, b = 2)$
- $Beta(a = 3, b = 3)$
- $Beta(a = 6, b = 6)$
- $Beta(a = 21, b = 21)$

Each of these priors reflects a belief that $\theta = 0.5$, but with varying degrees of (un)certainty. Given the general formula we developed above for the beta-binomial case, we just need to plug in the likelihood and the prior to get the posterior:

$$p(\theta|n, k) \propto p(k|n, \theta)p(\theta) \tag{2.23}$$

The four corresponding posterior distributions would be:

$$p(\theta \mid k, n) \propto [\theta^{80}(1-\theta)^{20}][\theta^{2-1}(1-\theta)^{2-1}] = \theta^{82-1}(1-\theta)^{22-1} \tag{2.24}$$

$$p(\theta \mid k, n) \propto [\theta^{80}(1-\theta)^{20}][\theta^{3-1}(1-\theta)^{3-1}] = \theta^{83-1}(1-\theta)^{23-1} \tag{2.25}$$

$$p(\theta \mid k, n) \propto [\theta^{80}(1-\theta)^{20}][\theta^{6-1}(1-\theta)^{6-1}] = \theta^{86-1}(1-\theta)^{26-1} \tag{2.26}$$

$$p(\theta \mid k, n) \propto [\theta^{80}(1-\theta)^{20}][\theta^{21-1}(1-\theta)^{21-1}] = \theta^{101-1}(1-\theta)^{41-1} \tag{2.27}$$

We can visualize each of these triplets of priors, likelihoods, and posteriors; see Figure 2.3.

Given some data and given a likelihood function, the tighter the prior, the greater the extent to which the posterior orients itself toward the prior. In general, we can say the following about the likelihood-prior-posterior relationship:

- The posterior distribution of a parameter is a compromise between the prior and the likelihood.
- For a given set of data, the greater the certainty in the prior, the more heavily will the posterior be influenced by the prior mean.
- Conversely, for a given set of data, the greater the *un*certainty in the prior, the more heavily will the posterior be influenced by the likelihood.

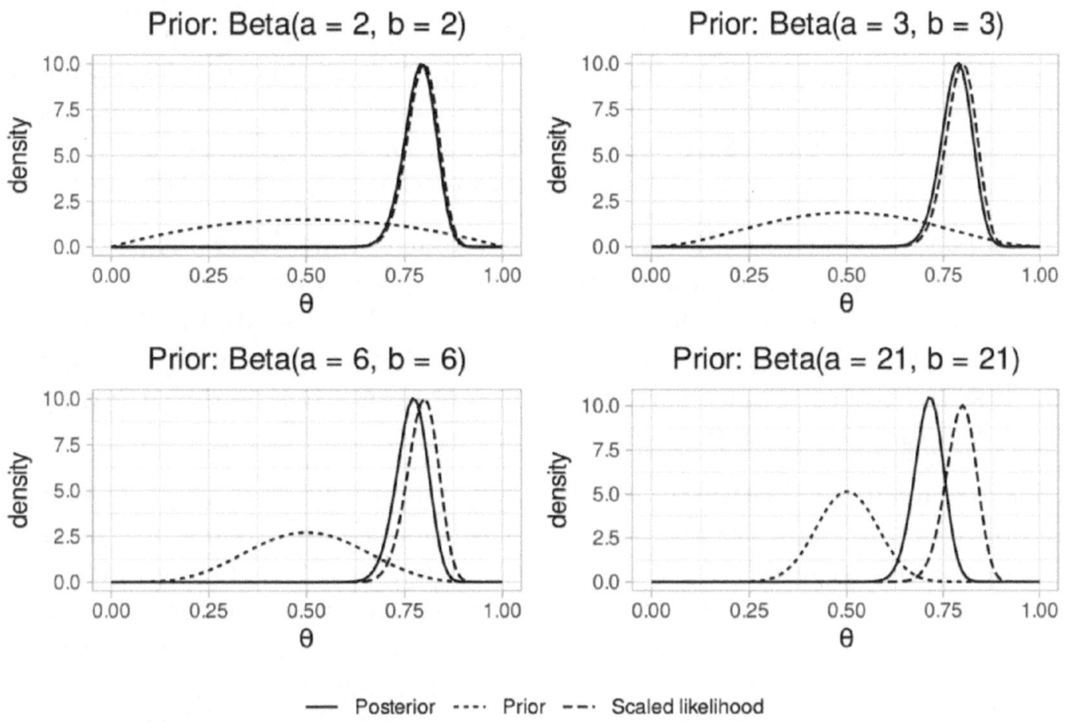

FIGURE 2.3: The (scaled) likelihood, prior, and posterior in the beta-binomial conjugate example, for different uncertainties in the prior. The likelihood is scaled to integrate to 1 to make its comparison easier.

Another important observation emerges if we increase the sample size (here, the number of trials) from 100 to, say, 1,000,000. Suppose we still get a sample mean of 0.8 here, so that $k = 800{,}000$. Now, the posterior mean will be influenced almost entirely by the sample mean. This is because, in the general form for the posterior $Beta(a+k, b+n-k)$ that we computed above, the n and k become very large relative to the a, b values, and dominate in determining the posterior mean.

Whenever we do a Bayesian analysis, it is good practice to check whether the parameter you are interested in estimating is sensitive to the prior specification. Such an investigation is called a *sensitivity analysis*. Later in this book, we will see many examples of sensitivity analyses in realistic data-analysis settings.

2.2.7 Incremental knowledge gain using prior knowledge

In the above example, we used an artificial example where we asked 100 subjects to complete the sentence shown at the beginning of the chapter, and then we counted the number of times that they produced "umbrella" vs. some other word as a continuation. Given 80 instances of "umbrella," and using a $Beta(4,4)$ prior, we derived the posterior to be $Beta(84,24)$. We could now use this posterior as our prior for the next study. Suppose that we were to carry out a second experiment, again with 100 subjects, and this time 60 produced "umbrella." We could now use our new prior ($Beta(84, 24)$) to obtain an updated posterior. We have $a = 84, b = 24, n = 100, k = 60$. This gives us as posterior: $Beta(a + k, b + n - k) = Beta(84 + 60, 24 + 100 - 60) = Beta(144, 64)$.

Now, if we were to pool all our data that we have from the two experiments, then we would have as data $n = 200, k = 140$. Suppose that we keep our initial prior of $a = 4, b = 4$. Then, our posterior would be $Beta(4+140, 4+200-140) = Beta(144, 64)$. This is exactly the same posterior that we got when first analyzed the first 100 subjects' data, derived the posterior, and then used that posterior as a prior for the next 100 subjects' data.

This toy example illustrates an important point that has great practical importance for cognitive science. One can incrementally gain information about a research question by using information from previous studies and deriving a posterior, and then use that posterior as a prior for the next study. For practical examples from psycholinguistics showing how information can be pooled from previous studies, see Jäger, Engelmann, and Vasishth (2017) and Nicenboim, Roettger, and Vasishth (2018). Vasishth and Engelmann (2022) illustrates an example of how the posterior from a previous study or collection of studies can be used as a prior to compute the posterior from new data. This approach allows us to build on the information available from previous work.

2.3 Summary

In this chapter, we learned how to use Bayes' rule in the specific case of a binomial likelihood, and a beta prior on the θ parameter in the likelihood function. Our goal in any Bayesian analysis will follow the path we took in this simple example: decide on an appropriate likelihood function, decide on priors for all the parameters involved in the likelihood function, and use this model (i.e., the likelihood and the priors) to derive the posterior distribution of each parameter. Then we draw inferences about our research question based on the posterior distribution of the parameter.

In the example discussed in this chapter, Bayesian analysis was easy. This was because we considered the simple conjugate case of the beta-binomial. In realistic data-analysis settings, our likelihood function will be very complex, and many parameters will be involved. Multiplying the likelihood function and the priors will become mathematically difficult or impossible. For such situations, we use computational methods to obtain samples from the posterior distributions of the parameters.

2.4 Further reading

Accessible introductions to conjugate Bayesian analysis are Lynch (2007), and Lunn et al. (2012). Somewhat more demanding discussions of conjugate analysis are in Lee (2012), Carlin and Louis (2008), Christensen et al. (2011), O'Hagan and Forster (2004), and Bernardo and Smith (2009).

Part II

Regression models with brms

3

Computational Bayesian data analysis

In the previous chapter, we learned how to analytically derive the posterior distribution of the parameters in our model. In practice, however, this is possible for only a very limited number of cases. Although the numerator in Bayes' rule, the unnormalized posterior, is easy to calculate (by multiplying the likelihood and the probability density/mass function), the denominator, the marginal likelihood, requires us to carry out an integration; see Equation (3.1).

$$
\begin{aligned}
p(\Theta|y) &= \frac{p(y|\Theta) \cdot p(\Theta)}{p(y)} \\
p(\Theta|y) &= \frac{p(y|\Theta) \cdot p(\Theta)}{\int_{\Theta} p(y|\Theta) \cdot p(\Theta) d\Theta}
\end{aligned}
\tag{3.1}
$$

Unless we are dealing with conjugate distributions, the solution will be extremely hard to derive or there will be no analytical solution. This was the major bottleneck of Bayesian analysis in the past, and required Bayesian practitioners to program an approximation method by themselves before they could even begin the Bayesian analysis. Fortunately, many of the probabilistic programming languages freely available today (see the next section for a listing) allow us to define our models without having to acquire expert knowledge about the relevant numerical techniques.

3.1 Deriving the posterior through sampling

Let's say that we want to derive the posterior of the model from section 2.2, that is, the posterior distribution of the cloze probability of "*umbrella*," θ, given the following data: a word (e.g., "*umbrella*") was answered 80 out of 100 times, and assuming a binomial distribution as the likelihood function, and $Beta(a = 4, b = 4)$ as a prior distribution for the cloze probability. If we can generate samples[1] from the posterior distribution of θ, instead of an analytically derived posterior distribution, given enough samples we will have a

[1]In the brms package (Bürkner 2017), samples from the posterior are referred to as draws.

DOI: 10.1201/9780429342646-5

good approximation of the posterior distribution. Obtaining samples from the posterior will be the only viable option in the models that we will discuss in this book. By "obtaining samples," we are talking about a situation analogous to when we use `rbinom()` or `rnorm()` to obtain samples from a particular distribution. For more details about sampling algorithms, see the section Further Reading (section 3.10).

Thanks to probabilistic programming languages, it will be relatively straightforward to get these samples, and we will discuss how we will do it in more detail in the next section. For now let's assume that we used some probabilistic programming language to obtain 20,000 samples from the posterior distribution of the cloze probability, θ: 0.812, 0.814, 0.824, 0.799, 0.81, 0.755, 0.818, 0.809, 0.859, 0.763, 0.746, 0.717, 0.736, 0.789, 0.771, 0.81, 0.76, 0.758, 0.791, 0.76, ... Figure 3.1 shows that the approximation of the posterior looks quite similar to the analytically derived posterior. The difference between the analytically computed and approximated mean and variance are 0.00009 and 0.000005, respectively.

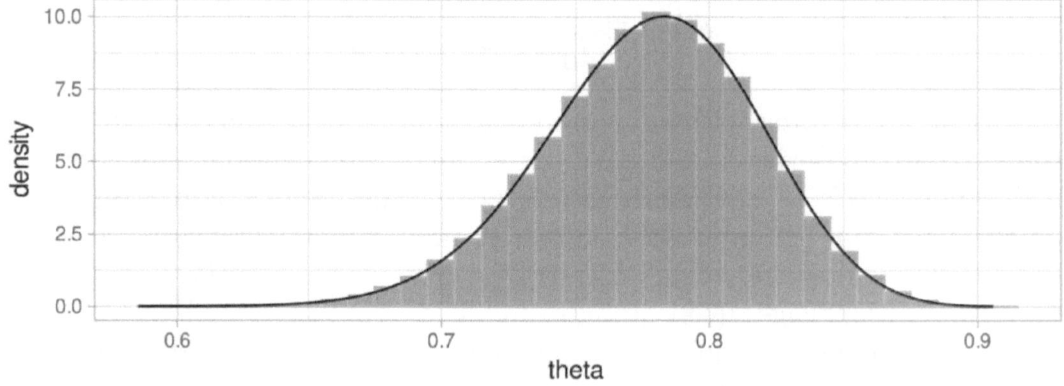

FIGURE 3.1: Histogram of the samples of θ from the posterior distribution generated via sampling. The black line shows the density plot of the analytically derived posterior.

3.2 Bayesian Regression Models using Stan: brms

The surge in popularity of Bayesian statistics is closely tied to the increase in computing power and the appearance of probabilistic programming languages, such as WinBUGS (Lunn et al. 2000), JAGS (Plummer 2016), PyMC3 (Salvatier, Wiecki, and Fonnesbeck 2016), Turing (Ge, Xu, and Ghahramani 2018), and Stan (Carpenter et al. 2017); for a historical review, see Plummer (2022).

These probabilistic programming languages allow the user to define models without having to deal (for the most part) with the complexities of the sampling process. However, they require learning a new language since the user has to fully specify the statistical model using a particular syntax.[2] Furthermore, some knowledge of the sampling process is needed to correctly parameterize the models and to avoid convergence issues (these topics will be covered in detail in chapter 9).

There are some alternatives that allow Bayesian inference in R without having to fully specify the model "by hand." The R packages `rstanarm` (Goodrich et al. 2018) and `brms` (Bürkner 2024) provide Bayesian equivalents of many popular R model-fitting functions, such as (g)lmer (Bates et al. 2015) and many others; both `rstanarm` and `brms` use Stan as the back-end for estimation and sampling. The package R-INLA (Lindgren and Rue 2015) allows for fitting a limited selection of likelihood functions and priors in comparison to `rstanarm` and `brms` (R-INLA can fit models that can be expressed as latent Gaussian models). This package uses the integrated nested Laplace approximation (INLA) method for carrying out Bayesian inference, rather than a sampling algorithm as in the other probabilistic languages listed above. Another alternative is JASP (JASP Team 2019), which provides a graphical user interface for both frequentist and Bayesian modeling, and is intended to be an open-source alternative to SPSS.

We will focus on `brms` in this part of the book. This is because it can be useful for a smooth transition from frequentist models to their Bayesian equivalents. The package `brms` is not only powerful enough to satisfy the statistical needs of many cognitive scientists, it has the added benefit that the Stan code can be inspected (with the `brms` functions `make_stancode()` and `make_standata()`), allowing the user to customize their models or learn from the code produced internally by `brms` to eventually transition to writing the models entirely in Stan. In the introduction to Stan in chapter 8, we implement in Stan the models presented in the current and the following chapters.

3.2.1 A simple linear model: A single subject pressing a button repeatedly (a finger tapping task)

To illustrate the basic steps for fitting a model using `brms`, consider the following example of a finger tapping task (for a review, see Hubel et al. 2013). Suppose that a subject first sees a blank screen. Then, after a certain amount of time (say 200 ms), the subject sees a cross in the middle of a screen, and as soon as they see the cross, they tap on the space bar as fast as they can until the experiment is over (361 trials). The dependent measure here is the time it

[2]The Python package PyMC3 and the Julia library Turing are recent exceptions since they are fully integrated into their respective languages.

takes in milliseconds from one press of the space bar to the next one. The data in each trial are therefore finger tapping times in milliseconds. Suppose that the research question is: how long does it take for this particular subject to press a key?

Let's model the data with the following assumptions:

1. There is a true (unknown) underlying time, μ ms, that the subject needs to press the space bar.
2. There is some noise in this process.
3. The noise is normally distributed (this assumption is questionable given that finger tapping as also response times are generally skewed; we will fix this assumption later).[3]

This means that the likelihood for each observation n will be:

$$t_n \sim Normal(\mu, \sigma) \tag{3.2}$$

where $n = 1, \ldots, N$, and t is the dependent variable (finger tapping times in milliseconds). The variable N indexes the total number of data points. The symbol μ indicates the *location* of the normal distribution function; the location parameter shifts the distribution left or right on the horizontal axis. For the normal distribution, the location is also the mean of the distribution. The symbol σ indicates the *scale* of the distribution; as the scale decreases, the distribution gets narrower. This compressing approaches a spike (all the probability mass get concentrated near one point) as the scale parameter approaches zero. For the normal distribution, the scale is also its standard deviation.

The reader may have encountered the model shown in Equation (3.2) in the form shown in Equation (3.3):

$$t_n = \mu + \varepsilon_n, \text{ where } \varepsilon_n \stackrel{iid}{\sim} Normal(0, \sigma) \tag{3.3}$$

When the model is written in this way, it should be understood as meaning that each data point t_n has some variability around a mean value μ, and that variability has standard deviation σ. The term "iid" (independent and identically distributed) implies that the residuals are independently generated (they are not correlated with any of the other residual values). The statement

[3]We refer to the time it takes for a subject to respond or react to a stimulus as response time (response and reaction times are often used interchangeably, cf. Luce 1991). In this case, however, there are no stimuli other than the cross on the screen, and the subject only taps the space bar as soon as they see the cross.

of the model in Equation (3.3) is exactly the same as saying that the observed data points t_n are iid and are coming from the $Normal(\mu, \sigma)$ distribution.

For a frequentist model that will give us the maximum likelihood estimate (the sample mean) of the time it takes to press the space bar, this would be enough information to write the formula in R, `t ~ 1`, and plug it into the function `lm()` together with the data: `lm(t ~ 1, data)`. The meaning of the `1` here is that `lm()` will estimate the intercept in the model, which is the estimate of μ in our example. If the reader is completely unfamiliar with linear models, the references in section 4.5 will be helpful.

For a Bayesian linear model, we will also need to define priors for the two parameters in our model. Let's say that we know for sure that the time it takes to press a key will be positive and lower than a minute (or 60000 ms), but we don't want to make a commitment regarding which values are more likely. We encode what we know about the noise in the task in σ: we know that this parameter must be positive and we'll assume that any value below 2000 ms is equally likely. These priors are in general strongly discouraged: A flat (or very wide) prior will almost never be the best approximation of what we know. Prior specification is discussed in detail in the online chapter E.

In this case, even if we know very little about the task, we know that pressing the spacebar will take at most a couple of seconds. We'll start with flat priors in this section for pedagogical purposes; the next sections will already show more realistic uses of priors.

$$\begin{aligned} \mu &\sim Uniform(0, 60000) \\ \sigma &\sim Uniform(0, 2000) \end{aligned} \tag{3.4}$$

First, load the data frame `df_spacebar` from the `bcogsci` package:

```
data("df_spacebar")
df_spacebar
```

```
## # A tibble: 361 x 2
##       t trial
##   <int> <int>
## 1   141     1
## 2   138     2
## 3   128     3
## # i 358 more rows
```

It is always a good idea to plot the data before doing anything else; see Figure 3.2. As we suspected, the data look a bit skewed, but we ignore this for the moment.

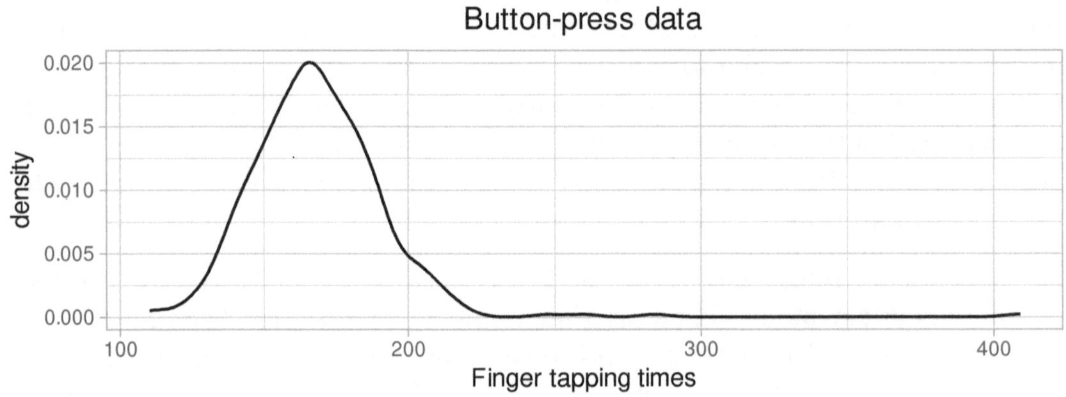

FIGURE 3.2: Visualizing the button-press data.

3.2.1.1 Specifying the model in `brms`

Fit the model defined by Equations (3.2) and (3.4) with `brms` in the following way.

```
fit_press <-
  brm(t ~ 1,
      data = df_spacebar,
      family = gaussian(),
      prior =
 c(prior(uniform(0, 60000), class = Intercept, lb = 0, ub = 60000),
   prior(uniform(0, 2000), class = sigma, lb = 0, ub = 2000)),
      chains = 4,
      iter = 2000,
      warmup = 1000)
```

The `brms` code has some differences from a model fit with `lm()`. At this beginning stage, we'll focus on the following options:

1. The term `family = gaussian()` makes it explicit that the underlying likelihood function is a normal distribution (Gaussian and normal are synonyms). This detail is implicit in the R function `lm()`. Other linking functions are possible, exactly as in the `glm()` function in R. The default for `brms` that corresponds to the `lm()` function is `family = gaussian()`.
2. The term `prior` takes as argument a list of prior specifications. Although this specification of priors is optional, the researcher should

always explicitly specify each prior. Otherwise, `brms` will define priors by default, which may or may not be appropriate for the research area. In cases where the distribution has a restricted coverage, that is, not every value is valid (e.g., smaller than 0 or larger than 60,000 are not valid for the intercept), we need to set lower and upper boundaries with `lb` and `ub`.[4]

3. The term `chains` refers to the number of independent runs for sampling (by default four).
4. The term `iter` refers to the number of iterations that the sampler makes to sample from the posterior distribution of each parameter (by default `2000`).
5. The term `warmup` refers to the number of iterations from the start of sampling that are eventually discarded (by default half of `iter`).

The last three options, `chains`, `iter`, `warmup` determine the behavior of the sampling algorithm: the No-U-Turn Sampler (NUTS; Hoffman and Gelman 2014) extension of Hamiltonian Monte Carlo (Duane et al. 1987; Neal 2011). We will discuss sampling in a bit more depth in chapter 8, but the basic process is explained next.

3.2.1.2 Sampling and convergence in a nutshell

The following is a gross oversimplification of the sampling process: The code specification starts (by default) four chains independently from each other. Each chain "searches" for samples of the posterior distribution in a multidimensional space, where each parameter corresponds to a dimension. The shape of this space is determined by the priors and the likelihood. The chains start at random locations, and in each iteration they take one sample each for all the parameters. When sampling begins, the samples may or may not belong to the posterior distributions of the parameters. Eventually, the chains end up in the vicinity of the posterior distribution, and from that point onward the samples will belong to the posterior.

Thus, when sampling begins, the samples from the different chains can be far from each other, but at some point they will "converge" and start delivering samples from the posterior distributions. Although there are no guarantees that the number of iterations we run the chains for will be sufficient for obtaining samples from the posteriors, the default values of `brms` (and Stan) are in many cases sufficient to achieve convergence. When the default number of iterations

[4]There is an additional problem here. Although the parameter for the intercept is assigned a uniform distribution bounded between 0 and 60,000 ms, the sampler might start sampling from an initial value outside this range producing warnings. The sampler can start from an initial value that is outside the 0–60,000 range because the initial value is chosen randomly (unless the user specifies an initial value explicitly).

does not suffice, brms (actually, Stan) will print out warnings, with suggestions for fixing the convergence problems. If all the chains converge to the same distribution, by removing the "warmup" samples, we make sure that we do not get samples from the initial path to the posterior distributions. The default in brms is that half of the total number of iterations in each chain (which default to 2000) will count as "warmup." So, if one runs a model with four chains and the default number of iterations, we will obtain a total of 4000 samples from the four chains, after discarding the warmup iterations.

Figure 3.3(a) shows the path of the chains from the warmup phase onward. Such plots are called traceplots. The warmup is shown only for illustration purposes; generally, one should only inspect the chains after the point where convergence has (presumably) been achieved (i.e., after the dashed line). After convergence has occurred, a visual diagnostic check is that chains should look like a "fat hairy caterpillar." Compare the traceplots of our model in Figure 3.3(a) with the traceplots of a model that did not converge, shown in Figure 3.3(b).

Traceplots are not always diagnostic as regards convergence. The traceplots might look fine, but the model may not have converged. Fortunately, Stan automatically runs several diagnostics with the information from the chains, and if there are no warnings after fitting the model and the traceplots look fine, we can be reasonably sure that the model converged, and assume that our samples are from the true posterior distribution. However, it is necessary to run more than one chain (preferably four), with a couple of thousands of iterations (at least) in order for the diagnostics to work.

3.2.1.3 Output of brms

Once the model has been fit (and assuming that we got no warning messages about convergence problems), we can print out the samples of the posterior distributions of each of the parameters using as_draws_df() (which stores metadata about the chains) or with as.data.frame():

```
as_draws_df(fit_press) %>%
  head(3)
```

```
## # A draws_df: 3 iterations, 1 chains, and 5 variables
##    b_Intercept sigma Intercept lprior   lp__
## 1          169    26       169    -19 -1683
## 2          168    25       168    -19 -1683
## 3          167    24       167    -19 -1684
## # ... hidden reserved variables {'.chain', '.iteration', '.draw'}
```

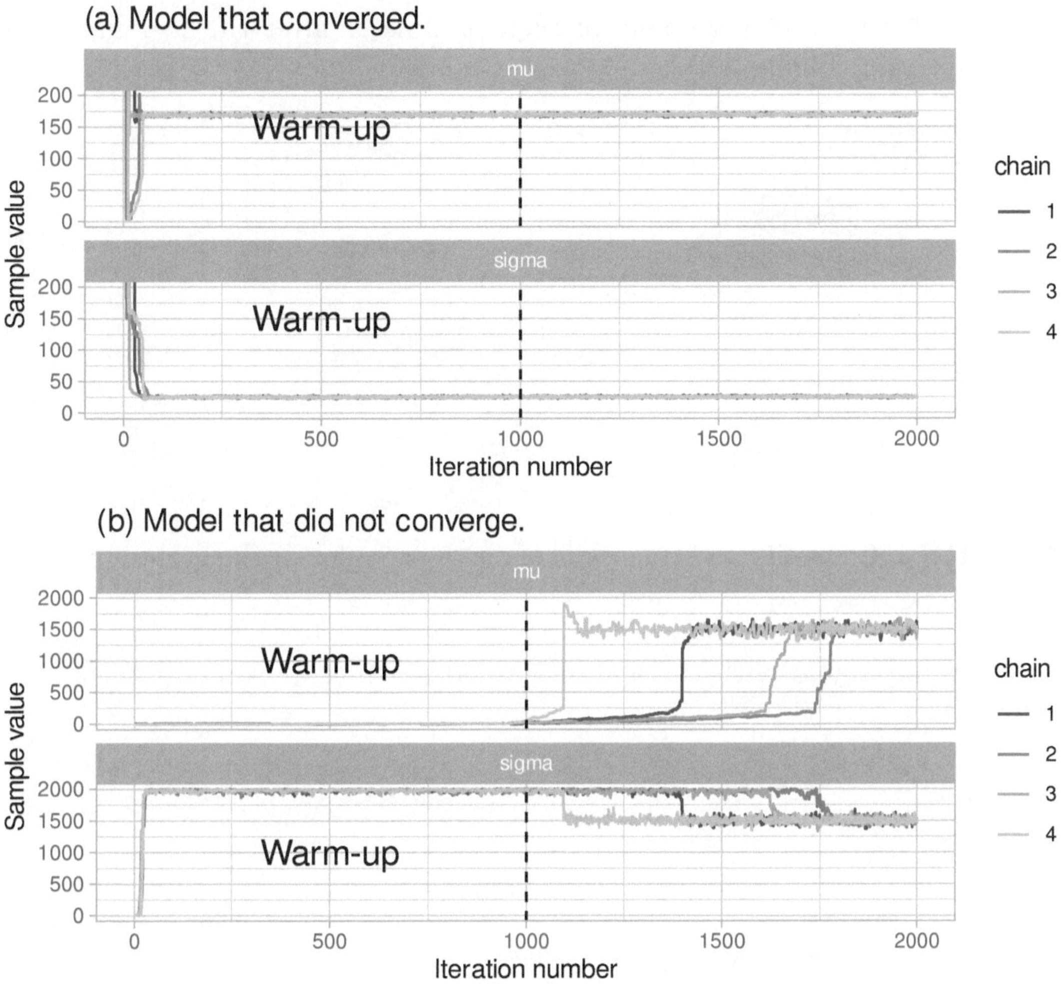

FIGURE 3.3: (a) Traceplots of our `brms` model for the button-pressing data. All the chains start from initial values above `200` and are outside of the plot. (b) Traceplots of a model that **did not** converge. We can diagnose the non-convergence by the observing that the chains do not overlap—each chain seems to be sampling from a different distribution.

The term `b_Intercept` in the `brms` output corresponds to our μ. We can ignore the last three columns: `Intercept` is an auxiliary intercept assuming centered predictors, which coincides here with `b_Intercept` and is discussed in online section A.3, `lprior` is the log-density of the (joint) prior distribution and it is there for compatibility with the package `priorsense` (`https://github.com/n-kall/priorsense`), and `lp` is not really part of the posterior, it's the log-density of the unnormalized posterior for each iteration (`lp` is discussed in online section B.1, which sould be read in the context of chapter 8).

Plot the density and traceplots of each parameter after the warmup with `plot(fit_press)` (Figure 3.4).

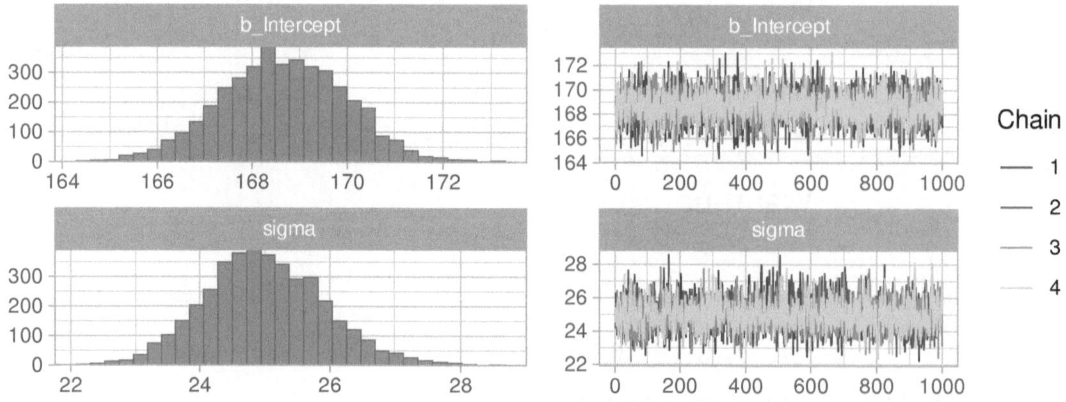

FIGURE 3.4: Density and traceplots of our `brms` model for the button-pressing data.

Printing the object with the `brms` fit provides a nice, if somewhat verbose, summary:

```
fit_press
# posterior_summary(fit_press) is also useful
```

```
##  Family: gaussian
##   Links: mu = identity; sigma = identity
## Formula: t ~ 1
##    Data: df_spacebar (Number of observations: 361)
##   Draws: 4 chains, each with iter = 2000; warmup = 1000; thin = 1;
##          total post-warmup draws = 4000
##
## Regression Coefficients:
##           Estimate Est.Error l-95% CI u-95% CI Rhat Bulk_ESS Tail_ESS
## Intercept   168.63      1.30   166.10   171.21 1.00     4058     2648
##
## Further Distributional Parameters:
##       Estimate Est.Error l-95% CI u-95% CI Rhat Bulk_ESS Tail_ESS
## sigma    24.99      0.93    23.24    26.88 1.00     3291     2790
##
## Draws were sampled using sampling(NUTS). For each parameter, Bulk_ESS
## and Tail_ESS are effective sample size measures, and Rhat is the potential
## scale reduction factor on split chains (at convergence, Rhat = 1).
```

The `Estimate` is just the mean of the posterior samples, `Est.Error` is the standard deviation of the posterior and the CIs mark the lower and upper bounds of the 95% credible intervals (to distinguish credible intervals from frequentist confidence intervals, the former will be abbreviated as CrIs):

```
as_draws_df(fit_press)$b_Intercept %>% mean()
```

```
## [1] 169
```

```
as_draws_df(fit_press)$b_Intercept %>% sd()
```

```
## [1] 1.3
```

```
as_draws_df(fit_press)$b_Intercept %>%
  quantile(c(0.025, .975))
```

```
##  2.5% 97.5%
##   166   171
```

Furthermore, the summary provides the `Rhat`, `Bulk_ESS`, and `Tail_ESS` of each parameter. R-hat compares the between- and within-chain estimates of each parameter. R-hat is larger than 1 when chains have not mixed well, one can only rely on the model if the R-hats for all the parameters are less than 1.05. (R warnings will appear otherwise). Bulk ESS (bulk effective sample size) is a measure of sampling efficiency in the bulk of the posterior distribution, that is the effective sample size for the mean and median estimates, whereas tail ESS (tail effective sample size) indicates the sampling efficiency at the tails of the distribution, that is the minimum of effective sample sizes for 5% and 95% quantiles. The effective sample size is generally smaller than the number of post-warmup samples, because the samples from the chains are not independent (they are correlated to some extent), and carry less information about the posterior distribution in comparison to independent samples. In some cases, however, the effective sample size is actually larger than the number of post-warmup samples. This might happen for parameters with a normally distributed posterior (in the unconstrained space, see online section B.1) and low dependence on the other parameters (Vehtari et al. 2021). A very low effective sample size indicates sampling problems (and is accompanied by R warnings) and in general appears together with chains that are not properly mixed. As a rule of thumb, Vehtari et al. (2021) suggest that a minimum effective sample size of 400 is required for statistical summaries.

We see that we can fit our model without problems, and we get some posterior distributions for our parameters. However, we should ask ourselves the following questions before we can interpret the posterior distributions of the model:

1. What information are the priors encoding? Do the priors make sense?
2. Does the likelihood assumed in the model make sense for the data?

We'll try to answer these questions by looking at the *prior and posterior predictive distributions*, and by doing sensitivity analyses. This is explained in the sections that follow.

3.3 Prior predictive distribution

We had defined the following priors for our linear model:

$$\begin{aligned} \mu &\sim \mathit{Uniform}(0, 60{,}000) \\ \sigma &\sim \mathit{Uniform}(0, 2000) \end{aligned} \tag{3.5}$$

These priors encode assumptions about the kind of data we would expect to see in a future study. To understand these assumptions, we are going to generate data from the model; such distribution of data, which is generated entirely by the prior distributions, is called the prior predictive distribution. Generating prior predictive distributions repeatedly helps us to check whether the priors make sense. What we want to know here is, do the priors generate realistic-looking data?

Formally, we want to know the density $p(\cdot)$ of data points $y_{pred_1}, \dots, y_{pred_N}$ from a data set $\boldsymbol{y_{pred}}$ of length N, given a vector of priors $\boldsymbol{\Theta}$ and our likelihood $p(\cdot|\boldsymbol{\Theta})$; (in our example, $\boldsymbol{\Theta} = \langle \mu, \sigma \rangle$). The prior predictive density is written as follows:

$$\begin{aligned} p(\boldsymbol{y_{pred}}) &= p(y_{pred_1}, \dots, y_{pred_n}) \\ &= \int_{\boldsymbol{\Theta}} p(y_{pred_1}|\boldsymbol{\Theta}) \cdot p(y_{pred_2}|\boldsymbol{\Theta}) \cdots p(y_{pred_N}|\boldsymbol{\Theta}) p(\boldsymbol{\Theta}) \, d\boldsymbol{\Theta} \end{aligned} \tag{3.6}$$

In essence, the vector of parameters is integrated out (see section 1.7). This yields the probability distribution of possible data sets given the priors and the likelihood, *before any observations are taken into account.*

The integration can be carried out computationally by generating samples from the prior distribution.

Here is one way to generate prior predictive distributions:

Repeat the following many times:

1. Take one sample from each of the priors.
2. Plug those samples into the probability density/mass function used as the likelihood in the model to generate a data set $y_{pred_1}, \dots, y_{pred_n}$.

Each sample is an imaginary or potential data set.

Create a function that does this:

```
normal_predictive_distribution <-
  function(mu_samples, sigma_samples, N_obs) {
    # empty data frame with headers:
    df_pred <- tibble(trialn = numeric(0),
                      t_pred = numeric(0),
                      iter = numeric(0))
    # i iterates from 1 to the length of mu_samples,
    # which we assume is identical to
    # the length of the sigma_samples:
    for (i in seq_along(mu_samples)) {
      mu <- mu_samples[i]
      sigma <- sigma_samples[i]
      df_pred <- bind_rows(df_pred,
                           tibble(trialn = seq_len(N_obs),
                                  # seq_len generates 1, 2,..., N_obs
                                  t_pred = rnorm(N_obs, mu, sigma),
                                  iter = i))
    }
    df_pred
  }
```

The following code produces 1000 samples of the prior predictive distribution of the model that we defined in section 3.2.1. This means that it will produce 361,000 predicted values (361 predicted observations for each of the 1000 simulations). Although this approach works, it's quite slow (a couple of seconds). See the online section A.1 for a more efficient version of this function. Section 3.7.2 will show that it's possible to use `brms` to sample from the priors, ignoring the `t` in the data by setting `sample_prior = "only"`. However, since `brms` still depends on Stan's sampler, which uses Hamiltonian Monte Carlo, the prior sampling process can also fail to converge, especially when one uses very uninformative priors, like the ones used in this example. In contrast, our function above, which uses `rnorm()`, cannot have convergence issues and will always produce multiple sets of prior predictive data $y_{pred_1}, \dots, y_{pred_n}$.

The code below uses the `tic()` and `toc()` functions from the `tictoc` package to print out the time it takes to run the code.

```
N_samples <- 1000
N_obs <- nrow(df_spacebar)
mu_samples <- runif(N_samples, 0, 60000)
sigma_samples <- runif(N_samples, 0, 2000)
tic()
prior_pred <-
  normal_predictive_distribution(mu_samples = mu_samples,
                                 sigma_samples = sigma_samples,
                                 N_obs = N_obs)
toc()
```

```
## 1.746 sec elapsed
```

```
prior_pred
```

```
## # A tibble: 361,000 x 3
##    trialn t_pred  iter
##     <dbl>  <dbl> <dbl>
## 1       1 16710.     1
## 2       2 16686.     1
## 3       3 17245.     1
## # i 360,997 more rows
```

Figure 3.5 shows the first 18 samples of the prior predictive distribution (i.e., 18 independently generated prior predicted data sets) with the code below.

```
prior_pred %>%
  filter(iter <= 18) %>%
  ggplot(aes(t_pred)) +
  geom_histogram(aes(y = after_stat(density))) +
  xlab("predicted t (ms)") +
  theme(axis.text.x = element_text(angle = 40,
                                   vjust = 1,
                                   hjust = 1)) +
  scale_y_continuous(limits = c(0, 0.0005),
                     breaks = c(0, 0.00025, 0.0005),
                     name = "density") +
  facet_wrap(~iter, ncol = 3)
```

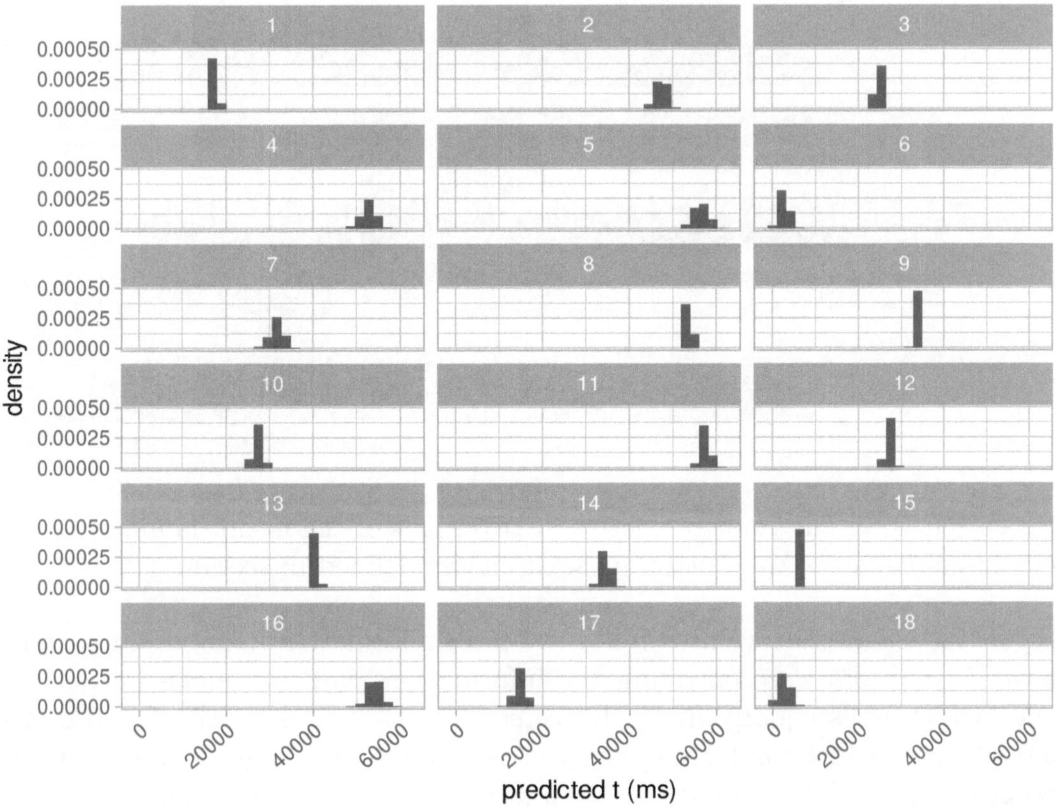

FIGURE 3.5: Eighteen samples from the prior predictive distribution of the model defined in section 3.2.1.

The prior predictive distribution in Figure 3.5 shows data sets generated from the model that are not realistic: Apart from the fact that the data sets show that finger tapping times distributions are symmetrical–and we know from prior experience with such data that they are generally right-skewed–some data sets present finger tapping times that are unrealistically long. Worse yet, if we inspect enough samples from the prior predicted data, it will become clear that a few data sets have negative finger tapping time values.

We can also look at the distribution of summary statistics in the prior predictive data. Even if we don't know beforehand what the data should look like, it's very likely that we have some expectations for possible mean, minimum, or maximum values. For example, in the button-pressing example, it seems reasonable to assume that average finger tapping times are between 200 and 600 ms; finger tapping times are very unlikely to be below 50 ms, and even long lapses of attention won't be greater than a couple of seconds.[5] Three distributions of summary statistics are shown in Figure 3.6.

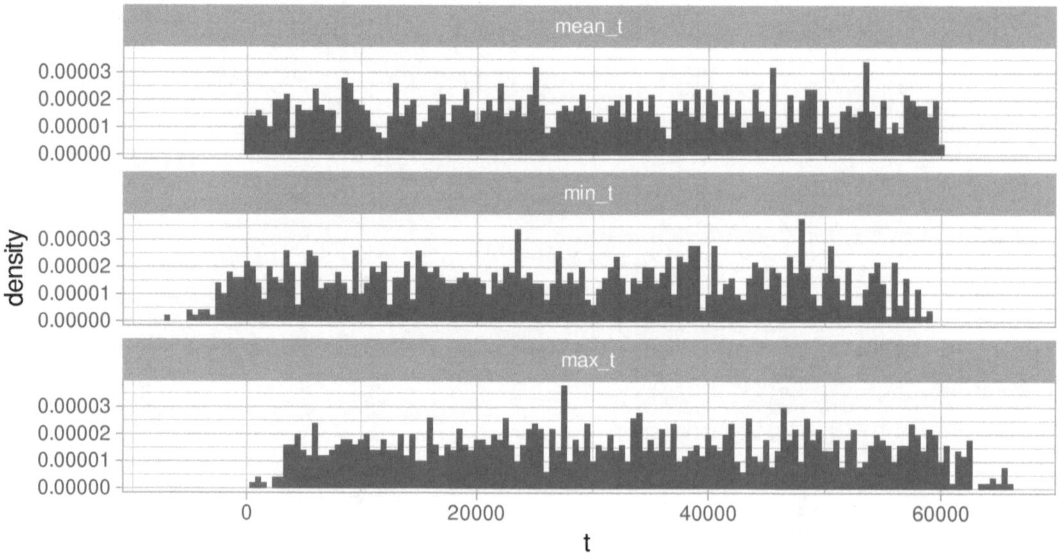

FIGURE 3.6: The prior predictive distributions of the mean, minimum, and maximum values of the button-pressing model defined in section 3.2.1.

Figure 3.6 shows that we used much less prior information than what we could have: Our priors were encoding the information that any mean between 0 and 60,000 ms is equally likely. It seems clear that a value close to 0 or to 60,000 ms would be extremely surprising. This wide range of mean values occurs because of the uniform prior on μ. Similarly, maximum values are quite "uniform," spanning a much wider range than what one would expect. Finally, in the distribution of minimum values, negative finger tapping times occur. This might seem surprising (our prior for μ excluded negative values), but the reason that negative values appear is that the prior is interpreted together with the likelihood (Gelman, Simpson, and Betancourt 2017), and the likelihood is a

[5]We'll see later how to generate prior predictive distributions of statistics such as mean, minimum, or maximum value in section 3.7.2 using `brms` and `pp_check()`.

normal distribution, which will allow for negative samples even if the location parameter μ is constrained to have only positive values.

To summarize the above discussion, the priors used in the example are clearly not very realistic given what we might know about finger tapping times for such a button-pressing task. This raises the question: what priors should we have chosen? In the next section, we consider this question.

3.4 The influence of priors: sensitivity analysis

For most cases that we will encounter in this book, there are four main classes of priors that we can choose from. Among Bayesians, there is no fixed nomenclature for classifying different kinds of priors. For this book, we have chosen specific names for each type of prior, but this is just a convention that we follow for consistency. There are also other classes of prior that we do not discuss in this book. An example is improper priors such as $Uniform(-\infty, +\infty)$, which are not proper probability distributions because the area under the curve does not sum to one.

When thinking about priors, the reader should not get hung up on what precisely the name is for a particular type of prior; they should rather focus on what that prior means in the context of the research problem.

3.4.1 Flat, uninformative priors

One option is to choose priors that are as uninformative as possible. The idea behind this approach is to let the data "speak for itself" and to not bias the statistical inference with "subjective" priors. There are several issues with this approach: First, the prior is as subjective as the likelihood, and in fact, different choices of likelihood might have a much stronger impact on the posterior than different choices of priors. Second, uninformative priors are in general unrealistic because they give equal weight to all values within the support of the prior distribution, ignoring the fact that usually there is some minimal information about the parameters of interest. Usually, at the very least, the order of magnitude is known (response times or finger tapping times will be in milliseconds and not days, EEG signals some microvolts and not volts, etc.). Third, uninformative priors make the sampling slower and might lead to convergence problems. Unless there is a large amount of data, it would be wise to avoid such priors. Fourth, it is not always clear which parameterization of a given distribution the flat priors should be assigned to. For example, the normal distribution is sometimes defined based on its standard deviation (σ), variance (σ^2), or precision ($1/\sigma^2$): a flat prior for the standard deviation is not flat for the precision of the distribution. Although it is sometimes possible to find an uninformative prior that is invariant under a change of parameters (also called Jeffreys priors; Jaynes 2003, sec. 6.15; Jeffreys 1939, Chapter 3), this is not always the case. Finally, if Bayes factors need to be computed, uninformative priors can lead to very misleading conclusions (chapter 13).

In the button-pressing example discussed in this chapter, an example of a flat, uninformative prior would be $\mu \sim \mathit{Uniform}(-10^{20}, 10^{20})$. On the millisecond scale, this is a very strange prior to use for a parameter representing mean button-pressing time: it allows for impossibly large positive values, and it also allows negative button-pressing times, which is of course impossible. It is technically possible to use such a prior, but it wouldn't make much sense.

3.4.2 Regularizing priors

If there does not exist much prior information (and if this information cannot be worked out through reasoning about the problem), and there is enough data (what "enough" means here will presently become clear when we look at specific examples), it is fine to use *regularizing priors.* These are priors that down-weight extreme values (that is, they provide regularization), they are usually not very informative, and mostly let the likelihood dominate in determining the posteriors. These priors are theory-neutral; that is, they usually do not bias the parameters to values supported by any prior belief or theory. The idea behind this type of prior is to help to stabilize computation. These priors are sometimes called *weakly informative* or *mildly informative* priors in the Bayesian literature. For many applications, they perform well, but as discussed in chapter 13, they tend to be problematic if Bayes factors need to be computed.

In the button-pressing example, an example of a regularizing prior would be $\mu \sim \mathit{Normal}_+(0, 1000)$. This is a normal distribution prior truncated at 0 ms, and allows a relatively constrained range of positive values for button-pressing times (roughly, up to 2000 ms or so). This is a regularizing prior because it rules out negative button-pressing times and down-weights extreme values over 2000 ms. Here, one could assume a non-truncated prior like $\mathit{Normal}(0, 1000)$. This could still be seen as a regularizing prior even though we don't expect μ to be negative; the data would ensure that the posterior distribution has positive values (because we would have no negative button-pressing times in the data).

3.4.3 Principled priors

The idea here is to have priors that encode all (or most of) the theory-neutral information that the researcher has. Since one generally knows what one's data do and do not look like, it is possible to build priors that truly reflect the properties of potential data sets, using prior predictive checks. In this book, many examples of this class of priors will come up.

In the button-pressing data, an example of a principled prior would be $\mu \sim Normal_+(250, 100)$. This prior is not overly restrictive, but represents a guess about plausible button-pressing times. Prior predictive checks using principled priors should produce realistic distributions of the dependent variable.

3.4.4 Informative priors

There are cases where a lot of prior knowledge exists. In general, unless there are very good reasons for having relatively informative priors (see chapter 13), it is not a good idea to let the priors have too much influence on the posterior. An example where informative priors would be important is when investigating a language-impaired population from which we can't get many subjects, but a lot of previously published papers exist on the research topic.

In the button-pressing data, an informative prior could be based on a meta-analysis of previously published or existing data, or the result of prior elicitation from an expert (or multiple experts) on the topic under investigation. An example of an informative prior would be $\mu \sim Normal_+(200, 20)$. This prior will have some influence on the posterior for μ, especially when one has relatively sparse data.

These four options constitute a continuum. The uniform prior from the last model (section 3.2.1) falls between flat, uninformative and regularizing priors. In practical data analysis situations, we are mostly going to choose priors that fall between regularizing and principled. Informative priors, in the sense defined above, will be used only relatively rarely; but they become more important to consider when doing Bayes factor analyses (chapter 13).

3.5 Revisiting the button-pressing example with different priors

What would happen if even wider priors were used for the model defined previously (in section 3.2.1)? Suppose that every value between -10^6 and 10^6 ms is assumed to be equally likely for the location parameter μ. This prior is clearly unrealistic and actually makes no sense at all: we are not expecting negative finger tapping times. Regarding the standard deviation, one could assume that any value between 0 and 10^6 is equally likely.[6] The likelihood remains unchanged.

$$\begin{aligned} \mu &\sim \mathit{Uniform}(-10^6, 10^6) \\ \sigma &\sim \mathit{Uniform}(0, 10^6) \end{aligned} \tag{3.7}$$

```
# The default settings are used when they are not set explicitly:
# 4 chains, with half of the iterations (set as 3000) as warmup.
fit_press_unif <- brm(t ~ 1,
                      data = df_spacebar,
                      family = gaussian(),
                      prior = c(prior(uniform(-10^6, 10^6),
                                      class = Intercept,
                                      lb = -10^6,
                                      ub = 10^6),
                                prior(uniform(0, 10^6),
                                      class = sigma,
                                      lb = 0,
                                      ub = 10^6)),
                      iter = 3000,
                      control = list(adapt_delta = .99,
                                     max_treedepth = 15))
```

Even with these extremely unrealistic priors, which require us to change the number of iterations,`iter`, and `adapt_delta` and `max_treedepth` default values to achieve convergence, the output of the model is virtually identical to the previous one (see Figure 3.7).

```
fit_press_unif
```

```
## ...
## Population-Level Effects:
##           Estimate Est.Error l-95% CI u-95% CI Rhat Bulk_ESS Tail_ESS
## Intercept   168.64      1.32   166.01   171.20 1.00     4010     3112
##
## Family Specific Parameters:
##       Estimate Est.Error l-95% CI u-95% CI Rhat Bulk_ESS Tail_ESS
## sigma    25.01      0.92    23.22    26.81 1.01      812      751
```

[6]Even though, in theory, one could use even wider priors, in practice, these are the widest priors that achieve convergence using `brms`.

```
##
## ...
```

Next, consider what happens if very informative priors are used. Assume that mean values very close to 400 ms are the most likely, and that the standard deviation of the finger tapping times is very close to 100 ms. Given that this is a model of button-pressing times, such an informative prior seems wrong—200 ms seems like a more realistic mean button-pressing time, not 400 ms. You can check this by doing an experiment yourself and looking at the recorded times; software like Linger (`http://tedlab.mit.edu/~dr/Linger/`) makes it easy to set up such an experiment.

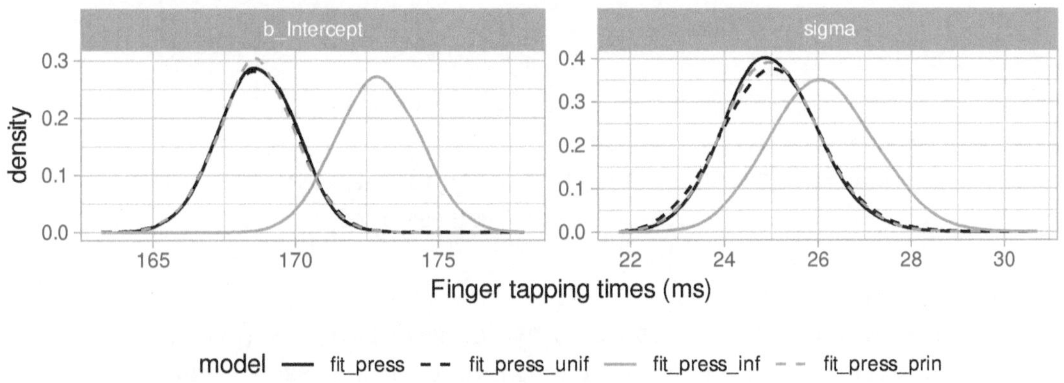

FIGURE 3.7: Comparison of the posterior distributions from the model with "realistically" bounded uniform prior distribution (though still not recommended), `fit_press`, against the model with extremely unrealistically wide priors, `fit_press_unif`, the model with misspecified informative priors, `fit_press_inf`, and the model with principled priors, `fit_press_prin`. All the posteriors virtually overlap except for the posterior of `fit_press_inf`, which is shifted by a few milliseconds.

The $Normal_+$ notation indicates a normal distribution truncated at zero such that only positive values are allowed (see online section A.2 discusses this type of distribution in detail). Even though the prior for the standard deviation is restricted to be positive, we are not required to add `lb = 0` to the prior, and it is automatically taken into account by `brms`.

$$\begin{aligned} \mu &\sim Normal(400, 10) \\ \sigma &\sim Normal_+(100, 10) \end{aligned} \tag{3.8}$$

```
fit_press_inf <- brm(t ~ 1,
  data = df_spacebar,
```

```
  family = gaussian(),
  prior = c(prior(normal(400, 10), class = Intercept),
            # `brms` knows that SDs need to be bounded
            # to exclude values below zero:
            prior(normal(100, 10), class = sigma)))
```

Despite these unrealistic but informative priors, the likelihood mostly dominates and the new posterior means and credible intervals are just a couple of milliseconds away from the previous estimates (see Figure 3.7):

```
fit_press_inf
```

```
## ...
## Population-Level Effects:
##           Estimate Est.Error l-95% CI u-95% CI Rhat Bulk_ESS Tail_ESS
## Intercept   172.92      1.38   170.28   175.65 1.00     3253     2990
##
## Family Specific Parameters:
##       Estimate Est.Error l-95% CI u-95% CI Rhat Bulk_ESS Tail_ESS
## sigma    26.07      1.01    24.21    28.16 1.00     2536     2577
##
## ...
```

As a final example of a sensitivity analysis, choose some principled priors. Assuming that we have some prior experience with previous similar experiments, suppose the mean response time is expected to be around 200 ms, with a 95% probability of the mean ranging from 0 to 400 ms. This uncertainty is perhaps unreasonably large, but one might want to allow a bit more uncertainty than one really thinks is reasonable (this kind of conservativity in allowing somewhat more uncertainty is sometimes called Cromwell's rule in Bayesian statistics; see O'Hagan and Forster 2004, sec. 3.19). In such a case, one can decide on the prior $Normal(200, 100)$. Given that the experiment involves only one subject and the task is very simple, one might not expect the residual standard deviation σ to be very large: As an example, one can settle on a location of 50 ms for a truncated normal distribution, but still allow for relatively large uncertainty: $Normal_+(50, 50)$. The prior specifications are summarized below.

$$\begin{aligned} \mu &\sim Normal(200, 100) \\ \sigma &\sim Normal_+(50, 50) \end{aligned} \tag{3.9}$$

Why are these priors principled? The designation "principled" here largely depends on our domain knowledge. The online chapter E discusses how one can use domain knowledge when specifying priors.

One can achieve a better understanding of what a particular set of priors implies by visualizing the priors graphically, and carrying out prior predictive checks. These steps are skipped here, but these issues are discussed in detail in the online chapters E and F. These chapters will give more detailed information about choosing priors and on developing a principled workflow for Bayesian data analysis.

```
fit_press_prin <-
  brm(t ~ 1,
      data = df_spacebar,
      family = gaussian(),
      prior = c(prior(normal(200, 100), class = Intercept),
                prior(normal(50, 50), class = sigma)))
```

The new estimates are virtually the same as before (see Figure 3.7):

```
fit_press_prin
```

```
## ...
## Population-Level Effects:
##           Estimate Est.Error l-95% CI u-95% CI Rhat Bulk_ESS Tail_ESS
## Intercept   168.64      1.32   166.10   171.16 1.00     3832     2782
##
## Family Specific Parameters:
##       Estimate Est.Error l-95% CI u-95% CI Rhat Bulk_ESS Tail_ESS
## sigma    24.99      0.93    23.23    26.93 1.00     3324     2515
##
## ...
```

The above examples of using different priors should not be misunderstood to mean that priors never matter. When there is enough data, the likelihood will dominate in determining the posterior distributions. What constitutes "enough" data is also a function of the complexity of the model; as a general rule, more complex models require more data.

Even in cases where there is enough data and the likelihood dominates in determining the posteriors, regularizing, principled priors (i.e., priors that are more consistent with our a priori beliefs about the data) will in general speed-up model convergence.

In order to determine the extent to which the posterior is influenced by the priors, it is a good practice to carry out a sensitivity analysis: try different priors and either verify that the posterior doesn't change drastically, or report how the posterior is affected by some specific priors (for examples from psycholinguistics, see Vasishth et al. 2013; Vasishth and Engelmann 2022). Chapter 13 will demonstrate that sensitivity analysis becomes crucial for reporting Bayes factors; even in cases where the choice of priors does not affect the posterior distribution, it generally affects the Bayes factor.

3.6 Posterior predictive distribution

The posterior predictive distribution is a collection of data sets generated from the model (the likelihood and the priors). Having obtained the posterior distributions of the parameters after taking into account the data, the posterior distributions can be used to generate future data from the model. In other words, given the posterior distributions of the parameters of the model, the posterior predictive distribution gives us some indication of what future data might look like.

Once the posterior distributions $p(\boldsymbol{\Theta} \mid \boldsymbol{y})$ are available, the predictions based on these distributions can be generated by integrating out the parameters:

$$p(\boldsymbol{y_{pred}} \mid \boldsymbol{y}) = \int_{\boldsymbol{\Theta}} p(\boldsymbol{y_{pred}}, \boldsymbol{\Theta} \mid \boldsymbol{y}) \, d\boldsymbol{\Theta} = \int_{\boldsymbol{\Theta}} p(\boldsymbol{y_{pred}} \mid \boldsymbol{\Theta}, \boldsymbol{y}) p(\boldsymbol{\Theta} \mid \boldsymbol{y}) \, d\boldsymbol{\Theta} \tag{3.10}$$

Assuming that past and future observations are conditionally independent[7] given $\boldsymbol{\Theta}$, i.e., $p(\boldsymbol{y_{pred}} \mid \boldsymbol{\Theta}, \boldsymbol{y}) = p(\boldsymbol{y_{pred}} \mid \boldsymbol{\Theta})$, the above equation can be written as:

$$p(\boldsymbol{y_{pred}} \mid \boldsymbol{y}) = \int_{\boldsymbol{\Theta}} p(\boldsymbol{y_{pred}} \mid \boldsymbol{\Theta}) p(\boldsymbol{\Theta} \mid \boldsymbol{y}) \, d\boldsymbol{\Theta} \tag{3.11}$$

In Equation (3.11), we are conditioning $\boldsymbol{y_{pred}}$ only on $\boldsymbol{y}$, we do not condition on what we don't know ($\boldsymbol{\Theta}$); the unknown parameters have been integrated out. This posterior predictive distribution has important differences from predictions obtained with the frequentist approach. The frequentist approach gives a point estimate of each predicted observation given the maximum likelihood estimate of $\boldsymbol{\Theta}$ (a point value), whereas the Bayesian approach gives a distribution of

[7] Two events A and B are defined to be conditionally independent given some event E if $P(A \cap B|E) = P(A|E)P(B|E)$. See chapter 2 of Blitzstein and Hwang (2014) for examples and more discussion of conditional independence.

values for each predicted observation. As with the prior predictive distribution, the integration can be carried out computationally by generating samples from the posterior predictive distribution. The same function that we created before, `normal_predictive_distribution()`, can be used here. The only difference is that instead of sampling `mu` and `sigma` from the priors, the samples come from the posterior.

```
N_obs <- nrow(df_spacebar)
mu_samples <- as_draws_df(fit_press)$b_Intercept
sigma_samples <- as_draws_df(fit_press)$sigma
normal_predictive_distribution(mu_samples = mu_samples,
                               sigma_samples = sigma_samples,
                               N_obs = N_obs)
```

```
## # A tibble: 1,444,000 x 3
##   trialn t_pred  iter
##    <dbl>  <dbl> <dbl>
## 1      1   155.     1
## 2      2   163.     1
## 3      3   209.     1
## # i 1,443,997 more rows
```

The `brms` function `posterior_predict()` is a convenient function that delivers samples from the posterior predictive distribution. Using the command `posterior_predict(fit_press)` yields the predicted finger tapping times in a matrix, with the samples as rows and the observations (data-points) as columns. (Bear in mind that if a model is fit with `sample_prior = "only"`, the dependent variable is ignored and `posterior_predict()` will yield samples from the prior predictive distribution).

The posterior predictive distribution can be used to examine the "descriptive adequacy" of the model under consideration (Gelman et al. 2014, Chapter 6; Shiffrin et al. 2008). Examining the posterior predictive distribution to establish descriptive adequacy is called posterior predictive checking. The goal here is to establish that the posterior predictive data look more or less similar to the observed data. Achieving descriptive adequacy means that the current data could have been generated by the model. Although passing a test of descriptive adequacy is not strong evidence in favor of a model, a major failure in descriptive adequacy can be interpreted as strong evidence against a model (Shiffrin et al. 2008). For this reason, comparing the descriptive adequacy of different models is not enough to differentiate between their relative performance.

When doing model comparison, it is important to consider the criteria that Roberts and Pashler (2000) define. Although Roberts and Pashler (2000) are more interested in process models and not necessarily Bayesian models, their criteria are important for any kind of model comparison. Their main point is that it is not enough to have a good fit to the data for a model to be convincing. One should check that the range of predictions that the model makes is reasonably constrained; if a model can capture any possible outcome, then the model fit to a particular data set is not so informative. In the Bayesian modeling context, although posterior predictive checks are important, they should been as only sanity checks to assess whether the model behavior is reasonable (for more on this point, see the online chapter F).

In many cases, one can simply use the plotting functions provided with `brms` (these act as wrappers for `bayesplot` functions). For example, the plotting function `pp_check()` takes as arguments the model, the number of predicted data sets, and the type of visualization, and it can display different visualizations of posterior predictive checks. In these type of plots, the observed data are plotted as y and predicted data as y_{rep}. Below, we use `pp_check()` to investigate how well the observed distribution of finger tapping times fit our model based on some number (11 and 100) of samples of the posterior predictive distributions (that is, simulated data sets); see Figures 3.8 and 3.9.

```
pp_check(fit_press, ndraws = 11, type = "hist")
```

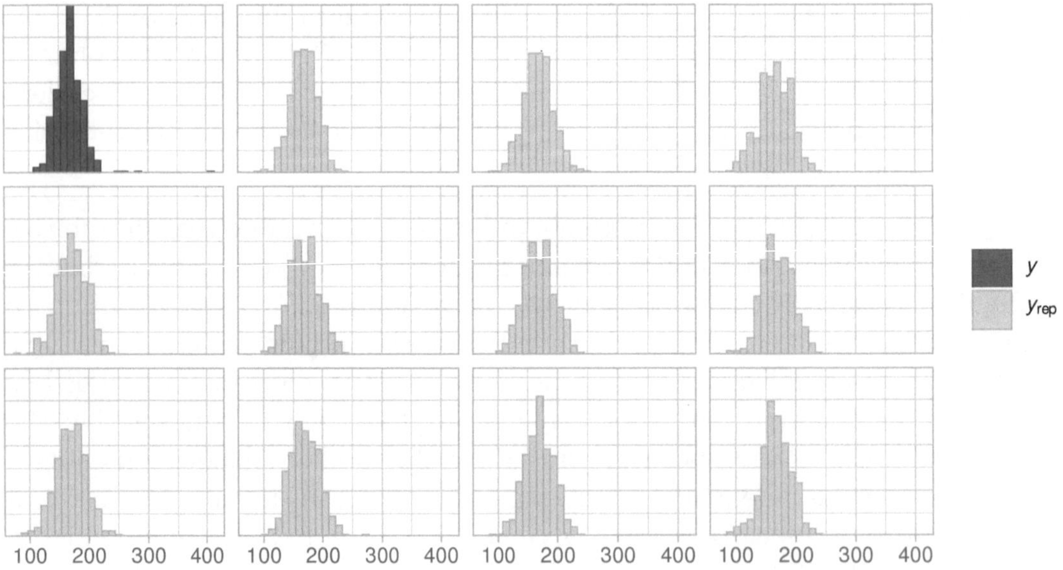

FIGURE 3.8: Histograms of eleven samples from the posterior predictive distribution of the model `fit_press` (y_{rep}).

```
pp_check(fit_press, ndraws = 100, type = "dens_overlay")
```

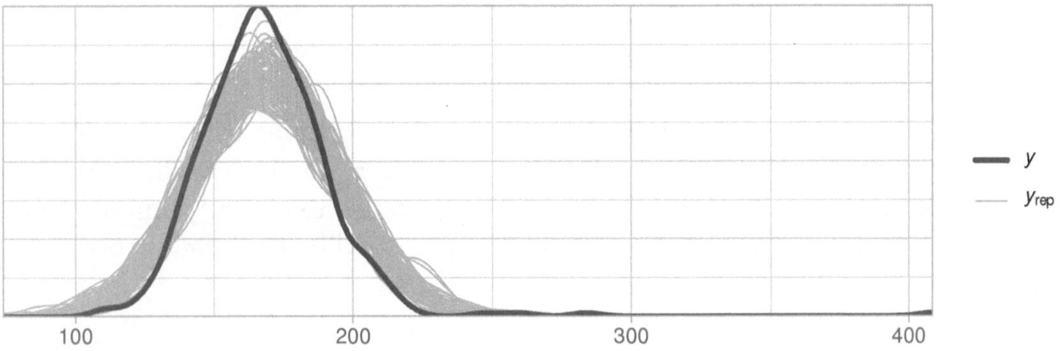

FIGURE 3.9: A posterior predictive check that shows the fit of the model `fit_press` in comparison to data sets from the posterior predictive distribution using an overlay of density plots.

The data is slightly skewed and has no values smaller than 100 ms, but the predictive distributions are centered and symmetrical; see Figures 3.8 and 3.9. This posterior predictive check shows a slight mismatch between the observed and predicted data. Can we build a better model? We'll come back to this issue in the next section.

3.7 The influence of the likelihood

Finger tapping times (and response times in general) are not usually normally distributed. A more realistic distribution is the log-normal. A random variable (such as time) that is log-normally distributed takes only positive real values and is right-skewed. Although other distributions can also produce data with such properties, the log-normal will turn out to be a pretty reasonable distribution for finger tapping times and response times.

3.7.1 The log-normal likelihood

If $\boldsymbol{y}$ is log-normally distributed, this means that $\log(\boldsymbol{y})$ is normally distributed.[8] The log-normal distribution is also defined using the parameters location, μ, and scale, σ, but these are on the log ms scale; they correspond to the mean and standard deviation of the logarithm of the data $\boldsymbol{y}$, $\log(\boldsymbol{y})$, which will be normally distributed. Thus, when we model some data $\boldsymbol{y}$ using the log-normal

[8] More precisely, $\log_e(\boldsymbol{y})$ or $\ln(\boldsymbol{y})$, but we'll write it as just $\log(\boldsymbol{y})$.

likelihood, the parameters μ and σ are on a different scale than the data $\boldsymbol{y}$. Equation (3.12) shows the relationship between the log-normal and the normal.

$$\begin{aligned} \log(\boldsymbol{y}) &\sim Normal(\mu, \sigma) \\ \boldsymbol{y} &\sim LogNormal(\mu, \sigma) \end{aligned} \tag{3.12}$$

We can obtain samples from the log-normal distribution, using the normal distribution by first setting an auxiliary variable z, so that $z = \log(y)$. This means that $z \sim Normal(\mu, \sigma)$. Then we can just use $exp(z)$ as samples from the $LogNormal(\mu, \sigma)$, since $\exp(z) = \exp(\log(y)) = y$. The code below produces Figure 3.10.

```
mu <- 6
sigma <- 0.5
N <- 500000
# Generate N random samples from a log-normal distribution
sl <- rlnorm(N, mu, sigma)
ggplot(tibble(samples = sl), aes(samples)) +
  geom_histogram(aes(y = after_stat(density)), binwidth = 50) +
  ggtitle("Log-normal distribution\n") +
  coord_cartesian(xlim = c(0, 2000))
# Generate N random samples from a normal distribution,
# and then exponentiate them
sn <- exp(rnorm(N, mu, sigma))
ggplot(tibble(samples = sn), aes(samples)) +
  geom_histogram(aes(y = after_stat(density)), binwidth = 50) +
  ggtitle("Exponentiated samples from\na normal distribution") +
  coord_cartesian(xlim = c(0, 2000))
```

3.7.2 Using a log-normal likelihood to fit data from a single subject pressing a button repeatedly

If we assume that finger tapping times are log-normally distributed, the likelihood function changes as follows:

$$t_n \sim LogNormal(\mu, \sigma) \tag{3.13}$$

But now the scale of the priors needs to change! Let's start with uniform priors for ease of exposition, even though, as we mentioned earlier, these are not really appropriate here. (More realistic priors are discussed below.)

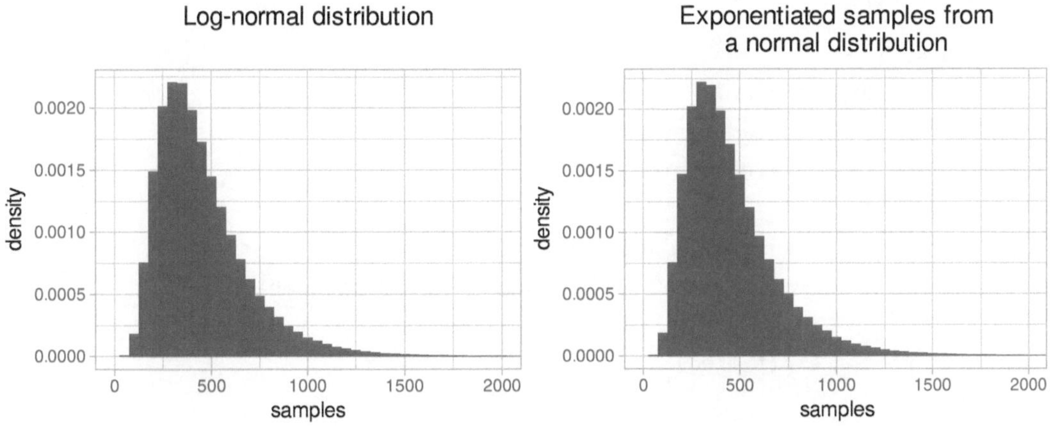

FIGURE 3.10: Two log-normal distributions with the same parameters generated by either generating samples from a log-normal distribution or exponentiating samples from a normal distribution.

$$\begin{aligned} \mu &\sim Uniform(0, 11) \\ \sigma &\sim Uniform(0, 1) \end{aligned} \tag{3.14}$$

Because the parameters are on a different scale than the dependent variable, their interpretation changes and it is more complex than if we were dealing with a linear model that assumes a normal likelihood (location and scale do not coincide with the mean and standard deviation of the log-normal):

- *The location* μ: In our previous linear model, μ represented the mean (in a normal distribution, the mean happens to be identical to the median and the mode). But now, the mean needs to be calculated by computing $\exp(\mu + \sigma^2/2)$. In other words, in the log-normal, the mean is dependent on both μ and σ. The median is just $\exp(\mu)$. Notice that the prior of μ is not on the milliseconds scale, but rather on the log milliseconds scale.
- *The scale* σ: This is the standard deviation of the normal distribution of $\log(\boldsymbol{y})$. The standard deviation of a log-normal distribution with *location* μ and *scale* σ will be $\exp(\mu + \sigma^2/2) \times \sqrt{\exp(\sigma^2) - 1}$. Unlike the normal distribution, the spread of the log-normal distribution depends on both μ and σ.

To understand the meaning of the priors on the millisecond scale, both the priors and the likelihood need to be taken into account. Generating a prior predictive distribution will help in interpreting the priors. This distribution can be generated by just exponentiating the samples produced by `normal_predictive_distribution()` (or, alternatively, edit the function by replacing `rnorm()` with `rlnorm()`).

```
N_samples <- 1000
N_obs <- nrow(df_spacebar)
mu_samples <- runif(N_samples, 0, 11)
sigma_samples <- runif(N_samples, 0, 1)
prior_pred_ln <-
  normal_predictive_distribution(mu_samples = mu_samples,
                                 sigma_samples = sigma_samples,
                                 N_obs = N_obs) %>%
  mutate(t_pred = exp(t_pred))
```

Next, plot the distribution of some representative statistics; see Figure 3.11.

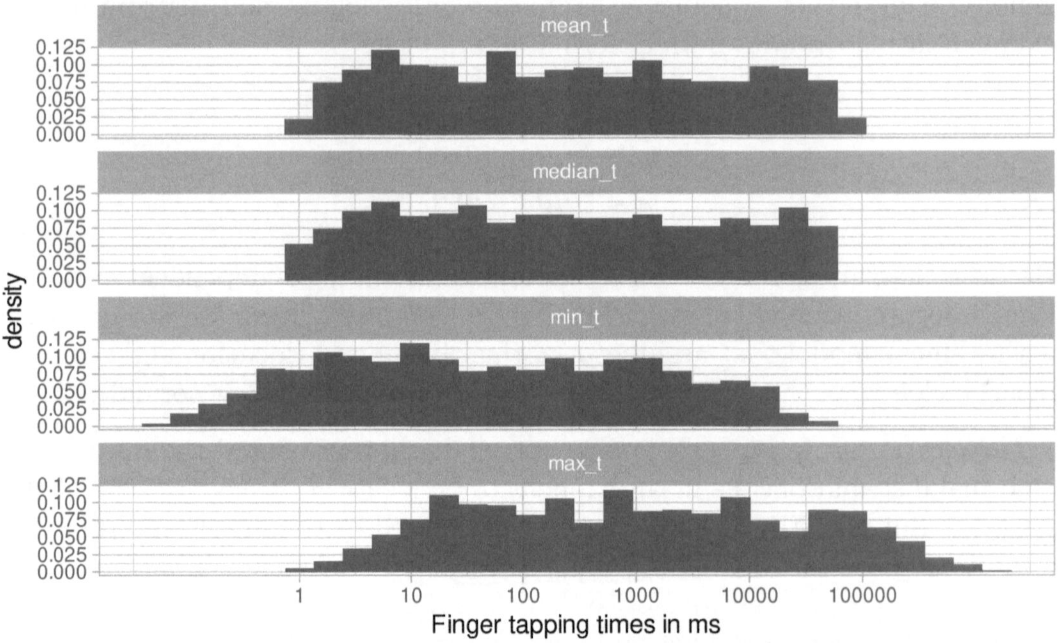

FIGURE 3.11: The prior predictive distribution of the mean, median, minimum, and maximum value of the log-normal model with priors defined in Equation (3.14), that is $\mu \sim Uniform(0, 11)$ and $\sigma \sim Uniform(0, 1)$. The x-axis shows values back-transformed from the log-scale.

We cannot generate negative values any more, since exp(any finite real number) > 0. These priors might work in the sense that the model might converge; but it would be better to have regularizing priors for the model. An example of regularizing priors:

$$\begin{aligned} \mu &\sim Normal(6, 1.5) \\ \sigma &\sim Normal_+(0, 1) \end{aligned} \tag{3.15}$$

The prior for σ here is a truncated distribution, and although its location is zero, this is not its mean. We can calculate its approximate mean from a large number of random samples of the prior distribution using the function `rtnorm()` from the package `extraDistr`. In this function, we have to set the parameter `a = 0` to express the fact that the normal distribution is truncated from the left at 0. (Online section A.2 discusses this type of distribution in detail):

```
mean(rtnorm(100000, 0, 1, a = 0))
```

```
## [1] 0.797
```

Even before generating the prior predictive distributions, we can calculate the range within which we are 95% sure that the expected median of the observations will lie. We do this by looking at what happens at two standard deviations away from the mean of the prior, μ, that is $6-2\times1.5$ and $6+2\times1.5$, and exponentiating these values:

```
c(lower = exp(6 - 2 * 1.5),
  higher = exp(6 + 2 * 1.5))
```

```
##  lower higher
##   20.1 8103.1
```

This means that the prior for μ is still not too informative (these are medians; the actual values generated by the log-normal distribution can be much more spread out). Next, plot the distribution of some representative statistics of the prior predictive distributions. `brms` allows one to sample from the priors, ignoring the observed data `t` , by setting `sample_prior = "only"` in the `brm` function.

If we want to use `brms` to generate prior predictive data in this manner even before we have any data, we do need to include a data frame with the relevant dependent variable (`y` in this case). Setting `sample_prior = "only"` will ignore the values of the dependent variable.

```
fit_prior_press_ln <-
  brm(t ~ 1,
      data = df_spacebar,
      family = lognormal(),
      prior = c(prior(normal(6, 1.5), class = Intercept),
                prior(normal(0, 1), class = sigma)),
      sample_prior = "only",
      control = list(adapt_delta = 0.9))
```

To avoid the warnings, increase the `adapt_delta` parameter's default value from 0.8 to 0.9 to simulate the data. Since Stan samples from the prior distributions in the same way that it samples from the posterior distribution, one should not ignore warnings; always ensure that the model converges. In that respect, the custom function `normal_predictive_distribution()` defined in section 3.3 has the advantage that it will always yield independent samples from the prior distribution and will not experience any convergence problems. This is because it just relies on the `rnorm()` function in R.

Plot the prior predictive distribution of means with the following code. In a prior predictive distribution, we generally want to ignore the data; this requires setting `prefix = "ppd"` in `pp_check()`.

To plot the distributions of minimum and maximum values, in the R code just replace `mean` with `min` or `max`. The distributions of the three statistics are displayed in Figure 3.12.

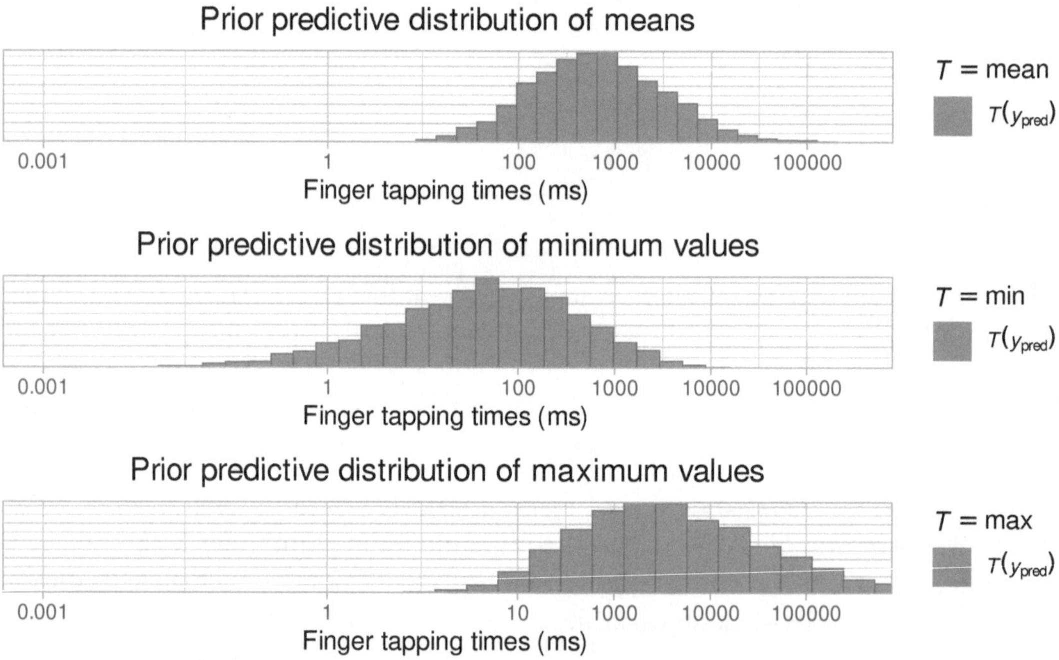

FIGURE 3.12: The prior predictive distributions of the mean, maximum, and minimum values of the log-normal model with priors defined in Equation (3.15). The prior predictive distributions are labeled y_{pred}. The x-axis shows values back-transformed from the log-scale.

Figure 3.12 shows that the priors used here are still quite uninformative. The tails of the prior predictive distributions that correspond to our normal priors shown in Figure 3.12 are even further to the right, reaching more extreme values than for the prior predictive distributions generated by uniform priors (shown in Figure 3.11). The new priors are still far from representing our prior knowledge. We could run more iterations of choosing priors and generating prior predictive distributions until we have priors that generate realistic data. However, given that the bulk of the distributions of the mean, maximum, and minimum values lies roughly in the correct order of magnitude, these priors are going to be acceptable. In general, summary statistics (e.g., mean, median, min, max) can be used to test whether the priors are in a plausible range. This can be done by defining, for the particular research problem under study, the extreme data that would be very implausible to ever observe (e.g., reading times at a word larger than one minute) and choosing priors such that such extreme finger tapping times occur only very rarely in the prior predictive distribution.

Next, fit the model; recall that both the distribution family and prior change in comparison to the previous example.

```
fit_press_ln <-
  brm(t ~ 1,
      data = df_spacebar,
      family = lognormal(),
      prior = c(prior(normal(6, 1.5), class = Intercept),
                prior(normal(0, 1), class = sigma)))
```

When we look at the summary of the posterior, the parameters are on the log-scale:

```
fit_press_ln
```

```
## ...
## Population-Level Effects:
##           Estimate Est.Error l-95% CI u-95% CI Rhat Bulk_ESS Tail_ESS
## Intercept     5.12      0.01     5.10     5.13 1.00     4185     2852
##
## Family Specific Parameters:
##       Estimate Est.Error l-95% CI u-95% CI Rhat Bulk_ESS Tail_ESS
## sigma     0.13      0.01     0.13     0.15 1.00     2785     2235
##
## ...
```

If the research goal is to find out how long it takes to press the space bar in milliseconds, we need to transform the μ (or `Intercept` in the model) to milliseconds. Because the median of the log-normal distribution is $\exp(\mu)$, the following returns the median estimate in milliseconds (recall that for the mean we would need to calculate $\exp(\mu + \sigma^2/2)$):

```
estimate_ms <- exp(as_draws_df(fit_press_ln)$b_Intercept)
```

To display the mean and 95% credible interval of these samples, type:

```
c(mean = mean(estimate_ms),
  quantile(estimate_ms, probs = c(.025, .975)))
```

```
##   mean   2.5% 97.5%
##    167    165   169
```

Next, check whether the predicted data sets look similar to the observed data set. See Figure 3.13; compare this with the earlier Figure 3.9.

```
pp_check(fit_press_ln, ndraws = 100)
```

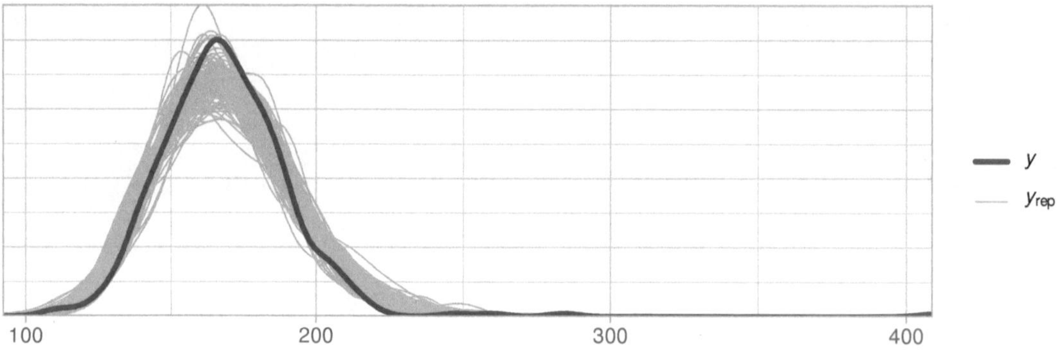

FIGURE 3.13: The posterior predictive distribution of `fit_press_ln`.

The key question is: Are the posterior predicted data now more similar to the observed data, compared to the case where we had a normal likelihood? According to Figure 3.13, it seems so, but it's not easy to tell.

Another way to examine the extent to which the predicted data looks similar to the observed data would be to look at the distribution of some summary statistic. As with prior predictive distributions, examine the distribution of representative summary statistics for the data sets generated by different models. However, in contrast with what occurs with prior predictive distributions, at this point we have a clear reference, our observations, and this means that we can compare the summary statistics with the observed statistics from our data. We suspect that the normal distribution would generate finger tapping times that are too fast (since the normal distribution is symmetrical) and that the log-normal distribution may capture the long tail better than the normal model. Based on this supposition, compute the distribution of minimum and maximum values for the posterior predictive distributions, and compare them with the minimum and maximum value, respectively. The function `pp_check()` implements this by specifying `stat` either `"min"` or `"max"` for both `fit_press`, and `fit_press_ln`; an example is shown below. The plots are shown in Figures 3.14 and 3.15.

```
pp_check(fit_press, type = "stat", stat = "min")
```

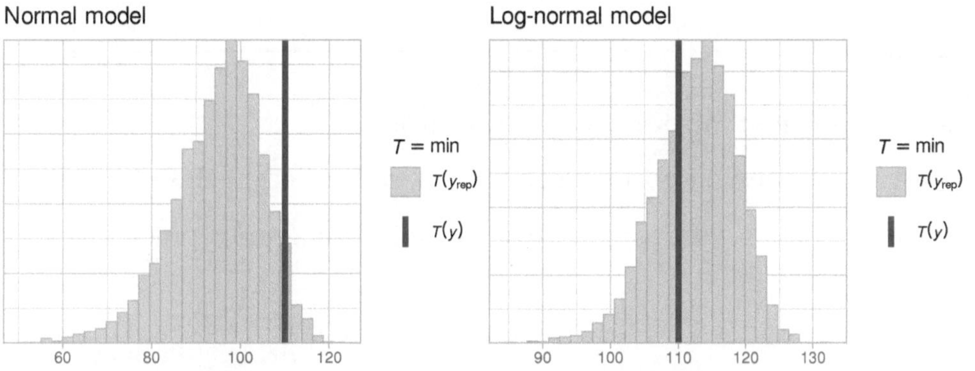

FIGURE 3.14: The distributions of minimum values in a posterior predictive check, using the normal and log-normal probability density functions. The minimum in the data is 110 ms.

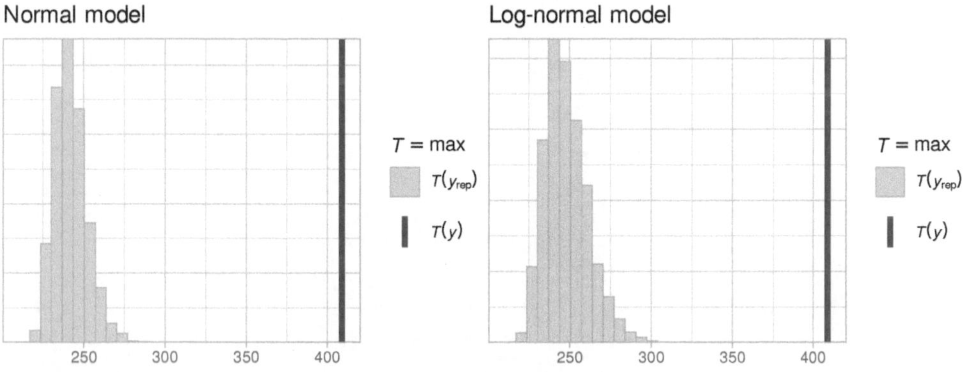

FIGURE 3.15: The distributions of maximum values in a posterior predictive check using the normal and log-normal. The maximum in the data is 409 ms.

Figure 3.14 shows that the log-normal likelihood does a slightly better job since the minimum value is contained in the bulk of the log-normal distribution and in the tail of the normal one. Figure 3.15 shows that both models are unable to capture the maximum value of the observed data. One explanation for this is that the approximately log-normally distributed observations in our data are being generated by the task of pressing as fast as possible, whereas the observations with long finger tapping times are possibly being generated by lapses of attention. If this assumption is correct, this would mean that the distribution of button pressing times is a mixture of two distributions; modeling such a mixture process involves more complex tools that we will take up in chapter 17.

This completes our introduction to `brms`. We are now ready to learn more about regression models.

3.8 List of the most important commands

Here is a list of the most important commands we learned in this chapter.

- The core `brms` function for fitting models, for generating prior predictive and posterior predictive data:

```
fit_press <-
  brm(t ~ 1,
      data = df_spacebar,
      family = gaussian(),
      prior = c(prior(uniform(0, 60000),
                      class = Intercept,
                      lb = 0,
                      ub = 60000),
                prior(uniform(0, 2000),
                      class = sigma,
                      lb = 0,
                      ub = 2000)),
      ## uncomment for prior predictive:
      ## sample_prior = "only",
      ## uncomment when dealing with divergent transitions
      ## control = list(adapt_delta = .9)
      ## default values for chains, iterations and warmup:
      chains = 4,
      iter = 2000,
      warmup = 1000)
```

- Extract samples from fitted model:

```
as_draws_df(fit_press)
```

- Basic plot of posteriors

```
plot(fit_press)
```

- Plot prior predictive/posterior predictive data

```
## Posterior predictive check:
pp_check(fit_press, ndraws = 100, type = "dens_overlay")
## Plot posterior predictive distribution of statistical summaries:
pp_check(fit_press, ndraws = 100, type = "stat", stat = "mean")
## Plot prior predictive distribution of statistical summaries:
pp_check(fit_press, ndraws = 100, type = "stat", stat = "mean",
         prefix = "ppd")
```

3.9 Summary

This chapter showed how to fit and interpret a Bayesian model with a normal likelihood. We looked at the effect of priors by investigating prior predictive distributions and by carrying out a sensitivity analysis. We also looked at the fit of the posterior, by inspecting the posterior predictive distribution (which gives us some idea about the descriptive adequacy of the model). We also showed how to fit a Bayesian model with a log-normal likelihood, and how to compare the predictive accuracy of different models.

3.10 Further reading

Sampling algorithms are discussed in detail in Gamerman and Lopes (2006). Also helpful are the sections on sampling from the short open-source book by Bob Carpenter, *Probability and Statistics: a simulation-based introduction* (`https://github.com/bob-carpenter/prob-stats`), and the sections on sampling algorithms in Lambert (2018) and Lynch (2007). The evolution of probabilistic programming languages for Bayesian inference is discussed in Štrumbelj et al.

(2024). Introductory linear modeling theory is covered in Dobson and Barnett (2011); more advanced treatments are in Montgomery, Peck, and Vining (2012) and Seber and Lee (2003). Generalized linear models are covered in detail in McCullagh and Nelder (2019). The reader may also benefit from our own freely available online lecture notes on linear modeling: `https://github.com/vasishth/LM`.

4

Bayesian regression models

We generally run experiments because we are interested in the relationship between two or more variables. A regression will tell us how our *dependent variable*, also called the *response* or *outcome variable* (e.g., pupil size, response times, accuracy, etc.) is affected by one or many *independent variables*, *predictors*, or *explanatory variables*. Predictors can be categorical (e.g., male or female), ordinal (first, second, third, etc.), or continuous (e.g., age). In this chapter we focus on simple regression models with different likelihood functions.

4.1 A first linear regression: Does attentional load affect pupil size?

Let us look at the effect of cognitive processing on human pupil size to illustrate the use of Bayesian linear regression models. Although pupil size is mostly related to the amount of light that reaches the retina or the distance to a perceived object, pupil sizes are also systematically influenced by cognitive processing: Increased cognitive load leads to an increase in the pupil size (for a review, see Mathot 2018).

For this example, we'll use data from one subject's pupil size of the control experiment by Wahn et al. (2016), averaged by trial. The data are available from `df_pupil` in the package `bcogsci`. In this experiment, the subject covertly tracks between zero and five objects among several randomly moving objects on a computer screen. Viewing zero targets is called a passive viewing condition.

This task is called multiple object tracking (or MOT; see Pylyshyn and Storm 1988). First, several objects appear on the screen for two seconds, and a subset of the objects are indicated as "targets" at the beginning. Then, the objects start moving randomly across the screen and become indistinguishable. After several seconds, the objects stop moving and the subject needs to indicate which objects were the targets. See Figure 4.1.[1] Our research goal is to examine

[1]The description of the procedure in the Wahn et al. (2016) paper is as follows: "At the beginning of a trial, all objects were shown for two seconds. Then, a subset of objects turned gray for two seconds, indicating the targets and then reverted back to look like all the other objects. After an additional 500 ms, objects started to move and participants were

DOI: 10.1201/9780429342646-6

how the number of moving objects being tracked–that is, how attentional load–affects pupil size.

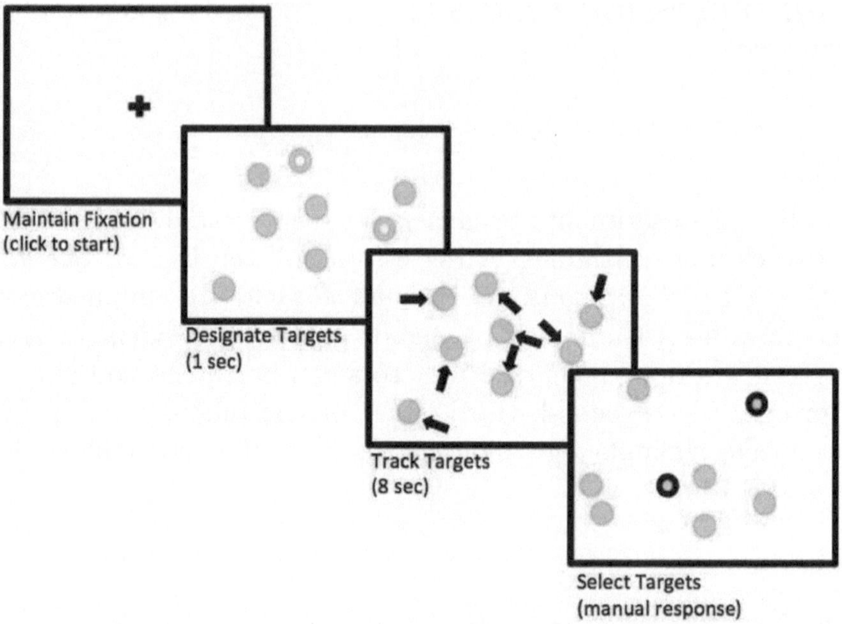

FIGURE 4.1: Flow of events in a trial where two objects need to be tracked. Adapted from Blumberg, Peterson, and Parasuraman (2015); licensed under CC BY 4.0 (`https://creativecommons.org/licenses/by/4.0/`).

4.1.1 Likelihood and priors

We will model pupil size as normally distributed, because we are not expecting a skew, and we have no further information available about the distribution of pupil sizes. (Given the units used here, pupil sizes cannot be of size zero or negative, so we know for sure that the choice of a normal likelihood is not exactly right.) For simplicity, assume a linear relationship between load and pupil size.

Let's summarize our assumptions:

1. There is some average pupil size, represented by α.
2. The increase of attentional load has a linear relationship with pupil size, determined by β.
3. There is some noise in this process, that is, variability around the true pupil size. This variability is represented by the scale σ.
4. The noise is normally distributed.

instructed to track the targets. While objects were moving, they repelled each other and bounced off the screen borders."

The generative probability density function will be as follows:

$$p\_size_n \sim Normal(\alpha + c\_load_n \cdot \beta, \sigma) \tag{4.1}$$

where n indicates the observation number with $n = 1, \dots, N$.

This means that the formula in `brms` will be `p_size ~ 1 + c_load`, where `1` represents the intercept, α, which doesn't depend on the predictor, and `c_load` is the predictor that is multiplied by β. The prefix `c_` will generally indicate that a predictor (in this case load) is centered (i.e., the mean of all the values is subtracted from each value). If load is centered, the intercept represents the pupil size at the average load in the experiment (because at the average load, the centered load is zero, yielding $\alpha + 0 \cdot \beta$). If load had not been centered (i.e., starts with no load, then one, two, etc.), then the intercept would represent pupil size when there is no load. We could fit a frequentist model with `lm(p_size ~ 1 + c_load, data_set)`; when we fit a Bayesian model using `brms`, the syntax is the same as in the `lm()` function, but we now have to specify priors for each of the parameters.

For setting plausible priors, some research needs to be done to find some information about pupil sizes. Although we might know that pupil diameters range between 2 to 4 mm in bright light to 4 to 8 mm in the dark (Spector 1990), this experiment was conducted with the Eyelink-II eyetracker which measures the pupils in arbitrary units (Hayes and Petrov 2016). If this is our first-ever analysis of pupil size, before setting up the priors, we'll need to look at some measures of pupil size. (If we have analyzed this type of data before from previously conducted experiments, we could also look at estimates from those studies). Fortunately, we have some measurements of the same subject with no attentional load for the first 100 ms, measured every 10 ms, in the data frame `df_pupil_pilot` from the package `bcogsci`: This will give us some idea about the order of magnitude of our dependent variable.

```
data("df_pupil_pilot")
df_pupil_pilot$p_size %>% summary()
```

```
##    Min. 1st Qu.  Median    Mean 3rd Qu.    Max.
##     852     856     862     861     866     868
```

With this information we can set a regularizing prior for α. Center the prior around 1000 to be in the right order of magnitude.[2] Since we don't know how

[2]The average pupil size will probably be higher than 800, since this measurement was with no load, but, in any case, the exact number won't matter, any mean for the prior between 500 and 1500 would be fine if the standard deviation is large.

much pupil sizes are going to vary by load yet, we include a rather wide prior by defining it as a normal distribution and setting its standard deviation as 500.

$$\alpha \sim Normal(1000, 500) \tag{4.2}$$

Given that our predictor load is centered, with the prior for α, we are saying that we suspect that the average pupil size for the average load in the experiment will be in a 95% credible interval limited by approximately $1000 \pm 2 \cdot 500 = [0, 2000]$ units. We can calculate this with more precision in R using the `qnorm()` function:

```
qnorm(c(.025, .975), mean = 1000, sd = 500)
```

```
## [1]   20 1980
```

We know that the measurements of the pilot data are strongly correlated because they were taken 10 milliseconds apart. For this reason, they won't give us a realistic estimate of how much the pupil size can vary. Accordingly, set up quite an uninformative prior for σ that encodes this lack of precise information: σ is surely larger than zero and has to be in the order of magnitude of the pupil size with no load.

$$\sigma \sim Normal_{+}(0, 1000) \tag{4.3}$$

With this prior for σ, we are saying that we expect that the standard deviation of the pupil sizes should be in the following 95% credible interval.

```
qtnorm(c(0.025, 0.975), mean = 0, sd = 1000, a = 0)
```

```
## [1]   31.3 2241.4
```

In order to compute the 95% credible interval, we used `qtnorm()` from the `extraDistr` package rather than `qnorm()`. As mentioned earlier, the relevant command specification is `qtnorm(n, mean, sd, a = 0)`; recall that `a = 0` indicates a truncated normal distribution, truncated at the left by zero.

The mean of $Normal_{+}$, a normal distribution truncated at zero so as to allow for only positive values, does not coincide with its location indicated with the parameter μ (and neither does the standard deviation coincide with the scale, σ); see online section A.2.

```
samples <- rtnorm(20000, mean = 0, sd = 1000, a = 0)
c(mean = mean(samples), sd = sd(samples))
```

```
## mean   sd
##  799  602
```

We still need to set a prior for β, the change in pupil size produced by the attentional load. Given that pupil size changes are not easily perceptible (we don't usually observe changes in pupil size in our day-to-day life), we expect them to be much smaller than the pupil size (which we assume has mean 1000 units), so we use the following prior:

$$\beta \sim Normal(0, 100) \tag{4.4}$$

With the prior of β, we are saying that we don't really know if increasing attentional load will increase or even decrease the pupil size (the prior distribution for β is centered at zero), but we do know that increasing one unit of load (that is, having one more object to track) will potentially change the pupil size in a way that is consistent with the following 95% credible interval.

```
qnorm(c(0.025, 0.975), mean = 0, sd = 100)
```

```
## [1] -196  196
```

That is, we don't expect changes in size that increase or decrease the pupil size more than 200 units for one unit increase in load.

The priors we have specified here are relatively uninformative; as mentioned earlier, this is because we are considering the situation where we don't have much prior experience with pupil size studies. In other settings, we might have more prior knowledge and experience; in that case, one could use somewhat more principled priors. We address this point in the online chapter on priors (online chapter E) and Bayesian workflow (online chapter F).

4.1.2 The `brms` model

Before fitting the `brms` model of the effect of load on pupil size, load the data and center the predictor `load`:

```
data("df_pupil")
(df_pupil <- df_pupil %>%
  mutate(c_load = load - mean(load)))
```

```
## # A tibble: 41 x 5
##    subj trial  load p_size c_load
##   <int> <int> <int>  <dbl>  <dbl>
## 1   701     1     2  1021. -0.439
## 2   701     2     1   951. -1.44
## 3   701     3     5  1064.  2.56
## # i 38 more rows
```

Now fit the brms model:

```
fit_pupil <-
  brm(p_size ~ 1 + c_load,
      data = df_pupil,
      family = gaussian(),
      prior = c(prior(normal(1000, 500), class = Intercept),
                prior(normal(0, 1000), class = sigma),
                prior(normal(0, 100), class = b, coef = c_load)))
```

The only difference from our previous models is that we now have a predictor in the formula and in the priors. Priors for predictors are indicated with `class = b`, and the specific predictor with `coef = c_load`. If we want to set the same priors to different predictors we can omit the argument `coef`. Even if we drop the `1` from the formula, `brm()` will fit the same model as when we specify `1` explicitly. If we really want to remove the intercept, this must be indicated with `0 +...` or `-1 +...`. Also see the online section A.3 for more details about the treatment of the intercept by brms. The priors are normal distributions for the intercept (α) and the slope (β), and a truncated normal distribution for the scale parameter σ, which coincides with the standard deviation (because the likelihood is a normal distribution). brms will automatically truncate the prior specification for σ and allow only positive values.

Next, inspect the output of the model. The posteriors and traceplots are shown in Figure 4.2; this figure is generated by typing:

```
plot(fit_pupil)
```

```
fit_pupil
```

```
## ...
## Population-Level Effects:
##           Estimate Est.Error l-95% CI u-95% CI Rhat Bulk_ESS Tail_ESS
## Intercept   700.94     20.48   660.50   740.11 1.00     3533     2602
```

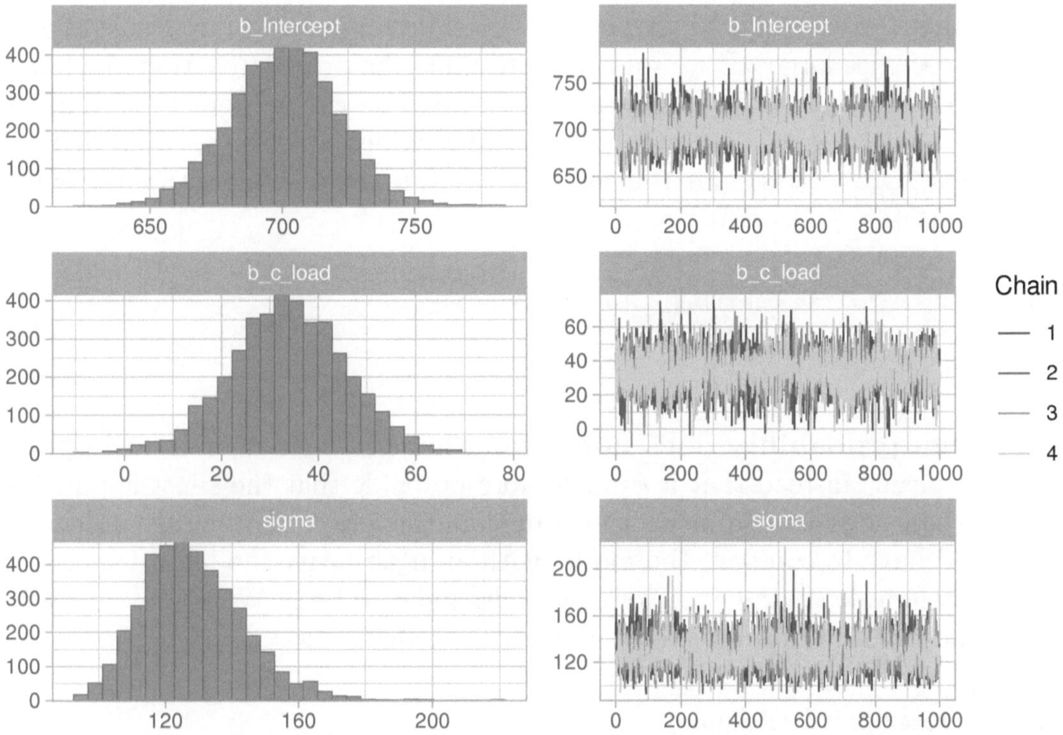

FIGURE 4.2: The posterior distributions of the parameters in the `brms` model `fit_pupil`, along with the corresponding traceplots.

```
## c_load        33.73     11.90     9.93     56.63 1.00     3439     2579
##
## Family Specific Parameters:
##       Estimate Est.Error l-95% CI u-95% CI Rhat Bulk_ESS Tail_ESS
## sigma   128.45     15.19   102.64   162.75 1.00     3421     2855
##
## ...
```

In the next section, we discuss how one can communicate the relevant information from the model.

4.1.3 How to communicate the results?

We want to answer the research question "What is the effect of attentional load on the subject's pupil size?" To answer this question, we'll need to examine what happens with the posterior distribution of β, which is printed out as `c_load` in the summary of `brms`. The summary of the posterior tells us that the most likely values of β will be around the mean of the posterior, 33.73, and we can be 95% certain that the value of β, given the model and the data, lies between 9.93 and 56.63.

The model tells us that as attentional load increases, the pupil size of the subject becomes larger. If we want to determine the probability that the pupil size increases rather than decreases, we can compute the proportion of samples that lie above zero. (The intercept and the slopes are always preceded by `b_` in `brms`. One can see all the names of parameters being estimated with `variables()`.)

```
mean(as_draws_df(fit_pupil)$b_c_load > 0)
```

```
## [1] 0.996
```

This high probability does not mean that the effect of load is non-zero. It means instead that it's much more probable that the effect is positive rather than negative. In order to claim that the effect is probably non-zero, we would have to compare the model with an alternative model in which the model assumes that the effect of load is assumed to be 0. We'll come back to this issue when we discuss model comparison in chapter 12.

4.1.4 Descriptive adequacy

Our model converged and we obtained posterior distributions for the parameters. However, there is no guarantee that our model is good enough to represent our data. We can use posterior predictive checks to investigate the descriptive adequacy of the model.

Sometimes it's useful to customize the posterior predictive check to visualize the fit of our model. We iterate over the different loads (e.g., 0 to 4), and we show the posterior predictive distributions based on 100 simulations for each load together with the observed pupil sizes in Figure 4.3. We don't have enough data to derive a strong conclusion: both the predictive distributions and our data look very widely spread out, and it's hard to tell if the distribution of the observations could have been generated by our model. For now we can say that it doesn't look too bad.

```
for (l in 0:4) {
  df_sub_pupil <- filter(df_pupil, load == l)
  p <- pp_check(fit_pupil,
                type = "dens_overlay",
                ndraws = 100,
                newdata = df_sub_pupil) +
    geom_point(data = df_sub_pupil, aes(x = p_size, y = 0.0001)) +
    ggtitle(paste("load: ", l)) +
```

```
    coord_cartesian(xlim = c(400, 1000))
  print(p)
}
```

In Figure 4.4, we look instead at the distribution of a summary statistic, such as mean pupil size by load. Figure 4.4 shows that the observed means for no load and for a load of one are falling in the tails of the distributions. Although our model predicts a monotonic increase in pupil size, the data might be indicating that the relevant difference is simply between no load and some load. However, given the uncertainty in the posterior predictive distributions and that the observed means are contained somewhere in the predicted distributions, it could be the case that with this model we are overinterpreting noise.

```
for (l in 0:4) {
  df_sub_pupil <- filter(df_pupil, load == l)
  p <- pp_check(fit_pupil,
                type = "stat",
                ndraws = 1000,
                newdata = df_sub_pupil,
                stat = "mean") +
    geom_point(data = df_sub_pupil, aes(x = p_size, y = 0.9)) +
    ggtitle(paste("load: ", l)) +
    coord_cartesian(xlim = c(400, 1000))
  print(p)
}
```

4.2 Log-normal model: Does trial affect response times?

Let us revisit the small experiment from section 3.2.1, where a subject repeatedly taps the space bar as fast as possible. Suppose that we want to know whether the subject tended to speed up (a practice effect) or slow down (a fatigue effect) while pressing the space bar. We'll use the same data set `df_spacebar` as before, and we'll center the column `trial`:

```
df_spacebar <- df_spacebar %>%
  mutate(c_trial = trial - mean(trial))
```

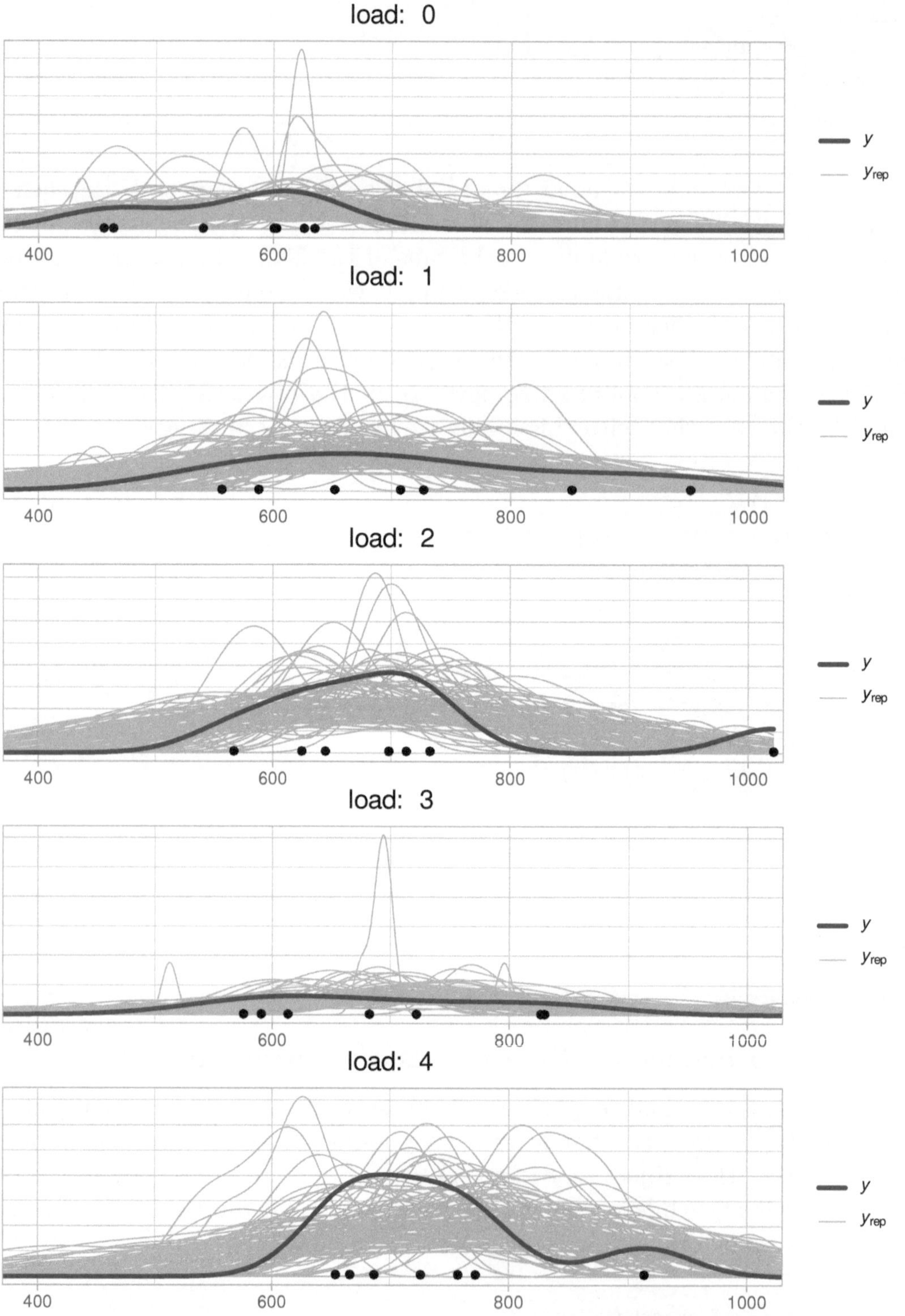

FIGURE 4.3: The plot shows 100 posterior predicted distributions with the label y_{rep}, the distribution of pupil size data in black with the label y, and the observed pupil sizes in black dots for the five levels of attentional load.

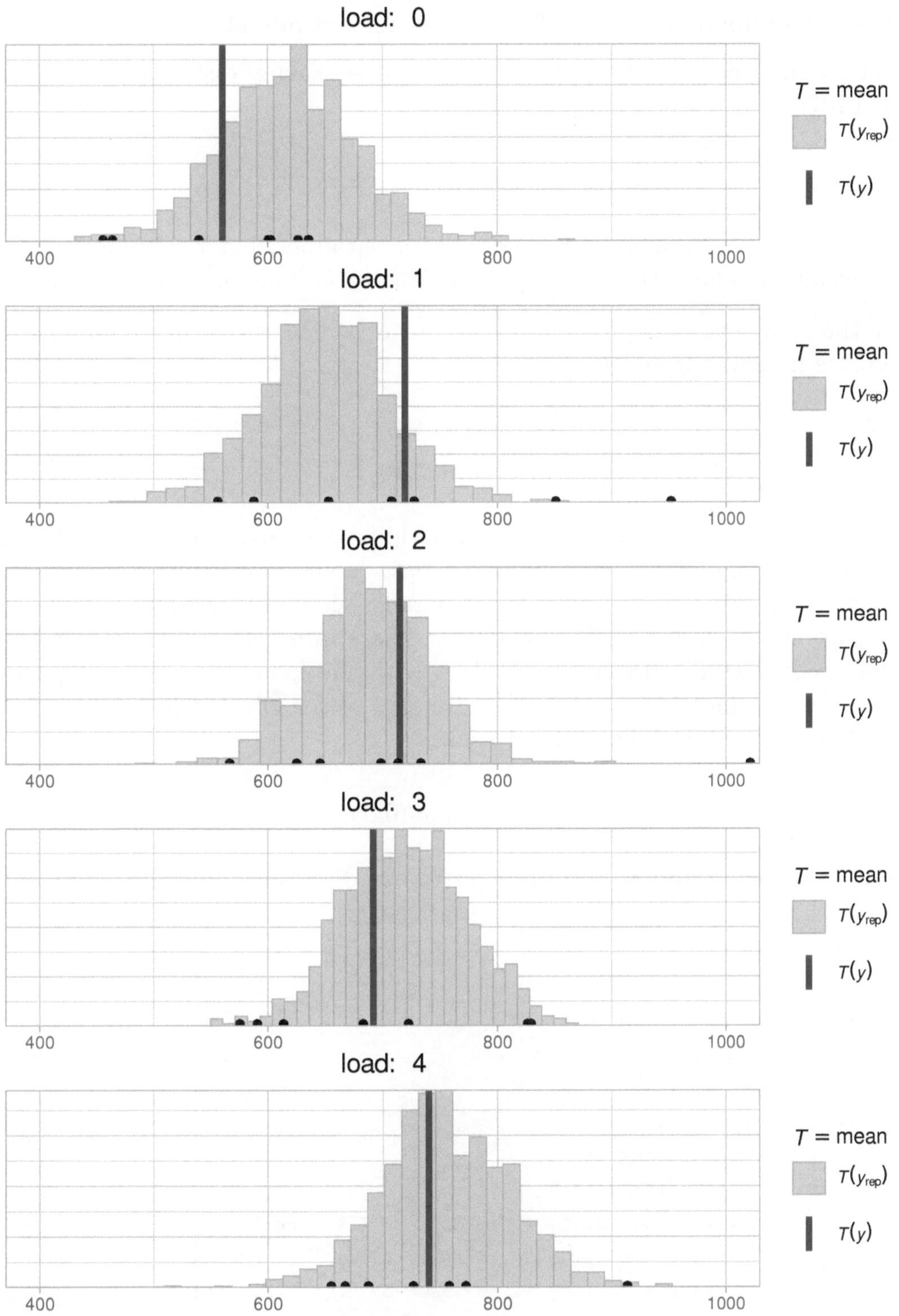

FIGURE 4.4: Shown are the distributions of posterior predicted means for each load level (in gray) and the corresponding means of observed pupil size (vertical black lines.

4.2.1 Likelihood and priors for the log-normal model

If we assume that finger-tapping times are log-normally distributed, the likelihood becomes:

$$t_n \sim LogNormal(\alpha + c\_trial_n \cdot \beta, \sigma) \tag{4.5}$$

where $n = 1, \dots, N$, and t is the dependent variable (finger tapping times in milliseconds). The variable N represents the total number of data points.

Use the same priors as in section 3.7.2 for α (which is equivalent to μ in the previous model) and for σ.

$$\begin{aligned} \alpha &\sim Normal(6, 1.5) \\ \sigma &\sim Normal_+(0, 1) \end{aligned} \tag{4.6}$$

We still need a prior for β. Effects are multiplicative rather than additive when we assume a log-normal likelihood, and that means that we need to take into account α in order to interpret β; for details, see online section A.4. We are going to try to understand how all our priors interact, by generating some prior predictive distributions. We start with the following prior centered in zero, a prior agnostic regarding the direction of the effect, which allows for a slowdown ($\beta > 0$) and a speedup ($\beta < 0$):

$$\beta \sim Normal(0, 1) \tag{4.7}$$

Here is our first attempt at a prior predictive distribution:

```
# Ignore the dependent variable,
# use a vector of ones a placeholder.
df_spacebar_ref <- df_spacebar %>%
  mutate(t = rep(1, n()))
fit_prior_press_trial <-
  brm(t ~ 1 + c_trial,
      data = df_spacebar_ref,
      family = lognormal(),
      prior = c(prior(normal(6, 1.5), class = Intercept),
                prior(normal(0, 1), class = sigma),
                prior(normal(0, 1), class = b, coef = c_trial)),
      sample_prior = "only",
      control = list(adapt_delta = .9))
```

In order to understand the type of data that we are assuming a priori with the prior of the parameter β, plot the median difference between the finger tapping times at adjacent trials. As the prior of β gets wider, larger differences are observed between adjacent trials. The objective of the prior predictive checks is to calibrate the prior of β to obtain a plausible range of differences. We are going to plot a distribution of medians because they are less affected by the variance in the prior predicted distribution than the distribution of mean differences; distributions of means will have much more spread. To make the distribution of means more realistic, we would also need to find a more accurate prior for the scale σ. (Recall that the mean of log-normal distributed values depend on both the location, μ and the scale, σ, of the distribution.) To plot the median effect, first define a function that calculates the difference between adjacent trials, and then apply the median to the result. We use that function in `pp_check` and show the result in Figure 4.5. As expected, the median effect is centered on zero (as is our prior), but we see that the distribution of possible medians for the effect is too widely spread out and includes values that are too extreme.

```
median_diff <- function(x) {
  median(x - lag(x), na.rm = TRUE)
}
pp_check(fit_prior_press_trial,
         type = "stat",
         stat = "median_diff",
  # show only prior predictive distributions
         prefix = "ppd",
  # each bin has a width of 500 ms
         binwidth = 500) +
  # cut the top of the plot to improve its scale
  coord_cartesian(ylim = c(0, 50))
```

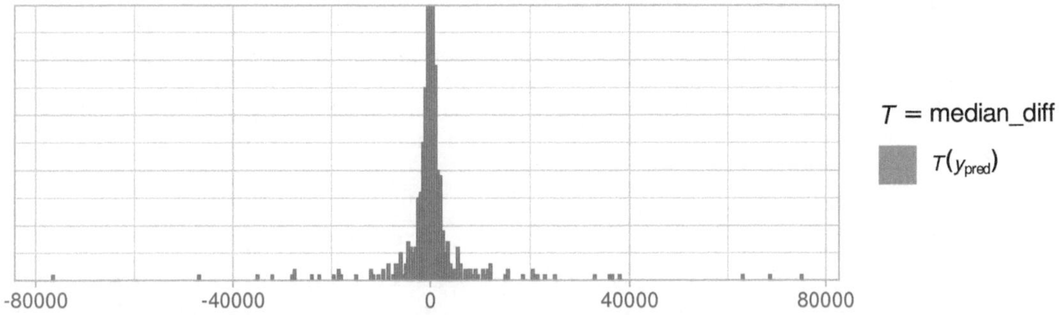

FIGURE 4.5: The prior predictive distribution of the median effect of the model defined in section 4.2 with $\beta \sim Normal(0, 1)$.

Repeat the same procedure with $\beta \sim \mathit{Normal}(0, 0.01)$; the resulting prior predictive distribution is shown in Figure 4.6. The prior predictive distribution shows us that the prior is still quite vague; it is, however at least in the right order of magnitude.

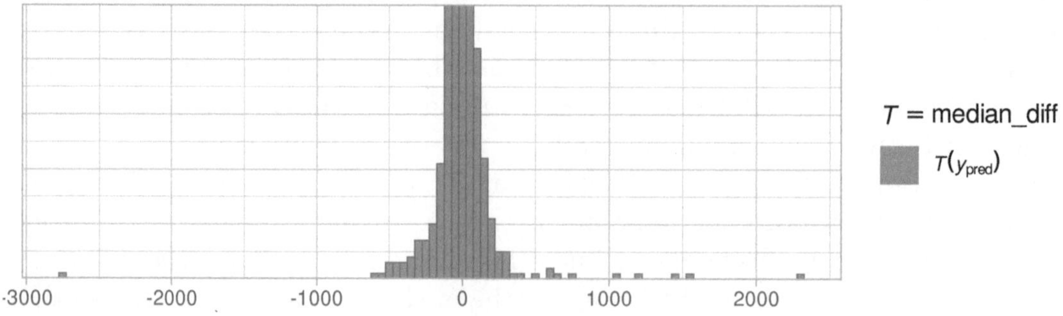

FIGURE 4.6: The prior predictive distribution of the median difference in finger tapping times between adjacent trials based on the model defined in section 4.2 with $\beta \sim \mathit{Normal}(0, 0.01)$.

Prior selection might look daunting and can be a lot of work. However, this work is usually done only the first time we start working with an experimental paradigm; besides, priors can be informed by the estimates from previous experiments (even maximum likelihood estimates from frequentist models can be useful). We will generally use very similar (or identical priors) for analyses dealing with the same type of task. When in doubt, a sensitivity analysis (see section 3.4) can tell us whether the posterior distribution depends unintentionally strongly on our prior selection. We also discuss prior selection in the online chapter E.

4.2.2 The `brms` model

We are now relatively satisfied with the priors for our model, and we can fit the model of the effect of trial on button-pressing time using `brms`. We need to specify that the family is `lognormal()`.

```
fit_press_trial <-
  brm(t ~ 1 + c_trial,
      data = df_spacebar,
      family = lognormal(),
      prior = c(prior(normal(6, 1.5), class = Intercept),
                prior(normal(0, 1), class = sigma),
                prior(normal(0, 0.01), class = b, coef = c_trial)))
```

Instead of printing out the complete output from the model, look at the estimates from the posteriors for the parameters α, β, and σ. These parameters are on the log scale:

```
posterior_summary(fit_press_trial,
                  variable = c("b_Intercept",
                               "b_c_trial",
                               "sigma"))
```

```
##              Estimate Est.Error   Q2.5    Q97.5
## b_Intercept 5.118404 0.0063173 5.1064 5.130800
## b_c_trial   0.000523 0.0000622 0.0004 0.000645
## sigma       0.123232 0.0047641 0.1148 0.133332
```

The posterior distributions can be plotted to obtain a graphical summary of all the parameters in the model (Figure 4.7):

```
plot(fit_press_trial)
```

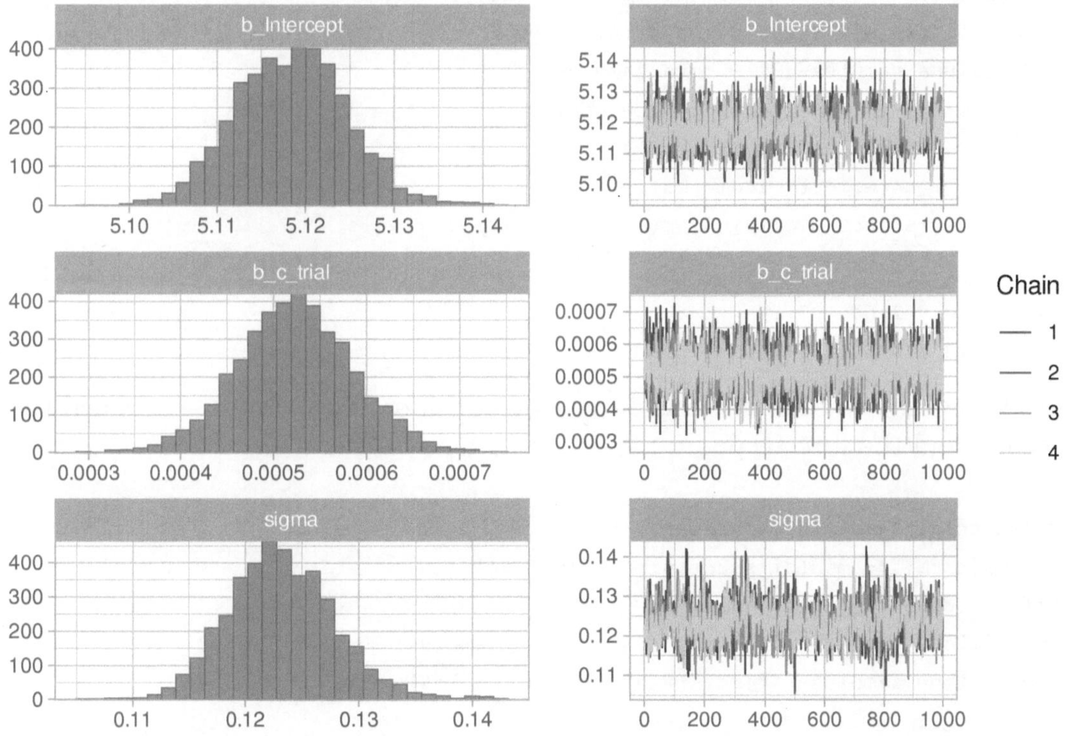

FIGURE 4.7: Posterior distributions of the model of the effect of trial on button-pressing.

Next, we turn to the question of what we can report as our results, and what we can conclude from the data.

4.2.3 How to communicate the results?

As shown above, the first step is to summarize the posteriors in a table or graphically (or both). If the research relates to the effect estimated by the model, the posterior of β can be summarized in the following way: $\hat{\beta} = 0.00052$, 95% CrI $= [0.0004, 0.00065]$.

The effect is easier to interpret in milliseconds. We can transform the estimates back to the millisecond scale from the log scale, but we need to take into account that the scale is not linear and that the effect between two button presses will differ depending on where we are in the experiment.

We will have a certain estimate if we consider the difference between response times in a trial at the middle of the experiment (when the centered trial number is zero) and the previous one (when the centered trial number is minus one).

```
alpha_samples <- as_draws_df(fit_press_trial)$b_Intercept
beta_samples <- as_draws_df(fit_press_trial)$b_c_trial
effect_middle_ms <- exp(alpha_samples) -
  exp(alpha_samples - 1 * beta_samples)
## ms effect in the middle of the expt
## (mean trial vs. mean trial - 1)
c(mean = mean(effect_middle_ms),
  quantile(effect_middle_ms, c(0.025, 0.975)))
```

```
##    mean   2.5%  97.5%
## 0.0874 0.0669 0.1078
```

We will obtain a different estimate if we consider the difference between the second and the first trial:

```
first_trial <- min(df_spacebar$c_trial)
second_trial <- min(df_spacebar$c_trial) + 1
effect_beginning_ms <-
  exp(alpha_samples + second_trial * beta_samples) -
  exp(alpha_samples + first_trial * beta_samples)
## ms effect from first to second trial:
c(mean = mean(effect_beginning_ms),
  quantile(effect_beginning_ms, c(0.025, 0.975)))
```

```
##    mean    2.5%  97.5%
## 0.0795 0.0623 0.0961
```

So far; we converted the estimates to obtain median effects, that's why we used $exp(\cdot)$. If we want to obtain mean effects, we need to take into account σ, since we need to calculate $exp(\cdot + \sigma^2/2)$. However, we can also use the built-in function `fitted()` which calculates mean effects. Consider again the difference between the second and the first trial this time using `fitted()`.

First, define for which observations we want to obtain the fitted values in millisecond scale. If we are interested in the difference between the second and first trial, create a data frame with their centered versions.

```
newdata_1 <- data.frame(c_trial = c(first_trial, second_trial))
```

Second, use `fitted()` on the `brms` object, including the new data, and setting the `summary` parameter to `FALSE`. The first column contains the posterior samples transformed into milliseconds of the first trial, and the second column of the second trial.

```
beginning <- fitted(fit_press_trial,
                    newdata = newdata_1,
                    summary = FALSE)
head(beginning, 3)
```

```
##      [,1] [,2]
## [1,]  152  152
## [2,]  154  154
## [3,]  154  154
```

The final step is to calculate the difference between trials, and report the mean and the 95% credible interval.

```
effect_beginning_ms <- beginning[, 2] - beginning[,1]
c(mean = mean(effect_beginning_ms),
  quantile(effect_beginning_ms, c(0.025, 0.975)))
```

```
##    mean    2.5%  97.5%
## 0.0801 0.0627 0.0968
```

Given that σ is much smaller than μ, σ doesn't have a large influence on the mean effects, and the mean and 95% CrI of the mean and median effects are quite similar.

We see that no matter how we calculate the trial effect, there is a slowdown. When reporting the results of these analyses, one should present the posterior mean and a credible interval, and then reason about whether the observed estimates are consistent with the prediction from the theory being investigated. The 95% credible interval used here is just a convention adopted from standard practice in psychology and related areas, where the 95% confidence interval is the default. One could, for example, have used an 80% CrI.

The practical relevance of the effect for the research question can be important too. For example, only after **100** button presses do we see a barely noticeable slowdown:

```
effect_100 <-
  exp(alpha_samples + 100 * beta_samples) -
  exp(alpha_samples)
c(mean = mean(effect_100),
  quantile(effect_100, c(0.025, 0.975)))
```

```
##  mean  2.5% 97.5%
##  8.97  6.83 11.14
```

We need to consider whether our uncertainty of this estimate and the estimated mean effect have any scientific relevance. Such relevance can be established by considering the previous literature, predictions from a quantitative model, or other expert domain knowledge. Sometimes, a quantitative meta-analysis is helpful; for examples, see Bürki, Alario, and Vasishth (2023), Cox et al. (2022), Bürki et al. (2020), Jäger, Engelmann, and Vasishth (2017), Mahowald et al. (2016), Nicenboim, Roettger, and Vasishth (2018), and Vasishth et al. (2013). We will discuss meta-analysis later in the book, in chapter 11.

Sometimes, researchers are only interested in establishing that an effect is present or absent; the magnitude and uncertainty of the estimate is of secondary interest. Here, the goal is to argue that there is **evidence** of a slowdown. The word evidence has a special meaning in statistics (Royall 1997), and in null hypothesis significance testing, a likelihood ratio test is the standard way to argue that one has evidence for an effect. In the Bayesian data analysis context, in order to answer such a question, a Bayes factor analysis must be carried out. We'll come back to this issue in the model comparison chapters 12–14.

4.2.4 Descriptive adequacy

We look now at the predictions of the model. Since we now know that trial effects are very small, let's examine the predictions of the model for differences in response times between 100 button presses. Similarly as for prior predictive checks, we define a function, `median_diff100()`, that calculates the median difference between a trial n and a trial $n + 100$. This time we'll compare the observed median difference against the range of predicted differences based on the model and the data rather than only the model as we did for the prior predictions. Below we use virtually the same code that we use for plotting prior predictive checks, but since we now use the fitted model, we'll obtain posterior predictive checks; this is displayed in Figure 4.8.

```
median_diff100 <- function(x) median(x - lag(x, 100), na.rm = TRUE)
pp_check(fit_press_trial,
         type = "stat",
         stat = "median_diff100")
```

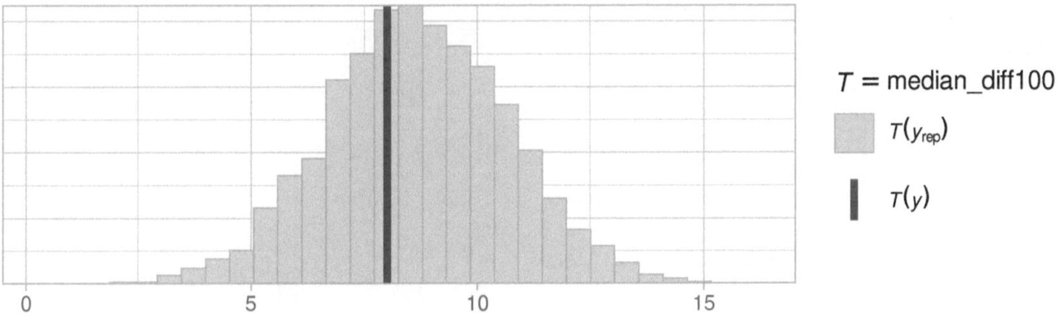

FIGURE 4.8: The posterior predictive distribution of the median difference in response times between a trial n and a trial $n + 100$ based on the model `fit_press_trial` and the observed data.

From Figure 4.8, we can conclude that model predictions for differences in response trials between trials are reasonable.

4.3 Logistic regression: Does set size affect free recall?

In this section, we will learn how the principles we have learned so far can naturally extend to *generalized* linear models (GLMs). We focus on one special case of GLMs that has wide application in linguistics and psychology, logistic regression.

As an example data set, we look at a study investigating the capacity level of working memory. The data are a subset of a data set created by Oberauer (2019). Each subject was presented word lists of varying lengths (2, 4, 6, and 8 elements), and then was asked to recall a word given its position on the list; see Figure 4.9. We will focus on the data from one subject.

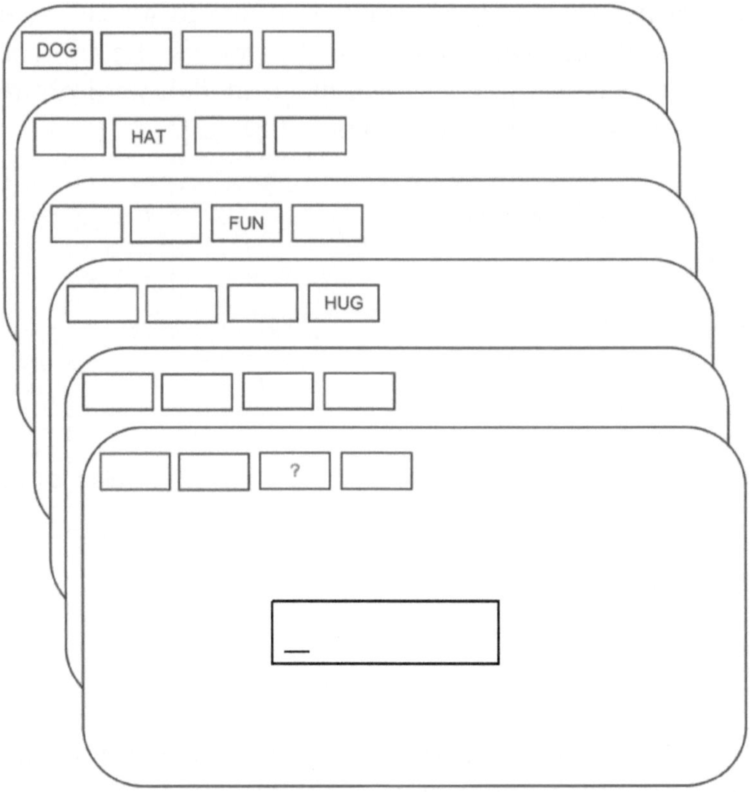

FIGURE 4.9: The flow of events in a trial with memory set size 4 and free recall. Adapted from Oberauer (2019); licensed under CC BY 4.0 (`https://creativecommons.org/licenses/by/4.0/`).

It is well-established that as the number of items to be held in working memory increases, performance, that is, accuracy, decreases (see Oberauer and Kliegl 2001, among others). We will investigate this claim with data from only one subject.

The data can be found in `df_recall` in the package `bcogsci`. The code below loads the data, centers the predictor `set_size`, and briefly explores the data set.

```
data("df_recall")
df_recall <- df_recall %>%
  mutate(c_set_size = set_size - mean(set_size))
# Set sizes in the data set:
```

```
df_recall$set_size %>%
  unique() %>% sort()
```

```
## [1] 2 4 6 8
```

```
# Trials by set size
df_recall %>%
  group_by(set_size) %>%
  count()
```

```
## # A tibble: 4 x 2
## # Groups:   set_size [4]
##   set_size     n
##      <int> <int>
## 1        2    23
## 2        4    23
## 3        6    23
## # i 1 more row
```

Here, the column `correct` records the incorrect vs. correct responses with `0` vs. `1`, respectively, and the column `c_set_size` records the centered memory set size; these latter scores have continuous values -3, -1, 1, and 3. These continuous values are centered versions of 2, 4, 6, and 8.

```
df_recall
```

```
## # A tibble: 92 x 8
##   subj  set_size correct trial session block tested c_set_size
##   <chr>    <int>   <int> <int>   <int> <int>  <int>      <dbl>
## 1 10           4       1     1       1     1      2         -1
## 2 10           8       0     4       1     1      8          3
## 3 10           2       1     9       1     1      2         -3
## # i 89 more rows
```

We want to model the trial-by-trial accuracy and examine whether the probability of recalling a word is related to the number of words in the set that the subject needs to remember.

4.3.1 The likelihood for the logistic regression model

Recall that the Bernoulli likelihood generates a 0 or 1 response with a particular probability θ. For example, one can generate simulated data for 10 observations, with a 50% probability of getting a 1 using `rbern()` from the package `extraDistr`.

```
rbern(10, prob = 0.5)
```

```
##  [1] 1 1 0 0 1 1 1 1 1 1
```

We can therefore define each dependent value `correct_n` in the data as being generated from a Bernoulli random variable with probability of success θ_n. Here, $n = 1, \dots, N$ indexes the observation, $correct_n$ is the dependent variable (0 indicates an incorrect recall and 1 a correct recall), and θ_n is the probability of correctly recalling a probe in a given observation n.

$$correct_n \sim Bernoulli(\theta_n) \tag{4.8}$$

Since the observed data are either 0 or 1, we shouldn't just fit a regression model using the normal (or log-normal) likelihood as we did in the preceding examples. Such a model would be inappropriate because it would assume that the observed data are real numbers ranging from $-\infty$ to $+\infty$ (or from 0 to $+\infty$).

The generalized linear modeling framework solves this problem by defining a *link function* $g(\cdot)$ that connects the linear model to the quantity to be estimated (here, the probabilities θ_n). The link function used for 0, 1 responses is called the *logit link*, and is defined as follows.

$$\eta_n = g(\theta_n) = \log\left(\frac{\theta_n}{1-\theta_n}\right) \tag{4.9}$$

The term $\frac{\theta_n}{1-\theta_n}$ is called the *odds*.[3] The logit link function is therefore a log-odds; it maps probability values ranging from $(0, 1)$ to real numbers lying between $-\infty$ and $+\infty$. Figure 4.10 shows the logit link function, $\eta = g(\theta)$, and the inverse logit, $\theta = g^{-1}(\eta)$, which is called the *logistic function*; the relevance of this logistic function will become clear in a moment.

The linear model is now fit not to the 0,1 responses as the dependent variable, but to η_n, i.e., log-odds, as the dependent variable:

[3] Odds are defined to be the ratio of the probability of success to the probability of failure. For example, the odds of obtaining a one in a fair six-sided die are $\frac{1/6}{1-1/6} = 1/5$. The odds of obtaining a heads in a fair coin are 1/1. Do not confuse this technical term with the day-to-day usage of the word "odds" to mean probability.

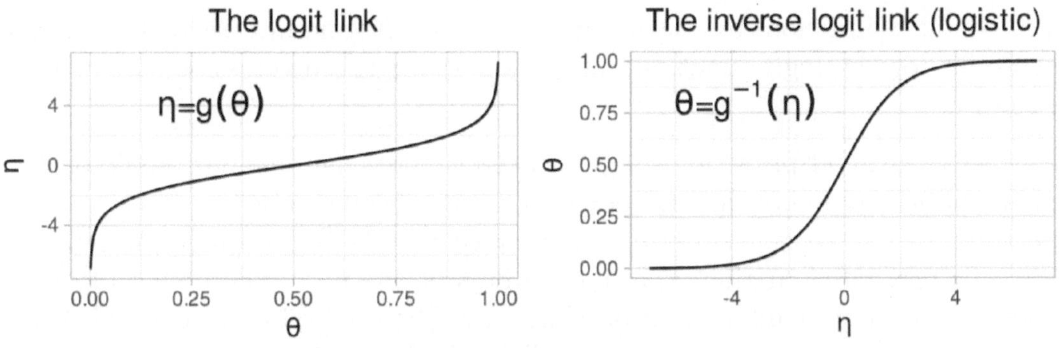

FIGURE 4.10: The logit and inverse logit (logistic) function.

$$\eta_n = \log\left(\frac{\theta_n}{1-\theta_n}\right) = \alpha + \beta \cdot c\_set\_size \tag{4.10}$$

Unlike linear models, the model is defined so that there is no residual error term (ε) in this model. Once η_n is estimated, one can solve the above equation for θ_n (in other words, we compute the inverse of the logit function and obtain the estimates on the probability scale). This gives the above-mentioned logistic regression function:

$$\theta_n = g^{-1}(\eta_n) = \frac{\exp(\eta_n)}{1+\exp(\eta_n)} = \frac{1}{1+exp(-\eta_n)} \tag{4.11}$$

The last equality in the equation above arises by dividing both the numerator and denominator by $\exp(\eta_n)$.

In summary, the generalized linear model with the logit link fits the following Bernoulli likelihood:

$$correct_n \sim Bernoulli(\theta_n) \tag{4.12}$$

The model is fit on the log-odds scale, $\eta_n = \alpha + c\_set\_size_n \cdot \beta$. Once η_n has been estimated, the inverse logit or the logistic function is used to compute the probability estimates $\theta_n = \frac{\exp(\eta_n)}{1+\exp(\eta_n)}$. An example of this calculation will be shown in the next section.

4.3.2 Priors for the logistic regression

In order to decide on priors for α and β, we need to take into account that these parameters do not represent probabilities or proportions, but *log-odds*, the x-axis in Figure 4.10 (right-hand side figure). As shown in the figure, the relationship between log-odds and probabilities is not linear.

There are two functions in R that implement the logit and inverse logit functions: `qlogis(p)` for the logit function and `plogis(x)` for the inverse logit or logistic function.

Now we need to set priors for α and β. Given that we centered our predictor, the intercept, α, represents the log-odds of correctly recalling one word in a random position for the average set size of five (since $5 = \frac{2+4+6+8}{4}$), which, incidentally, was not presented in the experiment. This is one case where the intercept doesn't have a clear interpretation if we leave the prediction uncentered: With non-centered set size, the intercept will be the log-odds of recalling one word in a set of *zero* words, which obviously makes no sense.

The prior for α will depend on how difficult the recall task is. We could assume that the probability of recalling a word for an average set size, α, is centered at 0.5 (a 50/50 chance) with a great deal of uncertainty. The R command `qlogis(0.5)` tells us that 0.5 corresponds to zero in log-odds space. How do we include a great deal of uncertainty? We could look at Figure 4.10, and decide on a standard deviation of 4 in a normal distribution centered at zero:

$$\alpha \sim Normal(0, 4) \tag{4.13}$$

Let's plot this prior in log-odds and in probability scale by drawing random samples.

```
samples_log_odds <- tibble(log_odds = rnorm(100000, 0, 4))
samples_p <- tibble(p = plogis(rnorm(100000, 0, 4)))
ggplot(samples_log_odds, aes(log_odds)) +
  geom_density()
ggplot(samples_p, aes(p)) +
  geom_density()
```

Figure 4.11 shows that our prior assigns more probability mass to extreme probabilities of recall than to intermediate values. Clearly, this is not what we intended.

We could try several values for the standard deviation of the prior until we find a prior that makes sense for us. Reducing the standard deviation to 1.5 seems to make sense as shown in Figure 4.12.

$$\alpha \sim Normal(0, 1.5) \tag{4.14}$$

We need to decide now on the prior for β, the effect in log-odds of increasing the set size. We could choose a normal distribution centered on zero, reflecting

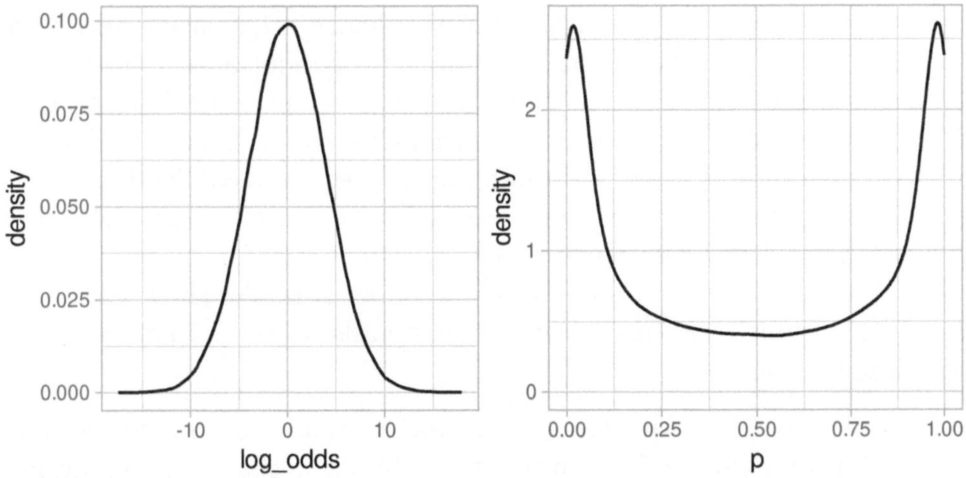

FIGURE 4.11: The prior for $\alpha \sim Normal(0, 4)$ in log-odds and in probability space.

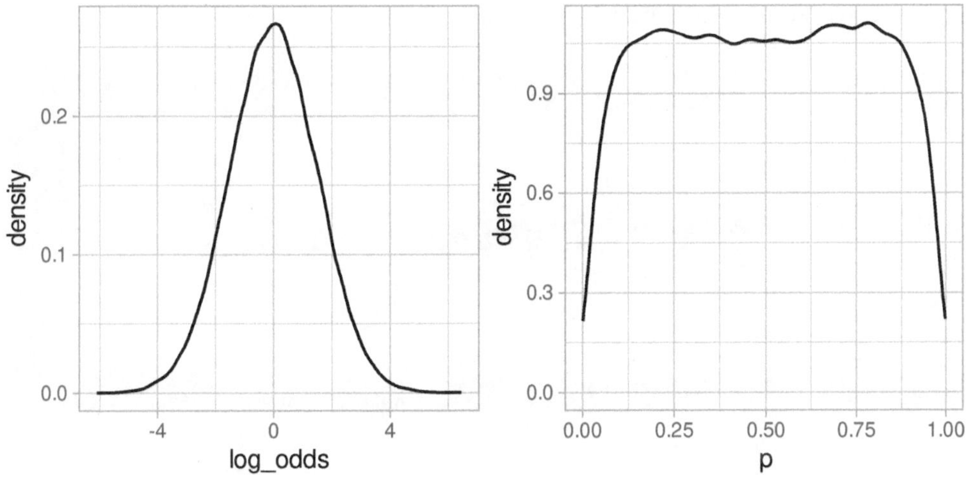

FIGURE 4.12: Prior for $\alpha \sim Normal(0, 1.5)$ in log-odds and in probability space.

our lack of any commitment regarding the direction of the effect. Let's get some intuitions regarding different possible standard deviations for this prior, by testing the following distributions as priors:

(a) $\beta \sim Normal(0, 1)$
(b) $\beta \sim Normal(0, .5)$
(c) $\beta \sim Normal(0, .1)$
(d) $\beta \sim Normal(0, .01)$
(e) $\beta \sim Normal(0, .001)$

In principle, we could produce the prior predictive distributions using `brms` with `sample_prior = "only"` and then `predict()`. However, as mentioned before, `brms` also uses Stan's Hamiltonian sampler for sampling from the priors, and this can lead to convergence problems when the priors are too uninformative (as in this case). We solve this issue by performing prior predictive checks directly in R using the `r*` family of functions (e.g., `rnorm()`, `rbinom()`, etc.) together with loops. This method is not as simple as using the convenient functions provided by `brms`, but it is very flexible and can be very powerful. We show the prior predictive distributions in Figure 4.13; for the details on the implementation in R, see online section A.5.

Figure 4.13 shows that, as expected, the priors are centered on zero. We see that the distribution of possible accuracies for the prior that has a standard deviation of 1 is problematic: There is too much probability concentrated near 0 and 1 for set sizes of 2 and 8. It's hard to tell the differences between the other priors, and it might be more useful to look at the predicted differences in accuracy between set sizes in Figure 4.14.

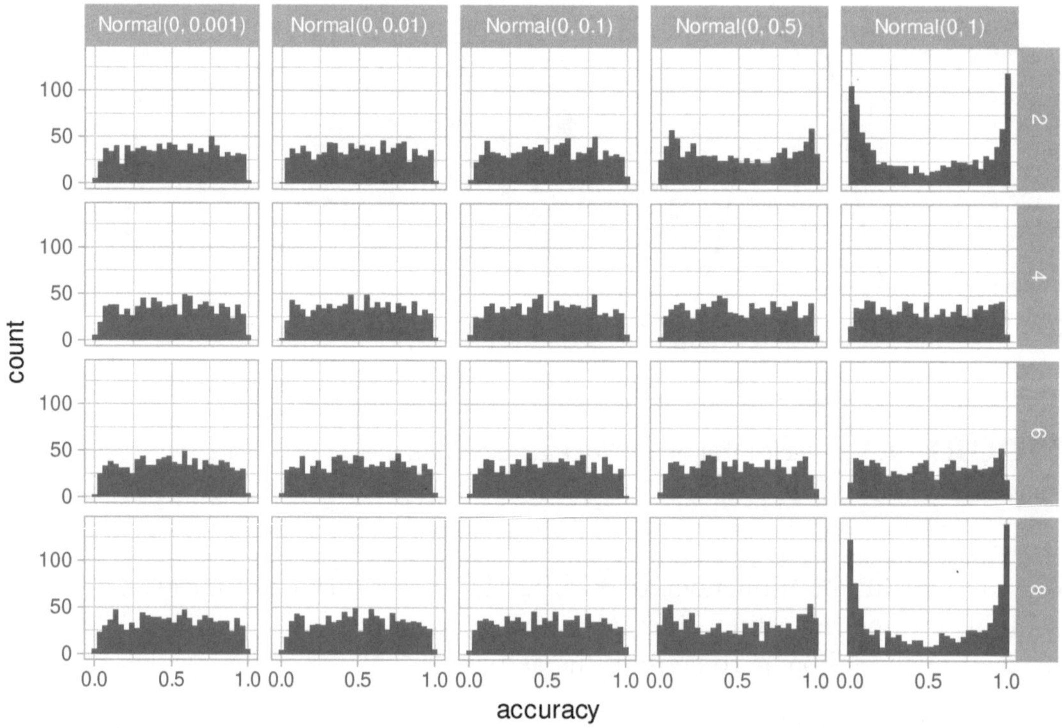

FIGURE 4.13: The prior predictive distributions of mean accuracy of the model defined in section 4.3, for different set sizes and different priors for β.

If we are not sure whether the increase of set size could produce something between a null effect and a relatively large effect, we can choose the prior with

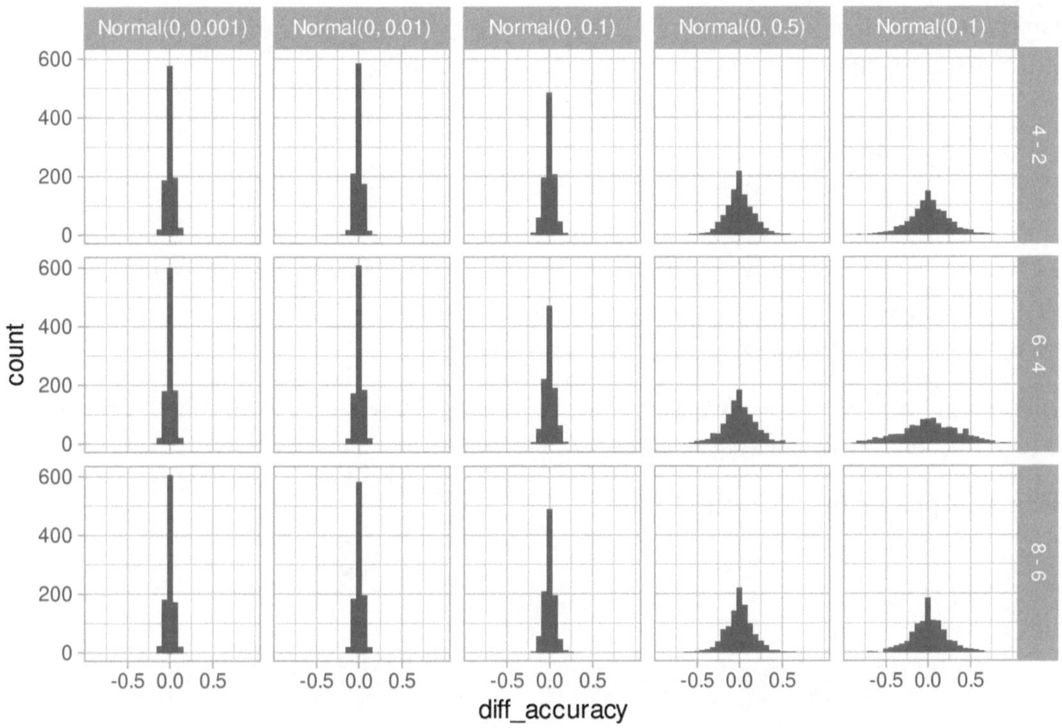

FIGURE 4.14: The prior predictive distributions of differences in mean accuracy between set sizes of the model defined in section 4.3 for different priors for β.

a standard deviation of 0.1. Under this reasoning, we settle on the following priors:

$$\begin{aligned} \alpha &\sim Normal(0, 1.5) \\ \beta &\sim Normal(0, 0.1) \end{aligned} \tag{4.15}$$

4.3.3 The `brms` model

Having decided on the likelihood, the link function, and the priors, the model can now be fit using `brms`. We need to specify that the family is `bernoulli()`, and the link is `logit`.

```
fit_recall <-
  brm(correct ~ 1 + c_set_size,
      data = df_recall,
      family = bernoulli(link = logit),
      prior = c(prior(normal(0, 1.5), class = Intercept),
                prior(normal(0, 0.1), class = b, coef = c_set_size)))
```

Next, look at the summary of the posteriors of each of the parameters. Keep in mind that the parameters are in log-odds space:

```
posterior_summary(fit_recall,
                  variable = c("b_Intercept", "b_c_set_size"))
```

```
##              Estimate Est.Error   Q2.5   Q97.5
## b_Intercept     1.920    0.3024  1.366  2.5311
## b_c_set_size   -0.182    0.0793 -0.337 -0.0236
```

Inspecting `b_c_set_size`, we see that increasing the set size has a detrimental effect on recall, as we suspected.

Plot the posteriors as well (Figure 4.15):

```
plot(fit_recall)
```

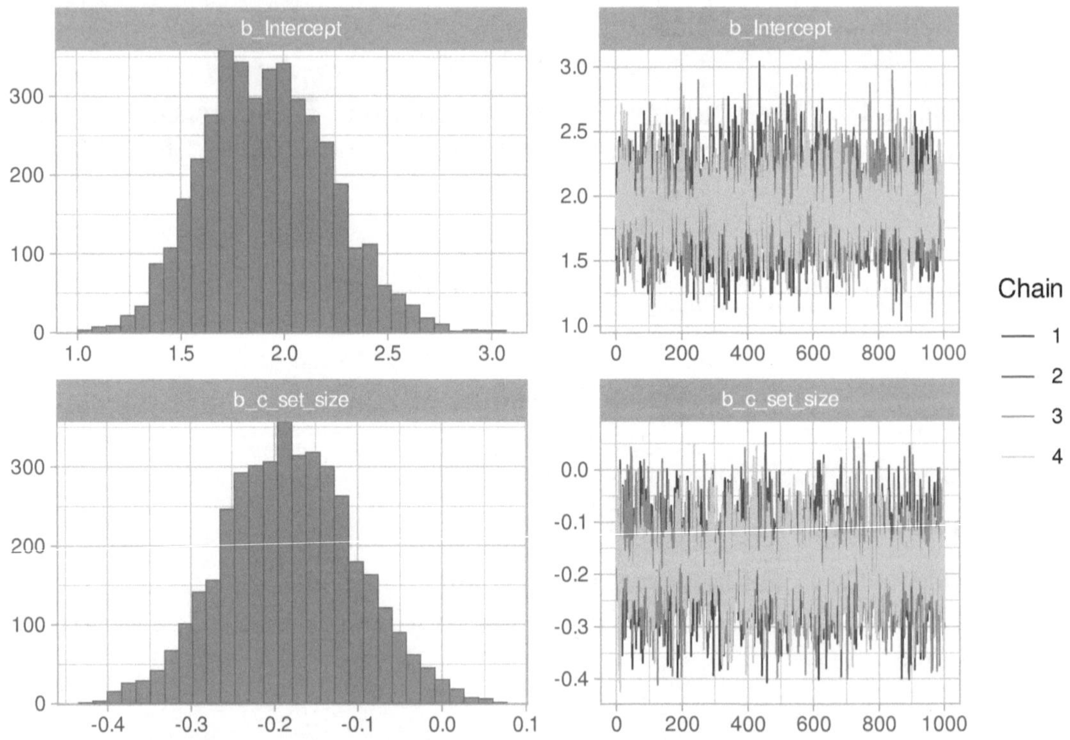

FIGURE 4.15: The posterior distributions of the parameters in the `brms` model `fit_recall`, along with the corresponding traceplots.

Next, we turn to the question of what we can report as our results, and what we can conclude from the data.

4.3.4 How to communicate the results?

Here, we are in a situation analogous to the one we saw earlier with the log-normal model. If we want to talk about the effect estimated by the model in log-odds space, we summarize the posterior of β in the following way: $\hat{\beta} = -0.182$, 95% CrI = $[-0.337, -0.024]$.

However, the effect is easier to understand in proportions rather than in log-odds. Let's look at the average accuracy for the task first:

```
alpha_samples <- as_draws_df(fit_recall)$b_Intercept
av_accuracy <- plogis(alpha_samples)
c(mean = mean(av_accuracy), quantile(av_accuracy, c(0.025, 0.975)))
```

```
##  mean  2.5% 97.5%
## 0.868 0.797 0.926
```

As before, to transform the effect of our manipulation to an easier to interpret scale (i.e., proportions), we need to take into account that the scale is not linear, that the effect of increasing the set size depends on the average accuracy, and the set size that we start from.

We can do the following calculation, similar to what we did for the trial effects experiment, to find out the decrease in accuracy in proportions or probability scale:

```
beta_samples <- as_draws_df(fit_recall)$b_c_set_size
effect_middle <- plogis(alpha_samples) -
  plogis(alpha_samples - beta_samples)
c(mean = mean(effect_middle),
  quantile(effect_middle, c(0.025, 0.975)))
```

```
##     mean     2.5%    97.5%
## -0.01878 -0.03674 -0.00252
```

Notice the interpretation here, if we increase the set size from the average set size minus one to the average set size, we get a reduction in the accuracy of recall of -0.019, 95% CrI = $[-0.037, -0.003]$. Recall that the average set size, 5, was not presented to the subject! We could alternatively look at the decrease in accuracy from a set size of 2 to 4:

```
four <- 4 - mean(df_recall$set_size)
two <- 2 - mean(df_recall$set_size)
```

```
effect_4m2 <-
  plogis(alpha_samples + four * beta_samples) -
  plogis(alpha_samples + two * beta_samples)
c(mean = mean(effect_4m2),
  quantile(effect_4m2, c(0.025, 0.975)))
```

```
##      mean     2.5%    97.5%
## -0.02944 -0.05302 -0.00489
```

We can also back-transform to probability scale using the function `fitted()` rather than using `plogis()`. One advantage is that this will work regardless of the type of link function; in this section we only discussed the logit link, but other link functions are possible to use in generalized linear models (e.g., the probit link; see Dobson and Barnett 2011).

Since set size is the only predictor and it is centered, for estimating the average accuracy, we can consider an imaginary observation where the `c_set_size` is zero, which corresponds to an average set size. (If there were more centered predictors, we would need to set all of them to zero). Now we can use the `summary` argument provided by `fitted()`.

```
fitted(fit_recall,
       newdata = data.frame(c_set_size = 0),
       summary = TRUE)[,c("Estimate", "Q2.5","Q97.5")]
```

```
## Estimate     Q2.5    Q97.5
##    0.868    0.797    0.926
```

For estimating the difference in accuracy from the average set size minus one to the average set size, and from a set size of two to four, first, define `newdata` with these set sizes.

```
new_sets <- data.frame(c_set_size = c(0, -1, four, two))
set_sizes <- fitted(fit_recall,
                    newdata = new_sets,
                    summary = FALSE)
```

Then calculate the appropriate differences considering that column one of `set_sizes` corresponds to the average set size, column two to the average set size minus one, and so forth.

```
effect_middle <- set_sizes[, 1] - set_sizes[, 2]
effect_4m2 <- set_sizes[, 3] - set_sizes[, 4]
```

Finally, calculate the summaries.

```
c(mean = mean(effect_middle), quantile(effect_middle,
                                       c(0.025, 0.975)))
```

```
##     mean     2.5%    97.5%
## -0.01878 -0.03674 -0.00252
```

```
c(mean = mean(effect_4m2), quantile(effect_4m2,
                                       c(0.025, 0.975)))
```

```
##     mean     2.5%    97.5%
## -0.02944 -0.05302 -0.00489
```

As expected, we get exactly the same values with `fitted()` as when we calculate them "by hand."

4.3.5 Descriptive adequacy

One potentially useful aspect of posterior distributions is that we could also make predictions for other conditions not presented in the actual experiment, such as set sizes that weren't tested. We could then carry out another experiment to investigate whether our model was right using another experiment. To make predictions for other set sizes, we extend our data set, adding rows with set sizes of 3, 5, and 7. To be consistent with the data of the other set sizes in the experiment, we add 23 observations for each new set size (this is the number of observations by set size in the data set). Something important to notice is that **we need to center our predictor based on the original mean set size**. This is because we want to maintain our interpretation of the intercept. We extend the data as follows, and we summarize the data and plot it in Figure 4.16.

```
df_recall_ext <- df_recall %>%
  bind_rows(tibble(set_size = rep(c(3, 5, 7), 23),
                   c_set_size = set_size -
                     mean(df_recall$set_size),
                   correct = 0))
# nicer label for the facets:
```

```
set_size <- paste("set size", 2:8) %>%
  setNames(-3:3)
pp_check(fit_recall,
         type = "stat_grouped",
         stat = "mean",
         group = "c_set_size",
         newdata = df_recall_ext,
         facet_args = list(ncol = 1, scales = "fixed",
                           labeller = as_labeller(set_size)),
         binwidth = 0.02)
```

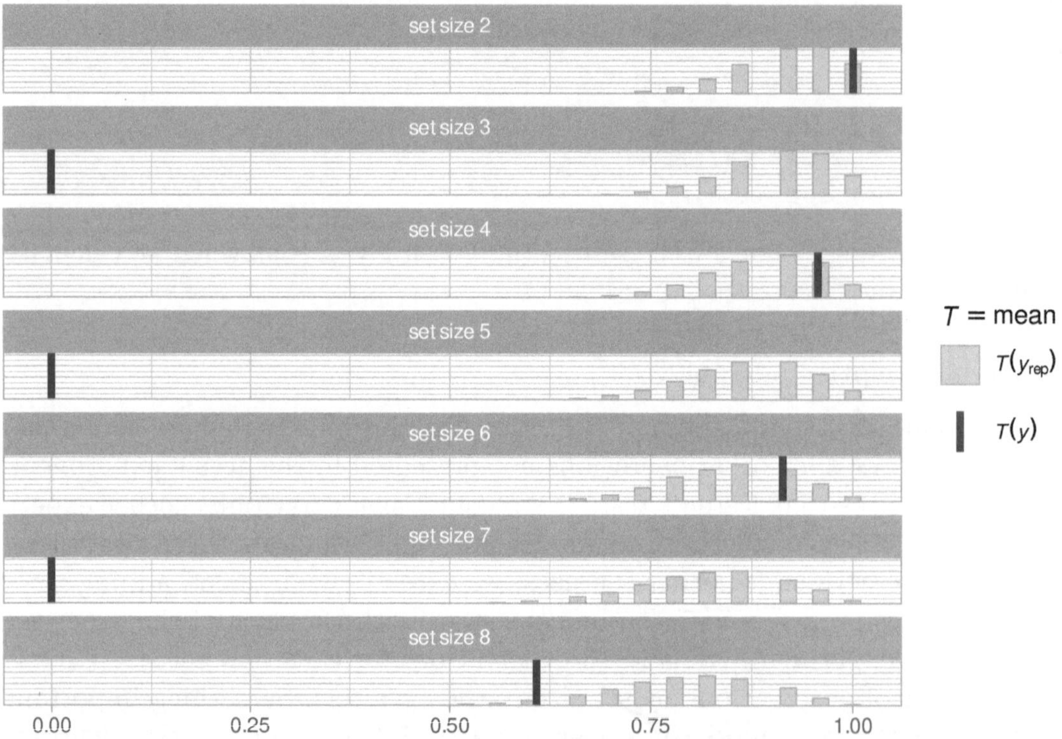

FIGURE 4.16: The distributions of posterior predicted mean accuracies for tested set sizes (2, 4, 6, and 8) and untested ones (3, 5, and 7) are labeled with y_{rep}. The observed mean accuracy, y, are only relevant for the tested set sizes (2, 4, 6, and 8); the "observed" accuracies of the untested set sizes are represented as 0.

We could now gather new data in an experiment that also shows set sizes of 3, 5, and 7. These data would be *held out* from the model `fit_recall`, since the model was fit when those data were not available. Verifying that the new observations fit in our already generated posterior predictive distribution would be a way to test genuine predictions from our model.

Having seen how we can fit simple regression models, we turn to hierarchical models in the next chapter.

4.4 Summary

In this chapter, we learned how to fit simple linear regression models and to fit and interpret models with a log-normal likelihood and logistic regression models. We investigated the prior specification for the models, using prior predictive checks, and the descriptive adequacy of the models using posterior predictive checks.

4.5 Further reading

Linear regression is discussed in several classic textbooks; these have largely a frequentist orientation, but the basic theory of linear modeling presented there can easily be extended to the Bayesian framework. An accessible textbook is by Dobson and Barnett (2011). Other useful textbooks on linear modeling are Harrell Jr. (2015), Faraway (2016), Fox (2015), and Montgomery, Peck, and Vining (2012).

5

Bayesian hierarchical models

Usually, experimental data in cognitive science contain "clusters." These are natural groups that contain observations that are more similar within the clusters than between them. The most common examples of clusters in experimental designs are subjects and experimental items (e.g., words, pictures, objects that are presented to the subjects). These clusters arise because we have multiple (repeated) observations for each subject and for each item. Such data are sometimes also called dependent data, because the observations within a cluster are not independent. If we want to incorporate this grouping structure in our analysis, we generally use a hierarchical model (also called multi-level or mixed models, Pinheiro and Bates 2000). This kind of clustering and hierarchical modeling arises as a consequence of the idea of *exchangeability*.

5.1 Exchangeability and hierarchical models

Exchangeability is the Bayesian analog of the phrase "independent and identically distributed" that appears regularly in classical (i.e., frequentist) statistics. Some connections and differences between exchangeability and the frequentist concept of independent and identically distributed (iid) are detailed in online section A.6.

Informally, the idea of exchangeability is as follows. Suppose we assign a numerical index to each of the levels of a group (e.g., to each subject). When the levels are exchangeable, we can reassign the indices arbitrarily and lose no information; that is, the joint distribution will remain the same, because we don't have any different prior information for each cluster (here, for each subject). In hierarchical models, we treat the levels of the group as exchangeable, and observations within each level in the group as also exchangeable. We generally include predictors at the level of the observations, those are the predictors that correspond to the experimental manipulations (e.g., attentional load, trial number, cloze probability, etc.); and maybe also at the group level, these are predictors that indicate characteristics of the levels in the group (e.g., the working memory capacity score of each subject). Then the conditional

DOI: 10.1201/9780429342646-7

distributions given these explanatory variables would be exchangeable; that is, our predictors incorporate all the information that is not exchangeable, and when we factor the predictors out, the observations or units in the group are exchangeable. This is the reason why the item number is an appropriate cluster, but trial number is not: In the first case, if we permute the numbering of the items there is no loss of information because the indexes are exchangeable, all the information about the items is incorporated as predictors in the model. In the second case, the numbering of the trials include information that will be lost if we treat them as exchangeable. For example, consider the case where, as trial numbers increase, subjects get more experienced or fatigued. Even if we are not interested in the specific cluster-level estimates, hierarchical models allow us to generalize to the underlying population (subjects, items) from which the clusters in the sample were drawn. Of course, this generalization is only possible if the samples are representative (e.g., if the subjects and items are randomly drawn); strictly speaking, this condition often does not hold in experimental fields like psychology and linguistics. For more on exchangeability, consult the further reading at the end of the chapter.

Exchangeability is important in Bayesian statistics because of a theorem called the Representation Theorem (de Finetti 1931). This theorem states that if a sequence of random variables is exchangeable, then the prior distributions on the parameters in a model are a necessary consequence; priors are not merely an arbitrary addition to the frequentist modeling approach that we are familiar with. For a detailed discussion on the Representation Theorem, see chapter 4 of O'Hagan and Forster (2004).

Furthermore, exchangeability has been shown (Bernardo and Smith 2009) to be mathematically equivalent to assuming a hierarchical structure in the model. The argument goes as follows. Suppose that the parameters for each level in a group are μ_i, where the levels are labeled $i = 1, \dots, I$. An example of groups is subjects. Suppose also that the data y_n, where $n = 1, \dots, N$ are observations from these subjects (e.g., pupil size measurements). The data are assumed to be generated as

$$y_n \sim Normal(\mu_{subj[n]}, \sigma) \tag{5.1}$$

The notation $subj[n]$, which roughly follows Gelman and Hill (2007), identifies the subject index. Suppose that each of 20 subjects provide 50 observations. If the data are ordered by subject id, then $subj[1]$ to $subj[50]$ corresponds to a subject with id $i = 1$, $subj[51]$ to $subj[100]$ corresponds to a subject with id $i = 2$, and so on.

We can code this representation in a straightforward way in R:

```
N_subj <- 20
N_obs_subj <- 50
N <- N_subj * N_obs_subj
df <- tibble(row = 1:N,
             subj = rep(1:N_subj, each = N_obs_subj))
df
```

```
## # A tibble: 1,000 x 2
##      row  subj
##    <int> <int>
## 1      1     1
## 2      2     1
## 3      3     1
## # i 997 more rows
```

```
# Example:
c(df$subj[1], df$subj[2], df$subj[51])
```

```
## [1] 1 1 2
```

If the data y_n are exchangeable, the parameters μ_i are also exchangeable. The fact that the μ_i are exchangeable can be shown (Bernardo and Smith 2009) to be mathematically equivalent to assuming that they come from a common distribution, for example:

$$\mu_i \sim Normal(\mu, \tau) \tag{5.2}$$

To make this more concrete, assume some completely arbitrary true values for the parameters and generate observations based on a hierarchical process in R.

```
mu <- 100
tau <- 15
sigma <- 4
mu_i <- rnorm(N_subj, mu, tau)
df_h <- mutate(df, y = rnorm(N, mu_i[subj], sigma))
df_h
```

```
## # A tibble: 1,000 x 3
##      row  subj     y
##    <int> <int> <dbl>
```

```
## 1      1     1  87.3
## 2      2     1  90.7
## 3      3     1  87.5
## # i 997 more rows
```

The parameters μ and τ, called hyperparameters, are unknown and have prior distributions (hyperpriors) defined for them. This fact leads to a hierarchical relationship between the parameters: there is a common parameter μ for each of the levels of a group (each of the subjects), and the parameters μ_i are assumed to be generated as a function of this common parameter μ. Here, τ represents between-group (here, between-subject) variability and σ represents within-group (within-subject) variability. The three parameters have priors defined for them. The first two priors below are called hyperpriors.

- $p(\mu)$
- $p(\tau)$
- $p(\sigma)$

In such a model, information about μ_i comes from two sources:

(a) from each of the observed y_n corresponding to the respective $\mu_{subj[n]}$ parameter, and

(b) from the parameters μ and τ that led to all the other y_k (where $k \neq n$) being generated.

This is illustrated in Figure 5.1.

Fit this model in `brms` in the following way. For now, the prior distributions are arbitrary.

```
fit_h <- brm(y ~ 1 + (1 | subj), df_h,
             prior =
               c(prior(normal(50, 100), class = Intercept),
                 prior(normal(2, 5), class = sigma),
                 prior(normal(10, 10), class = sd)),
             # increase iterations, adapt_delta to avoid convergence issues
             iter = 5000,
             control = list(adapt_delta = .99,
                            max_treedepth = 12))
```

```
fit_h
```

```
## ...
```

```
## Group-Level Effects:
## ~subj (Number of levels: 20)
##               Estimate Est.Error l-95% CI u-95% CI Rhat Bulk_ESS
## sd(Intercept)    15.52      2.65    11.43    22.09 1.00      482
##               Tail_ESS
## sd(Intercept)      847
##
## Population-Level Effects:
##           Estimate Est.Error l-95% CI u-95% CI Rhat Bulk_ESS Tail_ESS
## Intercept   102.21      3.49    95.13   108.94 1.01      389      714
##
## Family Specific Parameters:
##       Estimate Est.Error l-95% CI u-95% CI Rhat Bulk_ESS Tail_ESS
## sigma     4.02      0.09     3.84     4.20 1.01     1547     2419
##
## ...
```

In this output, `Intercept` corresponds to the posterior of μ, `sigma` to σ, and `sd(Intercept)` to τ. There is more information in the `brms` object: we can also get the posteriors for each level of our group (i.e., for each subject). However, rather than estimating μ_i, `brms` estimates the adjustments to μ, u_i, named `r_subj[i,Intercept]`, so that $\mu_i = \mu + u_i$. See the code below.

```
# Extract the posterior estimates of u_i
u_i_post <- as_draws_df(fit_h) %>%
  select(starts_with("r_subj"))
# Extract the posterior estimate of mu
mu_post <- as_draws_df(fit_h)$b_Intercept
# Build the posterior estimate of mu_i
mu_i_post <- mu_post + u_i_post
colMeans(mu_i_post) %>% unname()
```

```
##  [1]  91.8  96.3 123.2 101.0 102.0 126.4 107.0  80.3  90.3  93.4 118.9
## [12] 104.5 106.0 101.0  92.7 126.5 107.7  71.4 110.3  93.0
```

```
# Compare with true values
mu_i
```

```
##  [1]  91.6  96.5 123.4 101.1 101.9 125.7 106.9  81.0  89.7  93.3 118.4
## [12] 105.4 106.0 101.7  91.7 126.8 107.5  70.5 110.5  92.9
```

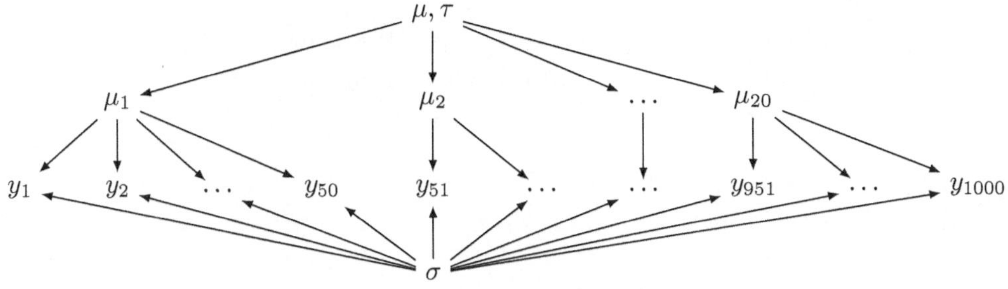

FIGURE 5.1: A directed acyclic graph illustrating a hierarchical model (partial pooling).

There are two other configurations possible that do not involve this hierarchical structure and which represent two alternative, extreme scenarios.

One of these two configurations is called the *complete pooling model*, Here, the data y_n are assumed to be generated from a single distribution:

$$y_n \sim Normal(\mu, \sigma). \tag{5.3}$$

This model is an intercept only regression similar to what we saw in chapter 3.

Generate simulated observations in a vector `y` based on arbitrary true values in R in the following way.

```
sigma <- 4
mu <- 100
df_cp <- mutate(df, y = rnorm(N, mu, sigma))
df_cp
```

```
## # A tibble: 1,000 x 3
##     row  subj     y
##   <int> <int> <dbl>
## 1     1     1  99.3
## 2     2     1  99.3
## 3     3     1 106.
## # i 997 more rows
```

Fit it in `brms`.

```
fit_cp <- brm(y ~ 1, df_cp,
              prior =
```

```
             c(prior(normal(50, 200), class = Intercept),
               prior(normal(2, 5), class = sigma)))
```

```
fit_cp
```

```
## ...
## Population-Level Effects:
##           Estimate Est.Error l-95% CI u-95% CI Rhat Bulk_ESS Tail_ESS
## Intercept   100.15      0.13    99.90   100.40 1.00     2733     2263
##
## Family Specific Parameters:
##       Estimate Est.Error l-95% CI u-95% CI Rhat Bulk_ESS Tail_ESS
## sigma     4.00      0.09     3.83     4.18 1.00     3020     2617
##
## ...
```

The configuration of the complete pooling model is illustrated in Figure 5.2.

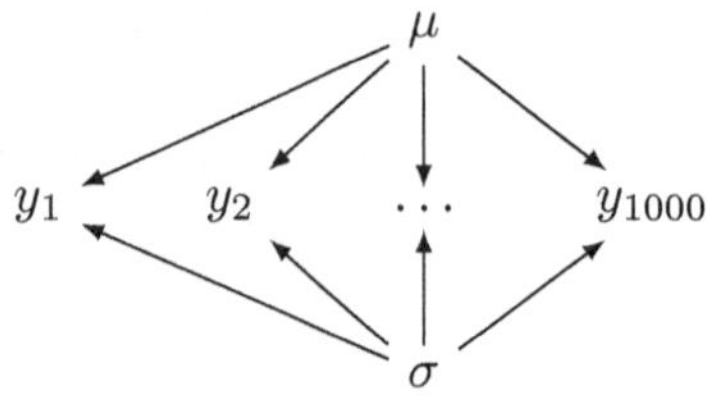

FIGURE 5.2: A directed acyclic graph illustrating a complete pooling model.

The other configuration is called the *no pooling model*; here, each y_n is assumed to be generated from an independent distribution:

$$y_n \sim Normal(\mu_{subj[n]}, \sigma) \tag{5.4}$$

Generate simulated observations from the no pooling model in R with arbitrary true values.

```
sigma <- 4
mu_i <- c(156, 178, 95, 183, 147, 191, 67, 153, 129, 119, 195,
          150, 172, 97, 110, 115, 78, 126, 175, 80)
df_np <- mutate(df, y = rnorm(N, mu_i[subj], sigma))
df_np
```

```
## # A tibble: 1,000 x 3
##     row  subj     y
##   <int> <int> <dbl>
## 1     1     1  157.
## 2     2     1  160.
## 3     3     1  164.
## # i 997 more rows
```

Fit the no pooling model in `brms`. By using the formula `0 + factor(subj)`, the common intercept can be removed and the model is forced to estimate one intercept for each `subj` separately. The column `subj` is converted to a factor so that `brms` does not interpret it as a number.

```
fit_np <- brm(y ~ 0 + factor(subj), df_np,
              prior =
                c(prior(normal(0, 200), class = b),
                  prior(normal(2, 5), class = sigma)))
```

The summary shows now the 20 estimates of μ_i as `b_factorsubj` and σ. (We ignore `lp__` and `lprior`.)

```
fit_np %>% posterior_summary()
```

```
##                 Estimate Est.Error   Q2.5  Q97.5
## b_factorsubj1     156.23    0.5542 155.15 157.32
## b_factorsubj2     176.73    0.5766 175.63 177.85
## b_factorsubj3      94.40    0.5538  93.32  95.48
## b_factorsubj4     183.09    0.5617 181.95 184.20
## b_factorsubj5     145.87    0.5432 144.78 146.94
## b_factorsubj6     191.09    0.5497 190.03 192.20
## b_factorsubj7      67.07    0.5535  65.98  68.18
## b_factorsubj8     152.44    0.5788 151.32 153.59
## b_factorsubj9     129.11    0.5414 128.04 130.15
## b_factorsubj10    119.74    0.5435 118.66 120.78
## b_factorsubj11    195.12    0.5722 194.00 196.23
## b_factorsubj12    150.09    0.5538 148.99 151.21
## b_factorsubj13    172.41    0.5664 171.30 173.52
## b_factorsubj14     95.42    0.5522  94.36  96.49
## b_factorsubj15    110.18    0.5492 109.11 111.24
## b_factorsubj16    115.27    0.5524 114.22 116.36
## b_factorsubj17     78.52    0.5634  77.41  79.63
## b_factorsubj18    125.62    0.5472 124.51 126.68
```

```
## b_factorsubj19    175.20    0.5409    174.17    176.25
## b_factorsubj20     81.15    0.5657     80.03     82.26
## sigma               3.91    0.0871      3.74      4.09
## lprior           -131.51    0.0112   -131.53   -131.49
## lp__            -2912.05    3.2170  -2919.29  -2906.75
```

Unlike the hierarchical model, now there is no common distribution that generates the μ_i parameters. This is illustrated in Figure 5.3.

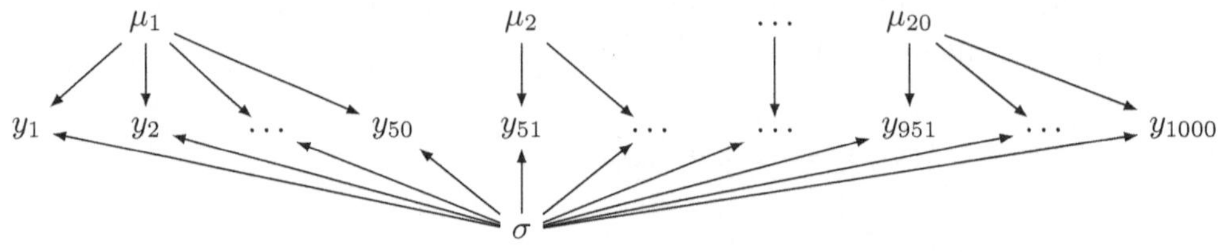

FIGURE 5.3: A directed acyclic graph illustrating a no pooling model.

The hierarchical model lies between these two extremes and for this reason is sometimes called a *partial pooling model.* One way that the hierarchical model is often described is that the estimates μ_i "borrow strength" from the parameter μ (which represents the grand mean in the above example).

An important practical consequence of partial pooling is the idea of "borrowing strength from the mean": if we have very sparse data from a particular member of a group (e.g., missing data from a particular subject), the estimate μ_i of that particular group member n is determined by the parameter μ. In other words, when the data are sparse for group member n, the posterior estimate μ_i is determined largely by the posterior on μ. In this sense, even the frequentist hierarchical modeling software in R, `lmer()` from the package `lme4`, is essentially Bayesian in formulation (except of course that there is no prior on μ).

So far we have focused on the structure of μ, the location parameter of the likelihood. We could even have partial pooling, complete pooling or no pooling with respect to σ, the scale parameter of the likelihood. More generally, any parameter of a likelihood can have any of these kinds of pooling.

In the coming sections, we will be looking at each of these models with more detail and using realistic examples.

5.2 A hierarchical model with a normal likelihood: The N400 effect

Event-related potentials (ERPs) allow scientists to observe electrophysiological responses in the brain measured by means of electroencephalography (EEG) that are time-locked to a specific event (i.e., the presentation of the stimuli). A very robust ERP effect in the study of language is the N400. Words with low predictability are accompanied by an *N400 effect* in comparison with high-predictable words, this is a relative negativity that peaks around 300–500 ms after word onset over central parietal scalp sites (first reported in Kutas and Hillyard 1980, for semantic anomalies, and in 1984 for low predictable word; for a review, see Kutas and Federmeier 2011). The N400 is illustrated in Figure 5.4.

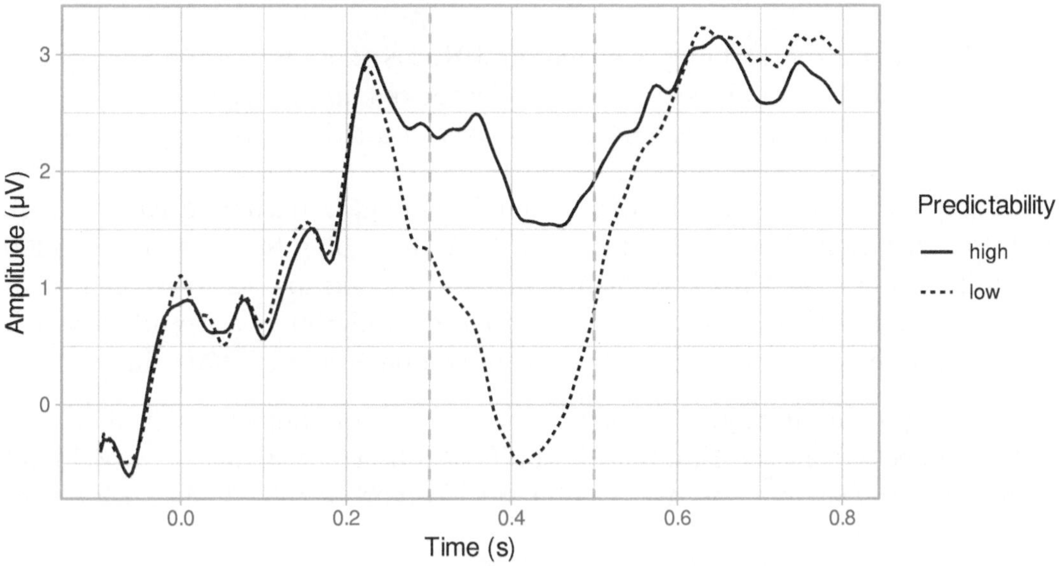

FIGURE 5.4: Typical ERPs for the grand average across the N400 spatial window (central parietal electrodes: Cz, CP1, CP2, P3, Pz, P4, POz) for high and low predictability nouns (specifically from the constraining context of the experiment reported in Nicenboim, Vasishth, and Rösler 2020). The x-axis indicates time in seconds and the y-axis indicates voltage in microvolts (unlike many EEG/ERP plots, negative polarity is plotted downward).

For example, in (1) below, the continuation '*dog*' has lower predictability than the continuation '*paint*,' and thus we would expect a more negative signal, that is, an N400 effect, in '*dog*' in (b) in comparison with '*paint*' in (a). Predictability is often measured with a cloze task (see section 1.4).

1. Example from Kutas and Hillyard (1984)
   a. Don't touch the wet paint.
   b. Don't touch the wet dog.

The EEG data are typically recorded in tens of electrodes every couple of milliseconds, but for our purposes (i.e., for learning about Bayesian hierarchical models), we can safely ignore the complexity of the data. A common way to simplify the high-dimensional EEG data when we are dealing with the N400 is to focus on the average amplitude of the EEG signal at the typical spatio-temporal window of the N400 (for example, see Frank et al. 2015).

For this example, we are going to focus on the N400 effect for critical nouns from a subset of the data of Nieuwland et al. (2018). Nieuwland et al. (2018) presented a replication attempt of an original experiment of DeLong, Urbach, and Kutas (2005) with sentences like (2).

2. Example from DeLong, Urbach, and Kutas (2005)
   a. The day was breezy so the boy went outside to fly a kite.
   b. The day was breezy so the boy went outside to fly an airplane.

We'll ignore the goal of original experiment (DeLong, Urbach, and Kutas 2005), and its replication (Nieuwland et al. 2018). We are going to focus on the N400 at the final nouns in the experimental stimuli. In example (2), the final noun '*kite*' has higher predictability than '*airplane*,' and thus we would expect a more negative signal in '*airplane*' in (b) in comparison with '*kite*' in (a).

To speed up computation, we restrict the data set of Nieuwland et al. (2018) to the subjects from the Edinburgh lab. This subset of the data can be found in `df_eeg` in the `bcogsci` package. Center the cloze probability before using it as a predictor.

```
data("df_eeg")
(df_eeg <- df_eeg %>%
  mutate(c_cloze = cloze - mean(cloze)))
```

```
## # A tibble: 2,863 x 7
##     subj cloze  item  n400 cloze_ans     N c_cloze
##    <dbl> <dbl> <dbl> <dbl>     <dbl> <dbl>   <dbl>
## 1      1 0         1  7.08         0    44  -0.476
## 2      1 0.03      2 -0.68         1    44  -0.446
## 3      1 1         3  1.39        44    44   0.524
## # i 2,860 more rows
```

```
# Number of subjects
df_eeg %>%
  distinct(subj) %>%
  count()
```

```
## # A tibble: 1 x 1
##       n
##   <int>
## 1    37
```

One convenient aspect of using averages of EEG data is that they are roughly normally distributed. This allows us to use the normal likelihood. Figure 5.5 shows the distribution of the data.

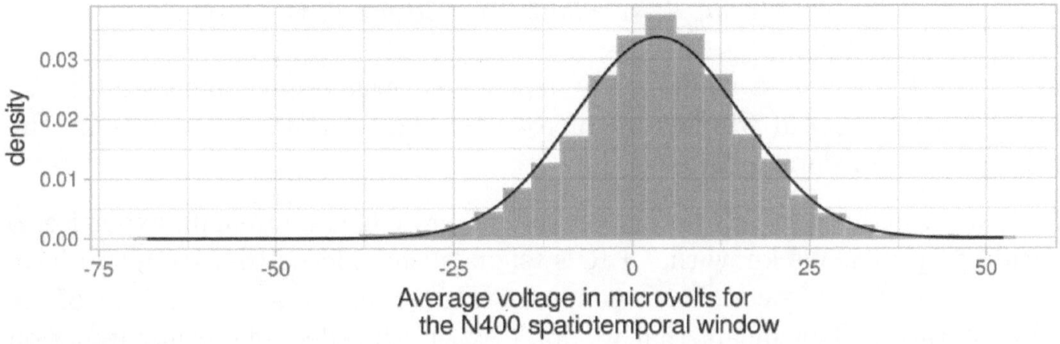

FIGURE 5.5: Histogram of the N400 averages for every trial, overlaid is a density plot of a normal distribution.

5.2.1 Complete pooling model (M_{cp})

We'll start from the simplest model which is basically the linear regression we encountered in the preceding chapter.

5.2.1.1 Model assumptions

This model, call it M_{cp}, makes the following assumptions.

1. The EEG averages for the N400 spatiotemporal window are normally distributed.
2. Observations are independent.
3. There is a linear relationship between cloze and the EEG signal for the trial.

This model is incorrect for these data due to assumption (2) being violated.

With the last assumption, we are saying that the difference in the average signal when we compare nouns with cloze probability of 0 and 0.1 is the same as the difference in the signal when we compare nouns with cloze values of 0.1 and 0.2 (or 0.9 and 1). This is just an assumption, and it may not necessarily be the case in the actual data. This means that we are going to get a posterior for β conditional on the assumption that the linear relationship holds. Even if the assumption approximately holds, we still don't know how much we deviate from this assumption.

We can now decide on a likelihood and priors.

5.2.1.2 Likelihood and priors

A normal likelihood seems reasonable for these data:

$$signal_n \sim Normal(\alpha + c\_cloze_n \cdot \beta, \sigma) \tag{5.5}$$

where $n = 1, \dots, N$, and $signal$ is the dependent variable (average signal in the N400 spatiotemporal window in microvolts). The variable N represents the total number of data points.

As always, we need to rely on our previous knowledge and domain expertise to decide on priors. We know that ERPs (signals time-locked to a stimulus) have mean amplitudes of a couple of microvolts: This is easy to see in any plot of the EEG literature. This means that we don't expect the effect of our manipulation to exceed, say, $10\mu V$. As before, a priori we'll assume that effects can be negative or positive. We can quantify our prior knowledge regarding plausible values of β as normally distributed, centered at zero with a standard deviation of $10\mu V$. (Other values such as $5\mu V$ would have been equally reasonable, since this smaller standard deviation would entail that 95% of the prior mass probability is between -10 and $10\mu V$.)

If the signal for each ERP is *baselined*, that is, the mean signal of a time window before the time window of interest is subtracted from the time window of interest, then the mean signal would be relatively close to 0. Since we know that the ERPs were baselined in this study, we expect that the grand mean of our signal should be relatively close to zero. Our prior for α is then also normally distributed and centered at zero with a standard deviation of $10\mu V$.

The prior for the standard deviation σ of our signal distribution requires some thought. We know that EEG signals are quite noisy and that the standard deviation must be higher than zero. Our prior for σ is a truncated normal distribution with location zero and scale 50. Recall that since we truncate the distribution, the parameters location and scale do not correspond to the mean and standard deviation of the new distribution; see online section A.2.

We can draw random samples from this truncated distribution and calculate their mean and standard deviation:

```
samples <- rtnorm(20000, mean = 0, sd = 50, a = 0)
c(mean = mean(samples), sd = sd(samples))
```

```
## mean   sd
## 39.9 30.2
```

So we are essentially saying that we assume a priori that we will find the true standard deviation of the signal in the following interval with 95% probability:

```
quantile(samples, probs = c(0.025, 0.975))
```

```
##   2.5%  97.5%
##   1.51 112.07
```

```
# Analytically:
# c(qtnorm(0.025, 0, 50, a = 0), qtnorm(0.975, 0, 50, a = 0))
```

To sum up, we are going to use the following priors:

$$\begin{aligned}\alpha &\sim Normal(0, 10)\\ \beta &\sim Normal(0, 10)\\ \sigma &\sim Normal_{+}(0, 50)\end{aligned} \tag{5.6}$$

A model such as M_{cp} is sometimes called a *fixed-effects* model: all the parameters are fixed in the sense that do not vary from subject to subject or from item to item. A similar frequentist model would correspond to fitting a simple linear model using the `lm()` function: `lm(n400 ~ 1 + c_cloze, data = df_eeg)`.

We fit this model in `brms` as follows (the default family is `gaussian()` so we can omit it). As with the `lm()` function in R, by default an intercept is fitted and thus `n400 ~ c_cloze` is equivalent to `n400 ~ 1 + c_cloze`:

```
fit_N400_cp <-
  brm(n400 ~ c_cloze,
      prior = c(prior(normal(0, 10), class = Intercept),
                prior(normal(0, 10), class = b, coef = c_cloze),
                prior(normal(0, 50), class = sigma)),
      data = df_eeg)
```

For now, check the summary and plot the posteriors of the model (Figure 5.6).

```
fit_N400_cp
```

```
## ...
## Population-Level Effects:
##           Estimate Est.Error l-95% CI u-95% CI Rhat Bulk_ESS Tail_ESS
## Intercept     3.67      0.22     3.25     4.11 1.00     4081     2788
## c_cloze       2.24      0.55     1.18     3.32 1.00     3461     3056
##
## Family Specific Parameters:
##       Estimate Est.Error l-95% CI u-95% CI Rhat Bulk_ESS Tail_ESS
## sigma    11.82      0.16    11.52    12.13 1.00     4471     2929
##
## ...
```

```
plot(fit_N400_cp)
```

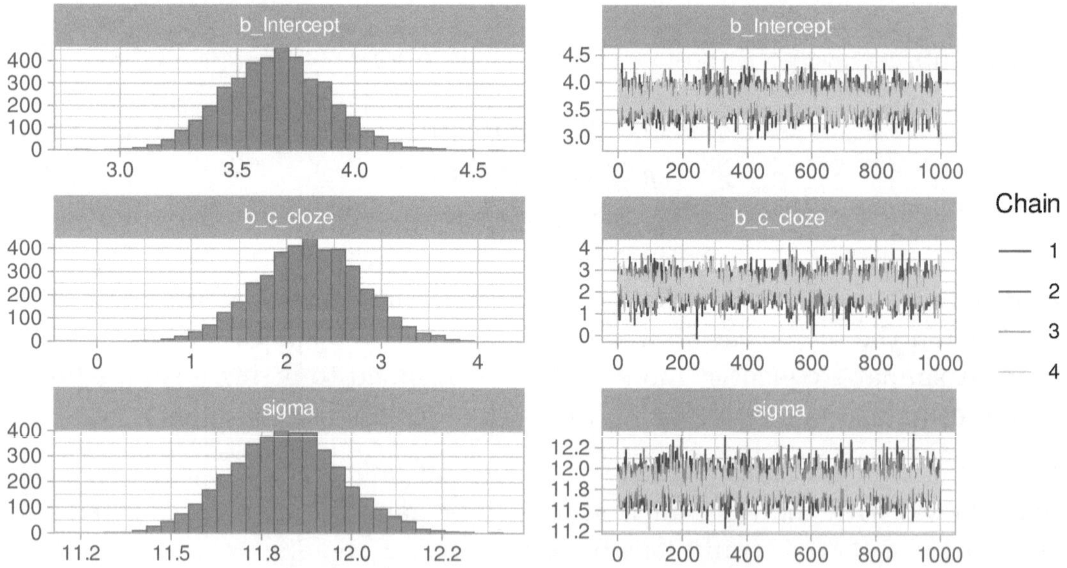

FIGURE 5.6: Posterior distributions of the complete pooling model, `fit_N400_cp`.

5.2.2 No pooling model (M_{np})

One of the assumptions of the previous model is clearly wrong: observations are not independent, they are clustered by subject (and also by the specific item, but we'll ignore this until section 5.2.4). It is reasonable to assume that EEG signals are more similar within subjects than between them. The following model assumes that each subject is completely independent from each other.[1]

5.2.2.1 Model assumptions

1. EEG averages for the N400 spatio-temporal window are normally distributed.
2. Every subject's model is fit independently of the other subjects; the subjects have no parameters in common (an exception is the standard deviation, σ; this is the same for all subjects in Equation (5.7)).
3. There is a linear relationship between cloze and the EEG signal for the trial.

What likelihood and priors can we choose here?

5.2.2.2 Likelihood and priors

The likelihood is a normal distribution as before:

$$signal_n \sim Normal(\alpha_{subj[n]} + c\_cloze_n \cdot \beta_{subj[n]}, \sigma) \tag{5.7}$$

As before, n represents each observation, that is, the nth row in the data frame, which has N rows, and now the index i identifies the subject. The notation $subj[n]$, which roughly follows Gelman and Hill (2007), identifies the subject index; for example, if $subj[10] = 3$, then the 10th row of the data frame is from subject 3.

We define the priors as follows:

$$\begin{aligned} \alpha_i &\sim Normal(0, 10) \\ \beta_i &\sim Normal(0, 10) \\ \sigma &\sim Normal_+(0, 50) \end{aligned} \tag{5.8}$$

In `brms`, such a model can be fit by removing the common intercept with the formula `n400 ~ 0 + factor(subj) + c_cloze:factor(subj)`.

This formula forces the model to estimate one intercept and one slope for *each*

[1]For simplicity, we assume that they share the same standard deviation.

level of `subj`.[2] The by-subject intercepts are indicated with `factor(subj)` and the by-subject slopes with `c_cloze:factor(subj)`.

It's very important to specify that `subject` should be treated as a factor and not as a numerical variable; we don't assume that subject number 3 will show 3 times more positive (or negative) average signal than subject number 1! The model fits 37 independent intercepts and 37 independent slopes. By setting a prior to `class = b` and omitting `coef`, we are essentially setting identical priors to all the intercepts and slopes of the model. The parameters are independent from each other; it is only our previous knowledge (or prior beliefs) about their possible values (encoded in the priors) that is identical. We can set different priors to each intercept and slope, but that will mean setting 74 priors!

```
fit_N400_np <- brm(n400 ~ 0 + factor(subj) + c_cloze:factor(subj),
                   prior = c(prior(normal(0, 10), class = b),
                             prior(normal(0, 50), class = sigma)),
                   data = df_eeg)
```

For this model, printing a summary means printing the 75 parameters ($\alpha_{1,...,37}$, $\beta_{1,...,37}$, and σ). We could do this as always by printing out the model results: just type `fit_N400_np`.

It may be easier to understand the output of the model by plotting $\beta_{1,..,37}$ using `bayesplot`. (`brms` also includes a wrapper for this function called `stanplot`.) We can take a look at the internal names that `brms` gives to the parameters with `variables(fit_N400_np)`; they are `b_factorsubj`, then the subject index and then `:c_cloze`. The code below changes the subject labels back to their original numerical indices and plots them in Figure 5.7. The subjects are ordered by the magnitude of their mean effects.

The model M_{np} does not estimate a unique population-level effect; instead, there is a different effect estimated for each subject. However, given the posterior means from each subject, it is still possible to calculate the average of these estimates $\hat{\beta}_{1,...,I}$, where I is the total number of subjects:

```
# parameter name of beta by subject:
ind_effects_np <- paste0("b_factorsubj",
                         unique(df_eeg$subj), ":c_cloze")
```

[2]If we don't remove the intercept, that is, if we use the formula `n400 ~ 1 + factor(subj) + c_cloze:factor(subj)`, with `factor(subj)` we are going estimate the deviation between the first subject and each of the other subjects. This is because the default contrast coding for the subjects, treated as a population-level or fixed effect, is treatment contrasts. See chapters 6 and 7.

```
beta_across_subj <- as.data.frame(fit_N400_np) %>%
  #removes the meta data from the object
  select(all_of(ind_effects_np)) %>%
  rowMeans()

# Calculate the average of these estimates
(grand_av_beta <- tibble(mean = mean(beta_across_subj),
                         lq = quantile(beta_across_subj, c(0.025)),
                         hq = quantile(beta_across_subj, c(0.975))))
```

```
## # A tibble: 1 x 3
##    mean    lq    hq
##   <dbl> <dbl> <dbl>
## 1  2.17  1.18  3.18
```

In Figure 5.7, the 95% credible interval of this overall mean effect is plotted as two vertical lines together with the effect of cloze probability for each subject (ordered by effect size). Here, rather than using a plotting function from `brms`, we can extract the summary of by-subject effects, reorder them by magnitude, and then plot the summary with a custom plot using `ggplot2`.

```
# make a table of beta's by subject
beta_by_subj <- posterior_summary(fit_N400_np,
                                  variable = ind_effects_np) %>%
  as.data.frame() %>%
  mutate(subject = 1:n()) %>%
  ## reorder plot by magnitude of mean:
  arrange(Estimate) %>%
  mutate(subject = factor(subject, levels = subject))
```

The code below generates Figure 5.7.

```
ggplot(beta_by_subj,
       aes(x = Estimate, xmin = Q2.5, xmax = Q97.5, y = subject)) +
  geom_point() +
  geom_errorbarh() +
  geom_vline(xintercept = grand_av_beta$mean) +
  geom_vline(xintercept = grand_av_beta$lq, linetype = "dashed") +
  geom_vline(xintercept = grand_av_beta$hq, linetype = "dashed") +
  xlab("By-subject effect of cloze probability in microvolts")
```

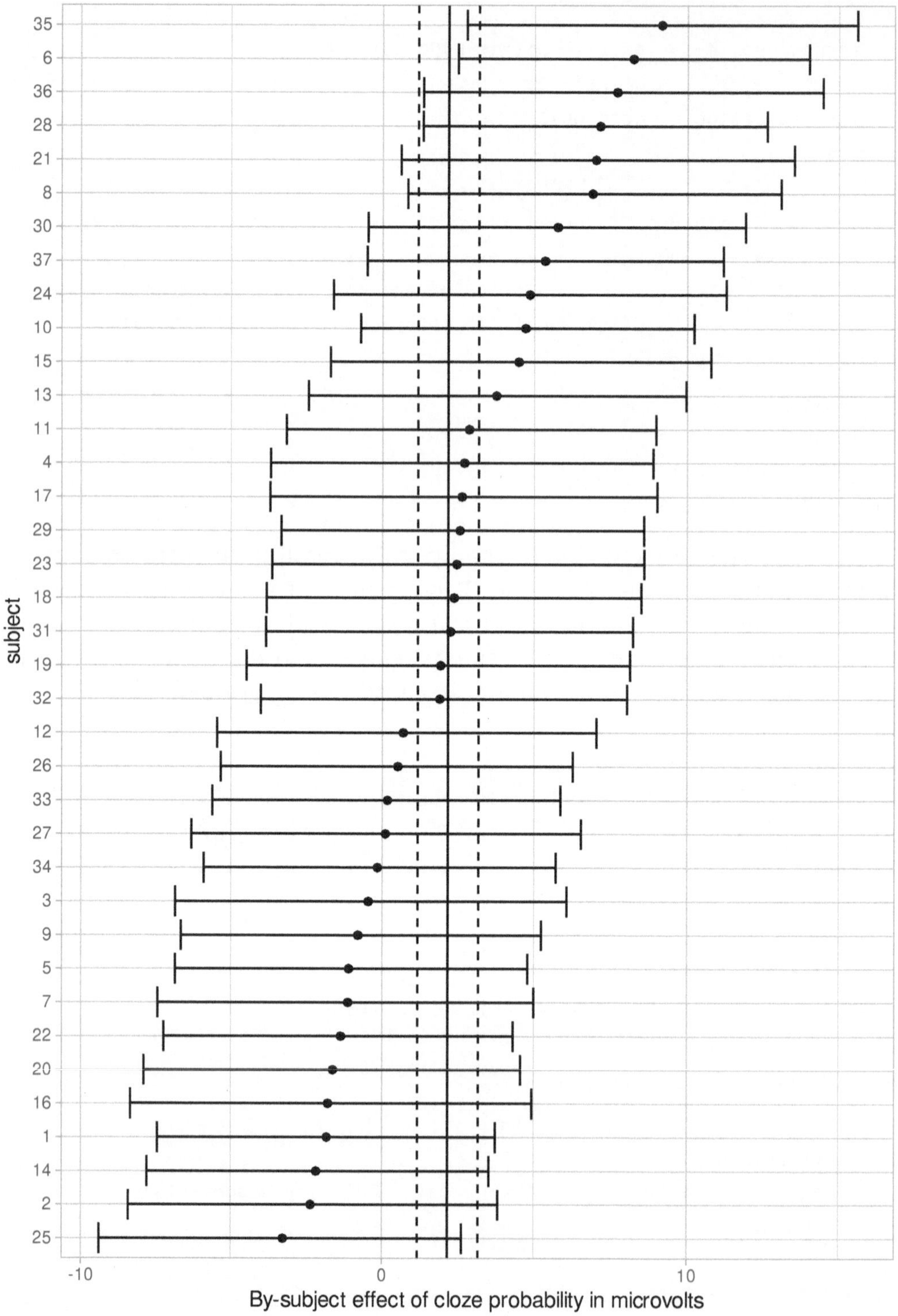

FIGURE 5.7: 95% credible intervals of the effect of cloze probability for each subject according to the no pooling model, `fit_N400_np`. The solid vertical line represents the mean over all the subjects; and the broken vertical lines mark the 95% credible interval for this mean.

5.2.3 Varying intercepts and varying slopes model (M_v)

One major problem with the no-pooling model is that we fit each subject's data ignoring the information available in the other subjects' data. Depending on the characteristics of the data set, the Bayesian no-pooling model might either underfit or overfit the individual subjects' data. It will ignore what can be learned from all the data taken together and might end up in one of these two situations: if there is not enough data for some subjects, it will underfit—relying too much on the priors and learning too little from the data. Alternatively, if there is enough data, it might overfit, that is, it might learn "too much" from the individual subjects (the estimates from some subjects might be quite extreme and unrepresentative), leading us to overinterpret noisy estimates from the subjects' data (Type M error; see Gelman and Carlin 2014). The model can be modified to explicitly assume that the subjects have an overall effect common to all the subjects, with the individual subjects deviating from this common effect.

In the model that we fit next, we will assume that there is an overall effect that is common to the subjects and, importantly, that all subjects' parameters originate from one common (normal) distribution. This model specification will result in the estimation of posteriors for each subject being also influenced by what we know about all the subjects together. We begin with a hierarchical model with uncorrelated varying intercepts and slopes. The analogous frequentist model can be fit using `lmer()` from the package `lme4`, using `(1 + c_cloze || subj)` or, equivalently, `(c_cloze || subj)` for the by-subject random effects.

5.2.3.1 Model assumptions

1. EEG averages for the N400 spatio-temporal window are normally distributed.
2. Each subject deviates to some extent (this is made precise below) from the grand mean and from the mean effect of predictability. This implies that there is some between-subject variability in the individual-level intercept and slope adjustments by subject.
3. There is a linear relationship between cloze and the EEG signal.

5.2.3.2 Likelihood and priors

The likelihood now incorporates the assumption that both the intercept and slope are adjusted by subject.

$$signal_n \sim Normal(\alpha + u_{subj[n],1} + c\_cloze_n \cdot (\beta + u_{subj[n],2}), \sigma) \tag{5.9}$$

$$
\begin{aligned}
\alpha &\sim \mathit{Normal}(0, 10) \\
\beta &\sim \mathit{Normal}(0, 10) \\
u_1 &\sim \mathit{Normal}(0, \tau_{u_1}) \\
u_2 &\sim \mathit{Normal}(0, \tau_{u_2}) \\
\tau_{u_1} &\sim \mathit{Normal}_+(0, 20) \\
\tau_{u_2} &\sim \mathit{Normal}_+(0, 20) \\
\sigma &\sim \mathit{Normal}_+(0, 50)
\end{aligned}
\tag{5.10}
$$

In this model each subject has their own intercept adjustment, $u_{subj,1}$, and slope adjustment, $u_{subj,2}$.[3] If $u_{subj,1}$ is positive, the subject will have a more positive EEG signal than the grand mean average. If $u_{subj,2}$ is positive, the subject will have a more positive EEG response to a change of one unit in `c_cloze` than the overall mean effect (i.e., there will be a more positive effect of cloze probability on the N400). The parameters u are sometimes called random effects and thus a model with fixed effects (α and β) and random effects is called a mixed model. However, random effects have different meanings in different contexts. To avoid ambiguity, `brms` calls these random-effects parameters *group-level* effects. Since we are estimating α and u at the same time and we assume that the average of the u's is 0 (since it is assumed to be normally distributed with mean 0), what is common between the subjects, the grand mean, is estimated as the intercept α, and the deviations of individual subjects' means from this grand mean are the adjustments u_1. Similarly, the mean effect of cloze is estimated as β, and the deviations of individual subjects' mean effects of cloze from β are the adjustment u_2. The standard deviations of these two adjustment terms, τ_{u_1} and τ_{u_2}, respectively, represent between subject variability.

Thus, the model M_v has three *standard deviations*: σ, τ_{u_1} and τ_{u_2}. In statistics, it is conventional to talk about variances (the square of these standard deviations); for this reason, these standard deviations are also (confusingly) called *variance components*. The variance components τ_{u_1} and τ_{u_2} characterize between-subject variability, and the variance component σ characterizes within-subject variability.

The by-subject adjustments u_1 and u_2 are parameters in the model, and therefore have priors defined on them. By contrast, in the frequentist `lmer` model, the adjustments u_1 and u_2 are not parameters; they are called conditional modes; see Bates et al. (2015).

---

[3]The intercept adjustment is often called u_0 in statistics books, where the intercept might be called α (or sometimes also β_0), and thus u_1 refers to the adjustment to the slope. However, in this book, we start the indexing with 1 to be consistent with the Stan language.

Parameters that appear in the prior specifications for parameters, such as τ_u, are often called *hyperparameters*,[4] and the priors on such hyperparameters are called *hyperpriors*. Thus, the parameter u_1 has $Normal(0, \tau_{u_1})$ as a prior; τ_{u_1} is a hyperparameter, and the hyperprior on τ_{u_1} is $Normal(0, 20)$.[5]

We know that in general, in EEG experiments, the standard deviations for the by-subject adjustments are smaller than the standard deviation of the observations (which is the within-subjects standard deviation). That is, usually the between-subject variability in the intercepts and slopes is smaller than the within-subjects variability in the data. For this reason, reducing the scale of the truncated normal distribution to 20 (in comparison to 50) seems reasonable for the priors of the τ parameters. As always, we can do a sensitivity analysis to verify that our priors are reasonably uninformative (if we intended them to be uninformative).

For now, we are assuming that there is no relationship (no correlation) between the by-subject intercept and slope adjustments u_1 and u_2; this lack of correlation is indicated in `brms` using the double pipe `||`. The double pipe is also used in the same way in `lmer()` from the package `lme4` (in fact `brms` bases its syntax on that of the `lme4` package).

In `brms`, we need to specify hyperpriors for τ_{u_1} and τ_{u_2}; these are called `sd` in `brms`, to distinguish these standard deviations from the standard deviation of the residuals σ. As with the population-level effects, the by-subjects intercept adjustments are implicitly fit for the group-level effects and thus `(c_cloze || subj)` is equivalent to `(1 + c_cloze || subj)`. If we don't want an intercept we need to explicitly indicate this with `(0 + c_cloze || subj)` or `(-1 + c_cloze || subj)`. Such a removal of the intercept is not normally done.

```
prior_v <-
  c(prior(normal(0, 10), class = Intercept),
    prior(normal(0, 10), class = b, coef = c_cloze),
    prior(normal(0, 50), class = sigma),
    prior(normal(0, 20), class = sd, coef = Intercept, group = subj),
    prior(normal(0, 20), class = sd, coef = c_cloze, group = subj))

fit_N400_v <- brm(n400 ~ c_cloze + (c_cloze || subj),
                  prior = prior_v,
                  data = df_eeg)
```

[4]Another potential source of confusion here is that *hyperparameters* is also used in the machine learning literature with a different meaning.

[5]One could in theory keep going deeper and deeper, defining hyper-hyperpriors etc., but the model would quickly become impossible to fit.

When we print a `brms` fit, we first see the summaries of the posteriors of the standard deviation of the by-group intercept and slopes, τ_{u_1} and τ_{u_2} as `sd(Intercept)` and `sd(c_cloze)`, and then, as with previous models, the population-level effects, α and β as `Intercept` and `c_cloze`, and the scale of the likelihood, σ, as `sigma`. The full summary can be printed out by typing:

```
fit_N400_v
```

Because the above command will result in some pages of output, it is easier to understand the summary graphically (Figure 5.8). Rather than the wrapper `plot()`, we use the original function of the package `bayesplot`, `mcmc_dens()`, to only show density plots. We extract the first 5 parameters of the model with `variables(fit_N400_v)[1:5]`.

```
mcmc_dens(fit_N400_v, pars = variables(fit_N400_v)[1:5])
```

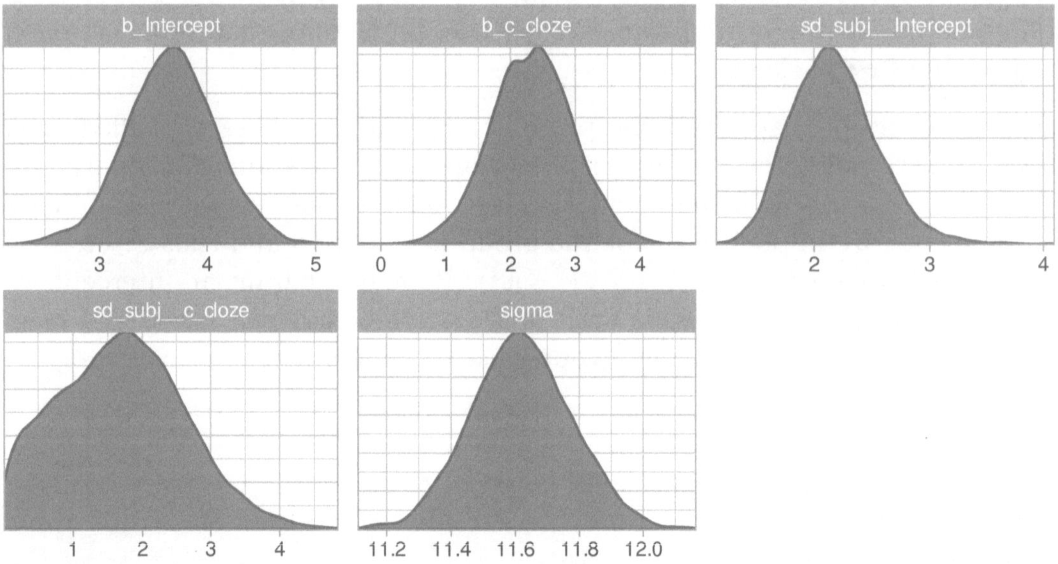

FIGURE 5.8: Posterior distributions of the parameters in the model `fit_N400_v`.

Because we estimated how the population-level effect of cloze is adjusted for each subject, we could examine how each subject is being affected by the manipulation. For this we do the following, and we plot it in Figure 5.9. These are adjustments, $u_{1,2}, u_{2,2}, ..., u_{37,2}$, and not the effect of the manipulation by subject, $\beta + [u_{1,2}, u_{2,2}, ..., u_{37,2}]$. The code below produces Figure 5.9.

```
# make a table of u_2s
ind_effects_v <- paste0("r_subj[", unique(df_eeg$subj),
                        ",c_cloze]")
u_2_v <- posterior_summary(fit_N400_v, variable = ind_effects_v) %>%
  as_tibble() %>%
  mutate(subj = 1:n()) %>%
  ## reorder plot by magnitude of mean:
  arrange(Estimate) %>%
  mutate(subj = factor(subj, levels = subj))
# We plot:
ggplot(u_2_v,
       aes(x = Estimate, xmin = Q2.5, xmax = Q97.5, y = subj)) +
  geom_point() +
  geom_errorbarh() +
  xlab("By-subject adjustment to the slope in microvolts")
```

There is an important difference between the no-pooling model and the varying intercepts and slopes model we just fit. The no-pooling model fits each individual subject's intercept and slope independently for each subject. By contrast, the varying intercepts and slopes model takes *all* the subjects' data into account in order to compute the population-level (or fixed) effects α and β; and the model "shrinks" (Pinheiro and Bates 2000) the by-subject intercept and slope adjustments toward the population-level effects estimates. In Figure 5.10, we can see the shrinkage of the estimates in the varying intercepts model by comparing them with the estimates of the no pooling model (M_{np}).

```
b_c_cloze <- posterior_summary(fit_N400_v, variable = "b_c_cloze")
ggplot(by_subj_df,
       aes(ymin = Q2.5, ymax = Q97.5, x = subj,
           y = Estimate, color = model, shape = model)) +
  geom_errorbar(position = position_dodge(1)) +
  geom_point(position = position_dodge(1)) +
  # We'll also add the mean and 95% CrI of the overall difference
  # to the plot:
  geom_hline(yintercept = b_c_cloze[, "Estimate"]) +
  geom_hline(yintercept = b_c_cloze[, "Q2.5"],
             linetype = "dotted", linewidth = .5) +
  geom_hline(yintercept = b_c_cloze[, "Q97.5"],
             linetype = "dotted", linewidth = .5) +
```

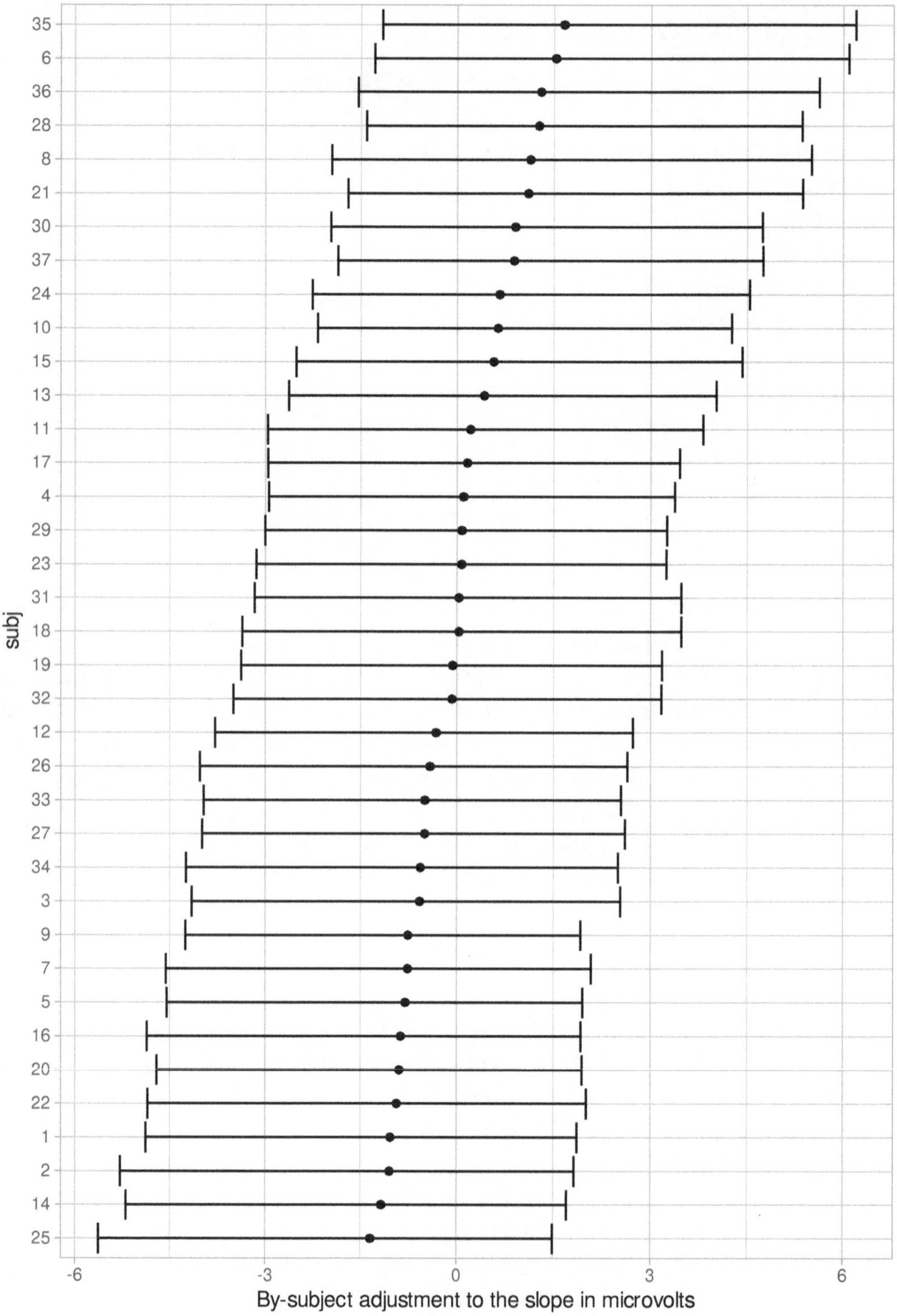

FIGURE 5.9: 95% credible intervals of adjustments to the effect of cloze probability for each subject ($u_{1..37,2}$) according to the varying intercept and varying slopes model, `fit_N400_v`. To obtain the effect of cloze probability for each subject, we would need to add the estimate of β to each adjustment.

```
xlab("N400 effect of predictability") +
coord_flip()
```

5.2.4 Correlated varying intercept varying slopes model (M_h)

The model $\boldsymbol{M_v}$ allowed for differences in intercepts (mean voltage) and slopes (effects of cloze) across subjects, but it has the implicit assumption that these varying intercepts and varying slopes are independent. It is in principle possible that subjects showing more negative voltage may also show stronger effects (or weaker effects). Next, we fit a model that allows a correlation between the intercepts and slopes. We model the correlation between varying intercepts and slopes by defining a variance-covariance matrix $\boldsymbol{\Sigma}$ between the by-subject varying intercepts and slopes, and by assuming that both adjustments (intercept and slope) come from a multivariate (in this case, a bivariate) normal distribution.

- In $\boldsymbol{M_h}$, we model the EEG data with the following assumptions:

  1. EEG averages for the N400 spatio-temporal window are normally distributed.
  2. Some aspects of the mean signal voltage and of the effect of predictability depend on the subject, and these two might be correlated, i.e., we assume group-level intercepts and slopes, and allow a correlation between them by-subject.
  3. There is a linear relationship between cloze and the EEG signal for the trial.

The likelihood remains identical to the model $\boldsymbol{M_v}$, which assumes no correlation between group-level intercepts and slopes (section 5.2.3):

$$signal_n \sim Normal(\alpha + u_{subj[n],1} + c\_cloze_n \cdot (\beta + u_{subj[n],2}), \sigma) \tag{5.11}$$

The correlation is indicated in the priors on the adjustments for intercept u_1 and slopes u_2.

- Priors:

$$\begin{aligned} \alpha &\sim Normal(0, 10) \\ \beta &\sim Normal(0, 10) \\ \sigma &\sim Normal_+(0, 50) \\ \begin{pmatrix} u_{i,1} \\ u_{i,2} \end{pmatrix} &\sim \mathcal{N}\left(\begin{pmatrix} 0 \\ 0 \end{pmatrix}, \boldsymbol{\Sigma_u} \right) \end{aligned} \tag{5.12}$$

In this model, a bivariate normal distribution generates the varying intercepts

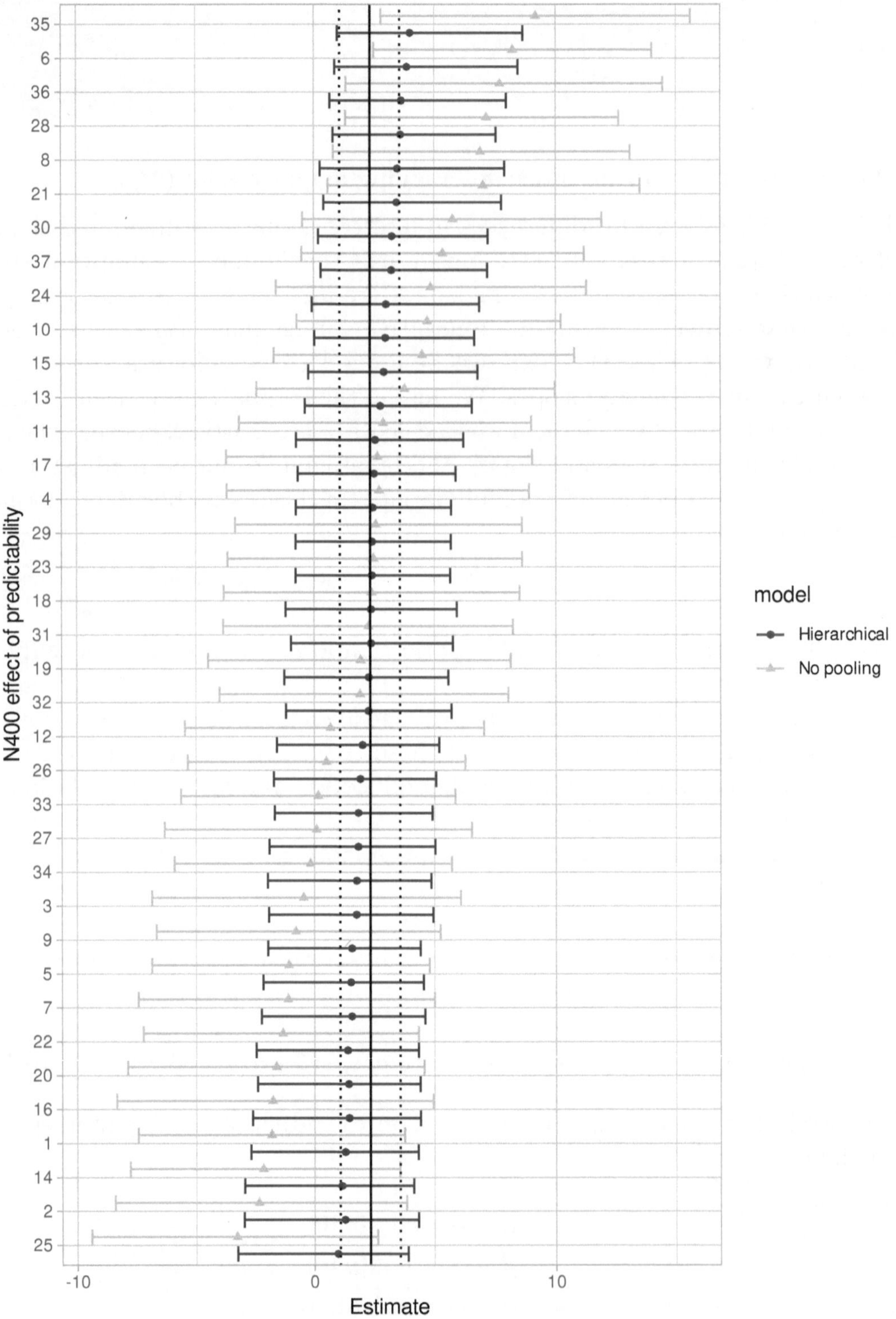

FIGURE 5.10: This plot compares the estimates of the effect of cloze probability for each subject between (i) the no pooling, `fit_N400_np` and (ii) the varying intercepts and varying slopes model, `fit_N400_v`.

and varying slopes $\mathbf{u}$; this is an $n\times 2$ matrix. The variance-covariance matrix $\boldsymbol{\Sigma}_u$ defines the standard deviations of the varying intercepts and varying slopes, and the correlation between them. Recall from section 1.6.2 that the diagonals of the variance-covariance matrix contain the variances of the correlated random variables, and the off-diagonals contain the covariances. In this example, the covariance $Cov(u_1, u_2)$ between two variables u_1 and u_2 is defined as the product of their correlation ρ and their standard deviations τ_{u_1} and τ_{u_2}. In other words, $Cov(u_1, u_2) = \rho_u \tau_{u_1} \tau_{u_2}$.

$$\boldsymbol{\Sigma}_u = \begin{pmatrix} \tau_{u_1}^2 & \rho_u \tau_{u_1} \tau_{u_2} \\ \rho_u \tau_{u_1} \tau_{u_2} & \tau_{u_2}^2 \end{pmatrix} \tag{5.13}$$

In order to specify a prior for Σ_u, we need priors for the standard deviations, τ_{u_1}, and τ_{u_2}, and also for their correlation, ρ_u. We can use the same priors for τ as before. For the correlation matrix that contains ρ_u, we use the LKJ prior. The basic idea of the LKJ prior on the correlation matrix is that as its parameter, η (*eta*), increases, it will favor correlations closer to zero.[6] At $\eta = 1$, the LKJ correlation distribution is uninformative (if there is a single correlation parameter, this is similar to $Beta(1, 1)$, scaled to the interval $[-1, 1]$), at $\eta < 1$, it favors extreme correlations (similar to $Beta(a < 1, b < 1)$). We set $\eta = 2$ so that we don't favor extreme correlations, and we still represent our lack of knowledge through the wide spread of the prior between -1 and 1. Thus, $\eta = 2$ gives us a regularizing, relatively uninformative or mildly informative prior.

Figure 5.11 shows a visualization of different parameterizations of the LKJ prior.

$$\begin{aligned} \tau_{u_1} &\sim Normal_+(0, 20) \\ \tau_{u_2} &\sim Normal_+(0, 20) \\ \begin{bmatrix} 1 & \rho_u \\ \rho_u & 1 \end{bmatrix} &\sim LKJcorr(2) \end{aligned} \tag{5.14}$$

In our `brms` model, we allow a correlation between the by-subject intercepts and slopes by using a single pipe `|` instead of the double pipe `||` that we used previously. This convention follows the syntax used in the frequentist `lmer()` function. As before, the varying intercepts are implicitly fit.

Because we have a new parameter, the correlation ρ_u, we need to add a new prior for this correlation. In `brms`, this is achieved by adding a prior for the

[6]This is because an LKJ correlation distribution with a large η corresponds to a correlation matrix with values close to zero in the lower and upper triangles.

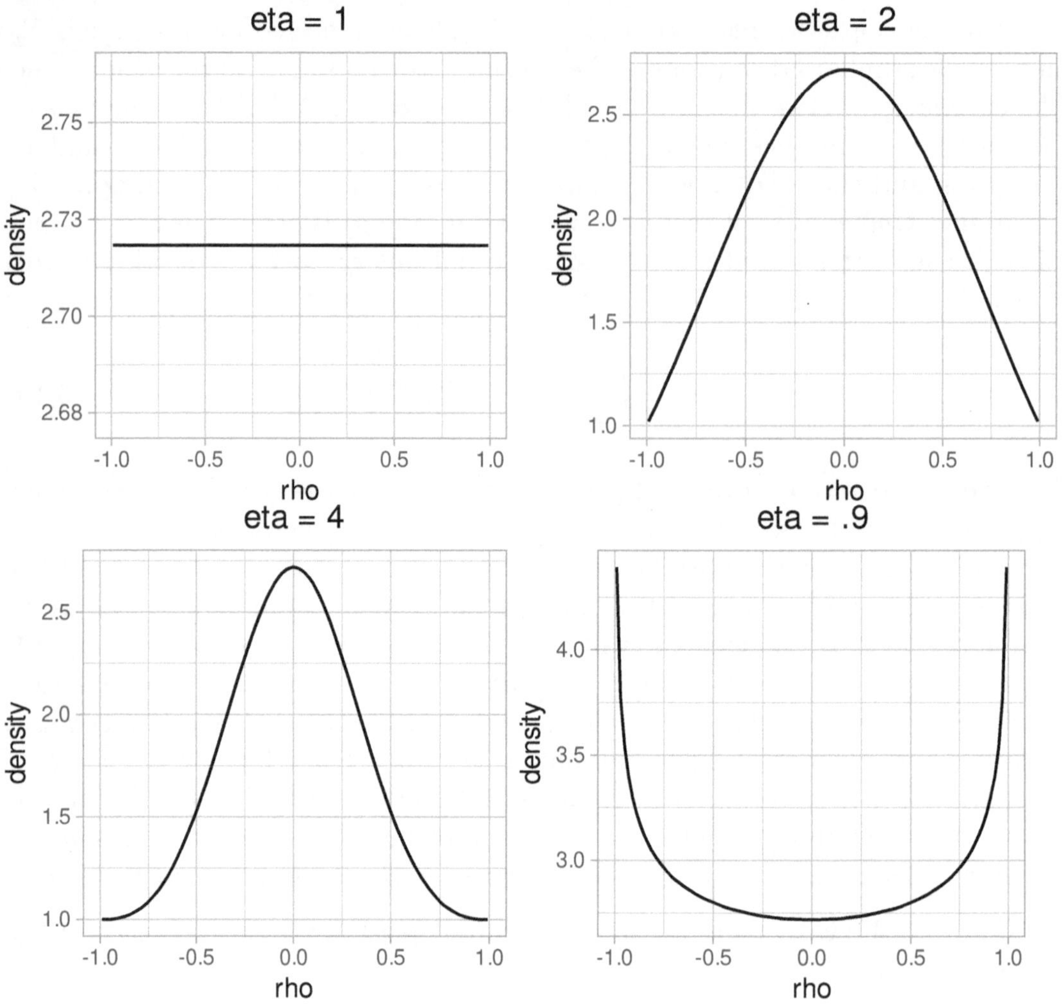

FIGURE 5.11: Visualization of the LKJ correlation distribution prior with four different values of the η parameter.

parameter type `cor`. This is not obvious from the `brms` syntax, but the LKJ prior is being defined on the correlation matrix, not the correlation ρ_u.

```
prior_h <- c(prior(normal(0, 10), class = Intercept),
             prior(normal(0, 10), class = b, coef = c_cloze),
             prior(normal(0, 50), class = sigma),
             prior(normal(0, 20),
                   class = sd, coef = Intercept,
                   group = subj),
             prior(normal(0, 20),
                   class = sd, coef = c_cloze,
```

```
                group = subj),
          prior(lkj(2), class = cor, group = subj))

fit_N400_h <- brm(n400 ~ c_cloze + (c_cloze | subj),
                  prior = prior_h,
                  data = df_eeg)
```

The estimates do not change much in comparison with the varying intercept and varying slope model, probably because the estimation of the correlation is quite poor (i.e., there is a lot of uncertainty in the posterior). While the inclusion of the correlation parameter is justified, this results, in our experience, in a slower convergence of the models (see also the discussion at the end of the section 5.2.6). As before, the estimates are shown graphically, in Figure 5.12. As always, one can access the complete summary by typing `fit_N400_h`.

```
plot(fit_N400_h, nvariables = 6)
```

We are now half-way to what is sometimes called the "maximal" hierarchical model (Barr et al. 2013). This usually refers to a model with all the by-participant and by-items group-level variance components *allowed by the experimental design* and *a full variance covariance matrix* for all the group-level parameters. Not all variance components are allowed by the experimental design; in particular, between-group manipulations cannot have variance components. For example, even if we assume that the working memory capacity of the subjects might affect the N400, we cannot measure how working memory affects the subjects differently.

When we refer to a full variance-covariance matrix, we mean a variance-covariance matrix where all the elements (variances and covariances) are non-zero. In our previous model, for example, the variance-covariance matrix $\boldsymbol{\Sigma_u}$ was full because no element was zero. If we assume no correlation between group-level intercept and slope, there would need to be zeros in the diagonal of the matrix, and this would render the model identical to $\boldsymbol{M_v}$ defined in section 5.2.3. If we also assume that the bottom right element (τ^2) is zero, the model would turn into a varying intercept model (in `brms` formula `n400 ~ c_cloze + (1 | subj)`); and if we assume that the variance-covariance matrix has zeros in every cell, the model would turn into a complete pooling model, $\boldsymbol{M_{cp}}$, as defined in section 5.2.1.

As we will see in section 5.2.6 and in the online chapter F, "maximal" is probably a misnomer for Bayesian models, since this mostly refers to limitations of the popular frequentist package for fitting models, `lme4`.

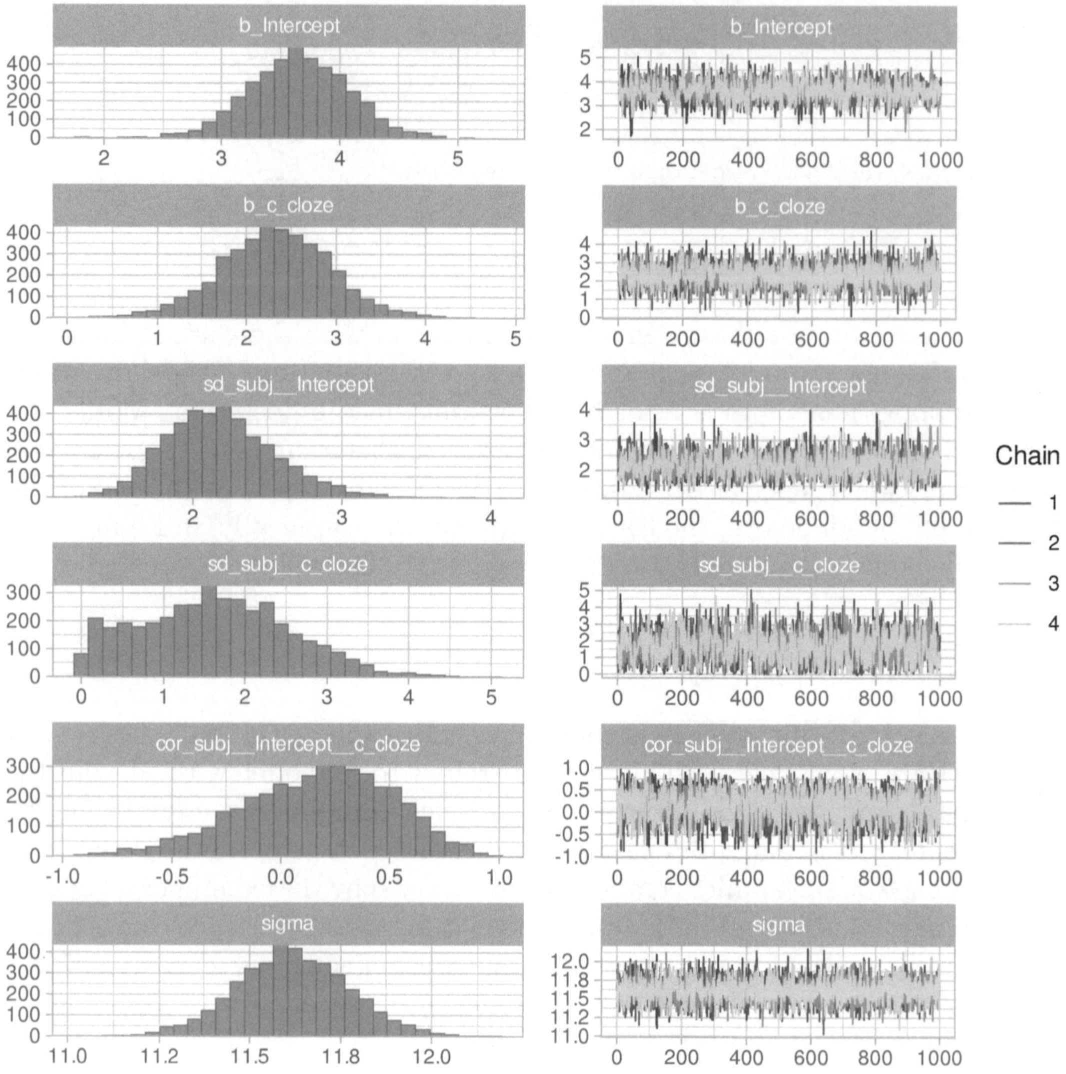

FIGURE 5.12: The posteriors of the parameters in the model `fit_N400_h`.

The next section spells out a model with full variance-covariance matrix for both subjects and items-level effects.

5.2.5 By-subjects and by-items correlated varying intercept varying slopes model (M_{sih})

Our new model, M_{sih}, will allow for differences in intercepts (mean voltage) and slopes (effects of predictability) across subjects and across items. In typical Latin square designs, subjects and items are said to be *crossed random effects*—each subject sees exactly one instance of each item. Here we assume a possible correlation between varying intercepts and slopes by subjects, as well as a correlation between varying intercepts and slopes by items.

- In M_{sih}, we model the EEG data with the following assumptions:

  1. EEG averages for the N400 spatio-temporal window are normally distributed.
  2. Some aspects of the mean signal voltage and of the effect of predictability depend on the subject, i.e., we assume group-level intercepts, and slopes, and a correlation between them by-subject.
  3. Some aspects of the mean signal voltage and of the effect of predictability depend on the item, i.e., we assume group-level intercepts, and slopes, and a correlation between them by-item.
  4. There is a linear relationship between cloze and the EEG signal for the trial.

- Likelihood:

$$
\begin{aligned}
signal_n \sim Normal(\alpha + u_{subj[n],1} + w_{item[n],1} + \\
c\_cloze_n \cdot (\beta + u_{subj[n],2} + w_{item[n],2}), \sigma)
\end{aligned}
\tag{5.15}
$$

- Priors:

$$
\begin{aligned}
\alpha &\sim Normal(0, 10) \\
\beta &\sim Normal(0, 10) \\
\sigma &\sim Normal_+(0, 50) \\
\begin{pmatrix} u_{i,1} \\ u_{i,2} \end{pmatrix} &\sim \mathcal{N}\left(\begin{pmatrix} 0 \\ 0 \end{pmatrix}, \boldsymbol{\Sigma_u} \right) \\
\begin{pmatrix} w_{j,1} \\ w_{j,2} \end{pmatrix} &\sim \mathcal{N}\left(\begin{pmatrix} 0 \\ 0 \end{pmatrix}, \boldsymbol{\Sigma_w} \right)
\end{aligned}
\tag{5.16}
$$

The index j represents the item id; this is analogous to the index i for subjects. Similarly, $item[n]$ indicates the item that corresponds to the observation in the n-th row of the data frame.

We have hyperparameters and hyperpriors as before:

$$
\begin{aligned}
\boldsymbol{\Sigma_u} &= \begin{pmatrix} \tau_{u_1}^2 & \rho_u \tau_{u_1} \tau_{u_2} \\ \rho_u \tau_{u_1} \tau_{u_2} & \tau_{u_2}^2 \end{pmatrix} \\
\boldsymbol{\Sigma_w} &= \begin{pmatrix} \tau_{w_1}^2 & \rho_w \tau_{w_1} \tau_{w_2} \\ \rho_w \tau_{w_1} \tau_{w_2} & \tau_{w_2}^2 \end{pmatrix}
\end{aligned}
\tag{5.17}
$$

$$
\begin{aligned}
\tau_{u_1} &\sim Normal_+(0, 20)\\
\tau_{u_2} &\sim Normal_+(0, 20)\\
\begin{bmatrix} 1 & \rho_u \\ \rho_u & 1 \end{bmatrix} &\sim LKJcorr(2)\\
\tau_{w_1} &\sim Normal_+(0, 20)\\
\tau_{w_2} &\sim Normal_+(0, 20)\\
\begin{bmatrix} 1 & \rho_w \\ \rho_w & 1 \end{bmatrix} &\sim LKJcorr(2)
\end{aligned}
\tag{5.18}
$$

We set identical priors for by-items group-level effects as for by-subject group-level effects, but this is only because we don't have any differentiated prior information about subject-level vs. item-level variation. However, bear in mind that the estimation for items is completely independent from the estimation for subjects. Although we wrote many more equations than before, the `brms` model is quite straightforward to extend.

We assign the priors to the by-subject and by-item parameters first:

```
prior_sih_full <-
  c(prior(normal(0, 10), class = Intercept),
    prior(normal(0, 10), class = b, coef = c_cloze),
    prior(normal(0, 50), class = sigma),
    prior(normal(0, 20),
          class = sd, coef = Intercept,
          group = subj),
    prior(normal(0, 20),
          class = sd, coef = c_cloze,
          group = subj),
    prior(lkj(2), class = cor, group = subj),
    prior(normal(0, 20),
          class = sd, coef = Intercept,
          group = item),
    prior(normal(0, 20),
          class = sd, coef = c_cloze,
          group = item),
    prior(lkj(2), class = cor, group = item))
```

We can also simplify the priors, when we assign the same priors to the by-subject and by-item parameters:

```
prior_sih <-
  c(prior(normal(0, 10), class = Intercept),
    prior(normal(0, 10), class = b),
    prior(normal(0, 50), class = sigma),
    prior(normal(0, 20), class = sd),
    prior(lkj(2), class = cor))
```

Fit the model:

```
fit_N400_sih <- brm(n400 ~ c_cloze + (c_cloze | subj) +
                      (c_cloze | item),
                    prior = prior_sih,
                    data = df_eeg)
```

We have new group-level effects in the summary if we run `fit_N400_sih`, but again the estimate of the effect of cloze remains virtually unchanged (Figure 5.13).

```
fit_N400_sih
```

```
## ...
## Group-Level Effects:
## ~item (Number of levels: 80)
##                        Estimate Est.Error l-95% CI u-95% CI Rhat
## sd(Intercept)              1.51      0.35     0.81     2.18 1.00
## sd(c_cloze)                2.31      1.03     0.26     4.30 1.01
## cor(Intercept,c_cloze)    -0.42      0.32    -0.90     0.32 1.00
##                        Bulk_ESS Tail_ESS
## sd(Intercept)              1459     1717
## sd(c_cloze)                1040     1236
## cor(Intercept,c_cloze)     2299     2353
##
## ~subj (Number of levels: 37)
##                        Estimate Est.Error l-95% CI u-95% CI Rhat
## sd(Intercept)              2.21      0.37     1.55     3.02 1.00
## sd(c_cloze)                1.47      0.89     0.08     3.28 1.00
## cor(Intercept,c_cloze)     0.13      0.35    -0.60     0.77 1.00
##                        Bulk_ESS Tail_ESS
## sd(Intercept)              1433     2064
## sd(c_cloze)                1131     1446
```

```
## cor(Intercept,c_cloze)     4318     2764
##
## Population-Level Effects:
##           Estimate Est.Error l-95% CI u-95% CI Rhat Bulk_ESS Tail_ESS
## Intercept     3.66      0.46     2.72     4.54 1.00     2108     2604
## c_cloze       2.31      0.70     0.92     3.63 1.00     3998     3250
##
## Family Specific Parameters:
##       Estimate Est.Error l-95% CI u-95% CI Rhat Bulk_ESS Tail_ESS
## sigma    11.49      0.16    11.18    11.79 1.00     5701     2832
##
## ...
```

```
plot(fit_N400_sih, nvariables = 9)
```

5.2.6 Beyond the maximal model–distributional regression models

We can use posterior predictive checks to verify that our last model can capture the entire signal distribution. This is shown in Figure 5.14.

```
pp_check(fit_N400_sih, ndraws = 50, type = "dens_overlay")
```

However, we know that in ERP studies, large levels of impedance between the recording electrodes and the skin tissue increase the noise in the recordings (Picton et al. 2000). Given that skin tissue is different between subjects, it could be the case that the level of noise varies by subject. It might be a good idea to verify that our model is good enough for capturing the by-subject variability. The code below produces Figure 5.15.

```
pp_check(fit_N400_sih,
         type = "dens_overlay_grouped",
         ndraws = 100,
         group = "subj") +
  xlab("Signal in the N400 spatiotemporal window") +
  theme(legend.position = "inside",
        legend.position.inside = c(.5,.08))
```

Figure 5.15 hints that we might be misfitting some subjects: Some of the by-subject observed distributions of the EEG signal averages look much tighter than their corresponding posterior predictive distributions (e.g., subjects 3, 5, 9, 10, 14), whereas some other by-subject observed distributions look wider

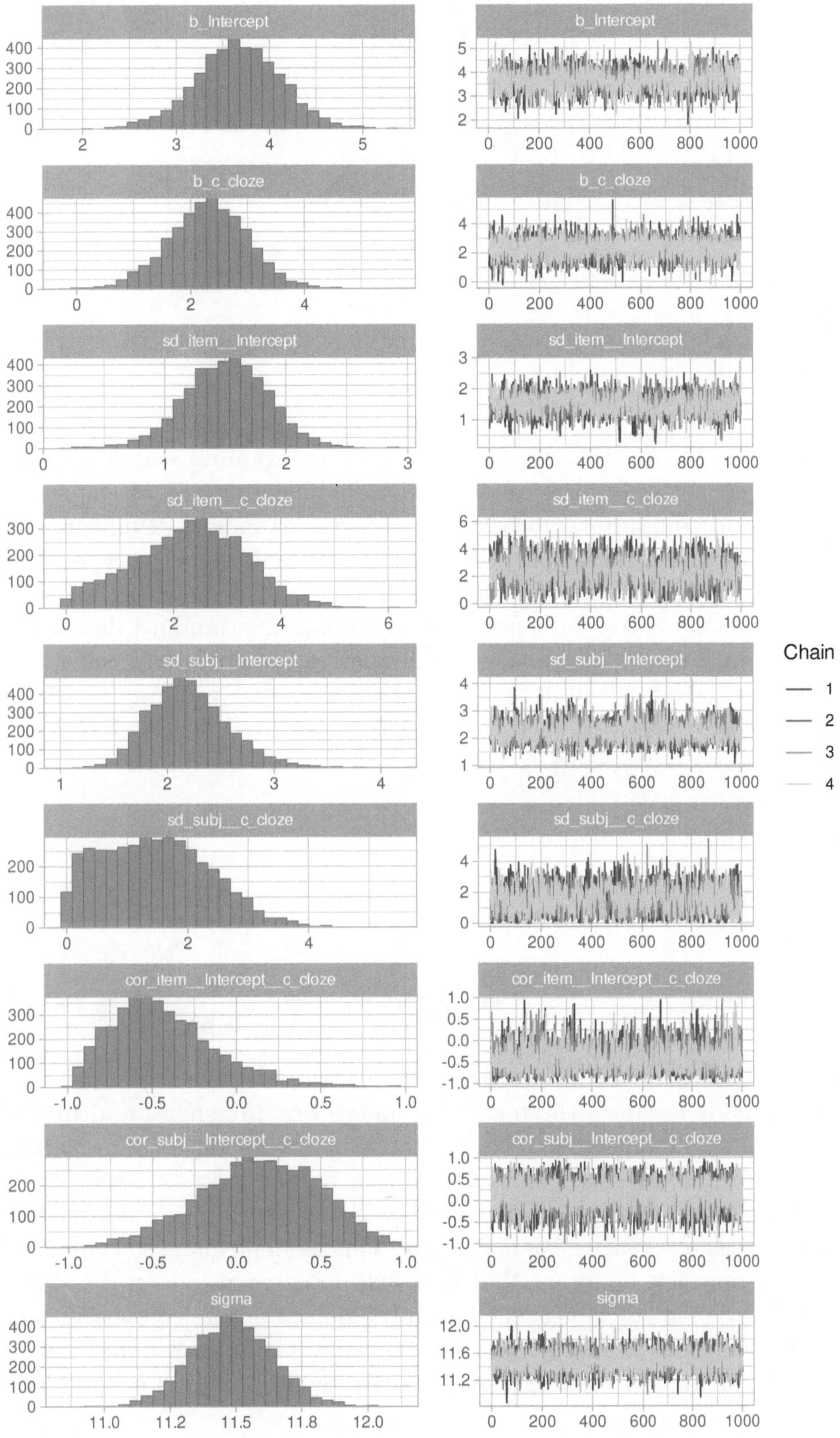

FIGURE 5.13: The posterior distributions of the parameters in the model `fit_N400_sih`.

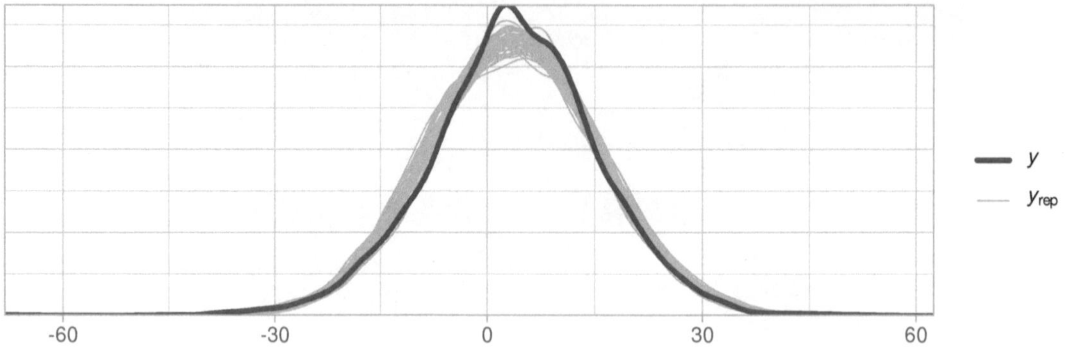

FIGURE 5.14: Overlay of densities from the posterior predictive distributions of the model `fit_N400_sih`.

(e.g., subjects 25, 26, 27). Another approach to examine whether we misfit the by-subject noise level is to plot posterior distributions of the standard deviations and compare them with the observed standard deviation. This is achieved in the following code, which groups the data by subject, and shows the distribution of standard deviations. The result is shown in Figure 5.16. It is clear now that, for some subjects, the observed standard deviation lies outside the bulk of the distribution of predictive standard deviations.

```
pp_check(fit_N400_sih,
         type = "stat_grouped",
         ndraws = 1000,
         group = "subj",
         stat = "sd",
         facet_args = list(scales = "fixed")) +
  scale_x_continuous(breaks = c(8, 12, 16), limits = c(7,19)) +
  theme(legend.position = "inside",
        legend.position.inside = c(.5,.08))
```

Why is our "maximal" hierarchical model misfitting the by-subject distribution of data? This is because, the maximal models are, in general and implicitly, models with the maximal group-level effect structure for the location parameter (e.g., the mean, μ, in a normal model). Other parameters (e.g., scale or shape parameters) are estimated as auxiliary parameters, and are assumed to be constant across observations and clusters. This assumption is so common that researchers may not be aware that it is just an assumption. In the Bayesian framework, it is easy to change such default assumptions if necessary. Changing the assumption that all subjects have the same residual standard deviation leads to the distributional regression model. Such models can be fit in `brms`;

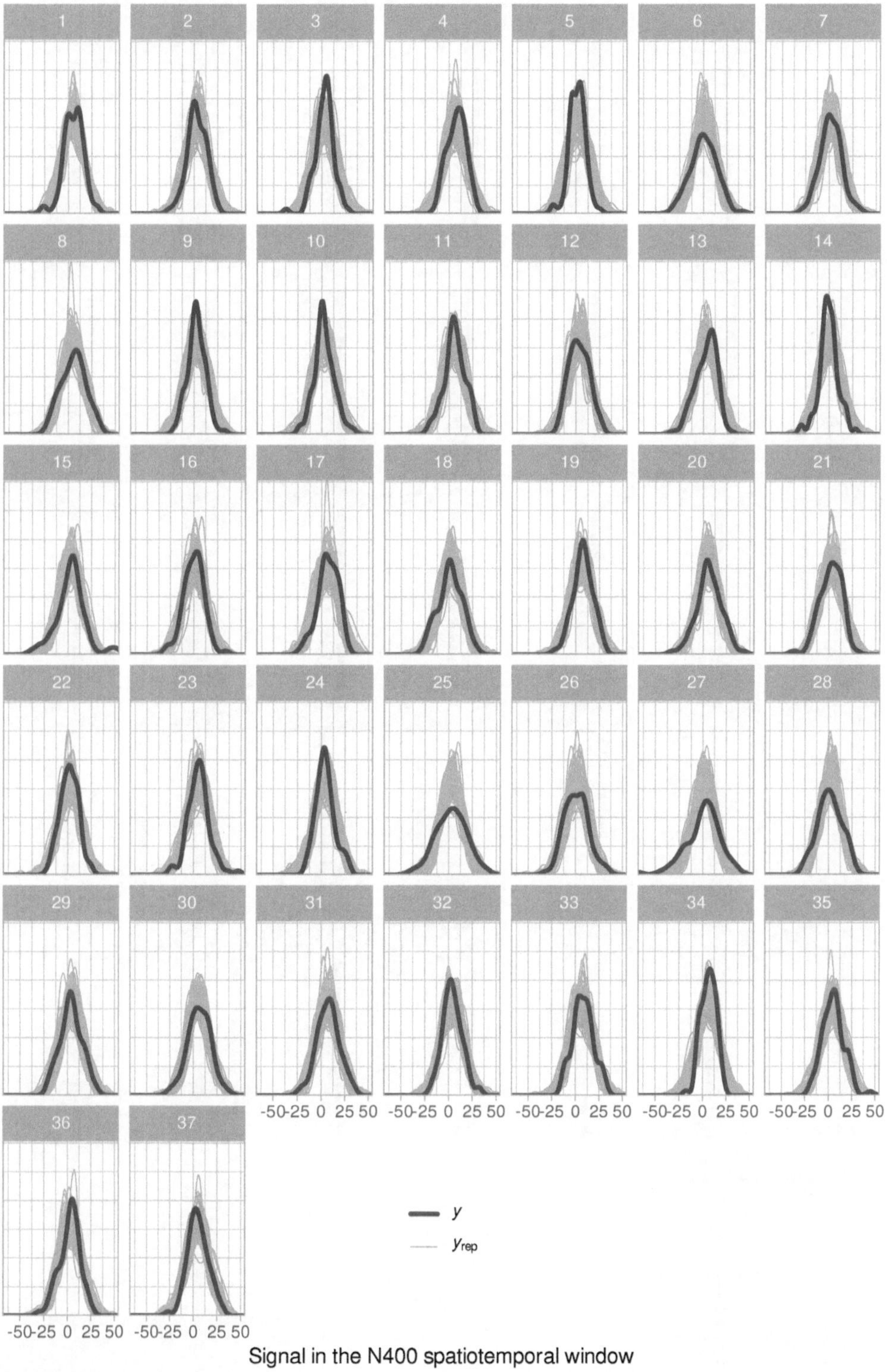

FIGURE 5.15: The plot shows 100 predicted distributions with the label y_{rep} and the distribution of the average signal data with the label y density plots for the 37 subjects that participated in the experiment.

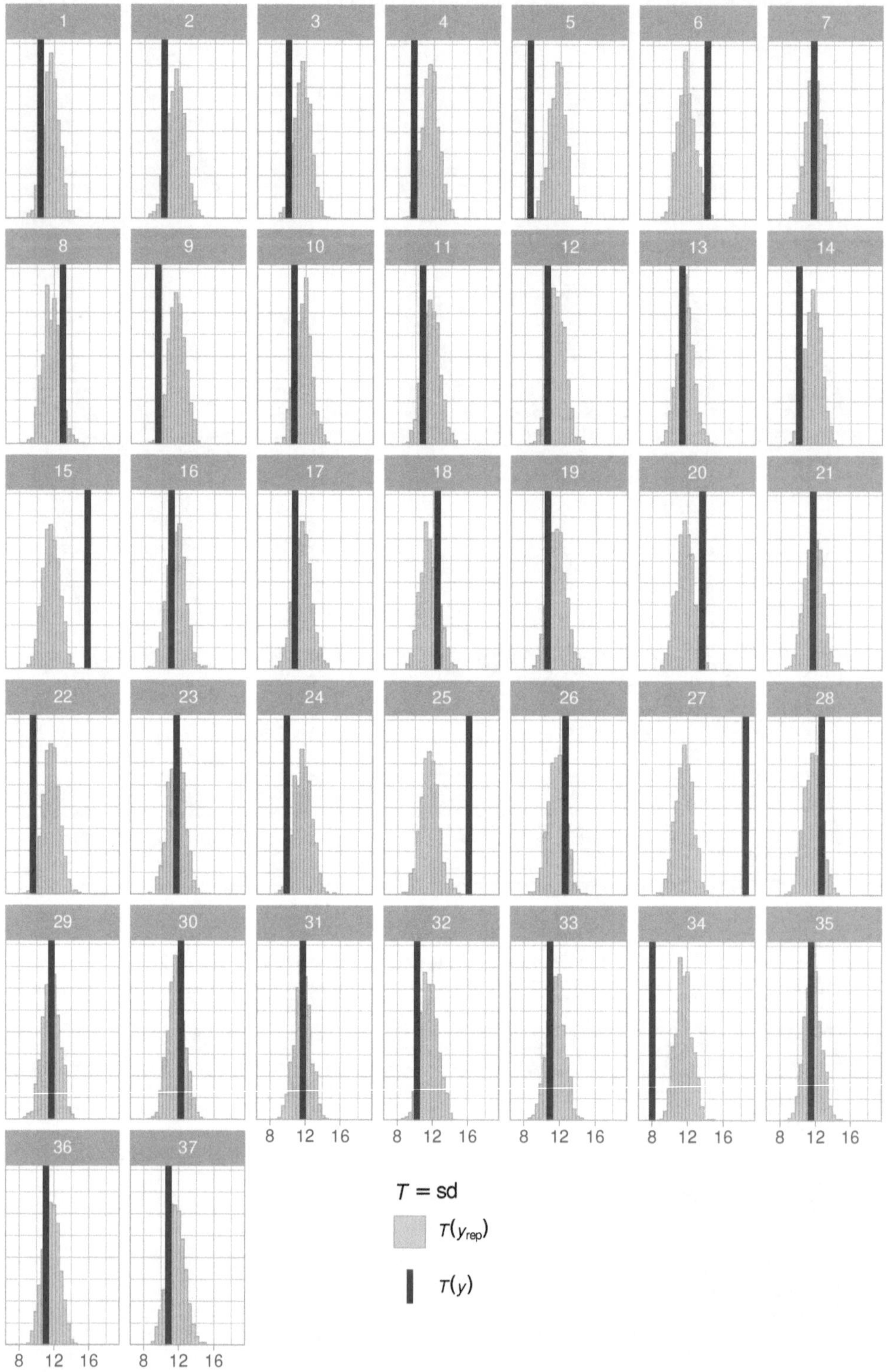

FIGURE 5.16: Distribution of posterior predicted standard deviations in gray and observed standard deviation in black lines by subject.

also see the `brms` vignette, https://cran.r-project.org/web/packages/brms/vignettes/brms_distreg.html.

We are going to change our previous likelihood, so that the scale σ also has a group-level effect structure. We exponentiate σ to make sure that the negative adjustments do not cause σ to become negative.

$$\begin{aligned} signal_n \sim & Normal(\alpha + u_{subj[n],1} + w_{item[n],1} + \\ & c\_cloze_n \cdot (\beta + u_{subj[n],2} + w_{item[n],2}), \sigma_n) \\ \sigma_n = & \exp(\sigma_\alpha + \sigma_{u_{subj[n]}}) \end{aligned} \tag{5.19}$$

We just need to add priors to our new parameters (that replace the old prior for σ). We set the prior to the intercept of the standard deviation, σ_α, to be similar to our previous σ. For the variance component of σ, τ_{σ_u}, we set rather uninformative hyperpriors. Recall that everything is exponentiated when it goes inside the likelihood; that is why we use log(50) rather than 50 for the standard deviation of the prior for σ.

$$\begin{aligned} \sigma_\alpha &\sim Normal(0, \log(50)) \\ \sigma_u &\sim Normal(0, \tau_{\sigma_u}) \\ \tau_{\sigma_u} &\sim Normal_+(0, 5) \end{aligned} \tag{5.20}$$

This model can be fit in `brms` using the internal function `brmsformula()` (or its shorter alias `bf()`). This is a powerful function that extends the formulas that we used so far allowing for setting a hierarchical regression to any parameter of a model. This will allow us to set a by-subject hierarchical structure to the parameter σ. We also need to set new priors; these priors are identified by `dpar = sigma`.

```
prior_s <-
  c(prior(normal(0, 10), class = Intercept),
    prior(normal(0, 10), class = b),
    prior(normal(0, 20), class = sd),
    prior(lkj(2), class = cor),
    prior(normal(0, log(50)), class = Intercept, dpar = sigma),
    prior(normal(0, 5),
          class = sd,
          group = subj,
          dpar = sigma))

fit_N400_s <- brm(
```

```
brmsformula(n400 ~ c_cloze + (c_cloze | subj) + (c_cloze | item),
            sigma ~ 1 + (1 | subj)),
  prior = prior_s, data = df_eeg)
```

Inspect the output below; notice that our estimate for the effect of cloze remains very similar to that of the model `fit_N400_sih`.

Compare the two models' estimates:

```
posterior_summary(fit_N400_sih, variable = "b_c_cloze")
```

```
##           Estimate Est.Error  Q2.5 Q97.5
## b_c_cloze     2.31     0.695 0.924  3.63
```

```
posterior_summary(fit_N400_s, variable = "b_c_cloze")
```

```
##           Estimate Est.Error Q2.5 Q97.5
## b_c_cloze      2.3      0.66    1  3.61
```

Nonetheless, Figure 5.17 shows that the fit of the model with respect to the by-subject variability is much better than before. Furthermore, Figure 5.18 shows that the observed standard deviations for each subject are well inside the posterior predictive distributions. The code below produces Figure 5.17.

```
pp_check(fit_N400_s,
         type = "dens_overlay_grouped",
         ndraws = 100,
         group = "subj") +
  xlab("Signal in the N400 spatiotemporal window") +
 theme(legend.position = "inside",
       legend.position.inside = c(.5,.08))
```

The model `fit_N400_s` raises the question: how much structure should we add to our statistical model? Should we also assume that σ can vary by items, and also by our experimental manipulation? Should we also have a maximal model for σ? Unfortunately, there are no clear answers that apply to every situation. The amount of complexity that we can introduce in a statistical model depends on (i) the answers we are looking for (we should include parameters that represent what we want to estimate), (ii) the size of the data at hand (more complex models require more data; this could mean more observations within subjects and/or the number of subjects), (iii) our computing power (as the

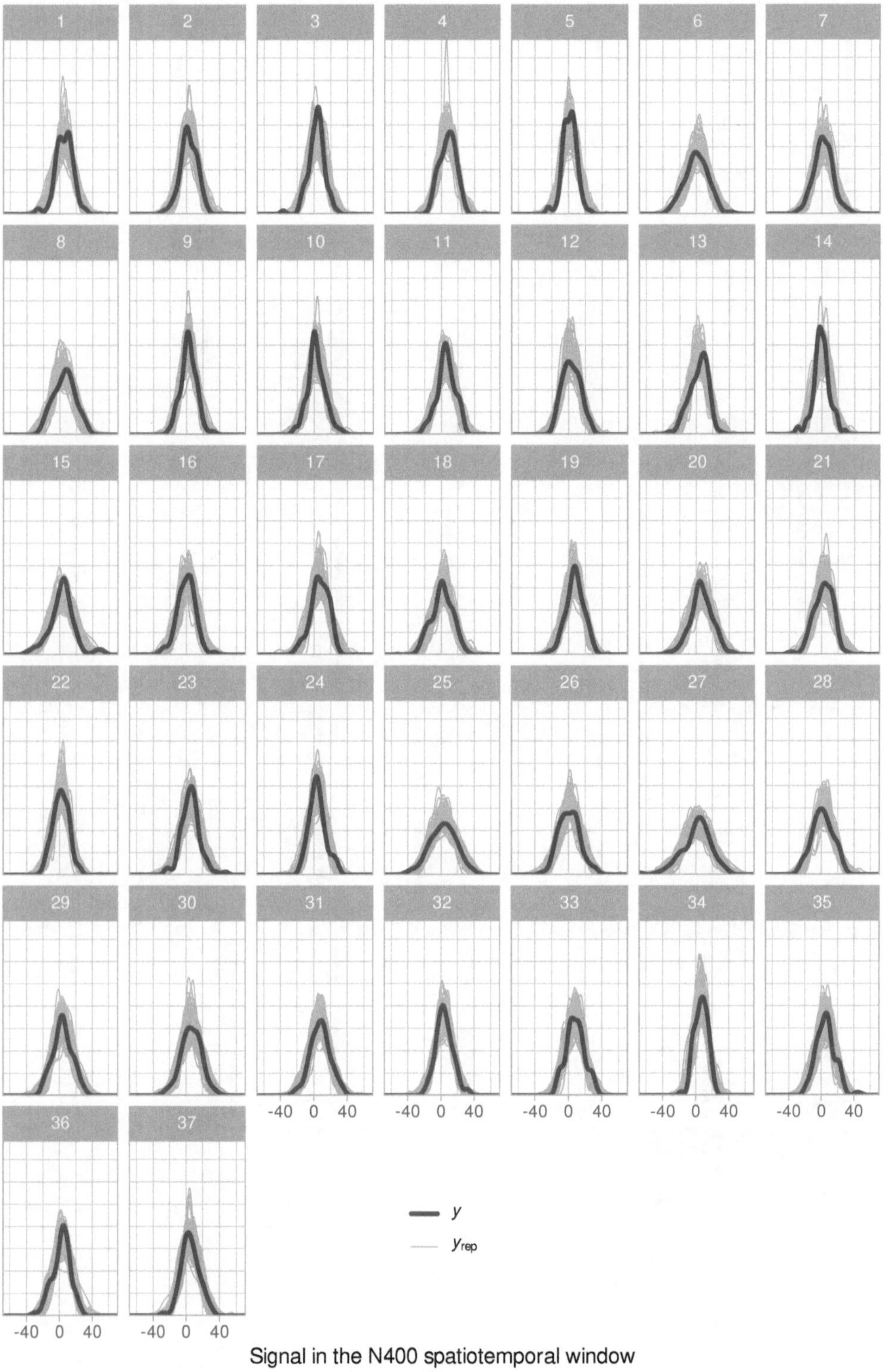

FIGURE 5.17: The gray density plots show 100 predicted distributions from a model that includes a hierarchical structure for σ. The black density plots show the distribution of the average signal data for the 37 subjects in the experiment.

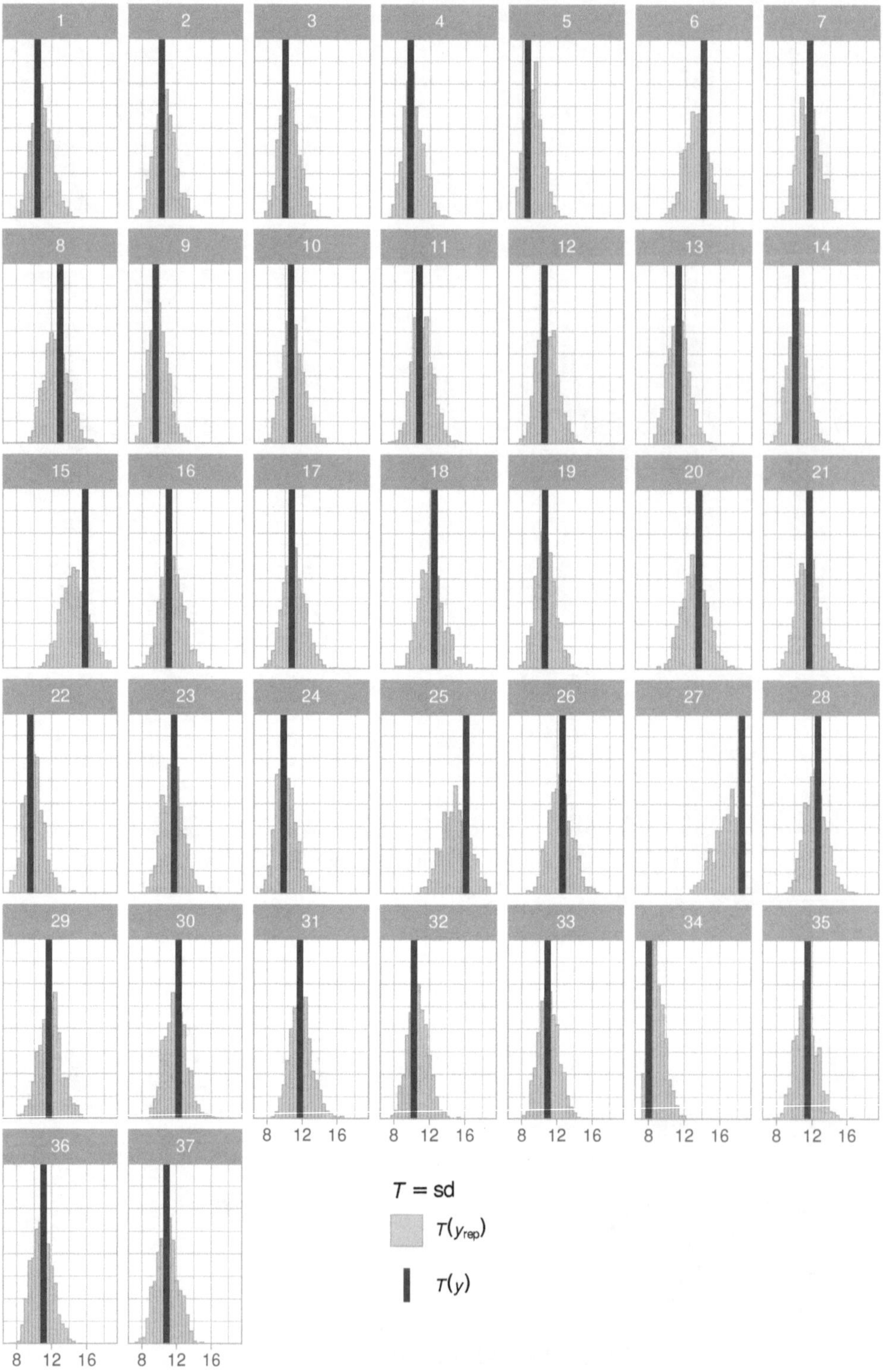

FIGURE 5.18: Shown (in gray) are the distributions of the posterior predicted standard deviations for each subject, generated from a model that includes a hierarchical structure for σ. Also shown are the observed mean standard deviations for each subject (black vertical lines).

complexity increases, models take more and more time to converge and require more computer power to finish the computations in a feasible time frame), and (iv) our domain knowledge.

Whether certain variance components should be included in a model also depends on whether there is evidence that they impact statistical inference (this could be investigated using model comparison techniques like Bayes factors). For example, it is well-known that estimating group-level effects for the location parameter can have a strong influence on the test statistics for the corresponding population-level effect (Barr et al. 2013; Schad et al. 2021). Ultimately, all models are approximations (that's in the best case; often, they are plainly wrong) and we need to think carefully about which aspects of our data we have to account and which aspects we can abstract away from.

In the context of cognitive modeling, McClelland (2009) argues that models should not focus on every single detail of the process they intend to explain. In order to understand a model, it needs to be simple enough. However, McClelland (2009) warns us that one must bear in mind that oversimplification does have an impact on what we can conclude from our analysis: A simplification can limit the phenomena that a model addresses, or can even lead to incorrect predictions. There is a continuum between purely statistical models (e.g., a linear regression) and computational cognitive models. For example, we can define "hybrid" models such as the linear ballistic accumulator (Brown and Heathcote 2008; and see Nicenboim 2018 for an implementation in Stan), where a great deal of cognitive detail is sacrificed for tractability. The conclusions of McClelland (2009) apply to any type of model in cognitive science: "Simplification is essential, but it comes at a cost, and real understanding depends in part on understanding the effects of the simplification."

5.3 A hierarchical log-normal model: The Stroop effect

Next, we illustrate some of the issues that arise with a log-normal likelihood in a hierarchical model. The data are from a Stroop task (Stroop 1935; for a review, see MacLeod 1991). We will analyze a subset of the data of 3337 subjects that participated in one variant of the Stroop task; this was part of a battery of tasks run in Ebersole et al. (2016).

For this variant of the Stroop task, subjects were presented with one word ("red," "blue," or "green") at the center of the screen, written in either red, blue, or green color. In one-third of the trials, the word matched the text color ("congruent" condition); in the remaining trials, it did not match ("incongruent" condition). Subjects were instructed to focus only on the color of the text and press `1` if the color was red, `2` if it was blue, and `3` if it was green. The incongruent condition is more difficult because subjects need to identify the color when it mismatches the word. For example, they might need to respond that the color is blue when the word "green" is written in blue; this is more challenging than the congruent condition, where the word "green" is written in green. This increased difficulty in the incongruent condition is known as the Stroop effect, which is extremely robust across task variations.

This task yields two measures: the accuracy of the decision made, and the time it took to respond. For the Stroop task, accuracy is usually almost at ceiling; to simplify the model, we will ignore accuracy. For cognitive models that incorporates accuracy and response times into a model to analyze these Stroop data, see Nicenboim (2018). Additionally, chapters 17 and 18 address response time and accuracy data.

5.3.1 A correlated varying intercept varying slopes log-normal model

If our theory only focuses on the difference between the response times for the "congruent" vs. "incongruent" condition, we can ignore the actual color presented and the word that was written. We can simply focus on whether a trial was congruent or incongruent. Define a predictor `c_cond` to represent these two conditions. For simplicity, we will also assume that all subjects share the same variance (as we saw in section 5.2.6, changing this assumption leads to distributional regression models).

The above assumptions mean that we are going to fit the data with the following likelihood. The likelihood function is identical to the one that we fit in section 5.2.4, except that here the location and scale are defined in a log-normal likelihood rather than a normal likelihood. Equation (5.21) states that we are dealing with a hierarchical model with by-subjects varying intercepts and varying slopes model:

$$rt_n \sim LogNormal(\alpha + u_{subj[n],1} + c\_cond_n \cdot (\beta + u_{subj[n],2}), \sigma) \tag{5.21}$$

In chapter 6, we will discuss the sum-contrast coding of the two conditions (`c_cond`). For now, it suffices to say that we assign a `+1` to `c_cond` for the "incongruent" condition, and a `-1` for the "congruent" condition (i.e., a sum-contrast coding). Under this contrast coding, if the posterior mean of the parameter β turns out to be positive, that would mean that the model predicts that the incongruent condition has slower response times than the congruent one. This is because on average the location of the log-normal likelihood for each condition will be as follows. In Equation (5.22), $\mu_{incongruent}$ refers to the location of the incongruent condition, and $\mu_{congruent}$ to the location of the congruent condition.

$$\begin{aligned} \mu_{incongruent} &= \alpha + 1 \cdot \beta \\ \mu_{congruent} &= \alpha + -1 \cdot \beta \end{aligned} \tag{5.22}$$

We could have chosen to do the opposite contrast coding assignments: -1 for the incongruent condition, and $+1$ for congruent condition. In that case, if the posterior mean of the parameter β turns out to be positive, that would mean that the incongruent condition has a *faster* response time than the congruent condition. Given that the Stroop effect is very robust, we do not expect such an outcome. In order to make the β parameter easier to interpret, we have chosen the contrast coding where a positive sign on the mean of β implies that the incongruent condition has slower response times.

As always, we need priors for all the parameters in our model. For the population-level parameters (or fixed effects), we use the same priors as we did when we were fitting a regression with a log-normal likelihood in section 3.7.2.

$$\begin{aligned} \alpha &\sim Normal(6, 1.5) \\ \beta &\sim Normal(0, 0.01) \\ \sigma &\sim Normal_+(0, 1) \end{aligned} \tag{5.23}$$

Here, β represents, on the log scale, the change in the intercept α as a function of the experimental manipulation. In this model, β will probably be larger in magnitude than for the model that examined the difference in pressing the spacebar for two consecutive trials in section 3.7.2. We might need to examine the prior for β with predictive distributions, but we will postpone that for now.

In contrast to our previous models, the intercept α is not the grand mean of the location. This is because the conditions were not balanced in the experiment (one-third of the conditions were congruent and two-thirds incongruent). The intercept could be interpreted here as the time (in log-scale) it takes to respond if we ignore the experimental manipulation. Next, we turn our attention to the prior specification for the group-level parameters (or random effects). If we assume a possible correlation between by-subject intercepts and slopes, our model will have the following structure. In particular, we have to define priors for the parameters in the variance-covariance matrix Σ_u.

$$\begin{pmatrix} u_{i,1} \\ u_{i,2} \end{pmatrix} \sim \mathcal{N}\left(\begin{pmatrix} 0 \\ 0 \end{pmatrix}, \boldsymbol{\Sigma_u} \right) \tag{5.24}$$

$$\boldsymbol{\Sigma}_u = \begin{pmatrix} \tau_{u_1}^2 & \rho_u \tau_{u_1} \tau_{u_2} \\ \rho_u \tau_{u_1} \tau_{u_2} & \tau_{u_2}^2 \end{pmatrix} \tag{5.25}$$

In practice, this means that we need priors for the by-subject standard deviations and correlations. For the variance components, we will set a similar prior as for σ. We don't expect the by-group adjustments to the intercept and slope to have more variance than the within-subject variance, so this prior will be quite conservative because it allows for a large range of prior uncertainty. We assign the same prior for the correlations as we did in section 5.2.5.

$$\begin{aligned} \tau_{u_1} &\sim Normal_+(0,1) \\ \tau_{u_2} &\sim Normal_+(0,1) \\ \begin{bmatrix} 1 & \rho_u \\ \rho_u & 1 \end{bmatrix} &\sim LKJcorr(2) \end{aligned} \tag{5.26}$$

We are now ready to fit the model. To speed up computation, we subset 50 subjects of the original data set; both the subsetted data and the original data set can be found in the package `bcogsci`. If we were analyzing these data for publication in a journal article or the like, we would obviously not subset the data.

We restrict ourselves to the correct trials only, and add a `c_cond` predictor, sum-coded as described earlier.

```
data("df_stroop")
(df_stroop <- df_stroop %>%
  mutate(c_cond = if_else(condition == "Incongruent", 1, -1)))
```

```
## # A tibble: 3,058 x 5
##     subj trial condition      RT c_cond
##    <dbl> <int> <chr>       <int>  <dbl>
## 1      1     0 Congruent    1484     -1
## 2      1     1 Incongruent  1316      1
## 3      1     2 Incongruent   628      1
## # i 3,055 more rows
```

Fit the model.

```
fit_stroop <- brm(RT ~ c_cond + (c_cond | subj),
                  family = lognormal(),
                  prior =
```

```
              c(prior(normal(6, 1.5), class = Intercept),
                prior(normal(0, 0.01), class = b),
                prior(normal(0, 1), class = sigma),
                prior(normal(0, 1), class = sd),
                prior(lkj(2), class = cor)),
            data = df_stroop)
```

We will focus on β (but you can verify that there is nothing surprising in the other parameters in the model `fit_stroop`).

```
posterior_summary(fit_stroop, variable = "b_c_cond")
```

```
##            Estimate Est.Error   Q2.5  Q97.5
## b_c_cond    0.0269   0.00554 0.0162 0.0375
```

As shown in Figure 5.19, if we overlay the density plots for the prior and posterior distributions of β, it becomes evident that the prior might have been too restrictive: the posterior is relatively far from the prior, and the prior strongly down-weights the values that the posterior is centered around. Such a strong discrepancy between the prior and posterior can be investigated with a sensitivity analysis.

```
sample_b_post <- as_draws_df(fit_stroop)$b_c_cond
# We generate samples from the prior as well:
N <- length(sample_b_post)
sample_b_prior <- rnorm(N, 0, 0.01)
samples <-
  tibble(sample = c(sample_b_post, sample_b_prior),
         distribution = c(rep("posterior", N), rep("prior", N)))
ggplot(samples, aes(x = sample, fill = distribution)) +
  geom_density(alpha = 0.5)
```

5.3.1.1 Sensitivity analysis

Here, the discrepancy evident in Figure 5.19 is investigated with a sensitivity analysis. We will examine what happens for the following priors for β. In the

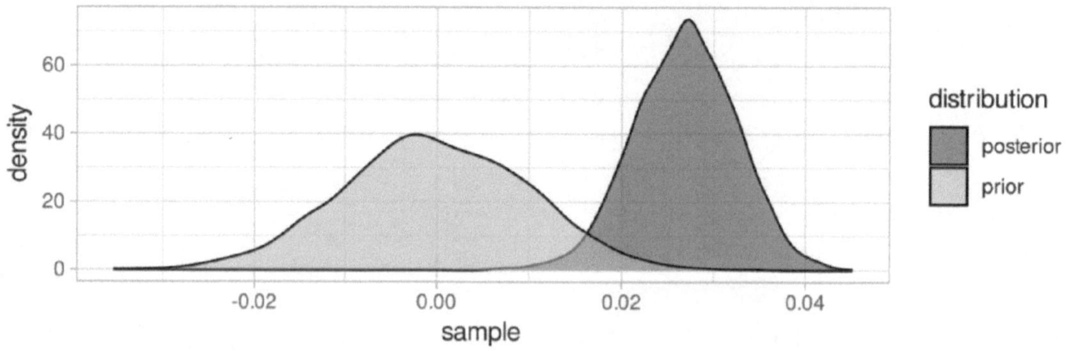

FIGURE 5.19: The discrepancy between the prior and the posterior distributions for the slope parameter in the model `fit_stroop`.

TABLE 5.1: The summary (mean and 95% credible interval) for the posterior distribution of the slope in the model (`fit_stroop`), given different priors on the slope parameter.

prior	Estimate	Q2.5	Q97.5
$Normal(0, 0.001)$	0.001	−0.001	0.003
$Normal(0, 0.01)$	0.027	0.016	0.037
$Normal(0, 0.05)$	0.037	0.025	0.049
$Normal(0, 0.1)$	0.037	0.025	0.049
$Normal(0, 1)$	0.037	0.025	0.049
$Normal(0, 2)$	0.038	0.025	0.050

models we fit below, all the other parameters have the same priors as in the model `fit_stroop`; we vary only the priors for β. The different priors are:

- $\beta \sim Normal(0, 0.001)$
- $\beta \sim Normal(0, 0.01)$
- $\beta \sim Normal(0, 0.05)$
- $\beta \sim Normal(0, 0.1)$
- $\beta \sim Normal(0, 1)$
- $\beta \sim Normal(0, 2)$

We can summarize the estimates of β given different priors as shown in Table 5.1.

It might be easier to see how much the posterior difference between conditions changes depending on the prior. In order to answer this question, we need to remember that the median difference between conditions ($MedianRT_{diff}$) can be calculated as the difference between the exponents of each condition's medians:

TABLE 5.2: A summary, under a range of priors, of the posterior distributions of the mean difference between the two conditions, back-transformed to the millisecond scale.

prior	mean diff (ms)	Q2.5	Q97.5
$Normal(0, 0.001)$	0.66	−1.53	2.82
$Normal(0, 0.01)$	30.11	18.01	41.91
$Normal(0, 0.05)$	41.50	28.07	55.55
$Normal(0, 0.1)$	41.76	28.41	55.73
$Normal(0, 1)$	41.98	28.56	55.71
$Normal(0, 2)$	42.16	28.29	56.29

$$\begin{aligned} MedianRT_{diff} &= MedianRT_{incongruent} - MedianRT_{congruent} \\ MedianRT_{diff} &= \exp(\alpha + \beta) - \exp(\alpha - \beta) \end{aligned} \tag{5.27}$$

Equation (5.27) gives us the posterior distributions of the median difference between conditions for the different models. We calculate the median difference rather than the mean difference because the mean depends on the parameter σ, but the median doesn't: The mean of a log-normal distribution is $\exp(\mu + \sigma^2/2)$, and the median is simply $\exp(\mu)$; also see section 3.7.2. Another reason for reporting the median difference is that it would be a better way to summarize a skewed reading time or response time distribution (the mean is very sensitive to extreme values).

Table 5.2 summarizes the posterior distributions under different priors using the means of the difference in medians, along with 95% credible intervals. It's important to realize that the use of the mean to summarize the posterior distribution is orthogonal to our use of the median to summarize the response times by condition: In the first case, we use the median to summarize a group of *observations*, and in the second case, we use the mean to summarize a group of *samples* from the posterior–we could have summarized the samples from the posterior with its median instead of the mean.

Table 5.2 shows us that the posterior changes substantially when we use wider priors. It seems that the posterior is relatively unaffected when we use priors with a standard deviation larger than 0.05. However, if we assume a priori that the effect of the manipulation must be small, we will end up obtaining a posterior that is consistent with that belief. When we include less information about the possible effect sizes by using a less informative prior, we allow the data to influence the posterior more. A sensitivity analysis is always an important component of a good-quality Bayesian analysis.

Which analysis should one report after carrying out a sensitivity analysis? In the above example, the priors ranging from $Normal(0, 0.05)$ to $Normal(0, 2)$ show rather similar posterior distributions for the mean difference. The most common approach in Bayesian analysis is to report the results of such relatively uninformative priors (e.g., one could report the posterior associated with the $Normal(0, 2)$ here), because this kind of prior allows for a broader range of possible effects and is relatively agnostic. However, if there is a good reason to use a prior from a previous analysis, then of course it makes sense to report the analysis with the informative prior alongside an analysis with an uninformative prior. Reporting only informative priors in a Bayesian analysis is generally not a good idea. The issue is transparency: the reader should know what the posterior looks like for both an informative and an uninformative prior.

Another situation where posterior distributions associated with multiple priors should be reported is when one is carrying out an adversarial sensitivity analysis (Spiegelhalter, Abrams, and Myles 2004): one can take a group of agnostic, enthusiastic, and adversarial or skeptical priors that, respectively, reflect a non-committal a priori position, an informed position based on the researcher's prior beliefs, and an adversarial position based on a scientific opponent's beliefs. In such a situation, analyses using all three priors can be reported, so that the reader can determine how different prior beliefs influence the posterior. For an example of such an adversarial analysis, see Vasishth and Engelmann (2022). Finally, when carrying out hypothesis testing using Bayes factors, the choice of the prior on the parameter of interest becomes critically important; in that situation, it is very important to report a sensitivity analysis, showing the Bayes factor as well as a summary of the posterior distributions (Schad et al. 2021); we return to this point in chapter 13, which covers Bayes factors.

5.4 Why fitting a Bayesian hierarchical model is worth the effort

Carrying out Bayesian data analysis clearly requires much more effort than fitting a frequentist model: we have to define priors, verify that our model works, and decide how to interpret the results. By comparison, fitting a linear mixed model using `lme4` consists of only a single line of code. But there is a hidden cost to the relatively high speed furnished by functions such as `lmer()`. First, the model fit using `lmer()` or the like makes many assumptions, but they are hidden from the user. This is not a problem for the knowledgeable modeler, but very dangerous for the naive user. A second conceptual problem is that the way frequentist models are typically used is to answer a binary question: is the effect "significant" or not? Although frequentist models can quickly answer the question whether there is evidence against the null hypothesis, this often is the wrong question given the actual research problem. For discussion, see Vasishth and Gelman (2021).

Nevertheless, it is natural to ask why one should bother to go through all the trouble of fitting a Bayesian model. An important reason is the flexibility in model specification. The approach we have presented here can be used to extend essentially any parameter of any model. This includes popular uses, such as logistic and Poisson regressions, and also useful models that are relatively rarely used in cognitive science, such as multi-logistic regression (e.g., accuracy in some task with more than two answers), ordered logistic (e.g., ratings, Bürkner and Vuorre 2019), models with a shifted log-normal distribution (see the online exercise G.10.1 and chapter 18, which deals with a log-normal race mode, and see Nicenboim, Logačev, et al. 2016; Rouder 2005), and distributional regression models (as shown in section 5.2.6). By contrast, a frequentist model, although easy to fit quickly, forces the user to use an inflexible canned model, which may not necessarily make sense for their data.

This flexibility allows us also to go beyond the statistical models discussed before, and to develop complex hierarchical computational process models that are tailored to specific phenomena. An example is computational cognitive models, which can be extended hierarchically in a straightforward way; see Lee (2011b) and Lee and Wagenmakers (2014). This flexibility is possible because, as we have seen with distributional regression models in section 5.2.6, any parameter can have a group-level effect structure. Some examples of hierarchical computational cognitive models in psycholinguistics are Logačev and Vasishth (2016), Nicenboim and Vasishth (2018), Vasishth, Chopin, et al. (2017), Vasishth, Jäger, and Nicenboim (2017), Lissón et al. (2021), Logačev and Dokudan (2021), Paape et al. (2021), and Paape and Vasishth (2022). The hierarchical Bayesian modeling approach can even be extended to process models that cannot be expressed as a likelihood function, although in such cases one may have to write one's own sampler; for an example from psycholinguistics, see Yadav et al. (2022) and Yadav et al. (2023). In chapters 15–18, we discuss and implement in Stan some relatively simple computational cognitive models.

5.5 Summary

This chapter presents two very commonly used classes of hierarchical model: those with normal and log-normal likelihoods. We saw several common variants of such models: varying intercepts, varying intercepts and varying slopes with or without a correlation parameter, and crossed random effects for subjects and items. We also experienced the flexibility of the Stan modeling framework through the example of a model that assumes a different residual standard deviation for each subject.

5.6 Further reading

Chapter 5 of Gelman et al. (2014) and O'Hagan and Forster (2004) provide rather technical but comprehensive treatments of exchangeability in Bayesian hierarchical models. Bernardo and Smith (2009) is a brief but useful article explaining exchangeability, and Lunn et al. (2012) also has a helpful discussion that we have drawn on in this chapter. Gelman and Hill (2007) is a comprehensive treatment of hierarchical modeling, although it uses WinBUGS. Yarkoni (2020) discusses the importance of modeling variability in variables that researchers clearly intend to generalize over (e.g., stimuli, tasks, or research sites), and how under-specification of group-level effects imposes strong constraints on the generalizability of results. Sorensen, Hohenstein, and Vasishth (2016) provides an introduction, using Stan, to the Laird-Ware style matrix formulation (Laird and Ware 1982) of hierarchical models; this formulation has the advantage of flexibility and efficiency when specifying models in Stan syntax.

6

Contrast coding

Whenever one uses a categorical variable as a predictor in a Bayesian regression model, for example when estimating the difference in a dependent variable among two or three experimental conditions, it becomes necessary to code the discrete factor levels into numeric predictor variables. This coding is termed *contrast coding*. For example, in the previous chapter (section 5.3), we coded two experimental conditions as -1 and $+1$; this is called a sum contrast. Contrasts are the values that we assign to predictor variables to encode specific comparisons between factor levels and to create predictor terms to estimate these comparisons in any type of regression, including Bayesian regressions.

Contrast coding in Bayesian models works more or less the same way as in frequentist models, and the same principles and tools can be used in both cases. As one important difference, the prior can matter in Bayesian models (especially when working with Bayes factors: Dablander et al. 2022; Rouder et al. 2012); the prior needs to be adapted to the scaling of contrasts. For example, given two conditions in an experiment, if we code a sum contrast (discussed below) with ± 1 coding, the parameter representing the difference between the two conditions will be half as large as the observed difference in the data, compared to the case when we code the contrasts with ± 0.5 coding. In the latter case, the parameter of interest does represent the observed difference in means between the conditions. The two contrast coding options (± 1 and ± 0.5) obviously imply different prior specifications for the parameter of interest.

This chapter will introduce contrast coding in the context of Bayesian models. The descriptions are in large parts taken from Schad et al. (2020) (which is published under a CC-BY 4.0 license: `https://creativecommons.org/licenses/by/4.0/`) and adapted for the current context.

Consider a situation where we want to estimate differences in a dependent variable between three levels of a factor. An example could be differences in response times between three levels of word class (noun, verb, adjective). We might be interested in whether word class influences response times. In frequentist statistics, one way to approach this question would be to run an ANOVA and compute an omnibus F-test for whether word class explains

DOI: 10.1201/9780429342646-8

response times.[1] However, if based on such omnibus approaches we find support for an influence of word class on response times, it remains unclear where this effect actually comes from, i.e., whether it originated from the nouns, verbs, or adjectives. This is problematic for inference because scientists typically have specific expectations about which groups differ from each other. In this chapter, we will show how to estimate specific comparisons directly in a Bayesian linear model. This gives the researcher a lot of control over Bayesian analyses. Specifically, we show how planned comparisons between specific conditions (groups) or clusters of conditions, are implemented as contrasts. This is a very effective way to align expectations with the statistical model. In Bayesian models, any specific comparisons can also be computed after the model is fit. Nevertheless, coding a priori expectations into contrasts for model fitting will make it much more straightforward to estimate certain comparisons between experimental conditions, and will allow us to perform Bayesian model comparisons using Bayes factors to provide evidence for or against very specific hypotheses.

For this and the next chapter, although knowledge of the matrix formulation of the linear model is not necessary, for a deeper understanding of contrast coding some exposure to the matrix formulation is desirable. We discuss the matrix formulation in the online section A.7.

6.1 Basic concepts illustrated using a two-level factor

We first consider the simplest case: suppose we want to compare the means of a dependent variable (DV) such as response times between two groups of subjects. A simulated data set is available in the package `bcogsci` as the data set `df_contrasts1`. The simulations assumed longer response times in condition $F1$ ($\mu_1 = 0.8$ sec) than $F2$ ($\mu_2 = 0.4$ sec). The data from the 10 simulated subjects are aggregated and summary statistics are computed for the two groups.

```
data("df_contrasts1")
df_contrasts1
```

[1] A Bayesian analog to the frequentist F-test is Bayesian model comparison (i.e., Bayes factors), where we might compare an alternative model including word class as a predictor term with a null model lacking this predictor. We will discuss such Bayesian model comparison using Bayes factors in chapter 13.

TABLE 6.1: Summary statistics per condition for the simulated data.

Factor	N data	Est. means	Std. dev.	Std. errors
F1	5	0.8	0.2	0.1
F2	5	0.4	0.2	0.1

```
## # A tibble: 10 x 3
##   F        DV    id
##   <fct> <dbl> <int>
## 1 F1    0.636     1
## 2 F1    0.841     2
## 3 F1    0.555     3
## # i 7 more rows
```

```
## [1] 0.6
```

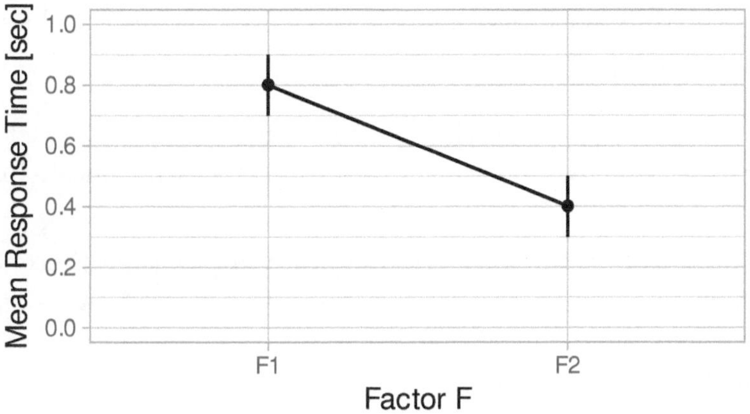

FIGURE 6.1: Means and standard errors of the simulated dependent variable (e.g., response times in seconds) in two conditions $F1$ and $F2$.

The results, displayed in Figure 6.1 and in Table 6.1, show that the assumed true condition means are exactly realized with the simulated data. The numbers are exact because the `mvrnorm()` function used here (see `?df_contrasts1`) ensures that the data are generated so that the sample mean yields the true means for each level. In real data sets, of course, the sample means will vary from experiment to experiment.

A simple Bayesian linear model of `DV` on F yields a straightforward estimate of the difference between the group means. We use relatively uninformative priors. The estimates for the population-level effects are presented below using the function `fixef()`:

```
fit_F <- brm(DV ~ F,
             data = df_contrasts1,
             family = gaussian(),
             prior = c(prior(normal(0, 2), class = Intercept),
                       prior(normal(0, 2), class = sigma),
                       prior(normal(0, 1), class = b)))
```

```
fixef(fit_F)
```

```
##           Estimate Est.Error  Q2.5 Q97.5
## Intercept     0.79      0.11  0.56  1.01
## FF2          -0.39      0.15 -0.70 -0.07
```

Comparing the means for each condition with the coefficients (`Estimates`) reveals that (i) the intercept (0.8) is the mean for condition $F1$, $\hat{\mu}_1$; and (ii) the slope (`FF2`: -0.4) is the difference between the estimated means for the two groups, $\hat{\mu}_2 - \hat{\mu}_1$ (Bolker 2018):

$$\begin{array}{lcl} \text{Intercept} = & \hat{\mu}_1 & = \text{estimated mean for } F1 \\ \text{Slope (FF2)} = & \hat{\mu}_2 - \hat{\mu}_1 & = \text{estim. mean for } F2 - \text{estim. mean for } F1 \end{array} \tag{6.1}$$

The new information is the 95% credible interval for the difference between the two groups.

6.1.1 Default contrast coding: Treatment contrasts

How does the function `brm()` arrive at these particular values for the intercept and slope? That is, why does the intercept assess the mean of condition $F1$ and how do we know the slope measures the difference in means between $F2$-$F1$? This result is a consequence of the default contrast coding of the factor F. R assigns treatment contrasts to factors and orders their levels alphabetically. The alphabetically first factor level (here: $F1$) is coded in R by default as 0 and the second level (here: $F2$) is coded as 1.[2] This becomes clear when we inspect the current contrast attribute of the factor using the `contrasts()` command:

[2]When using treatment coding, it's important to be aware that the intercept won't be centered. As mentioned in online section A.3, this means the prior one sets for the intercept in `brms` will actually be assigned to its centered version. Although this distinction may not have any impact when one uses wide priors, it's still worth keeping in mind.

```
contrasts(df_contrasts1$F)
```

```
##    F2
## F1  0
## F2  1
```

Why does this contrast coding yield these particular regression coefficients? Let's take a look at the regression equation. Let α represent the intercept and β_1 the slope. Then, the simple regression above expresses the belief that the expected response time $\hat{y}$ (or $E[Y]$) is a linear function of the factor F.

$$E[Y] = \alpha + \beta_1 x \tag{6.2}$$

So, if $x = 0$ (condition $F1$), the expectation is $\alpha + \beta_1 \cdot 0 = \alpha$; and if $x = 1$ (condition $F2$), the expectation is $\alpha + \beta_1 \cdot 1 = \alpha + \beta_1$.

Expressing the above in terms of the estimated coefficients:

$$\begin{array}{lccl} \text{estim. value for } F1 = & \hat{\mu}_1 = & \hat{\alpha} = & \text{Intercept} \\ \text{estim. value for } F2 = & \hat{\mu}_2 = & \hat{\alpha} + \hat{\beta}_1 = & \text{Intercept} + \text{Slope (FF2)} \end{array} \tag{6.3}$$

It is useful to think of such unstandardized regression coefficients as difference scores; they express the increase in the dependent variable y associated with a change in the independent variable x by one unit, such as going from 0 to 1 in this example. The difference between condition means is $0.4 - 0.8 = -0.4$, which is the estimated regression coefficient $\hat{\beta}_1$. The sign of the slope is negative because we have chosen to subtract the larger mean $F1$ score from the smaller mean $F2$ score.

6.1.2 Defining comparisons

The analysis of the regression equation demonstrates that in the treatment contrast the intercept assesses the average response in the baseline condition, whereas the slope estimates the difference between condition means. However, these are just verbal descriptions of what each coefficient assesses. Is it also possible to formally write down what each coefficient assesses?

From the perspective of parameter estimation, the slope represents the effect of main interest, so we consider this first. The treatment contrast specifies that the slope β_1 estimates the difference in means between the two levels of the factor F. This can formally be written as:

$$\beta_1 = \mu_{F2} - \mu_{F1} \tag{6.4}$$

or equivalently:

$$\beta_1 = -1 \cdot \mu_{F1} + 1 \cdot \mu_{F2} \tag{6.5}$$

The ± 1 weights in the parameter estimation directly express which means are compared by the treatment contrast.

The intercept in the treatment contrast estimates a quantity that is usually of little interest: it estimates the mean in condition $F1$. Formally, the parameter α estimates the following quantity:

$$\alpha = \mu_{F1} \tag{6.6}$$

or equivalently:

$$\alpha = 1 \cdot \mu_{F1} + 0 \cdot \mu_{F2}. \tag{6.7}$$

The fact that the intercept term formally estimates the mean of condition $F1$ is in line with our previous derivation (see Equation (6.1)).

In R, factor levels are ordered alphabetically and by default the first level is used as the baseline in treatment contrasts. Obviously, this default mapping will depend on the levels' alphabetical ordering. If a different baseline condition is desired, it is possible to re-order the levels. Here is one way of re-ordering the levels:

```
df_contrasts1$Fb <- factor(df_contrasts1$F, levels = c("F2", "F1"))
contrasts(df_contrasts1$Fb)
```

```
##    F1
## F2  0
## F1  1
```

This re-ordering did not change any data associated with the factor, only one of its attributes. With this new contrast attribute, a simple Bayesian model yields the following result.

```
fit_Fb <- brm(DV ~ Fb,
              data = df_contrasts1,
              family = gaussian(),
              prior = c(prior(normal(0, 2), class = Intercept),
                        prior(normal(0, 2), class = sigma),
                        prior(normal(0, 1), class = b)))
```

```
fixef(fit_Fb)
```

```
##           Estimate Est.Error Q2.5 Q97.5
## Intercept      0.4      0.11 0.17  0.63
## FbF1           0.4      0.16 0.07  0.72
```

The model now estimates different quantities. The intercept now codes the mean of condition $F2$, and the slope measures the difference in means between $F1$ minus $F2$. This represents an alternative coding of the treatment contrast.

These model posteriors do not allow us to claim that we have evidence for the hypothesis that the effect of factor F is different from zero. If the research focus is on such hypothesis testing, Bayesian hypothesis tests can be carried out using Bayes factors, by comparing a model containing a contrast of interest with a model lacking this contrast. We will discuss details of Bayesian hypothesis testing based on Bayes factors in chapter 13. For now, our focus is on obtaining estimates of the relevant comparison(s); we are interested in situations where we have two or more conditions.

6.1.3 Sum contrasts

Treatment contrasts are only one of many options. It is also possible to use sum contrasts, which code one of the conditions as -1 and the other as $+1$, effectively *centering* the effects at the grand mean (GM, i.e., the mean of the two group means). Here, we rescale the contrast to values of -0.5 and $+0.5$, which makes the estimated treatment effect the same as for treatment coding and are easier to interpret.

To define this contrast in a linear regression, one way is to use the `contrasts` function (another way is to define a column containing $+0.5$ and -0.5 for the corresponding levels of the factor).

```
contrasts(df_contrasts1$F) <- c(-0.5, +0.5)
fit_mSum <- brm(DV ~ F,
                data = df_contrasts1,
                family = gaussian(),
                prior = c(prior(normal(0, 2), class = Intercept),
                          prior(normal(0, 2), class = sigma),
                          prior(normal(0, 1), class = b)))
```

```
fixef(fit_mSum)
```

```
##              Estimate Est.Error  Q2.5 Q97.5
## Intercept       0.60      0.08  0.45  0.76
## F1             -0.39      0.15 -0.69 -0.10
```

Here, the slope ($F1$) again codes the difference between the groups associated with the first and second factor levels. It has the same value as in the treatment contrast. One important difference from the treatment contrast is that the intercept now represents the estimate of the average of condition means for $F1$ and $F2$, that is, the grand mean. For the scaled sum contrast:

$$\begin{array}{lll} \text{Intercept} = & (\hat{\mu}_1 + \hat{\mu}_2)/2 & = \text{estimated mean of } F1 \text{ and } F2 \\ \text{Slope (F1)} = & \hat{\mu}_2 - \hat{\mu}_1 & = \text{estim. mean of } F2 - \text{estim. mean for} F1 \end{array} \tag{6.8}$$

Why does the intercept assess the grand mean and why does the slope estimate the group difference? This is the result of rescaling the sum contrast. The first factor level ($F1$) was coded as -0.5, and the second factor level ($F2$) as $+0.5$:

```
contrasts(df_contrasts1$F)
```

```
##    [,1]
## F1 -0.5
## F2  0.5
```

Look again at the regression equation to better understand what computations are performed. Again, α represents the intercept, β_1 represents the slope, and the predictor variable x represents the factor F. The regression equation is written as:

$$E[Y] = \alpha + \beta_1 x \tag{6.9}$$

The group of $F1$ subjects is then coded as -0.5, and the response time for the group of $F1$ subjects is $\alpha + \beta_1 \cdot x_1 = 0.6 + (-0.4) \cdot (-0.5) = 0.8$. By contrast, the $F2$ group is coded as $+0.5$. By implication, the mean of the $F2$ group must be $\alpha + \beta_1 \cdot x_1 = 0.6 + (-0.4) \cdot 0.5 = 0.4$. Expressed in terms of the estimated coefficients:

$$\begin{array}{llll} \text{estim. value for } F1 = & \hat{\mu}_1 = & \hat{\alpha} - 0.5 \cdot \hat{\beta}_1 = & \text{Intercept} - 0.5 \cdot \text{Slope (F1)} \\ \text{estim. value for } F2 = & \hat{\mu}_2 = & \hat{\alpha} + 0.5 \cdot \hat{\beta}_1 = & \text{Intercept} + 0.5 \cdot \text{Slope (F1)} \end{array} \tag{6.10}$$

The unstandardized regression coefficient is a difference score: Taking a step of one unit on the predictor variable x, e.g., from -0.5 to $+0.5$, reflecting a step from condition $F1$ to $F2$, changes the dependent variable from 0.8 (for condition $F1$) to 0.4 (condition $F2$). This reflects a difference of $0.4-0.8=-0.4$; this is again the estimated regression coefficient $\hat{\beta}_1$. Moreover, as mentioned above, the intercept now assesses the grand mean, i.e., the unweighted mean of the means for conditions $F1$ and $F2$: it is in the middle between condition means for $F1$ and $F2$.

So far we gave verbal statements about what is estimated by the intercept and the slope in the case of the scaled sum contrast. It is possible to write these statements as formal parameter estimates. In scaled sum contrasts, the slope parameter β_1 assesses the following quantity:

$$\beta_1 = -1 \cdot \mu_{F1} + 1 \cdot \mu_{F2} \tag{6.11}$$

This estimates the same quantity as the slope in the treatment contrast. The intercept now assesses a different quantity: the average of the two conditions $F1$ and $F2$:

$$\alpha = 1/2 \cdot \mu_{F1} + 1/2 \cdot \mu_{F2} = \frac{\mu_{F1} + \mu_{F2}}{2} \tag{6.12}$$

In balanced data, i.e., in data sets where there are no missing data points, the average of the two conditions $F1$ and $F2$ is the grand mean. In unbalanced data sets, where there are missing values, this average is the weighted grand mean. To illustrate this point, consider an example with fully balanced data and two equal group sizes of 5 subjects for each group $F1$ and $F2$. Here, the grand mean is also the mean across all subjects. Next, consider a highly simplified unbalanced data set, where in condition $F1$ two observations of the dependent variable are available with values of 2 and 3, and where in condition $F2$ only one observation of the dependent variable is available with a value of 4. In this data set, the mean across all subjects is $\frac{2+3+4}{3} = \frac{9}{3} = 3$. However, the (weighted) grand mean as assessed in the intercept in a model using sum contrasts for factor F would first compute the mean for each group separately (i.e., $\frac{2+3}{2} = 2.5$, and 4), and then compute the mean across conditions $\frac{2.5+4}{2} = \frac{6.5}{2} = 3.25$. The grand mean of 3.25 is different from the mean 3 that was computed using all the subjects' data all at once.

To summarize, treatment contrasts and sum contrasts are two possible ways to parameterize the difference between two groups; they generally estimate different quantities. Treatment contrasts compare one or more means against a baseline condition, whereas sum contrasts compare a condition's mean to the grand mean (which in the two-group case also implies estimating the difference between the two group means). One question that comes up here is: how does one know what quantities are estimated by a given set of contrasts? (In the context of Bayes factors, the question would be: what hypothesis test does the contrast coding encode?) This question will be discussed in detail below for the general case of any arbitrary contrast coding.

6.1.4 Cell means parameterization and posterior comparisons

One alternative option is to use what is called the cell means parameterization (this coding is also called "one-hot encoding" in the context of machine learning). In this approach, one does not estimate an intercept term, and then differences between factor levels. Instead, each free parameter is used to simply estimate the mean of one of the factor levels. As a consequence, no comparisons between condition means are estimated, but simply the mean of each experimental condition is estimated. Cell means parameterization is specified by explicitly removing the intercept term (which is added automatically in `brms`) by adding a -1 in the regression formula:

```
fit_mCM <- brm(DV ~ -1 + F,
               data = df_contrasts1,
               family = gaussian(),
               prior = c(prior(normal(0, 2), class = sigma),
                         prior(normal(0, 2), class = b)))
```

```
fixef(fit_mCM)
```

```
##     Estimate Est.Error Q2.5 Q97.5
## FF1      0.8      0.11 0.57  1.02
## FF2      0.4      0.11 0.18  0.61
```

Now, the regression coefficients (see the column labeled `Estimate`) estimate the mean of the first factor level (0.8) and the mean of the second factor level (0.4). This cell means parameterization usually does not allow us to make inferences about the hypotheses of interest using Bayes factors, as these hypotheses usually relate to differences between conditions rather than to whether each condition differs from zero.

The cell means parameterization provides a good example demonstrating an advantage of Bayesian data analysis: In Bayesian models, it is possible to use the posterior samples to compute new estimates that were not directly contained in the fitted model. To implement this, we first extract the posterior samples from the `brm()` model object:

```
df_postSamp <- as_draws_df(fit_mCM)
```

In a second step, we can then compute comparisons from these posterior samples. For example, we can compute the difference between conditions $F2$ and $F1$. To do so, we simply take the posterior samples for each condition and compute their difference.

```
df_postSamp  <- df_postSamp %>%
  mutate(b_dif = b_FF2 - b_FF1)
```

This provides a posterior sample of the difference between conditions. It is possible to investigate this posterior sample by looking at its mean and 95% credible interval:

```
c(Estimate = mean(df_postSamp$b_dif),
  quantile(df_postSamp$b_dif, p = c(0.025, 0.975)))
```

```
## Estimate     2.5%    97.5%
##  -0.3959  -0.7096  -0.0695
```

The above summary provides the same estimate (roughly -0.4) that we obtained previously when using the treatment contrast or the scaled sum contrast. (Note that if the priors are sufficiently non-informative, the results will be similar. However, given that the priors differ in comparison to a model estimating the difference directly, the posterior may also differ between models.)

Thus, Bayesian models provide a lot of flexibility in computing new comparisons post-hoc from the posterior samples and in obtaining their posterior distributions. However, what these posterior computations do not provide directly are inferences on null hypotheses. That is, just by looking at the credible intervals, we cannot make inferences about whether a null hypothesis can be rejected; an explicit hypothesis test is needed to answer such a question (see chapter 13).

6.2 The hypothesis matrix illustrated with a three-level factor

Consider an example with the three word classes nouns, verbs, and adjectives. We load simulated data from a lexical decision task with response times as dependent variable. The research question is: do response times differ as a

TABLE 6.2: Summary statistics per condition for the simulated data.

Factor	N data	Est. means	Std. dev.	Std. errors
adjectives	4	400.2	19.9	9.9
nouns	4	500.0	20.0	10.0
verbs	4	450.2	20.0	10.0

function of the between-subject factor word class with three levels: nouns, verbs, and adjectives? Here, just to illustrate the case of a three-level factor, we make the arbitrary assumption that nouns have the longest response times and adjectives the shortest response times. Word class is specified as a between-subject factor. In cognitive science experiments, word class will usually vary within subjects and between items. Because the within- or between-subjects status of an effect is independent of its contrast coding, we assume the manipulation to be between subjects for ease of exposition. The concepts presented here extend to repeated measures designs that are often analyzed using hierarchical Bayesian (linear mixed) models.

Load and display the simulated data.

```
data("df_contrasts2")
head(df_contrasts2)
```

```
## # A tibble: 6 x 3
##   F        DV    id
##   <fct> <int> <int>
## 1 nouns   476     1
## 2 nouns   517     2
## 3 nouns   491     3
## # i 3 more rows
```

As shown in Table 6.2, the estimated means reflect our assumptions about the true means in the data simulation: Response times are longest for nouns and shortest for adjectives. In the following sections, we use this data set to illustrate sum contrasts. Furthermore, we will use an additional data set to illustrate repeated, Helmert, polynomial, and custom contrasts. In practice, usually only one set of contrasts is selected when the expected pattern of means is formulated during the design of the experiment.

6.2.1 Sum contrasts

We begin with sum contrasts. Suppose that the expectation is that nouns are responded to slower and adjectives are responded to faster than the grand mean response time. Then, the research question could be: By how much do nouns differ from the grand mean and by how much do adjectives differ from the grand mean? And are the responses slower or faster than the grand mean? We want to estimate the following two quantities:

$$\beta_1 = \mu_1 - \frac{\mu_1 + \mu_2 + \mu_3}{3} = \mu_1 - GM \tag{6.13}$$

and

$$\beta_2 = \mu_2 - \frac{\mu_1 + \mu_2 + \mu_3}{3} = \mu_2 - GM \tag{6.14}$$

β_1 can also be written as:

$$\beta_1 = \frac{2}{3}\mu_1 - \frac{1}{3}\mu_2 - \frac{1}{3}\mu_3 \tag{6.15}$$

Here, the weights $2/3, -1/3, -1/3$ are informative about how to combine the condition means to estimate the linear model coefficient.

β_2 is also rewritten as:

$$\beta_2 = \mu_2 - \frac{\mu_1 + \mu_2 + \mu_3}{3} \tag{6.16}$$

$$\Leftrightarrow \beta_2 = -\frac{1}{3}\mu_1 + \frac{2}{3}\mu_2 - \frac{1}{3}\mu_3 \tag{6.17}$$

Here, the weights are $-1/3, 2/3, -1/3$, and they again indicate how to combine the condition means for estimating the regression coefficient.

6.2.2 The hypothesis matrix

The weights of the condition means are not only useful for defining parameter estimates. They also provide the starting step in a very powerful method which allows the researcher to generate the contrasts that are needed to estimate these comparisons in a linear model. What we did so far is to explain some different contrast codings that exist and the comparisons that are estimated by these contrasts. Given a particular data set, if the goal is to estimate certain comparisons, then the procedure would be to check whether any of the contrasts that we encountered estimate these comparisons of interest. Sometimes it suffices to use one of these existing contrasts. At other times, our research questions may not correspond exactly to any of the contrasts in the default set of standard contrasts provided in R. For these cases, or for more complex designs, it is very useful to know how contrast matrices are created. Indeed, a relatively simple procedure exists in which we write our comparisons formally, extract the weights of the condition means from the comparisons, and then automatically generate the correct contrast matrix that we need in order to estimate these comparisons in a linear model. Using this powerful method, it is not necessary to find a match to a contrast matrix provided by the family of functions in R starting with the prefix `contr`.[3] Instead, it is possible to simply define the comparisons that one wants to estimate and to obtain the correct contrast matrix for these in an automatic procedure. Here, for pedagogical reasons, we show some examples of how to apply this procedure in cases where the comparisons *do* correspond to some of the existing contrasts.

Defining a custom contrast matrix involves four steps:

1. Write down the estimated comparisons
2. Extract the weights and write them into what we will call a *hypothesis matrix* (and can also be viewed as a *comparison matrix*)
3. Apply the *generalized matrix inverse* to the hypothesis matrix to create the contrast matrix
4. Assign the contrast matrix to the factor and run the (Bayesian) model

The term hypothesis matrix is used here because contrast coding is often done to carry out an explicit hypothesis test; but one could of course use contrast coding just to compute the estimates of an effect and their uncertainty, without doing a hypothesis test.

Let us apply this four-step procedure to our example of the sum contrast. The first step is to write down the estimated comparisons, as was done in section 6.2.1. The second step involves writing down the weights that each comparison gives to the condition means. The weights for the first comparison are `wH01 = c(+2/3, -1/3, -1/3)`, and the weights for the second comparison are `wH02 = c(-1/3, +2/3, -1/3)`.

Before writing these into a hypothesis matrix, we also define the estimated quantity for the intercept term. The intercept parameter estimates the mean across all conditions:

[3] These are the functions `contr.treatment()`, `contr.sum()`, `contr.poly()`, `contr.helmert()`, and `contr.sdif()`. The last contrast is available in the `MASS` package.

$$\alpha = \frac{\mu_1 + \mu_2 + \mu_3}{3} \tag{6.18}$$

$$\alpha = \frac{1}{3}\mu_1 + \frac{1}{3}\mu_2 + \frac{1}{3}\mu_3 \tag{6.19}$$

This estimate has weights of 1/3 for all condition means. The weights from all three model parameters that were defined are now combined and written into a matrix that we refer to as the *hypothesis matrix* (`Hc`):

```
HcSum <-
  rbind(cH00 = c(adjectives = 1 / 3, nouns = 1 / 3, verbs = 1 / 3),
        cH01 = c(adjectives = 2 / 3, nouns = -1 / 3, verbs = -1 / 3),
        cH02 = c(adjectives = -1 / 3, nouns = 2 / 3, verbs = -1 / 3))
fractions(t(HcSum))
```

```
##            cH00 cH01 cH02
## adjectives  1/3  2/3 -1/3
## nouns       1/3 -1/3  2/3
## verbs       1/3 -1/3 -1/3
```

Each set of weights is first entered as a row into the matrix (command `rbind()`). We switch rows and columns of the matrix for easier readability using the command `t()` (this transposes the matrix). The command `fractions()` from the `MASS` package turns the decimals into fractions to improve readability.

Now that the condition weights have been written into the hypothesis matrix, the third step of the procedure is implemented: a matrix operation called the *generalized matrix inverse*[4] is used to obtain the contrast matrix that is needed to estimate these comparisons in a linear model.

Use the function `ginv2()` from the `bcogsci` package for the next step. It is similar to `ginv()` from `MASS` but provides nicer formatting of the output.

Applying the generalized inverse to the hypothesis matrix results in the new matrix `XcSum`. This is the contrast matrix X_c that estimates exactly those comparisons that were specified earlier:

```
(XcSum <- ginv2(HcSum))
```

[4]At this point, there is no need to understand in detail what this means. We refer the interested reader to Schad et al. (2020). For a quick overview, we recommend a vignette explaining the generalized inverse in the `matlib` package (`https://cran.r-project.org/web/packages/matlib/vignettes/ginv.html`; Friendly, Fox, and Chalmers 2020).

```
##            cH00 cH01 cH02
## adjectives  1    1    0
## nouns       1    0    1
## verbs       1   -1   -1
```

This contrast matrix corresponds exactly to the sum contrasts described above. In the case of the sum contrast, the contrast matrix looks very different from the hypothesis matrix. The contrast matrix in sum contrasts codes with +1 the condition that is to be compared to the grand mean. The condition that is never compared to the grand mean is coded as -1. Without knowing the relationship between the hypothesis matrix and the contrast matrix, the meaning of the coefficients is completely opaque.

To verify this custom-made contrast matrix, it is compared to the sum contrast matrix as generated by the R function `contr.sum()` in the `stats` package. The resulting contrast matrix is identical to the result when adding the intercept term, a column of ones, to the contrast matrix:

```
fractions(cbind(1, contr.sum(3)))
```

```
##   [,1] [,2] [,3]
## 1  1    1    0
## 2  1    0    1
## 3  1   -1   -1
```

In order to estimate model parameters, step four in our procedure involves assigning sum contrasts to the factor F in our example data, and fitting a (Bayesian) linear model. This allows us to estimate the regression coefficients associated with each contrast. We compare these to the data shown above (Table 6.2) to test whether the regression coefficients actually correspond to the differences of condition means, as intended. To define the contrast, it is necessary to remove the intercept term, as this is automatically added by the modeling function `brm()`.

```
contrasts(df_contrasts2$F) <- XcSum[, 2:3]
fit_Sum <- brm(DV ~ F,
               data = df_contrasts2,
               family = gaussian(),
               prior = c(prior(normal(500, 100), class = Intercept),
                         prior(normal(0, 100), class = sigma),
                         prior(normal(0, 100), class = b)))
```

```
fixef(fit_Sum)
```

```
##           Estimate Est.Error  Q2.5 Q97.5
## Intercept    450.4      6.97 436.7 464.3
## FcH01        -49.2      9.75 -68.4 -29.8
## FcH02         49.1     10.05  29.4  70.0
```

The (Bayesian) linear model regression coefficients show the grand mean response time of 450 ms in the intercept. Remember that the first regression coefficient `FcH01` was designed to estimate the extent to which adjectives are responded to faster than the grand mean. The regression coefficient `FcH01` (`Estimate`) of -50 reflects the difference between adjectives (400 ms) and the grand mean of 450 ms. The second estimate of interest tells us the extent to which response times for nouns differ from the grand mean. The fact that the second regression coefficient `FcH02` is close to 50 indicates that the response time for nouns is (50 ms) slower than the grand mean of 450 ms. Although the nouns are estimated to have 50 ms longer reading times than the grand mean, the reading times for adjectives are 50 ms faster than the grand mean.

We have now not only derived contrasts, parameter estimates, and comparisons for the sum contrast, we have also used a powerful and highly general procedure that is used to generate contrasts for many kinds of different comparisons and experimental designs.

6.2.3 Generating contrasts: The `hypr` package

To work with the four-step procedure, i.e., to flexibly design contrasts to estimate specific comparisons, we have developed the R package `hypr` (Rabe et al. 2020). This package allows the researcher to specify the desired comparisons, and based on these comparisons, it automatically generates contrast matrices that allow us to estimate these comparisons in linear models. The functions available in this package thus considerably simplify the implementation of the four-step procedure outlined above. Because `hypr` was originally written with the frequentist framework in mind, the comparisons are expressed as null hypotheses. In the Bayesian framework, these should be treated as comparisons between (bundles of) condition means. To illustrate the functionality of the `hypr` package, we will use the two comparisons that we had defined and analyzed in the previous section:

$$\beta_1 = \mu_1 - \frac{\mu_1 + \mu_2 + \mu_3}{3} = \mu_1 - GM \tag{6.20}$$

and

$$\beta_2 = \mu_2 - \frac{\mu_1 + \mu_2 + \mu_3}{3} = \mu_2 - GM \tag{6.21}$$

These estimates are effectively comparisons between condition means or between bundles of condition means. That is, both μ_1 and μ_2 are compared to the grand mean. These two comparisons can be directly entered into R using the `hypr()` function from the `hypr` package. To do so, we use some labels to indicate factor levels. E.g., `adjectives`, `nouns`, and `verbs` can represent factor levels μ_1, μ_2, and μ_3. The first comparison specifies that μ_1 is compared to $\frac{\mu_1+\mu_2+\mu_3}{3}$. This can be written as a formula in R: `adjectives ~ (adjectives + nouns + verbs)/3`. The second comparison is that μ_2 is compared to $\frac{\mu_1+\mu_2+\mu_3}{3}$, which can be written as `nouns ~ (adjectives + nouns + verbs)/3`.

```
HcSum <- hypr(b1 = adjectives ~ (adjectives + nouns + verbs) / 3,
              b2 = nouns ~ (adjectives + nouns + verbs) / 3,
              levels = c("adjectives", "nouns", "verbs"))
HcSum
```

```
## hypr object containing 2 null hypotheses:
## H0.b1: 0 = (2*adjectives - nouns - verbs)/3
## H0.b2: 0 = (2*nouns - adjectives - verbs)/3
##
## Call:
## hypr(b1 = ~2/3 * adjectives - 1/3 * nouns - 1/3 * verbs, b2 = ~2/3 *
##     nouns - 1/3 * adjectives - 1/3 * verbs, levels = c("adjectives",
## "nouns", "verbs"))
##
## Hypothesis matrix (transposed):
##            b1   b2
## adjectives  2/3 -1/3
## nouns      -1/3  2/3
## verbs      -1/3 -1/3
##
## Contrast matrix:
##            b1 b2
## adjectives  1  0
## nouns       0  1
## verbs      -1 -1
```

The results show that the comparisons between condition means have been re-written into a form where 0 is coded on the left side of the equation, and the condition means together with associated weights are written on the right side of the equation. This presentation makes it easy to see the weights of the condition means to code a certain comparison. The next part of the results shows the hypothesis matrix, which contains the weights from the condition means. Thus, `hypr` takes formulas coding comparisons between condition means as input, and automatically extracts the corresponding weights and encodes them into the hypothesis matrix. `hypr` moreover applies the generalized matrix inverse to obtain the contrast matrix from the hypothesis matrix. The different steps correspond exactly to the steps we had carried out manually in the preceding section. `hypr` automatically performs these steps for us. We can now extract the contrast matrix by a simple function call:

```
contr.hypothesis(HcSum)
```

```
##              b1 b2
## adjectives    1  0
## nouns         0  1
## verbs        -1 -1
## attr(,"class")
## [1] "hypr_cmat" "matrix"    "array"
```

We can assign this contrast to our factor as we did before.

```
contrasts(df_contrasts2$F) <- contr.hypothesis(HcSum)
```

Now, we could again run the same model. However, since the contrast matrix is now the same as used before, the results would also be exactly the same, and we therefore skip the model fitting for brevity.

The `hypr` package can be used to create contrasts for Bayesian models, where the focus lies on estimation of contrasts that code comparisons between condition means or bundles of condition means. (Of course, one can use contrast coding for carrying out hypothesis tests using the Bayes factor; see chapter 13; in this case, the priors specified can lead to very different conclusions.) Thus, the comparison that one specifies implies the estimation of a difference between condition means or bundles of condition means. We see this in the output of the `hypr()` function (see the first section of the results); these formulate the comparison in a way that also illustrates the estimation of model parameters. That is, the comparison (expressed in the `hypr` package's syntax) $\mu_1 \sim \frac{\mu_1+\mu_2+\mu_3}{3}$ corresponds to a parameter estimate of `b1 = 2/3*m1 - 1/3*m2 - 1/3*m3`, where $m1$ to $m3$ are the means for each of the conditions. The resulting contrasts will then allow us to estimate the specified differences between condition means or bundles of condition means.

6.3 Other types of contrasts: Illustration with a factor of four levels

Here, we introduce repeated difference, Helmert, and polynomial contrasts.[5]

Consider an experiment with one between-subject factor with four levels (the discussion below does not depend on the factor being between-subjects–the same logic will hold for within-subjects designs). We load a corresponding data set, which contains simulated data containing response times with a four-level between-subject factor. The sample sizes for each level and the means and standard errors are shown in Table 6.3, and the means and standard errors are also shown graphically in Figure 6.2.

```
data("df_contrasts3")
```

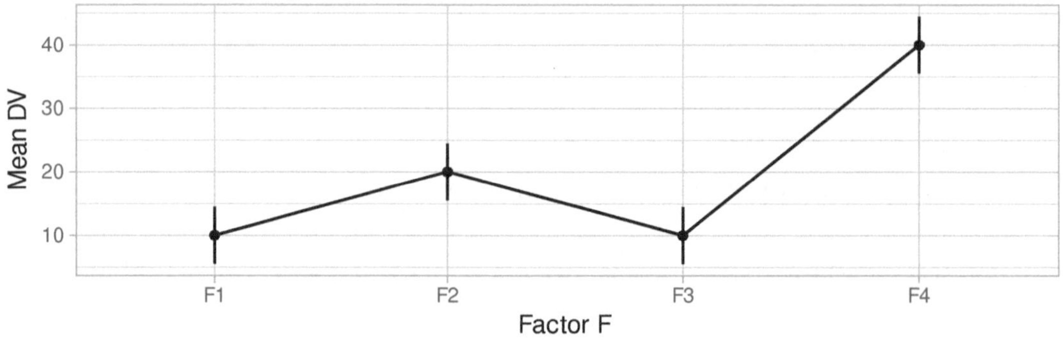

FIGURE 6.2: The means and error bars (showing standard errors) for a simulated data set with one between-subjects factor with four levels.

We assume that the four factor levels $F1$ to $F4$ reflect levels of word frequency, including the levels `low`, `medium-low`, `medium-high`, and `high` frequency words, and that the dependent variable reflects response time.

Qualitatively, the simulated pattern of results is similar to empirically observed values for word frequency effects on single fixation durations in eye tracking (see Figure 5.4 in Heister, Würzner, and Kliegl 2012). Normally, one may

[5]If one is trying to set a general prior for differences between means, then the function `bayestestR::contr.equalprior()` can be used. The `bayestestR::contr.equalprior()` function can be used when one aims to set equal marginal priors for differences between means across all levels of a factor. This avoids the unequal prior distributions that can result from other contrast methods (e.g., `stats::contr.treatment()`) and supports correct Bayes factor estimation in multi-level factors. It is particularly useful for factors with more than two levels, following the orthogonal-normal contrasts described in Rouder et al. (2012, 363). For more information, see `https://easystats.github.io/bayestestR/reference/contr.equalprior.html`.

TABLE 6.3: Summary statistics per condition for the simulated data.

Factor	N data	Est. means	Std. dev.	Std. errors
F1	5	10.0	10.0	4.5
F2	5	20.0	10.0	4.5
F3	5	10.0	10.0	4.5
F4	5	40.0	10.0	4.5

expect the pattern to be monotonically decreasing, i.e., higher word frequency should lead to faster reading times. However, we choose this more complex pattern because we think it may be helpful in understanding how polynomial and monotonic contrasts work (see below).

6.3.1 Repeated contrasts

Arguably, the most popular contrast psychologists and psycholinguists are interested in is the comparison between neighboring levels of a factor. This type of contrast is called the repeated contrast. In our example, our research question might be whether the frequency level `low` leads to slower response times than frequency level `medium-low`, whether frequency level `medium-low` leads to slower response times than frequency level `medium-high`, and whether frequency level `medium-high` leads to slower response times than frequency level `high`.

Repeated contrasts are used to implement these comparisons. Consider first how to derive the contrast matrix for repeated contrasts, starting out by specifying the comparisons that are to be estimated. Importantly, this again applies the general strategy of how to translate (any) comparisons between groups or conditions into a set of contrasts, yielding a powerful tool of great value in many research settings. We follow the four-step procedure outlined above.

The first step is to specify our comparisons and to write them down in a way such that their weights can be extracted easily. For a four-level factor, the three comparisons are:

$$\beta_{2-1} = -1 \cdot \mu_1 + 1 \cdot \mu_2 + 0 \cdot \mu_3 + 0 \cdot \mu_4 \tag{6.22}$$

$$\beta_{3-2} = 0 \cdot \mu_1 - 1 \cdot \mu_2 + 1 \cdot \mu_3 + 0 \cdot \mu_4 \tag{6.23}$$

$$\beta_{4-3} = 0 \cdot \mu_1 + 0 \cdot \mu_2 - 1 \cdot \mu_3 + 1 \cdot \mu_4 \tag{6.24}$$

Here, the μ_x are the mean response times in condition x. Each regression coefficient gives weights to the different condition means. For example, the first estimate (β_{2-1}) estimates the difference between condition mean for $F2$ (μ_2) minus the condition mean for $F1$ (μ_1), but ignores condition means for $F3$ and $F4$ (μ_3, μ_4). μ_1 has a weight of -1, μ_2 has a weight of $+1$, and μ_3 and μ_4 have weights of 0.

We can write these comparisons into `hypr`:

```
HcRep <- hypr(c2vs1 = F2 ~ F1,
              c3vs2 = F3 ~ F2,
              c4vs3 = F4 ~ F3,
              levels = c("F1", "F2", "F3", "F4"))
HcRep
```

```
## hypr object containing 3 null hypotheses:
## H0.c2vs1: 0 = F2 - F1
## H0.c3vs2: 0 = F3 - F2
## H0.c4vs3: 0 = F4 - F3
##
## Call:
## hypr(c2vs1 = ~F2 - F1, c3vs2 = ~F3 - F2, c4vs3 = ~F4 - F3, levels = c("F1",
## "F2", "F3", "F4"))
##
## Hypothesis matrix (transposed):
##    c2vs1 c3vs2 c4vs3
## F1 -1     0     0
## F2  1    -1     0
## F3  0     1    -1
## F4  0     0     1
##
## Contrast matrix:
##    c2vs1 c3vs2 c4vs3
## F1 -3/4  -1/2  -1/4
## F2  1/4  -1/2  -1/4
## F3  1/4   1/2  -1/4
## F4  1/4   1/2   3/4
```

The hypothesis matrix shows exactly the weights that we had written down above. Moreover, we see the contrast matrix. In the case of the repeated contrast, the contrast matrix again looks very different from the hypothesis matrix. In this case, the contrast matrix looks a lot less intuitive than the hypothesis matrix, and if one did not know the associated hypothesis matrix, it seems unclear what the contrast matrix would actually test. To verify this custom-made contrast matrix, we compare it to the repeated contrast matrix as generated by the R function `contr.sdif()` in the `MASS` package (Ripley 2023). The resulting contrast matrix is identical to our result:

```
fractions(contr.sdif(4))
```

```
##    2-1  3-2  4-3
## 1 -3/4 -1/2 -1/4
## 2  1/4 -1/2 -1/4
## 3  1/4  1/2 -1/4
## 4  1/4  1/2  3/4
```

We can thus use either approach (`hypr()` or `contr.sdif()`) to obtain the contrast matrix in this case. Next, we apply the repeated contrasts to the factor F in the example data and run a linear model. This allows us to estimate the regression coefficients associated with each contrast. These are compared to the data in Figure 6.2 to test whether the regression coefficients actually correspond to the differences between successive condition means, as intended.

```
contrasts(df_contrasts3$F) <- contr.hypothesis(HcRep)
fit_Rep <- brm(DV ~ F,
               data = df_contrasts3,
               family = gaussian(),
               prior = c(prior(normal(20, 50), class = Intercept),
                         prior(normal(0, 50), class = sigma),
                         prior(normal(0, 50), class = b)))
```

```
fixef(fit_Rep)
```

```
##           Estimate Est.Error  Q2.5 Q97.5
## Intercept    19.99      2.45  15.2 24.86
## Fc2vs1        9.65      6.84  -4.3 22.91
## Fc3vs2       -9.31      7.01 -22.9  5.06
## Fc4vs3       29.25      6.94  15.4 43.15
```

The results show that as expected, the regression coefficients reflect the differences that were of interest: the regression coefficient (`Estimate`) `Fc2vs1` has a value of approximately 10, which corresponds to the difference between the condition mean for $F2$ (20) minus the condition mean for $F1$ (10), i.e., $20 - 10 = 10$. Likewise, the regression coefficient `Fc3vs2` has a value of approximately -10, which corresponds to the difference between the condition mean for $F3$ (10) minus the condition mean for $F2$ (20), i.e., $10 - 20 = -10$. Finally, the regression coefficient `Fc4vs3` has a value of roughly 30, which reflects the difference between condition $F4$ (40) minus condition $F3$ (10), i.e., $40 - 10 = 30$. Thus, the regression coefficients estimate differences between successive or neighboring condition means.

6.3.2 Helmert contrasts

Another common contrast is the Helmert contrast. In a Helmert contrast for our four-level factor, the first contrast compares level $F1$ versus $F2$. The second contrast compares level $F3$ to the average of the first two, i.e., `F3 ~ (F1+F2)/2`. The third contrast then compares level $F4$ to the average of the first three. We can code this contrast in `hypr`:

```
HcHel <- hypr(b1 = F2 ~ F1,
              b2 = F3 ~ (F1 + F2) / 2,
              b3 = F4 ~ (F1 + F2 + F3) / 3,
              levels = c("F1", "F2", "F3", "F4"))
HcHel
```

```
## hypr object containing 3 null hypotheses:
## H0.b1: 0 = F2 - F1
## H0.b2: 0 = F3 - 1/2*F1 - 1/2*F2
## H0.b3: 0 = F4 - 1/3*F1 - 1/3*F2 - 1/3*F3
##
## Call:
## hypr(b1 = ~F2 - F1, b2 = ~F3 - 1/2 * F1 - 1/2 * F2, b3 = ~F4 -
##     1/3 * F1 - 1/3 * F2 - 1/3 * F3, levels = c("F1", "F2", "F3",
## "F4"))
##
## Hypothesis matrix (transposed):
##     b1   b2   b3
## F1   -1 -1/2 -1/3
## F2    1 -1/2 -1/3
## F3    0    1 -1/3
## F4    0    0    1
##
## Contrast matrix:
##     b1   b2   b3
## F1 -1/2 -1/3 -1/4
## F2  1/2 -1/3 -1/4
## F3    0  2/3 -1/4
## F4    0    0  3/4
```

The classical Helmert contrast coded by the function `contr.helmert()` yields a similar but slightly different result:

```
contr.helmert(4)
```

```
##   [,1] [,2] [,3]
## 1   -1   -1   -1
## 2    1   -1   -1
## 3    0    2   -1
## 4    0    0    3
```

These contrasts are scaled versions of our custom Helmert contrast, i.e., the first column of our custom Helmert contrast has to be multiplied by 2 to get the classical version, the second column has to be multiplied by 3, and the fourth column has to be multiplied by 4 to get to our custom Helmert contrast. In our opinion, the custom Helmert contrast defined using the `hypr` function is more appropriate and intuitive to use. Probably the only reason the classical Helmert contrast uses these scaled differences is that the rescaling yields an easier contrast matrix, which consists of integers rather than fractions. The estimates from our custom Helmert contrast seem much more relevant in Bayesian approaches today. This is because in our custom Helmert contrast, the coefficients estimate the differences between (groups of) conditions (rather than scaled versions of these differences). That means that we can set the priors intuitively (as priors on differences between (groups of) conditions—rather than on scaled differences) and that we can accordingly interpret the posterior much more straightforwardly. We apply the custom Helmert contrast here:

```
contrasts(df_contrasts3$F) <- contr.hypothesis(HcHel)
fit_Hel <- brm(DV ~ F,
               data = df_contrasts3,
               family = gaussian(),
               prior = c(prior(normal(20, 50), class = Intercept),
                         prior(normal(0, 50), class = sigma),
                         prior(normal(0, 50), class = b)))
```

```
fixef(fit_Hel)
```

```
##           Estimate Est.Error   Q2.5 Q97.5
## Intercept    19.95      2.43  15.08 24.76
## Fb1           9.83      6.87  -4.21 23.28
## Fb2          -4.85      6.03 -16.52  6.97
```

```
## Fb3          26.31      5.57  14.90 37.39
```

When we fit the Bayesian model using our custom Helmert contrast, we can see that the estimates reflect the comparisons outlined above. The first estimate `Fb1` has a value of roughly 10, reflecting the difference between conditions $F1$ and $F2$. The second estimate `Fb2` has a value of 5, which reflects the difference between condition $F3$ (10) and the average of the first two conditions ($(10 + 20)/2 = 15$). The estimate `Fb3` reflects the difference between $F4(40)$ and the average of the first three, which is $(10 + 20 + 10)/3 = 13.3$, and is thus $40 - 13.3 = 26.7$.

Here, we can see that the generalized matrix inverse can be used to create contrasts that are not exactly equivalent to the ones provided by R per default (i.e., here `contr.helmert()`). Another example for this is discussed in online section A.8, which implements a treatment contrast, where the intercept captures the grand mean.

6.3.3 Contrasts in linear regression analysis: The design or model matrix

We have discussed how different contrasts are created from the hypothesis matrix. What we have not treated in detail is how exactly contrasts are used in a linear model. Here, we will see that the contrasts for a factor in a linear model are just the same thing as continuous numeric predictors (i.e., covariates) in a linear/multiple regression analysis. That is, contrasts are the way to encode discrete factor levels into numeric predictor variables to use in linear/multiple regression analysis, by encoding which differences between factor levels are estimated. The contrast matrix X_c that we have looked at so far has one entry (row) for each experimental condition. For use in a linear model, the contrast matrix is coded into a design or model matrix X, where each individual data point has one row. The design matrix X can be extracted using the function `model.matrix()`:

```
# contrast matrix:
(contrasts(df_contrasts3$F) <- contr.hypothesis(HcRep))
```

```
##    c2vs1 c3vs2 c4vs3
## F1 -0.75  -0.5 -0.25
## F2  0.25  -0.5 -0.25
## F3  0.25   0.5 -0.25
## F4  0.25   0.5  0.75
## attr(,"class")
## [1] "hypr_cmat" "matrix"    "array"
```

```
# design/model matrix:
covars <- model.matrix(~ 1 + F, df_contrasts3)
(covars <- as.data.frame(covars))
```

```
##    (Intercept) Fc2vs1 Fc3vs2 Fc4vs3
## 1            1  -0.75   -0.5  -0.25
## 2            1  -0.75   -0.5  -0.25
## 3            1  -0.75   -0.5  -0.25
## 4            1  -0.75   -0.5  -0.25
## 5            1  -0.75   -0.5  -0.25
## 6            1   0.25   -0.5  -0.25
## 7            1   0.25   -0.5  -0.25
## 8            1   0.25   -0.5  -0.25
## 9            1   0.25   -0.5  -0.25
## 10           1   0.25   -0.5  -0.25
## 11           1   0.25    0.5  -0.25
## 12           1   0.25    0.5  -0.25
## 13           1   0.25    0.5  -0.25
## 14           1   0.25    0.5  -0.25
## 15           1   0.25    0.5  -0.25
## 16           1   0.25    0.5   0.75
## 17           1   0.25    0.5   0.75
## 18           1   0.25    0.5   0.75
## 19           1   0.25    0.5   0.75
## 20           1   0.25    0.5   0.75
```

For each of the 20 subjects, four numbers are stored in this model matrix. They represent the three values of three predictor variables used to predict response times in the task. Indeed, this matrix is exactly the design matrix X commonly used in multiple regression analysis, where each column represents one numeric predictor variable (covariate), and the first column codes the intercept term.

To further illustrate this, the covariates are extracted from this design matrix and stored separately as numeric predictor variables in the data frame:

```
df_contrasts3[, c("Fc2vs1", "Fc3vs2", "Fc4vs3")] <- covars[, 2:4]
```

They are now used as numeric predictor variables in a multiple regression analysis:

```
fit_m3 <- brm(DV ~ Fc2vs1 + Fc3vs2 + Fc4vs3,
              data = df_contrasts3,
              family = gaussian(),
              prior = c(prior(normal(20, 50), class = Intercept),
                        prior(normal(0, 50), class = sigma),
                        prior(normal(0, 50), class = b)))
```

```
fixef(fit_m3)
```

```
##           Estimate Est.Error   Q2.5 Q97.5
## Intercept    19.95      2.42  15.22  24.7
## Fc2vs1        9.91      6.88  -3.44  23.2
## Fc3vs2       -9.51      6.86 -23.14   3.9
## Fc4vs3       29.44      6.96  15.45  43.0
```

The results show that the regression coefficients are the same as in the contrast-based analysis shown in the previous section (on repeated contrasts). This demonstrates that contrasts serve to code discrete factor levels into a linear/multiple regression analysis by numerically encoding comparisons between specific condition means, or groups of condition means.

6.3.4 Polynomial contrasts

Polynomial contrasts are another option for analyzing factors. Suppose that we expect a linear trend across conditions, where the response increases by a constant magnitude with each successive factor level. This could be the expectation when four levels of a factor reflect decreasing levels of word frequency (i.e., four factor levels: high, medium-high, medium-low, and low word frequency), where one expects the fastest response for high-frequency words, and successively slower responses for lower word frequencies. The effect for each individual level of a factor (e.g., as coded via a repeated contrast) may not be strong enough for detecting it in the statistical model. Specifying a linear trend in a polynomial contrast (see effect `F.L` below) allows us to pool the whole increase (across all four factor levels) into a single coefficient for the linear trend, increasing statistical sensitivity for estimating/detecting the increase. Such a specification constrains the estimate to one interpretable parameter, e.g., a linear increase across factor levels. The larger the number of factor levels, the more parsimonious can polynomial contrasts be compared to contrast-based specifications as introduced in the previous sections. Going beyond a linear trend, one may also have expectations about quadratic trends (see the estimate for `F.Q` below). For example, one may expect an increase only among very low frequency words, but no difference between high and medium-high frequency words.

Here is an example of how to code polynomial contrasts for a four-level factor. In this case, one can estimate a linear (`F.L`), a quadratic (`F.Q`), and a cubic (`F.C`) trend. If more factor levels are present, then higher order trends can be estimated.

```
Xpol <- contr.poly(4)
(contrasts(df_contrasts3$F) <- Xpol)
```

```
##            .L   .Q     .C
## [1,] -0.671  0.5 -0.224
## [2,] -0.224 -0.5  0.671
## [3,]  0.224 -0.5 -0.671
## [4,]  0.671  0.5  0.224
```

```
fit_Pol <- brm(DV ~ F,
               data = df_contrasts3,
               family = gaussian(),
               prior = c(prior(normal(20, 50), class = Intercept),
                         prior(normal(0, 50), class = sigma),
                         prior(normal(0, 50), class = b)))
```

```
fixef(fit_Pol)
```

```
##           Estimate Est.Error  Q2.5 Q97.5
## Intercept    20.01      2.45 15.21  25.0
## F.L          17.73      4.86  8.17  27.1
## F.Q           9.89      4.88  0.01  19.2
## F.C          13.41      4.97  3.37  23.3
```

In this example, condition means increase across factor levels in a linear fashion: if one draws a linear regression line through all condition means, this line will increase toward the right. However, there may also be quadratic and cubic trends.

However, one difficulty with the `contr.poly()` function is that the scaling of the predictor variables is not very clear, making it difficult to define priors. Therefore, an alternative can be to use custom polynomial contrasts:

```
Xpol2            <- data.frame(linear = c(F1 = -3,
                                          F2 = -1,
                                          F3 = 1,
                                          F4 = 3) / 2)
Xpol2$quadratic <- Xpol2$linear^2 - mean(Xpol2$linear^2)
Xpol2$cubic     <- Xpol2$linear^3 - mean(Xpol2$linear^3)
```

Because the linear and cubic trends are highly correlated, we orthogonalize the two:

```
Xpol2$cubic <- resid(lm(cubic ~ linear, Xpol2))
(contrasts(df_contrasts3$F) <- as.matrix(Xpol2))
```

```
##    linear quadratic cubic
## F1   -1.5         1  -0.3
## F2   -0.5        -1   0.9
## F3    0.5        -1  -0.9
## F4    1.5         1   0.3
```

```
fit_Pol2 <- brm(DV ~ F,
                data = df_contrasts3,
                family = gaussian(),
                prior = c(prior(normal(20, 50), class = Intercept),
                          prior(normal(0, 50), class = sigma),
                          prior(normal(0, 50), class = b)))
```

```
fixef(fit_Pol2)
```

```
##            Estimate Est.Error  Q2.5 Q97.5
## Intercept     20.02      2.39 15.32 24.85
## Flinear        7.98      2.20  3.58 12.28
## Fquadratic     4.95      2.46  0.08  9.89
## Fcubic         9.96      3.65  2.88 17.00
```

The effects are now on interpretable scales.

6.3.5 An alternative to contrasts: Monotonic effects

An alternative to specifying contrasts to estimate specific comparisons between factor levels is monotonic effects (`https://paul-buerkner.github.io/brms/articles/brms_monotonic.html`; Bürkner and Charpentier 2020). This simply assumes that the dependent variable increases (or decreases) in a monotonic fashion across levels of an ordered factor. In this kind of analysis, one does not define contrasts specifying differences between (groups of) factor levels. Instead, one estimates one parameter which captures the average increase (or decrease) in the dependent variable associated with two neighboring factor levels. Moreover, one estimates the percentages of the overall increase (or decrease) that is associated with each of the differences between neighboring factor levels (i.e., similar to simple difference contrasts, but measured in percentage increase, and assuming monotonicity, i.e., that the same increase or decrease is present for all simple differences).

To implement a monotonic analysis, we first code the factor F as being an ordered factor (i.e., `ordered = TRUE`; remember that the default ordering is alphabetical and may need to be changed). Then, we specify that we want to estimate a monotonic effect of F using the notation `mo(F)` in our call to `brms`:

```
df_contrasts3$F <- factor(df_contrasts3$F, ordered = TRUE)
fit_mo <- brm(DV ~ 1 + mo(F),
              data = df_contrasts3,
              family = gaussian(),
              prior = c(prior(normal(20, 50), class = Intercept),
                        prior(normal(0, 50), class = sigma),
                        prior(normal(0, 50), class = b)))
```

```
fit_mo
```

```
## ...
## Population-Level Effects:
##           Estimate Est.Error l-95% CI u-95% CI Rhat Bulk_ESS Tail_ESS
## Intercept     9.33      4.26     0.38    17.12 1.00     2161     2181
## moF           9.49      2.45     4.29    14.26 1.00     2011     1933
##
## Simplex Parameters:
##          Estimate Est.Error l-95% CI u-95% CI Rhat Bulk_ESS Tail_ESS
## moF1[1]      0.20      0.14     0.01     0.54 1.00     2299     1851
## moF1[2]      0.11      0.11     0.00     0.42 1.00     2552     1865
## moF1[3]      0.69      0.17     0.27     0.94 1.00     2472     1976
##
## Family Specific Parameters:
##        Estimate Est.Error l-95% CI u-95% CI Rhat Bulk_ESS Tail_ESS
## sigma     11.77      2.31     8.32    17.31 1.00     2157     2051
```

```
##
## ...
```

The results show that there is an overall positive population-level effect of the factor F with an estimate (`moF`) of 9.49, reflecting an average increase in the dependent variables of 9.49 with each level of F. The model summary shows estimates for the simplex parameters, which represent the ratios of the overall increase associated with F that can be attributed to each of the differences between neighboring factor levels. The results show that most of the increase is associated with `moF1[3]`, i.e., with the last difference, reflecting the difference between $F3$ and $F4$, whereas the other two differences (`moF1[1]`, reflecting the difference between $F1$ and $F2$, and `moF1[2]`, reflecting the difference between $F2$ and $F3$) are smaller. Comparing conditional effects (i.e., the condition means estimated by the model) between a model using polynomial contrasts and a model assuming monotonic effects makes it clear that the current model "forces" the effects to increase in a monotonic fashion; see Figure 6.3.

```
ppoly <- conditional_effects(fit_Pol)
pmon  <- conditional_effects(fit_mo)
plot(ppoly, plot = FALSE)[[1]] + ggtitle("Polynomial contrasts")
plot(pmon, plot = FALSE)[[1]] + ggtitle("Monotonic effects")
```

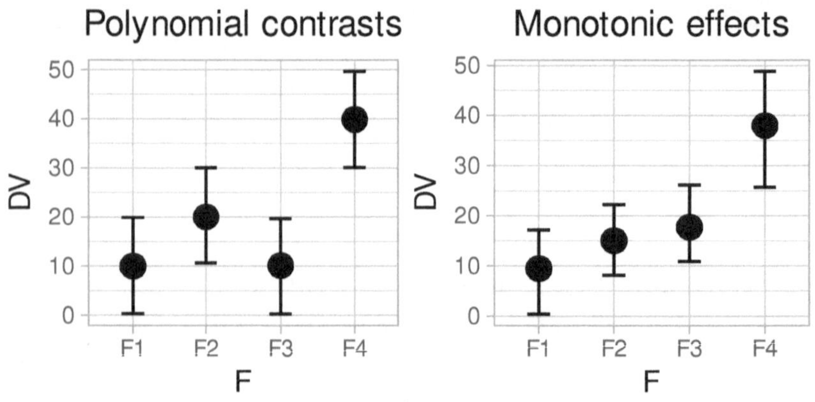

FIGURE 6.3: Conditional effects using the polynomial contrasts on the left side vs. assuming monotonic effects on the right side.

This is regardless of the information provided in the data; see the posterior predictive checks in Figure 6.4.

The monotonicity assumption is violated in the current data set, since the mean is larger in condition $F2$ than in condition $F3$. The monotonic model thus assumes this (negative) difference is due to chance; see Figure 6.4.

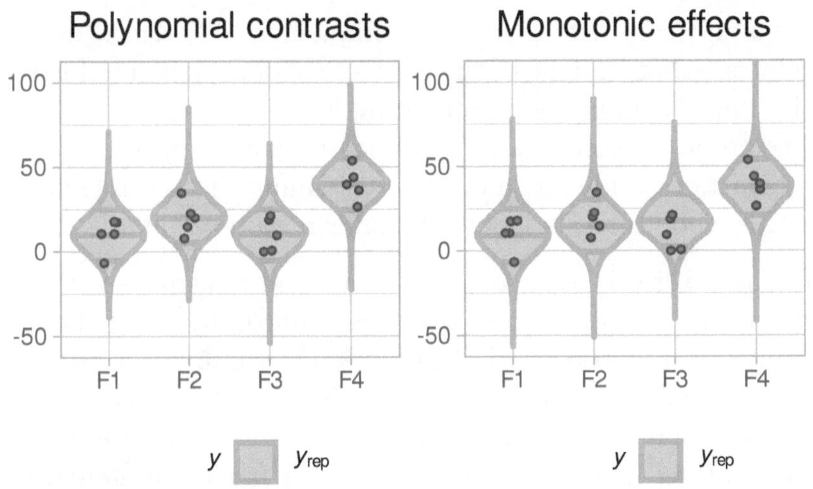

FIGURE 6.4: Posterior predictive distributions by condition using the polynomial contrasts on the left side vs. assuming monotonic effects on the right side.

Estimating such monotonic effects provides an alternative to the contrast coding we treat in the rest of this chapter. It may be relevant when the specific differences between factor levels are not of interest, but when instead the goal is to estimate the overall monotonic effect of a factor and when this overall effect is not well approximated by a simple linear trend.

6.4 What makes a good set of contrasts?

For a factor with I levels one can make only $I - 1$ linearly independent comparisons within a single model. For example, in a design with one factor with two levels, only one comparison is possible (between the two factor levels). The reason for this is that the intercept is also estimated. More generally, if we have a factor with I_1 and another factor with I_2 levels, then the total number of conditions is $I_1 \times I_2 = \nu$ (not $I_1 + I_2$), which implies a maximum of $\nu - 1$ contrasts.

For example, in a design with one factor with three levels, A, B, and C, in principle one could make three comparisons (A vs. B, A vs. C, B vs. C). However, after defining an intercept, only two means can be compared (or two linear combinations of means). Therefore, for a factor with three levels, we define two comparisons within one statistical model.

One critical precondition for contrasts is that they implement different comparisons that are not collinear; that is, none of the contrasts can be generated from the other contrasts by linear combination. For example, the contrast `c1 = c(1, 2, 3)` can be generated from the contrast `c2 = c(3, 4, 5)` simply by computing `c2 - 2`. Therefore, contrasts c1 and c2 cannot be used simultaneously. That is, each contrast needs to encode some independent information about the data.

There are (at least) three criteria to decide what a good contrast is. First, *orthogonal contrasts* have advantages as they estimate mutually independent comparisons in the data (see Dobson and Barnett 2011, sec. 6.2.5, p. 91 for a detailed explanation of orthogonality). Second, it is crucial that contrasts are defined in a way such that they answer the research questions. One way to accomplish this second point is to use the hypothesis matrix to generate contrasts (e.g., via the `hypr` package), as this ensures that one uses contrasts that exactly estimate the comparisons of interest in a given study. Third, in the Bayesian context, the scaling of contrasts is important, and has to be considered when defining priors: the scaling of the priors has to be tailored to the scaling of each individual contrasts. In situations where we define the same prior for all coefficients of a factor, the scaling of the contrasts has to be set accordingly.

6.4.1 Centered contrasts

Contrasts are often constrained to be centered, such that the individual contrast coefficients c_i for different factor levels i sum to 0: $\sum_{i=1}^{I} c_i = 0$. This has advantages when estimating interactions with other factors or covariates (we discuss interactions between factors in the next chapter). All contrasts discussed here are centered except for the treatment contrast, in which the contrast coefficients for each contrast do not sum to zero:

```
colSums(contr.treatment(4))
```

```
## 2 3 4
## 1 1 1
```

Other contrasts, such as repeated contrasts, are centered and the contrast coefficients for each contrast sum to 0:

```
colSums(contr.sdif(4))
```

```
## 2-1 3-2 4-3
##   0   0   0
```

The contrast coefficients mentioned above appear in the contrast matrix. The weights in the hypothesis matrix are always centered. This is also true for the treatment contrast. The reason is that they code comparisons between conditions or bundles of conditions. The only exception are the weights for the intercept, which are all the same and together always sum to 1 in the hypothesis matrix. This is done to ensure that when applying the generalized matrix inverse, the intercept results in a constant term with values of 1 in the contrast matrix. An important question concerns whether (or when) the intercept needs to be considered in the generalized matrix inversion and whether (or when) it can be ignored. This question is closely related to orthogonal contrasts, a concept we turn to below.

6.4.2 Orthogonal contrasts

Two centered contrasts c_1 and c_2 are orthogonal to each other if the following condition applies. Here, i is the i-th cell of the vector representing the contrast.

$$\sum_{i=1}^{I} c_{1,i} \cdot c_{2,i} = 0 \tag{6.25}$$

Whether contrasts are orthogonal can often be determined easily by computing the correlation between two contrasts. Orthogonal contrasts have a correlation of 0.

For example, coding two factors in a 2×2 design (we return to this case in a section on designs with two factors below) using sum contrasts, these sum contrasts and their interaction are orthogonal to each other:

```
(Xsum <- cbind(F1 = c(1, 1, -1, -1),
               F2 = c(1, -1, 1, -1),
               F1xF2 = c(1, -1, -1, 1)))
```

```
##      F1 F2 F1xF2
## [1,]  1  1     1
## [2,]  1 -1    -1
## [3,] -1  1    -1
## [4,] -1 -1     1
```

```
cor(Xsum)
```

```
##       F1 F2 F1xF2
## F1     1  0     0
```

```
## F2     0 1    0
## F1xF2  0 0    1
```

The correlations between the different contrasts (i.e., the off-diagonals) are exactly 0. Notice that sum contrasts coding one multi-level factor are not orthogonal to each other:

```
cor(contr.sum(4))
```

```
##      [,1] [,2] [,3]
## [1,]  1.0  0.5  0.5
## [2,]  0.5  1.0  0.5
## [3,]  0.5  0.5  1.0
```

Here, the correlations between individual contrasts, which appear in the off-diagonals, deviate from 0, indicating non-orthogonality. The same is also true for treatment and repeated contrasts:

```
cor(contr.sdif(4))
```

```
##       2-1   3-2   4-3
## 2-1 1.000 0.577 0.333
## 3-2 0.577 1.000 0.577
## 4-3 0.333 0.577 1.000
```

```
cor(contr.treatment(4))
```

```
##        2      3      4
## 2  1.000 -0.333 -0.333
## 3 -0.333  1.000 -0.333
## 4 -0.333 -0.333  1.000
```

Orthogonality of contrasts plays a critical role when computing the generalized inverse. In the inversion operation, orthogonal contrasts are converted independently from each other. That is, the presence or absence of another orthogonal contrast does not change the resulting weights. In fact, for orthogonal contrasts, applying the generalized matrix inverse to the hypothesis matrix simply furnishes a scaled version of the hypothesis matrix in the contrast matrix (for mathematical details see Schad et al. 2020).

In Bayesian models, scaling is always important, since we need to interpret the scale in order to define priors or interpret posteriors. Therefore, when working with contrasts in Bayesian models, the generalized matrix inverse is always a good procedure to use.

6.4.3 The role of the intercept in non-centered contrasts

A related question concerns whether the intercept needs to be considered when computing the generalized inverse for a contrast. This is of key importance when using the generalized matrix inverse to define contrasts: the resulting contrast matrix and also the definition of estimates can completely change between a situation where the intercept is explicitly considered or not considered, and can thus change the resulting estimates in possibly unintended ways. Thus, if the definition of the intercept is incorrect, the estimates of slopes may also be wrong.

More specifically, it turns out that considering the intercept is necessary for contrasts that are not centered. This is the case for treatment contrasts which are not centered; e.g., the treatment contrast for two factor levels `c1vs0 = c(0, 1)`: $\sum_i c_i = 0 + 1 = 1$. One can actually show that the formula to determine whether contrasts are centered (i.e., $\sum_i c_i = 0$) is the same formula as the formula to test whether a contrast is "orthogonal to the intercept." Remember that for the intercept, all contrast coefficients are equal to one: $c_{1,i} = 1$ (here, $c_{1,i}$ indicates the vector of contrast coefficients associated with the intercept). We enter these contrast coefficient values into the formula testing whether a contrast is orthogonal to the intercept (here, $c_{2,i}$ indicates the vector of contrast coefficients associated with some contrast for which we want to test whether it is "orthogonal to the intercept"): $\sum_i c_{1,i} \cdot c_{2,i} = \sum_i 1 \cdot c_{2,i} = \sum_i c_{2,i} = 0$. The resulting formula is: $\sum_i c_{2,i} = 0$, which is exactly the constraint that needs to be satisfied for a contrast to be centered. Because of this analogy, treatment contrasts can be viewed to be "not orthogonal to the intercept." This means that the intercept needs to be considered when computing the generalized inverse for treatment contrasts. As we have discussed above, when the intercept is included in the hypothesis matrix, the weights for this intercept term should sum to one, as this yields a column of ones for the intercept term in the contrast matrix.

We can see that considering the intercept makes a difference for the treatment contrast. First, we define the comparisons involved in a treatment contrast, where two experimental conditions `b` and `c` are each compared to a baseline condition `a` (`b~a` and `c~a`). In addition, we explicitly code the intercept term, which involves a comparison of the baseline to 0 (`a~0`). We take a look at the resulting contrast matrix:

```
hypr(int = a ~ 0, b1 = b ~ a, b2 = c ~ a)
```

```
## hypr object containing 3 null hypotheses:
## H0.int: 0 = a      (Intercept)
##  H0.b1: 0 = b - a
##  H0.b2: 0 = c - a
##
## Call:
## hypr(int = ~a, b1 = ~b - a, b2 = ~c - a, levels = c("a", "b",
## "c"))
##
## Hypothesis matrix (transposed):
##   int b1 b2
## a  1  -1 -1
## b  0   1  0
## c  0   0  1
##
## Contrast matrix:
##   int b1 b2
## a 1   0  0
## b 1   1  0
## c 1   0  1
```

```
contr.treatment(c("a", "b", "c"))
```

```
##   b c
## a 0 0
## b 1 0
## c 0 1
```

This shows a contrast matrix that we know from the treatment contrast. The intercept is coded as a column of ones. And each of the comparisons is coded as a 1 in the condition which is compared to the baseline, and a 0 in other conditions. The point here is that this gives us the contrast matrix that is expected and known for the treatment contrast.

We can also ignore the intercept in the specification of the comparisons:

```
hypr(b1 = m1 ~ m0, b2 = m2 ~ m0)
```

```
## hypr object containing 2 null hypotheses:
```

```
## H0.b1: 0 = m1 - m0
## H0.b2: 0 = m2 - m0
##
## Call:
## hypr(b1 = ~m1 - m0, b2 = ~m2 - m0, levels = c("m0", "m1", "m2"
## ))
##
## Hypothesis matrix (transposed):
##    b1 b2
## m0 -1 -1
## m1  1  0
## m2  0  1
##
## Contrast matrix:
##    b1   b2
## m0 -1/3 -1/3
## m1  2/3 -1/3
## m2 -1/3  2/3
```

Notice that the resulting contrast matrix now looks very different from the contrast matrix that we know from the treatment contrast. Indeed, this contrast also estimates a reasonable set of quantities. It again estimates whether the condition mean `m1` differs from the baseline and whether `m2` differs from baseline. However, the intercept now estimates the average dependent variable across all three conditions (i.e., the grand mean). This can be seen by explicitly adding a comparison of the average of all three conditions to 0:

```
hypr(int = (m0 + m1 + m2) / 3 ~ 0, b1 = m1 ~ m0, b2 = m2 ~ m0)
```

```
## hypr object containing 3 null hypotheses:
## H0.int: 0 = (m0 + m1 + m2)/3  (Intercept)
##  H0.b1: 0 = m1 - m0
##  H0.b2: 0 = m2 - m0
##
## Call:
## hypr(int = ~1/3 * m0 + 1/3 * m1 + 1/3 * m2, b1 = ~m1 - m0, b2 = ~m2 -
##     m0, levels = c("m0", "m1", "m2"))
##
## Hypothesis matrix (transposed):
##    int b1  b2
## m0 1/3  -1  -1
## m1 1/3   1   0
```

```
## m2 1/3   0   1
##
## Contrast matrix:
##       int  b1   b2
## m0      1 -1/3 -1/3
## m1      1  2/3 -1/3
## m2      1 -1/3  2/3
```

The last two columns of the resulting contrast matrix are now the same as when the intercept was ignored, which confirms that the two columns encode the same comparison.

6.5 Computing condition means from estimated contrasts

As mentioned earlier, one advantage of Bayesian modeling is that based on the posterior samples, it is possible to very flexibly compute new comparisons and estimates. Above (see section 6.1.4), we had discussed the case where the Bayesian model estimated the condition means instead of contrasts by removing the intercept from the `brms` model (the formula in `brms` was: `DV ~ -1 + F`). This allowed us to get posterior samples from each condition mean, and then to compute any possible comparison between condition means by subtracting the corresponding samples.

Importantly, posterior samples for the condition means can also be obtained after fitting a model with contrasts. We illustrate this here for the case of sum contrasts. Let's use our above example of a design where we assess response times (in milliseconds, `DV`) for three different word classes adjectives, nouns, and verbs, that is, for a 3-level factor ***F***. In the above example, factor ***F*** was coded using a sum contrast, where the first contrast coded the difference of adjectives from the grand mean, and the second contrast coded the difference of nouns from the grand mean. This was the corresponding contrast matrix:

```
contrasts(df_contrasts2$F) <- contr.hypothesis(HcSum)
contrasts(df_contrasts2$F)
```

```
##            b1 b2
## adjectives  1  0
## nouns       0  1
## verbs      -1 -1
```

We had estimated a `brms` model for this data. The posterior estimates show results for the intercept (which is estimated to be 450 ms) and for our two coded comparisons. The effect `FcH01` codes our first comparison that response times for adjectives differ from the grand mean, and show an estimate that response times for adjectives are about 50 ms shorter than the grand mean. Moreover, the effect `FcH02` codes our second comparison that response times for nouns differ from the grand mean, and show the estimate that response times for nouns are 50 ms longer than the grand mean.

```
fixef(fit_Sum)
```

```
##           Estimate Est.Error  Q2.5 Q97.5
## Intercept    450.4      6.97 436.7 464.3
## FcH01        -49.2      9.75 -68.4 -29.8
## FcH02         49.1     10.05  29.4  70.0
```

Of course other comparisons might be of interest to us as well. For example, we might be interested in estimating how strongly response times for verbs differ from the grand mean.

To do so, one possible first step is to obtain the posteriors for the response times in each of the three conditions. How can this be done? The first step is to again extract the posterior samples from the model:

```
df_postSamp_Sum <- as_draws_df(fit_Sum)
```

We can see the samples for our first contrast (`b_FcH01`) and for our second contrast (`b_FcH02`). How can we now compute the posterior samples for each of the condition means, i.e., for adjectives, nouns, and verbs? For this, we need to take another look at the contrast matrix.

```
contrasts(df_contrasts2$F)
```

```
##            b1 b2
## adjectives  1  0
## nouns       0  1
## verbs      -1 -1
```

It tells us how the condition means are computed. For adjectives (see the first row of the contrast matrix), we can see that the response time is computed by taking 1 times the coefficient for `b1` (i.e., `FcH01`) and 0 times the coefficient for `b2` (i.e., `FcH02`). Thus, response times for adjectives are simply the samples for the `b1` (i.e., `FcH01`) contrast. The contrast matrix does not show the intercept term, which is implicitly added. Thus, we also have to add the estimates for the intercept. Thus, the condition mean for adjectives is computed as follows:

```
df_postSamp_Sum <- df_postSamp_Sum %>%
  mutate(b_adjectives = b_Intercept + b_FcH01)
```

Similarly, we can obtain the posterior samples for the response times for nouns. The computation can be seen from the second row of the contrast matrix, which shows that the contrast `b1` (i.e., `FcH01`) has weight 0 times, whereas the contrast `b2` (i.e., `FcH02`) has weight 1. Adding the intercept thus gives:

```
df_postSamp_Sum <- df_postSamp_Sum %>%
  mutate(b_nouns = b_Intercept + b_FcH02)
```

Finally, we want to obtain posterior samples for the average response times for verbs. For verbs, the third row of the contrast matrix shows two times a -1. Thus, contrasts `b1` (i.e., `FcH01`) and `b2` (i.e., `FcH02`) have to be subtracted from the intercept:

```
df_postSamp_Sum$b_verbs <-
  df_postSamp_Sum$b_Intercept - df_postSamp_Sum$b_FcH01 -
  df_postSamp_Sum$b_FcH02
```

This yields posterior samples for the mean response times for verbs.

We can now look at the posterior means and 95% credible intervals for adjectives, nouns, and verbs by computing the means and quantiles across all computed samples.

```
postTab <- df_postSamp_Sum %>%
  # remove the meta-data:
  as.data.frame() %>%
  select(b_adjectives, b_nouns, b_verbs) %>%
  # transform from wide to long with tidyr:
  pivot_longer(cols = everything(),
               names_to = "condition",
               values_to = "samp") %>%
  group_by(condition) %>%
  summarize(post_mean = mean(samp),
            `2.5%` = quantile(samp, p = 0.025),
            `97.5%` = quantile(samp, p = 0.975))
```

```
postTab
```

```
## # A tibble: 3 x 4
##   condition    post_mean `2.5%` `97.5%`
##   <chr>            <dbl>  <dbl>   <dbl>
## 1 b_adjectives      401.   378.    425.
## 2 b_nouns           499.   475.    524.
## 3 b_verbs           451.   426.    475.
```

The results show that as expected the posterior mean for adjectives is 400 ms, for nouns it is 500 ms, and for verbs, the posterior mean is 450 ms. Moreover, we have now posterior credible intervals for each of these estimates.

In fact, `brms` has a very convenient built-in function that allows us to compute these nested effects automatically (`robust = FALSE` shows the posterior mean; by default `brms` shows the posterior median). Notice that you need to add a `[]` after the function call, otherwise `brms` will plot the results.

```
conditional_effects(fit_Sum, robust = FALSE)[]
```

```
## $F
##            F  DV cond__  effect1__ estimate__ se__ lower__ upper__
## 1 adjectives 450      1 adjectives        401 11.9     378     425
## 2      nouns 450      1      nouns        499 12.3     475     524
## 3      verbs 450      1      verbs        451 12.1     426     475
```

The same function allows us to visualize the effects, as shown in Figure 6.5.

```
conditional_effects(fit_Sum, robust = FALSE)
```

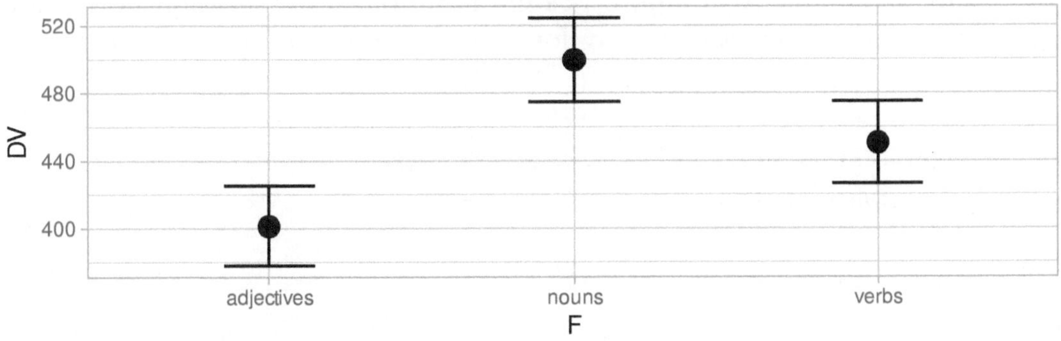

FIGURE 6.5: Estimated condition means computed from a `brms` model fitted with a sum contrast.

Coming back to our hand-crafted computations, the posterior samples can be used to compute additional comparisons. For example, we might be interested in how much response times for verbs differ from the grand mean. This can be computed based on the samples for the condition means: we first compute the grand mean from the three condition means, and then we compare this to the estimate for verbs.

```
df_postSamp_Sum <- df_postSamp_Sum %>%
  mutate(GM = (b_adjectives + b_nouns + b_verbs) / 3,
         b_FcH03 = b_verbs - GM)
c(post_mean = mean(df_postSamp_Sum$b_FcH03),
  quantile(df_postSamp_Sum$b_FcH03, p = c(0.025, 0.975)))
```

```
## post_mean      2.5%     97.5%
##     0.156   -19.714    20.205
```

The results show that reading times for verbs are quite the same as the grand mean, with a posterior mean estimate for the differences of nearly 0 ms, and with a 95% credible interval ranging between −20 and +20 ms.

The key message here is that based on the contrast matrix, it is possible to compute posterior samples for the condition means, and then to compute any arbitrary further comparisons or contrasts. We want to stress again that just obtaining the posterior distribution of a comparison does not allow us to argue that we have evidence for the effect; to argue that we have evidence for an effect being present/absent, we need Bayes factors. But the approach we outline above does allow us to obtain posterior means and credible intervals for arbitrary comparisons. However, different combinations of contrasts and priors may yield different results, especially when they are used to compute Bayes factors, due to their sensitivity for the priors.

We briefly show how to compute posterior samples for condition means for one more example contrast, namely for repeated contrasts. Here, the contrast matrix is:

```
contrasts(df_contrasts3$F) <- contr.hypothesis(HcRep)
contrasts(df_contrasts3$F)
```

```
##    c2vs1 c3vs2 c4vs3
## F1 -3/4  -1/2  -1/4
## F2  1/4  -1/2  -1/4
## F3  1/4   1/2  -1/4
## F4  1/4   1/2   3/4
```

The model estimates were:

```
fixef(fit_Rep)
```

```
##              Estimate Est.Error  Q2.5 Q97.5
## Intercept       19.99      2.45  15.2 24.86
## Fc2vs1           9.65      6.84  -4.3 22.91
## Fc3vs2          -9.31      7.01 -22.9  5.06
## Fc4vs3          29.25      6.94  15.4 43.15
```

We first obtain the posterior samples for the contrasts:

```
df_postSamp_Rep <- as.data.frame(fit_Rep)
```

Then we compute the posterior samples for condition $F1$. First, we have to add the intercept. Then, we can see in the contrast matrix that to compute the condition mean for $F1$, we have to add up all contrasts, using the weights `c(-3/4, -1/2, -1/4)` for each of the three contrasts (see first row of the contrast matrix). Thus, the posterior samples are computed as follows:

```
df_postSamp_Rep <- df_postSamp_Rep %>%
  mutate(b_F1 = b_Intercept +
           -3 / 4 * b_Fc2vs1 +
           -1 / 2 * b_Fc3vs2 +
           -1 / 4 * b_Fc4vs3)
```

The other condition means are computed correspondingly:

```
df_postSamp_Rep <- df_postSamp_Rep %>%
  mutate(b_F2 = b_Intercept +
           1 / 4 * b_Fc2vs1 +
           -1 / 2 * b_Fc3vs2 +
           -1 / 4 * b_Fc4vs3,
         b_F3 = b_Intercept +
           1 / 4 * b_Fc2vs1 +
           1 / 2 * b_Fc3vs2 +
           -1 / 4 *b_Fc4vs3,
         b_F4 = b_Intercept +
           1 / 4 * b_Fc2vs1 +
           1 / 2 * b_Fc3vs2 +
           3 / 4 * b_Fc4vs3)
```

Now we can look at the posterior means and credible intervals:

```
postTab <- df_postSamp_Rep %>%
  select(b_F1, b_F2, b_F3, b_F4) %>%
  pivot_longer(cols = everything(),
               names_to = "condition",
               values_to = "samp") %>%
  group_by(condition) %>%
  summarize(post_mean = round(mean(samp)),
            `2.5%` = round(quantile(samp, p = 0.025)),
            `97.5%` = round(quantile(samp, p = 0.975)))
```

```
print(postTab, n = 4)
```

```
## # A tibble: 4 x 4
##   condition post_mean `2.5%` `97.5%`
##   <chr>         <dbl>  <dbl>   <dbl>
## 1 b_F1             10      0      20
## 2 b_F2             20     10      30
## 3 b_F3             10      1      20
## 4 b_F4             40     30      50
```

We can verify that `brms` function return the same values:

```
conditional_effects(fit_Rep, robust = FALSE)[]
```

```
## $F
##    F DV cond__ effect1__ estimate__ se__ lower__ upper__
## 1 F1 20      1        F1       10.1 4.85   0.208    20.0
## 2 F2 20      1        F2       19.7 4.94   9.800    29.7
## 3 F3 20      1        F3       10.4 4.88   0.904    20.3
## 4 F4 20      1        F4       39.7 4.96  29.939    49.7
```

The posterior means reflect exactly the means in the data (for comparison see Figure 6.2 and Table 6.3). We now have posterior samples for each of the conditions and can compute posterior credible intervals as well as new comparisons between conditions.

6.6 Summary

Contrasts in Bayesian models work in exactly the same way as in frequentist models. Contrasts provide a way to tell the model how to code factors into numeric covariates. That is, they provide a way to define which comparisons between which condition means or bundles of condition means should be estimated in the Bayesian model. There are a number of default contrasts, like treatment contrasts, sum contrasts, repeated contrasts, and Helmert contrasts, that estimate specific comparisons between (groups of) condition means. A much more powerful procedure is to use the generalized matrix inverse, e.g., as implemented in the `hypr` package, to derive contrasts automatically after specifying the comparisons that a contrast should estimate. We have seen that in Bayesian models, it is quite straightforward to compute posterior samples for new contrasts post-hoc, after the model is fit. However, specifying precise contrasts is still of key importance when doing model comparisons (via Bayes factors) to answer the question of whether the data provide evidence for an effect of interest. If the effect of interest relates to a factor, then it has to be defined using contrast coding.

6.7 Further reading

A good discussion on contrast coding appears in Chapter 15 of Baguley (2012). A book-length treatment is by Rosenthal, Rosnow, and Rubin (2000). A brief discussion on contrast coding appears in Venables and Ripley (2002).

7

Contrast coding with two predictor variables

Chapter 6 provides a basic introduction to contrast coding in situations where there is one predictor variable, i.e., one factor, whose effects can be estimated using a specified contrast matrix. Here, we investigate how contrast coding generalizes to situations where there is more than one predictor variable. This could either be a situation where two factors are present or where one factor is paired with a continuous predictor variable, i.e., a covariate. We first discuss contrast coding for the case of two factors (for 2×2 designs; section 7.1) and then go on to investigate situations where one predictor is a factor and the other predictor is a covariate (section 7.2). One problem in the analysis of interactions occurs in situations where the model is not linear, but has some non-linear link function, for example in logistic models, or when assuming a log-normally distributed dependent variable. In these situations, the model makes predictions for each condition (i.e., design cell) at the latent level of the linear model. Sometimes it is important to translate these model predictions to the level of the observations (e.g., to probabilities in a logistic regression model). We will discuss how this can be implemented in section 7.3.

7.1 Contrast coding in a factorial 2×2 design

In section 6.3 of chapter 6, we used a data set with one 4-level factor. Here, we assume that the same four means come from an $\boldsymbol{A}(2) \times \boldsymbol{B}(2)$ between-subject-factor design rather than an $\boldsymbol{F}(4)$ between-subject-factor design. Load the simulated data and show summary statistics in Table 7.1 and in Figure 7.1. The means and standard deviations are exactly the same as in Figure 6.2 and in Table 6.3.

In order to carry out a 2×2 ANOVA-type (main effects and interaction) analysis, one needs sum contrasts in the linear model. (This is true for factors with two levels, but does not generalize to factors with more levels.) The results of such an analysis are shown in Table 7.2. (The analysis immediately below uses sum contrasts: $-1/+1$; however, scaled sum contrasts $-0.5/+0.5$ could also

DOI: 10.1201/9780429342646-9

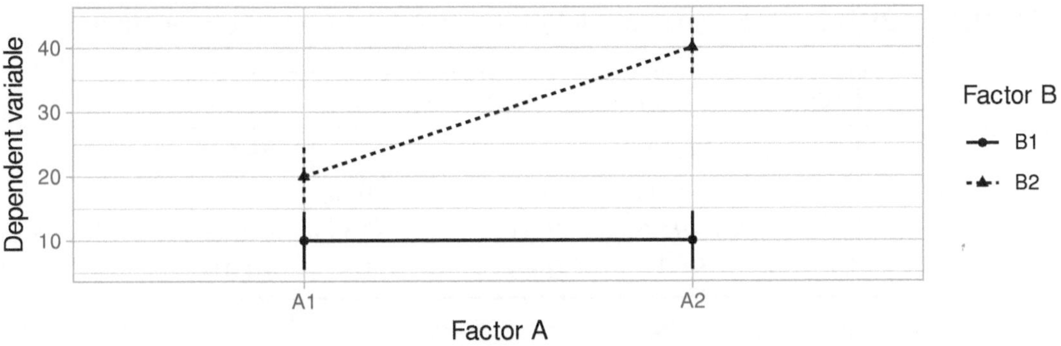

FIGURE 7.1: Means and error bars (showing standard errors) for a simulated data set with a two-by-two between-subjects factorial design.

TABLE 7.1: Summary statistics per condition for the simulated data.

Factor A	Factor B	N data	Means	Std. dev.	Std. errors
A1	B1	5	10	10	4.5
A1	B2	5	20	10	4.5
A2	B1	5	10	10	4.5
A2	B2	5	40	10	4.5

be used; in this case, the priors would need to be adapted to the respective scale.)

```
contrasts(df_contrasts4$A) <- contr.sum(2)
contrasts(df_contrasts4$B) <- contr.sum(2)
```

```
fit_AB.sum <- brm(DV ~ A * B,
                  data = df_contrasts4,
                  family = gaussian(),
                  prior = c(prior(normal(20, 50), class = Intercept),
                            prior(normal(0, 50), class = sigma),
                            prior(normal(0, 50), class = b)))
```

Next, we reproduce the $A(2) \times B(2)$ ANOVA, with contrasts specified for the corresponding one-way $F(4)$ ANOVA; that is, by treating the $2 \times 2 = 4$ condition means as four levels of a single factor F. In other words, we go back to the data frame simulated for the analysis of repeated contrasts (see chapter 6, section 6.3). We first define weights for condition means according to our hypotheses, invert this matrix, and use it as the contrast matrix for factor F. We define weights of 1/4 and −1/4. We do so because (a) we want to compare

TABLE 7.2: Bayesian linear model of a 2 x 2 design with sum contrasts.

Predictor	Estimate	Est. Error	2.5%	97.5%
Intercept	20.01	2.53	15.13	25.06
A1	−5.03	2.45	−9.88	−0.10
B1	−9.96	2.51	−15.04	−4.82
A1:B1	5.03	2.49	0.03	10.01

the mean of two conditions to the mean of two other conditions (e.g., factor $\boldsymbol{A}$ compares $\frac{F1+F2}{2}$ to $\frac{F3+F4}{2}$). Moreover, (b) we want coefficients to code half the difference between condition means, reflecting sum contrasts. Together $(a+b)$, this yields weights of $1/2 \cdot 1/2 = 1/4$. The resulting contrast matrix contains contrast coefficients of $+1$ or -1, showing that we successfully implemented sum contrasts. The results are identical to the previous models.

```
HcInt <-
  rbind(A = c(F1 = 1 / 4, F2 = 1 / 4, F3 = -1 / 4, F4 = -1 / 4),
        B = c(F1 = 1 / 4, F2 = -1 / 4, F3 = 1 / 4, F4 = -1 / 4),
        AxB = c(F1 = 1 / 4, F2 = -1 / 4, F3 = -1 / 4, F4 = 1 / 4))
t(fractions(HcInt))
```

```
##       A    B    AxB
## F1  1/4  1/4  1/4
## F2  1/4 -1/4 -1/4
## F3 -1/4  1/4 -1/4
## F4 -1/4 -1/4  1/4
```

```
(XcInt <- ginv2(HcInt))
```

```
##     A  B AxB
## F1  1  1  1
## F2  1 -1 -1
## F3 -1  1 -1
## F4 -1 -1  1
```

```
contrasts(df_contrasts3$F) <- XcInt
```

```
fit_F4.sum <- brm(DV ~ F,
                  data = df_contrasts3,
```

```
                 family = gaussian(),
                 prior = c(prior(normal(20, 50), class = Intercept),
                           prior(normal(0, 50), class = sigma),
                           prior(normal(0, 50), class = b)))
```

```
fixef(fit_F4.sum)
```

```
##           Estimate Est.Error   Q2.5 Q97.5
## Intercept    20.07      2.51  15.24 24.92
## FA           -5.01      2.56 -10.09 -0.01
## FB           -9.99      2.56 -15.10 -5.06
## FAxB          4.97      2.48  -0.11  9.97
```

This shows that it is possible to specify the contrasts not only for each factor (e.g., here in the 2 × 2 design) separately. Instead, one can also pool all experimental conditions (or design cells) into one large factor (here factor F with 4 levels), and specify the contrasts for the main effects and for the interactions in the resulting one large contrast matrix simultaneously.

In this approach, it can again be very useful to apply the `hypr` package to construct contrasts for a 2 × 2 design. The first parameter estimates the main effect A, i.e., it compares the average of $F1$ and $F2$ to the average of $F3$ and $F4$. The second parameter estimates the main effect B, i.e., it compares the average of $F1$ and $F3$ to the average of $F2$ and $F4$.

We code direct differences between the averages, i.e., we implement scaled sum contrasts instead of sum contrasts. This is shown below: the contrast matrix contains coefficients of $+1/2$ and $-1/2$ instead of $+1$ and -1. The interaction term estimates the difference between differences, i.e., the difference between $F1 - F2$ and $F3 - F4$.

```
hAxB <- hypr(A = (F1 + F2) / 2 ~ (F3 + F4) / 2,
             B = (F1 + F3) / 2 ~ (F2 + F4) / 2,
             AxB = (F1 - F2) ~ (F3 - F4))
hAxB
```

```
## hypr object containing 3 null hypotheses:
##   H0.A: 0 = (F1 + F2 - F3 - F4)/2
##   H0.B: 0 = (F1 + F3 - F2 - F4)/2
## H0.AxB: 0 = F1 - F2 - F3 + F4
##
## Call:
```

```
## hypr(A = ~1/2 * F1 + 1/2 * F2 - 1/2 * F3 - 1/2 * F4, B = ~1/2 *
##     F1 + 1/2 * F3 - 1/2 * F2 - 1/2 * F4, AxB = ~F1 - F2 - F3 +
##     F4, levels = c("F1", "F2", "F3", "F4"))
##
## Hypothesis matrix (transposed):
##       A    B    AxB
## F1  1/2  1/2    1
## F2  1/2 -1/2   -1
## F3 -1/2  1/2   -1
## F4 -1/2 -1/2    1
##
## Contrast matrix:
##       A    B    AxB
## F1  1/2  1/2  1/4
## F2  1/2 -1/2 -1/4
## F3 -1/2  1/2 -1/4
## F4 -1/2 -1/2  1/4
```

```
contrasts(df_contrasts3$F) <- contr.hypothesis(hAxB)
```

```
fit_F4hypr <- brm(DV ~ F,
                  data = df_contrasts3,
                  family = gaussian(),
                  prior = c(prior(normal(20, 50), class = Intercept),
                            prior(normal(0, 50), class = sigma),
                            prior(normal(0, 50), class = b)))
```

```
fixef(fit_F4hypr)
```

```
##           Estimate Est.Error   Q2.5  Q97.5
## Intercept     20.0      2.35  15.33  24.49
## FA           -10.0      4.67 -19.44  -0.93
## FB           -19.8      4.71 -28.99 -10.59
## FAxB          19.4      9.60   0.13  38.52
```

The results show that the estimates for the main effects have (approximately) double the size as compared to the ± 1 sum contrasts–this is the result of the scaling that we applied, i.e., the main effects now directly estimate the difference between averages. The interaction estimates the difference between differences. Importantly, if we change the scaling of the contrasts, we also need to adapt the scaling of the priors. If appropriate priors are used, then both contrasts would lead to the same hypothesis tests if one were doing hypothesis testing using Bayes factors. Thus, the `hypr` package can be used to code hypotheses in a 2×2 design.

Now the main effects and interaction can be directly interpreted as differences between averages and as differences between differences. If one wants the interaction term to be on the same scale as the main effects, it would need to be multiplied by 2; however, then its interpretation would not be straightforward any more (i.e., it is no longer "the difference between differences" but a scaled variant of this). Thus, putting the interaction on the same scale as the main effects has the advantage that it ensures that the prior has the same scaling across the different terms. However, it has the disadvantage that the interpretation of the interaction term is not straightforward.

Incidentally, an alternative way to code main effects and interactions is to use the `ifelse()` command in R. For example, if we want to use ±1/2 sum contrasts as in the above example, we can specify the contrasts for the main effects as vectors:

```
A <- ifelse(df_contrasts3$F %in% c("F1", "F2"), -1 / 2, 1 / 2)
B <- ifelse(df_contrasts3$F %in% c("F1", "F3"), -1 / 2, 1 / 2)
```

Now, defining the interaction is simply a matter of multiplying the two vectors:

```
AxB <- A * B
```

```
## rescale:
(AxB <- A * B * 2)
```

```
##  [1]  0.5  0.5  0.5  0.5  0.5 -0.5 -0.5 -0.5 -0.5 -0.5 -0.5 -0.5 -0.5
## [14] -0.5 -0.5  0.5  0.5  0.5  0.5  0.5
```

This kind of vector-based contrast coding is convenient for more complex designs, such as $2 \times 2 \times 2$ factorial designs. However, in such complex designs, the `ifelse()` command can become cumbersome, so a better approach is to use the `merge()` function in R, as shown below.

First, define a data frame that represents the main effects and interaction contrasts for the four conditions.

```
(contr_coding <- data.frame(Condition = c("F1", "F2", "F3", "F4"),
                            A = c(-1 / 2, -1 / 2, 1 / 2, 1 / 2),
```

```
                               B = c(-1 / 2, 1 / 2, -1 / 2, 1 / 2),
                               AxB = c(1 / 2, -1 / 2, -1 / 2, 1 / 2)))
```

```
##   Condition    A    B  AxB
## 1        F1 -0.5 -0.5  0.5
## 2        F2 -0.5  0.5 -0.5
## 3        F3  0.5 -0.5 -0.5
## 4        F4  0.5  0.5  0.5
```

Then, one can merge this data frame with the data frame that contains the data:

```
(df_contrasts3 <- merge(df_contrasts3,
                        contr_coding,
                        by.x = "F",
                        by.y = "Condition"))
```

```
##     F     DV id Fc2vs1 Fc3vs2 Fc4vs3    A    B  AxB
## 1  F1 17.547  1  -0.75   -0.5  -0.25 -0.5 -0.5  0.5
## 2  F1 -6.683  2  -0.75   -0.5  -0.25 -0.5 -0.5  0.5
## 3  F1 10.574  3  -0.75   -0.5  -0.25 -0.5 -0.5  0.5
## 4  F1 10.558  4  -0.75   -0.5  -0.25 -0.5 -0.5  0.5
## 5  F1 18.005  5  -0.75   -0.5  -0.25 -0.5 -0.5  0.5
## 6  F2 20.123  6   0.25   -0.5  -0.25 -0.5  0.5 -0.5
## 7  F2 14.773  7   0.25   -0.5  -0.25 -0.5  0.5 -0.5
## 8  F2 22.613  8   0.25   -0.5  -0.25 -0.5  0.5 -0.5
## 9  F2 34.713  9   0.25   -0.5  -0.25 -0.5  0.5 -0.5
## 10 F2  7.778 10   0.25   -0.5  -0.25 -0.5  0.5 -0.5
## 11 F3 18.936 11   0.25    0.5  -0.25  0.5 -0.5 -0.5
## 12 F3  9.518 12   0.25    0.5  -0.25  0.5 -0.5 -0.5
## 13 F3 21.227 13   0.25    0.5  -0.25  0.5 -0.5 -0.5
## 14 F3 -0.153 14   0.25    0.5  -0.25  0.5 -0.5 -0.5
## 15 F3  0.472 15   0.25    0.5  -0.25  0.5 -0.5 -0.5
## 16 F4 26.466 16   0.25    0.5   0.75  0.5  0.5  0.5
## 17 F4 36.243 17   0.25    0.5   0.75  0.5  0.5  0.5
## 18 F4 53.667 18   0.25    0.5   0.75  0.5  0.5  0.5
## 19 F4 39.651 19   0.25    0.5   0.75  0.5  0.5  0.5
## 20 F4 43.973 20   0.25    0.5   0.75  0.5  0.5  0.5
```

7.1.1 Nested effects

One can estimate effects that do not correspond directly to main effects and interaction of the traditional ANOVA. For example, in a 2×2 experimental design, where factor $\boldsymbol{A}$ codes word frequency (low/high) and factor $\boldsymbol{B}$ is part of speech (noun/verb), one can estimate the effect of word frequency within nouns and the effect of word frequency within verbs. Formally, $\boldsymbol{A}_{B1}$ versus $\boldsymbol{A}_{B2}$ are nested within levels of $\boldsymbol{B}$. Said differently, simple effects of factor $\boldsymbol{A}$ are estimated for each of the levels of factor $\boldsymbol{B}$. In this version, we estimate the main effect of part of speech ($\boldsymbol{B}$; as in traditional ANOVA). Instead of also estimating the second main effect word frequency, $\boldsymbol{A}$, and the interaction, we estimate (1) the differences between the two levels of word frequency $\boldsymbol{A}$ in the first level of $\boldsymbol{B}$ (i.e., nouns), and (2) the difference between the two levels of word frequency $\boldsymbol{A}$ in the second level of $\boldsymbol{B}$ (i.e., verbs). In other words, we estimate the differences between the two levels of $\boldsymbol{A}$ in each of the levels of $\boldsymbol{B}$. Often, researchers have hypotheses about these differences, and not about the interaction.

```
HcNes <-
  rbind(B = c(F1 = 1 / 2, F2 = -1 / 2, F3 = 1 / 2, F4 = -1 / 2),
        B1xA = c(F1 = -1, F2 = 0, F3 = 1, F4 = 0),
        B2xA = c(F1 = 0, F2 = -1, F3 = 0, F4 = 1))
t(fractions(HcNes))
```

```
##        B  B1xA B2xA
## F1   1/2    -1    0
## F2  -1/2     0   -1
## F3   1/2     1    0
## F4  -1/2     0    1
```

```
(XcNes <- ginv2(HcNes))
```

```
##        B  B1xA B2xA
## F1   1/2 -1/2     0
## F2  -1/2     0 -1/2
## F3   1/2  1/2     0
## F4  -1/2     0  1/2
```

```
contrasts(df_contrasts3$F) <- XcNes
```

```
fit_Nest <- brm(DV ~ F,
                data = df_contrasts3,
                family = gaussian(),
                prior = c(prior(normal(20, 50), class = Intercept),
                          prior(normal(0, 50), class = sigma),
                          prior(normal(0, 50), class = b)))
```

```
fixef(fit_Nest)
```

```
##           Estimate Est.Error  Q2.5 Q97.5
## Intercept    19.99      2.42  15.2 24.80
## FB          -19.93      5.12 -30.2 -9.81
## FB1xA         0.09      7.43 -14.7 14.39
## FB2xA        19.33      6.85   5.4 32.70
```

The regression coefficients estimate the grand mean, the difference for the main effect of part of speech (***B***) and the two differences (for ***A***; i.e., simple effects) within the two levels (noun and verb) of part of speech (***B***).

These custom nested contrasts' columns are scaled versions of the corresponding hypothesis matrix. This is the case because the columns are orthogonal. This illustrates one advantage of orthogonal contrasts for the interpretation of regression coefficients: the underlying comparisons being estimated are already clear from the contrast matrix.

There is also a built-in R formula specification for nested designs. The order of factors in the formula from left to right specifies a top-down order of nesting within levels, i.e., here factor ***A*** (word frequency) is nested within levels of the factor ***B*** (part of speech). This yields the same result as our previous result based on custom nested contrasts:

```
contrasts(df_contrasts4$A) <- c(-0.5, +0.5)
contrasts(df_contrasts4$B) <- c(+0.5, -0.5)
fit_Nest2 <- brm(DV ~ B / A,
                 data = df_contrasts4,
                 family = gaussian(),
                 prior = c(prior(normal(20, 50), class = Intercept),
                           prior(normal(0, 50), class = sigma),
                           prior(normal(0, 50), class = b)))
```

```
fixef(fit_Nest2)
```

```
##            Estimate Est.Error   Q2.5 Q97.5
## Intercept     19.96      2.37  15.25  24.7
## B1           -19.92      4.75 -29.45 -10.7
## BB1:A1        -0.01      6.90 -13.59  13.7
## BB2:A1        19.63      6.98   5.55  33.6
```

In cases such as these, where A_{B1} vs. A_{B2} are nested within levels of B, it is necessary to include the effect of B (part of speech) in the model, even if one is only interested in the effect of A (word frequency) within levels of B (part of speech). Leaving out factor B in this particular case of a between-subjects design would increase posterior uncertainty if the data are fully balanced, and could lead to biases in parameter estimation if the data are not fully balanced.

Again, we show how nested contrasts can be easily implemented using `hypr`:

```
hNest <- hypr(B = (F1 + F3) / 2 ~ (F2 + F4) / 2,
              B1xA = F3 ~ F1,
              B2xA = F4 ~ F2)
hNest
```

```
## hypr object containing 3 null hypotheses:
##    H0.B: 0 = (F1 + F3 - F2 - F4)/2
## H0.B1xA: 0 = F3 - F1
## H0.B2xA: 0 = F4 - F2
##
## Call:
## hypr(B = ~1/2 * F1 + 1/2 * F3 - 1/2 * F2 - 1/2 * F4, B1xA = ~F3 -
##     F1, B2xA = ~F4 - F2, levels = c("F1", "F2", "F3", "F4"))
##
## Hypothesis matrix (transposed):
##    B    B1xA B2xA
## F1  1/2   -1    0
## F2 -1/2    0   -1
## F3  1/2    1    0
## F4 -1/2    0    1
##
## Contrast matrix:
##    B    B1xA B2xA
## F1  1/2 -1/2    0
## F2 -1/2    0 -1/2
```

```
## F3  1/2  1/2    0
## F4 -1/2    0  1/2
```

```
contrasts(df_contrasts3$F) <- contr.hypothesis(hNest)
```

```
fit_NestHypr <-
  brm(DV ~ F,
      data = df_contrasts3,
      family = gaussian(),
      prior = c(prior(normal(20, 50), class = Intercept),
                prior(normal(0, 50), class = sigma),
                prior(normal(0, 50), class = b)))
```

```
fixef(fit_NestHypr)
```

```
##           Estimate Est.Error   Q2.5 Q97.5
## Intercept    19.98      2.42  15.24  24.9
## FB          -19.87      4.93 -29.45 -10.1
## FB1xA         0.16      6.83 -12.96  13.7
## FB2xA        19.57      7.16   4.97  33.1
```

Of course, we can also ask the reverse question: Are there differences for part of speech ($\boldsymbol{B}$) in the levels of word frequency ($\boldsymbol{A}$; in addition to estimating the main effect of word frequency, $\boldsymbol{A}$)? That is, do nouns differ from verbs for low-frequency words ($\boldsymbol{B}_{A1}$) and do nouns differ from verbs for high-frequency words ($\boldsymbol{B}_{A2}$)?

```
hNest2 <- hypr(A = (F1 + F2) / 2 ~ (F3 + F4) / 2,
               A1xB = F2 ~ F1,
               A2xB = F4 ~ F3)
hNest2
```

```
## hypr object containing 3 null hypotheses:
##    H0.A: 0 = (F1 + F2 - F3 - F4)/2
## H0.A1xB: 0 = F2 - F1
## H0.A2xB: 0 = F4 - F3
##
## Call:
## hypr(A = ~1/2 * F1 + 1/2 * F2 - 1/2 * F3 - 1/2 * F4, A1xB = ~F2 -
##     F1, A2xB = ~F4 - F3, levels = c("F1", "F2", "F3", "F4"))
```

```
##
## Hypothesis matrix (transposed):
##     A    A1xB A2xB
## F1  1/2   -1    0
## F2  1/2    1    0
## F3 -1/2    0   -1
## F4 -1/2    0    1
##
## Contrast matrix:
##     A    A1xB A2xB
## F1  1/2 -1/2    0
## F2  1/2  1/2    0
## F3 -1/2    0 -1/2
## F4 -1/2    0  1/2
```

```
contrasts(df_contrasts3$F) <- contr.hypothesis(hNest2)
```

```
fit_Nest2Hypr <-
  brm(DV ~ F,
      data = df_contrasts3,
      family = gaussian(),
      prior = c(prior(normal(20, 50), class = Intercept),
                prior(normal(0, 50), class = sigma),
                prior(normal(0, 50), class = b)))
```

```
fixef(fit_Nest2Hypr)
```

```
##           Estimate Est.Error   Q2.5 Q97.5
## Intercept    19.95      2.47  15.11 24.63
## FA           -9.81      4.92 -19.72 -0.04
## FA1xB         9.80      7.05  -4.35 23.75
## FA2xB        29.38      6.83  15.68 42.55
```

Regression coefficients estimate the grand mean, the difference for the main effect of word frequency ($\boldsymbol{A}$) and the two part of speech effects (for $\boldsymbol{B}$; i.e., simple effects) within levels of word frequency ($\boldsymbol{A}$).

An important issue to keep in mind is that if, using nested contrasts, one finds a large effect in one nested comparison (e.g., `FA2xB`) but no difference or a smaller effect in the other comparison (here, `FA1xB`), one cannot conclude that there is an interaction; to argue for an interaction, one would have to test for the main effects and interaction using the ANOVA contrast coding (Nieuwenhuis, Forstmann, and Wagenmakers 2011).

7.1.2 Interactions between contrasts

In a 2×2 experimental design, the results from sum contrasts are equivalent to typical ANOVA results that we see in frequentist analyses. This means that sum contrasts assess the main effects and the interactions. One interesting question that arises here is: what would happen in a 2×2 design if we had used treatment contrasts instead of sum contrasts? Is it still possible to meaningfully interpret the results from the treatment contrasts in a simple 2×2 design?

This leads us to a very important principle in interpreting results from contrasts: When interactions between contrasts are included in a model, then the results for one contrast actually depend on the specification of the other contrast(s) in the analysis! This may be counter-intuitive at first, but it is very important and essential to keep in mind when interpreting results from contrasts. How does this work in detail?

The general rule to remember is that the effect of one contrast measures its effect at the location 0 of the other contrast(s) in the analysis. This can be seen in the regression equation of a 2×2 design with factors A and B:

$$E[Y] = \alpha + \beta_A A + \beta_B B + \beta_{A \times B} A \times B \tag{7.1}$$

If we set the predictor B to zero, then the equation simplifies to:

$$E[Y] = \alpha + \beta_A A \tag{7.2}$$

Thus, now we can see the "pure" effect of A.

What does that mean practically? Let us consider the example that we use two treatment contrasts in a 2×2 design. Here are the results from the linear model

```
contrasts(df_contrasts4$A) <- c(0, 1)
contrasts(df_contrasts4$B) <- c(0, 1)
fit_treatm <- brm(DV ~ B * A,
                  data = df_contrasts4,
                  family = gaussian(),
                  prior = c(prior(normal(20, 50), class = Intercept),
```

```
                      prior(normal(0, 50), class = sigma),
                      prior(normal(0, 50), class = b)))
```

```
fixef(fit_treatm)
```

```
##           Estimate Est.Error   Q2.5 Q97.5
## Intercept     9.77      4.77   0.19  19.0
## B1           10.49      6.97  -2.79  24.8
## A1            0.39      6.72 -12.55  13.8
## B1:A1        19.09      9.63  -0.36  37.8
```

Let's take a look at the effect of factor ***A***. How can we interpret what this measures? This effect actually estimates the effect of factor ***A*** at the "location" where factor ***B*** is coded as 0. Factor ***B*** is coded as a treatment contrast, that is, it codes a zero at its baseline condition, which is ***B*1**. Thus, the effect of factor ***A*** estimates the effect of ***A*** nested within the baseline condition of ***B***, i.e., a simple effect. We take a look at the data presented in Figure 7.1, what this nested effect should be. Figure 7.1 shows that the effect of factor ***A*** nested in ***B*1** is 0. If we now compare this to the results from the linear model, it is indeed clear that the effect of factor ***A*** is exactly estimated as 0. As expected, when factor ***B*** is coded as a treatment contrast, the effect of factor ***A*** estimates the effect of ***A*** nested within the baseline level of factor ***B***.

Next, consider the effect of factor ***B***. According to the same logic, this effect estimates the effect of factor ***B*** at the "location" where factor ***A*** is 0. Factor ***A*** is also coded as a treatment contrast, that is, it codes its baseline condition ***A*1** as 0. The effect of factor ***B*** estimates the effect of ***B*** nested within the baseline condition of ***A***. Figure 7.1 shows that this effect should be 10.

How do we know what the "location" is, where a contrast applies? For the treatment contrasts discussed here, it is possible to reason this through because all contrasts are coded as 0 or 1. How can one derive the "location" in general? What we can do is to look at the comparisons that are estimated when using the treatment contrasts (or in case we use Bayes factors, which hypotheses are tested) in the presence of an interaction between them by using the generalized matrix inverse. We go back to the default treatment contrasts. Then we extract the contrast matrix from the design matrix:

```
contrasts(df_contrasts4$A) <- contr.treatment(2)
contrasts(df_contrasts4$B) <- contr.treatment(2)
XcTr <- df_contrasts4 %>%
```

```
  group_by(A, B) %>%
  summarise() %>%
  model.matrix(~ 1 + A * B, .) %>%
  as.data.frame() %>%
  as.matrix()
rownames(XcTr) <- c("A1_B1", "A1_B2", "A2_B1", "A2_B2")
XcTr
```

```
##       (Intercept) A2 B2 A2:B2
## A1_B1           1  0  0     0
## A1_B2           1  0  1     0
## A2_B1           1  1  0     0
## A2_B2           1  1  1     1
```

This shows the treatment contrast for factors ***A*** and ***B***, and their interaction. We can now assign this contrast matrix to a `hypr` object. `hypr` automatically converts the contrast matrix into a hypothesis matrix, such that we can read from the hypothesis matrix which comparisons are being estimated by the different contrasts.

```
htr <- hypr() # initialize empty hypr object
cmat(htr) <- XcTr # assign contrast matrix to hypr object
htr # look at the resulting hypothesis matrix
```

```
## hypr object containing 4 null hypotheses:
## H0.(Intercept): 0 = A1_B1                              (Intercept)
##          H0.A2: 0 = -A1_B1 + A2_B1
##          H0.B2: 0 = -A1_B1 + A1_B2
##       H0.A2:B2: 0 = A1_B1 - A1_B2 - A2_B1 + A2_B2
##
## Call:
## hypr(`(Intercept)` = ~A1_B1, A2 = ~-A1_B1 + A2_B1, B2 = ~-A1_B1 +
##     A1_B2, `A2:B2` = ~A1_B1 - A1_B2 - A2_B1 + A2_B2, levels = c("A1_B1",
## "A1_B2", "A2_B1", "A2_B2"))
##
## Hypothesis matrix (transposed):
##       (Intercept) A2 B2 A2:B2
## A1_B1  1          -1 -1  1
## A1_B2  0           0  1 -1
## A2_B1  0           1  0 -1
## A2_B2  0           0  0  1
```

```
##
## Contrast matrix:
##       (Intercept) A2 B2 A2:B2
## A1_B1 1           0  0  0
## A1_B2 1           0  1  0
## A2_B1 1           1  0  0
## A2_B2 1           1  1  1
```

The same result is obtained by applying the generalized inverse to the contrast matrix (this is what `hypr()` does as well). An important fact is that when we apply the generalized inverse to the contrast matrix, we obtain the corresponding hypothesis matrix (for details, see Schad et al. 2020).

```
t(ginv2(XcTr))
```

```
##       (Intercept) A2 B2 A2:B2
## A1_B1  1          -1 -1  1
## A1_B2  0           0  1 -1
## A2_B1  0           1  0 -1
## A2_B2  0           0  0  1
```

As discussed above, the effect of factor $\boldsymbol{A}$ estimates its effect nested within the baseline level of factor $\boldsymbol{B}$. Likewise, the effect of factor $\boldsymbol{B}$ estimates its effect nested within the baseline level of factor $\boldsymbol{A}$. The term $\boldsymbol{A} : \boldsymbol{B}$ always tests the interaction between both factors, irrespective of the centering of $\boldsymbol{A}$ and $\boldsymbol{B}$.

How does this work for sum contrasts? They do not have a baseline condition that is coded as 0. In sum contrasts, the average of the contrast coefficients is 0. Therefore, effects estimate the average effect across factor levels, i.e., they estimate a main effect. This is what is typically also tested in standard ANOVA. Let's look at the example shown in Table 7.2: given that factor $\boldsymbol{B}$ has a sum contrast, the main effect of factor $\boldsymbol{A}$ is estimated as the average across levels of factor $\boldsymbol{B}$. Figure 7.1 shows that the effect of factor $\boldsymbol{A}$ in level B1 is $10 - 10 = 0$, and in level $B2$ it is $20 - 40 = -20$. The average effect across both levels is $(0 - 20)/2 = -10$. Due to the sum contrast coding, we have to divide this by two, yielding an expected effect of $-10/2 = -5$. This is exactly what the effect of factor $\boldsymbol{A}$ measures (see Table 7.2, *Estimate* for $A1$).

Similarly, factor $\boldsymbol{B}$ estimates its effect at the location 0 of factor $\boldsymbol{A}$. Again, 0 is exactly the mean of the contrast coefficients from factor $\boldsymbol{A}$, which is coded as a sum contrast. Therefore, factor $\boldsymbol{B}$ estimates the effect of $\boldsymbol{B}$ averaged across factor levels of $\boldsymbol{A}$, i.e., the main effect of $\boldsymbol{B}$. For factor level $A1$, factor $\boldsymbol{B}$ has an effect of $10 - 20 = -10$. For factor level $A2$, factor $\boldsymbol{B}$ has an effect of $10 - 40 = -30$. The average effect is $(-10 - 30)/2 = -20$, which again needs to be divided by 2 due to the sum contrast. This yields exactly the estimate of −10 that is also reported in Table 7.2 (*Estimate* for B1).

Again, we look at the hypothesis matrix for the main effects and the interaction:

```
contrasts(df_contrasts4$A) <- contr.sum(2)
contrasts(df_contrasts4$B) <- contr.sum(2)
XcSum <- df_contrasts4 %>%
  group_by(A, B) %>%
  summarise() %>%
  model.matrix(~ 1 + A * B, .) %>%
  as.data.frame() %>%
  as.matrix()
rownames(XcSum) <- c("A1_B1", "A1_B2", "A2_B1", "A2_B2")

hsum <- hypr() # initialize empty hypr object
cmat(hsum) <- XcSum # assign contrast matrix to hypr object
hsum # look at the resulting hypothesis matrix
```

```
## hypr object containing 4 null hypotheses:
## H0.(Intercept): 0 = (A1_B1 + A1_B2 + A2_B1 + A2_B2)/4  (Intercept)
##          H0.A1: 0 = (A1_B1 + A1_B2 - A2_B1 - A2_B2)/4
##          H0.B1: 0 = (A1_B1 - A1_B2 + A2_B1 - A2_B2)/4
##       H0.A1:B1: 0 = (A1_B1 - A1_B2 - A2_B1 + A2_B2)/4
##
## Call:
## hypr(`(Intercept)` = ~1/4 * A1_B1 + 1/4 * A1_B2 + 1/4 * A2_B1 +
##     1/4 * A2_B2, A1 = ~1/4 * A1_B1 + 1/4 * A1_B2 - 1/4 * A2_B1 -
##     1/4 * A2_B2, B1 = ~1/4 * A1_B1 - 1/4 * A1_B2 + 1/4 * A2_B1 -
##     1/4 * A2_B2, `A1:B1` = ~1/4 * A1_B1 - 1/4 * A1_B2 - 1/4 *
##     A2_B1 + 1/4 * A2_B2, levels = c("A1_B1", "A1_B2", "A2_B1",
## "A2_B2"))
##
## Hypothesis matrix (transposed):
##       (Intercept) A1   B1   A1:B1
## A1_B1  1/4         1/4  1/4  1/4
## A1_B2  1/4         1/4 -1/4 -1/4
## A2_B1  1/4        -1/4  1/4 -1/4
## A2_B2  1/4        -1/4 -1/4  1/4
##
## Contrast matrix:
```

```
##        (Intercept) A1 B1 A1:B1
## A1_B1  1            1  1  1
## A1_B2  1            1 -1 -1
## A2_B1  1           -1  1 -1
## A2_B2  1           -1 -1  1
```

This shows that each of the effects now does not compute nested comparisons any more, but that they rather estimate their effect averaged across conditions of the other factor. The averaging involves using weights of 1/2 in the hypothesis matrix. Moreover, the regression coefficients in the sum contrast measure half the distance between conditions, leading to weights of $1/2 \cdot 1/2 = 1/4$ in the hypothesis matrix.

The general rule to remember from these examples is that when interactions between contrasts are estimated, what an effect of a factor estimates depends on the contrast coding of the other factors in the design! The effect of a factor estimates the effect nested within the location zero of the other contrast(s) in an analysis. If another contrast is centered, and zero is the average of this other contrasts' coefficients, then the contrast of interest estimates the average or main effect, averaged across the levels of the other factor. Importantly, this property, that the coding of the other factor determines the estimates of one factor, holds only when the interaction between two contrasts is included into a model. If the interaction is omitted and only effects are estimated, then there is no such influence.

This may be a very surprising result for interactions of contrasts. However, it is also essential to interpreting contrast coefficients involved in interactions. It is particularly relevant for the analysis of the default treatment contrast, where the main effects estimate nested effects rather than average effects.

7.2 One factor and one covariate

7.2.1 Estimating a group difference and controlling for a covariate

In this section we treat the case where there are again two predictor variables for one dependent variable, but where one predictor variable is a discrete factor, and the other is a continuous covariate. Let's assume we have measured some response time (RT), e.g., in a lexical decision task. We want to predict the response time based on each subject's IQ, and we expect that higher IQ leads to shorter response times. Moreover, we have two groups of each 30 subjects. These are coded as factor F, with factor levels $F1$ and $F2$. We assume that these two groups have obtained different training programs to optimize their response times in the task. Group $F1$ obtained a control training, whereas group $F2$ obtained training to improve lexical decisions. We want to estimate the extent to which the training for better lexical decisions in group $F2$ leads to shorter response times compared to the control group $F1$. This is our main question of interest here, i.e., the extent to which the training program in $F2$ leads to faster response times compared to the control group $F1$. We load the data, which is a simulated data set.

```
data("df_contrasts5")
```

Our main effect of interest is the factor *F*. We want to estimate its effect on response times and code it using scaled sum contrasts, such that negative parameter estimates would yield support for our hypothesis that response times are faster in the training group $F2$:

```
(contrasts(df_contrasts5$F) <- c(-0.5, +0.5))
```

```
## [1] -0.5  0.5
```

We run a `brms` model to estimate the effect of factor *F*, i.e., how strongly the response times in the two groups differ from each other.

```
fit_RT_F <- brm(RT ~ F,
                data = df_contrasts5,
                family = gaussian(),
                prior = c(prior(normal(200, 50), class = Intercept),
                          prior(normal(0, 50), class = sigma),
                          prior(normal(0, 50), class = b)))
```

```
fixef(fit_RT_F)
```

```
##           Estimate Est.Error  Q2.5  Q97.5
## Intercept    212.3      5.17 202.1 222.40
## F1           -23.7     10.32 -44.2  -3.19
```

We find (see model estimates and data shown in Figure 7.2) that response times in group $F2$ are roughly 25 ms faster than in group $F1$ (Estimate of -24). This suggests that as expected, the training program that group $F2$ obtained seems to be successful in speeding up response times. Recall that one cannot just look at the 95% credible interval and check whether zero is outside the interval to declare that we have found an effect. To make a discovery claim, we need to

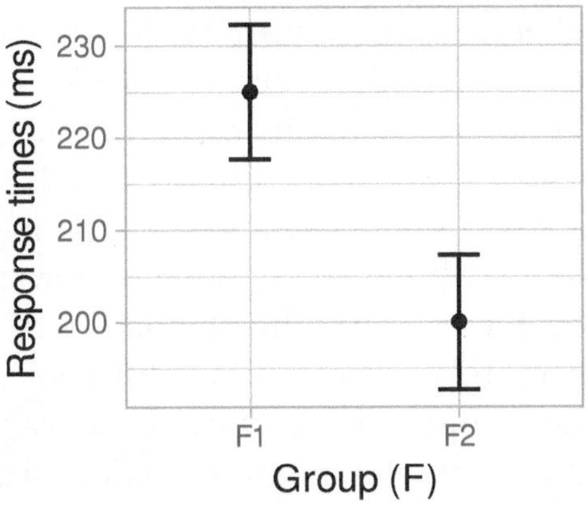

FIGURE 7.2: Means and error bars (showing standard errors) for a simulated data set of response times for two different groups of subjects, who have obtained a training in lexical decisions ($F2$) versus have obtained a control training ($F1$).

run a Bayes factor analysis on this data set to directly test this hypothesis, and this may or may not provide evidence for a difference in response times between groups.

Let's assume we have allocated subjects to the two groups randomly. Let's say that we also measured the IQ of each person using an IQ test. We did so because we expected that IQ could have an influence on response times, and we wanted to control for this influence. We now can check whether the two groups had the same average IQ.

```
df_contrasts5 %>%
  group_by(F) %>%
  summarize(M.IQ = mean(IQ))
```

```
## # A tibble: 2 x 2
##   F      M.IQ
##   <fct> <dbl>
## 1 F1       85
## 2 F2      115
```

Group $F2$ not only obtained additional training and had faster response times, group $F2$ also had a higher IQ (mean of 115) on average than group $F1$ (mean IQ = 85). Thus, the random allocation of subjects to the two groups seems to have created–by chance–a difference in IQs. Now we can ask the question: why might response times in group $F2$ be faster than in group $F1$? Is this because of the training program in $F2$? Or is this simply because the average

IQ in group $F2$ was higher than in group $F1$? To investigate this question, we add both predictor variables simultaneously in a `brms` model. Before we enter the continuous IQ variable, we center it, by subtracting its mean. Centering covariates is generally good practice. Moreover, it is often important to z-transform the covariate, i.e., to not only subtract the mean, but also to divide by its standard deviation (this can be done as follows: `df_contrasts5$IQ.s <- scale(df_contrasts5$IQ)`). The reason why this is often important is that the sampler doesn't work well if predictors have different scales. For the simple models we use here, the sampler works without z-transformation. However, for more realistic and more complex models, z-transformation of covariates is often very important. Importantly, when changing the scale of the predictor, then the scale of the prior has to be changed accordingly. Here, we set the same priors for the contrast and for the covariate IQ for convenience. In realistic models, these priors have to be set separately.

```
df_contrasts5 <- df_contrasts5 %>%
  mutate(c_IQ = IQ - mean(IQ))
fit_RT_F_IQ <- brm(RT ~ F + c_IQ,
                   data = df_contrasts5,
                   family = gaussian(),
                   prior = c(prior(normal(200, 50), class = Intercept),
                             prior(normal(0, 50), class = sigma),
                             prior(normal(0, 50), class = b)))
```

```
fixef(fit_RT_F_IQ)
```

```
##             Estimate Est.Error   Q2.5 Q97.5
## Intercept     212.33      4.79 202.99 221.9
## F1              6.47     13.31 -19.66  32.8
## c_IQ           -1.05      0.32  -1.67  -0.4
```

The results from the `brms` model now show that the difference in response times between groups (i.e., factor F) is not estimated to be -25 ms any more, but instead, the estimate is about $+7$ ms, and the 95% credible interval spans the range -20 to 33. Thus, there doesn't seem to be much reason to believe any more that the groups would differ. At the same time, we see that the predictor variable IQ shows a negative effect (Estimate $= -1$ with 95% credible interval: -1.7 to -0.4), suggesting that–as expected–response times seem to be faster in subjects with higher IQ.

This result can also be seen in Figure 7.3, which shows that response times decrease with increasing IQ, as suggested by the `brms` model. However, the

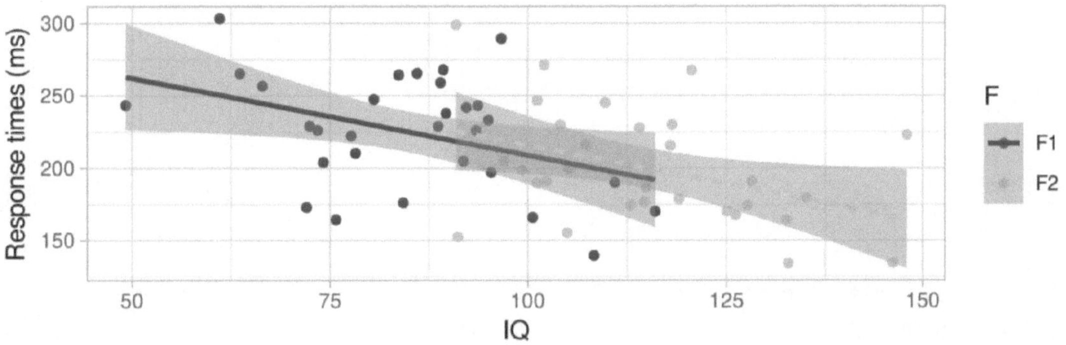

FIGURE 7.3: Response times as a function of individual IQ for two groups with a lexical decision training ($F2$) versus a control training ($F1$). Points indicate individual subjects, and lines with error bands indicate linear regression lines.

heights of the two regression lines do not differ from each other, consistent with the observation in the `brms` model that the effect of factor F did not seem to differ from zero. That is, factor F in the `brms` model estimates the difference in height of the regression line between both groups. That the height does not differ and the effect of F is estimated to be close to zero suggests that in fact group $F2$ showed faster response times not because of their additional training program. Instead, they had faster response times simply because their IQ was by chance higher on average compared to the control group $F1$. This analysis is the Bayesian equivalence of the frequentist "analysis of covariance" (ANCOVA), where it's possible to estimate a group difference after "controlling for" the influence of a covariate.

We can also see in Figure 7.3 that the two regression lines for the two groups are exactly parallel to each other. That is, the influence of IQ on response times seems to be exactly the same in both groups. This is actually an assumption in the ANCOVA analysis that needs to be checked in the data. That is, if we want to estimate the difference between groups after controlling for a covariate (here IQ), we have to investigate whether the influence of the covariate is the same in both groups. We can estimate this by including an interaction term between the factor and the covariate in the `brms` model:

```
fit_RT_FxIQ <-
  brm(RT ~ F * c_IQ,
      data = df_contrasts5,
      family = gaussian(),
      prior = c(prior(normal(200, 50), class = Intercept),
```

```
            prior(normal(0, 50), class = sigma),
            prior(normal(0, 50), class = b)))
```

```
fixef(fit_RT_FxIQ)
```

```
##              Estimate Est.Error   Q2.5  Q97.5
## Intercept      212.42      7.08 198.08 226.41
## F1               6.74     13.58 -20.23  33.42
## c_IQ            -1.06      0.32  -1.71  -0.44
## F1:c_IQ          0.01      0.67  -1.31   1.38
```

The estimate for the interaction (the term `F1:c_IQ`) is very small here (close to 0) and the 95% credible intervals clearly overlap with zero, showing that the two regression lines are estimated to be very similar, or parallel, to each other. If this is the case, then it is possible to correct for IQ when estimating the group difference. (From the perspective of causal inference, one might need additional requirements to draw strong conclusions here, such as measuring the covariate before the dependent variable, or reasoning based on theoretical plausibility.)

7.2.2 Estimating differences in slopes

We now take a look at a different data set.

```
data("df_contrasts6")
levels(df_contrasts6$F) <- c("simple", "complex")
```

This again contains data from response times (RT) in two groups. Let's assume the two groups have performed two different response time tasks, where one simple RT task doesn't rely on much cognitive processing (group "simple"), whereas the other task is more complex and depends on complex cognitive operations (group "complex"). We therefore expect that RTs in the simple task should be independent of IQ, whereas in the complex task, individuals with a high IQ should be faster in responding compared to individuals with low IQ. Thus, our primary hypothesis of interest states that the influence of IQ on RT differs between conditions. This means that we are interested in the difference between slopes. A slope in a linear regression assesses how strongly the dependent variable (here RT) changes with an increase of one unit on the covariate (here IQ), it thus assesses how "steep" the regression line is. Our hypothesis thus states that the regression lines differ between groups.

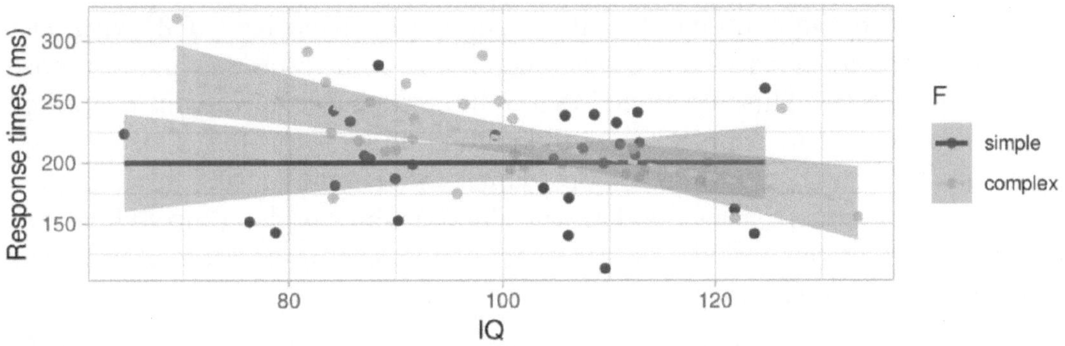

FIGURE 7.4: Response times as a function of individual IQ for two groups performing a simple versus a complex task. Points indicate individual subjects' responses and lines with error bands show the fitted linear regression lines.

The results, displayed in Figure 7.4, suggest that the data seem to support our hypothesis. For the subjects performing the complex task, response times seem to decrease with increasing IQ, whereas for subjects performing the simple task, response times seem to be independent of IQ. As stated before, our primary hypothesis relates to the difference in slopes. Statistically speaking, this is assessed in the interaction between the factor and the covariate. Thus, we run a `brms` model where the interaction is included. Importantly, we first use scaled sum contrasts for the group effect, and again center the covariate IQ.

```
contrasts(df_contrasts6$F) <- c(-0.5, +0.5)
df_contrasts6$c_IQ <- df_contrasts6$IQ - mean(df_contrasts6$IQ)
fit_RT_FxIQ2 <- brm(RT ~  F * c_IQ,
                    data = df_contrasts6,
                    family = gaussian(),
                    prior = c(prior(normal(200, 50), class = Intercept),
                              prior(normal(0, 50), class = sigma),
                              prior(normal(0, 50), class = b)))
```

```
fixef(fit_RT_FxIQ2)
```

```
##           Estimate Est.Error   Q2.5  Q97.5
## Intercept   209.97      4.88 200.33 219.39
## F1           19.14      9.59   0.18  38.41
## c_IQ         -0.80      0.33  -1.46  -0.15
## F1:c_IQ      -1.61      0.66  -2.87  -0.30
```

We can see that the main effect of IQ (term `c_IQ`) is negative (−0.8) with 95% credible intervals −1.5 to −0.2, suggesting that overall response times decrease with increasing IQ. This is qualified by the interaction term, which is estimated to be negative (−1.6), with 95% credible intervals −2.9 to −0.3. This suggests that the slope in the complex group (which was coded as +0.5 in the scaled sum contrast) is more negative than the slope in the simple group (which was coded as −0.5 in the scaled sum contrast). Thus, the interaction assesses the difference between slopes.

We can also run a model, where the nested slopes are estimated, i.e., the slope of IQ in the simple group and the slope of IQ in the complex group. This can be implemented by using the nested coding that we learned about in the previous section:

```
fit_RT_FnIQ2 <-
  brm(RT ~ F / c_IQ,
      data = df_contrasts6,
      family = gaussian(),
      prior = c(prior(normal(200, 50), class = Intercept),
                prior(normal(0, 50), class = sigma),
                prior(normal(0, 50), class = b)))
```

```
fixef(fit_RT_FnIQ2)
```

```
##                Estimate Est.Error   Q2.5  Q97.5
## Intercept        209.95      4.85 200.69 219.77
## F1                19.33      9.60   0.34  38.22
## Fsimple:c_IQ       0.01      0.47  -0.93   0.92
## Fcomplex:c_IQ     -1.61      0.47  -2.54  -0.71
```

Now we see that the slope of IQ in the simple group (`Fsimple:c_IQ`) is estimated to be 0, with credible intervals clearly including zero. By contrast, the slope in the complex group (`Fcomplex:c_IQ`) is estimated as -1.6 (95% CrI $= [-2.5, -0.7]$). This is consistent with our hypothesis that high IQ speeds up response times for the complex but not for the simple task. (To obtain evidence for this effect, we need Bayes factors, see chapter 13.) We can also see from the nested analysis that the difference in slopes between conditions is $-1.6 - 0.0 = -1.6$. This is exactly the value for the interaction term that we estimated in the previous model, demonstrating that interaction terms assess the difference between slopes; i.e., they estimate in how far the regression lines in the two conditions are parallel, with an estimate of 0 indicating perfectly parallel lines.

Notice that we can compute posterior samples for the nested slopes from the model with the interaction. That is, we can take the model that estimates main effects and the interaction, and compute posterior samples for the slope of IQ in the simple task and the slope of IQ in the complex task. First, we extract the posterior samples from the model.

```
df_postSamp_RT_FxIQ2 <- as_draws_df(fit_RT_FxIQ2)
```

Then, we take a look at the contrast coefficients for the group factor:

```
contrasts(df_contrasts6$F)
```

```
##         [,1]
## simple  -0.5
## complex  0.5
```

They show a value of -0.5 for the simple group. Thus, to compute the slope for the simple group we have to take the overall slope for `c_IQ` and subtract -0.5 times the estimate for the interaction:

```
df_postSamp_RT_FxIQ2 <- df_postSamp_RT_FxIQ2 %>%
  mutate(b_c_IQ_simple = b_c_IQ - 0.5 * `b_F1:c_IQ`)
```

Likewise, to estimate the slope for the complex group we have to take the overall slope for `c_IQ` and add $+0.5$ times the estimate for the interaction:

```
df_postSamp_RT_FxIQ2 <- df_postSamp_RT_FxIQ2 %>%
  mutate(b_c_IQ_complex = b_c_IQ + 0.5 * `b_F1:c_IQ`)
```

```
c(IQc.c_simple = mean(df_postSamp_RT_FxIQ2$b_c_IQ_simple),
  IQc.c_complex = mean(df_postSamp_RT_FxIQ2$b_c_IQ_complex))
```

```
##  IQc.c_simple IQc.c_complex
##       0.00398      -1.60336
```

The results show that the posterior means for the slope of `c_IQ` are 0 and -1.6 for the simple and the complex groups, as we had found above in the nested analysis. This is the case because the data are sufficiently informative compared to the priors. When the data are sparse or when working with Bayes factors, priors will have to be set carefully for each specific analysis.

In most situations one should always center covariates before including them into a model. If covariates are not centered, then the effects (here the effect for the factor) cannot be interpreted as main effects anymore.

One can also do analyses with interactions between a covariate and a factor, but by using different contrast codings. For example, if we use treatment contrasts for the factor, then the main effect of `c_IQ` assesses not the average slope of `c_IQ` across conditions, but instead the nested slope of `c_IQ` within the baseline group of the treatment contrast. The interaction still assesses the difference in slopes between groups. In a situation where there are more than two groups, when one estimates the interaction of contrasts with a covariate, then the contrasts define which slopes are compared with each other in the interaction terms. For example, when using sum contrasts in an example where the influence of IQ is measured on response times for nouns, verbs, and adjectives, then there are two interaction terms: these assess (1) whether the slope of IQ for nouns is different from the average slope across conditions, and (2) whether the slope of IQ for verbs is different from the average slope across conditions. If one uses repeated contrasts in a situation where the influence of IQ on response times is estimated for word frequency conditions "low," "medium-low," "medium-high," and "high," then there are three interaction terms (one for each contrast). The first interaction term estimates the difference in slopes between "low" and "medium-low" word frequencies, the second interaction term estimates the difference in slopes between "medium-low" and "medium-high" word frequencies, and the third interaction term estimates the difference in slopes between "medium-high" and "high" word frequency conditions. Thus, the logic of how contrasts specify certain comparisons between conditions extends directly to the situation where differences in slopes are estimated.

7.3 Interactions in generalized linear models (with non-linear link functions) and non-linear models

Next, we look at generalized linear models, where a linear predictor is passed through a non-linear link function to predict the dependent variable. Examples of generalized linear models include logistic regression models and models assuming a Poisson distribution. Even though a log-normal model is a linear model on a log-transformed dependent variable, the same techniques apply to this type of model since the logarithm transform is not linear. Here, we treat an example with a logistic model in a 2×2 factorial between-subject design. The logistic model has the following non-linear link function called the logistic function: $P(y = 1 \mid \eta) = \frac{1}{1+\exp(-\eta)}$, where η is the latent linear predictor. For

TABLE 7.3: Summary statistics per condition for the simulated data.

Factor A	Factor B	N data	Means
A1	B1	50	0.2
A1	B2	50	0.5
A2	B1	50	0.2
A2	B2	50	0.8

example, in our 2×2 factorial design with main effects A and B and their interaction, η is computed as a linear combination of the intercept plus the main effects and their interaction: $\eta = 1 + \beta_A x_A + \beta_B x_B + \beta_{A\times B} x_{A\times B}$.

Thus, there is a latent level of linear predictions (η), which are then passed through a non-linear link function to predict the probability that the observed data is a success ($P(y = 1)$). We will use this logistic model to analyze an example data set where the dependent variable is dichotomous, coded as either a 1 (indicating success) or a 0 (indicating failure).

We load a simulated data set where the dependent variable codes whether a subject performed a task successfully ($pDV = 1$) or not ($pDV = 0$). Moreover, the data set has two between-subject factors A and B. The means for each of the four conditions are shown in Table 7.3.

```
data("df_contrasts7")
```

To analyze this data, we use scaled sum contrasts, as we had done above for the 2×2 design with response times as the dependent variable; this allows us to interpret the coefficients directly as main effects. Next, we fit a `brms` model. The model specification is the same as the model with response times– with two differences: First, the `family` argument is now specified as `family = bernoulli(link = "logit")` to indicate the logistic model. We do not specify a prior for `sigma`, since there is no residual standard deviation in a logistic model.

```
contrasts(df_contrasts7$A) <- c(-0.5, +0.5)
contrasts(df_contrasts7$B) <- c(-0.5, +0.5)
fit_pDV_AB.sum <- brm(pDV ~ A * B,
                      data = df_contrasts7,
                      family = bernoulli(link = "logit"),
                      prior = c(prior(normal(0, 3), class = Intercept),
                                prior(normal(0, 3), class = b)))
```

```
fixef(fit_pDV_AB.sum)
```

```
##           Estimate Est.Error  Q2.5 Q97.5
## Intercept    -0.36      0.17 -0.70 -0.02
## A1            0.70      0.34  0.04  1.37
## B1            2.10      0.34  1.45  2.78
## A1:B1         1.35      0.66  0.08  2.64
```

The results from this analysis show that the estimates for the two main effects ($\boldsymbol{A}1$ and $\boldsymbol{B}1$) as well as the interaction ($\boldsymbol{A}1 : \boldsymbol{B}1$) are positive and the 95% credible intervals do not include zero. If we want to make a discovery claim, we would need to perform Bayes factor analyses to investigate the evidence that there is for each of the effects.

Next, we discuss how we can obtain model predictions for each of the four experimental conditions for this generalized linear model. To obtain such predictions, we first take a look at the contrast matrix. We simultaneously have contrasts for two main effects and one interaction:

We obtain the posterior samples for the estimates from the model:

```
df_postSamp_pDV <- as.data.frame(fit_pDV_AB.sum)
```

From these, we can compute the posterior samples for the linear predictions for each group. We see in the contrast matrix how we have to combine the posterior samples for the intercept, main effects, and interaction to obtain latent linear predictions for each condition. The first condition (design cell $\boldsymbol{A}1$, $\boldsymbol{B}1$) has a weight of 1 for the intercept, and then weights of -0.5 (for the main effect of $\boldsymbol{A}$), -0.5 (for the main effect of $\boldsymbol{B}$), and 0.25 (for the interaction). The posterior samples for the other conditions are computed accordingly.

```
df_postSamp_pDV <- df_postSamp_pDV %>%
  mutate(A1_B1 = b_Intercept - 0.5 * b_A1 - 0.5 * b_B1 +
           0.25 * `b_A1:B1`,
         A1_B2 = b_Intercept - 0.5 * b_A1 + 0.5 * b_B1 -
           0.25 * `b_A1:B1`,
         A2_B1 = b_Intercept + 0.5 * b_A1 - 0.5 * b_B1 -
           0.25 * `b_A1:B1`,
         A2_B2 = b_Intercept + 0.5 * b_A1 + 0.5 * b_B1 +
           0.25 * `b_A1:B1`)
```

Now, we have computed posterior samples for estimates of the latent linear predictor η for each experimental condition. We can look at the posterior means:

```
df_postSamp_pDV %>%
  select(A1_B1, A1_B2, A2_B1, A2_B2) %>%
  colMeans()
```

```
##   A1_B1   A1_B2   A2_B1   A2_B2
## -1.4223  0.0046 -1.3960  1.3813
```

This shows that these values are not on the probability scale. Instead, they are on the (log-odds) scale of the latent linear predictor η. For presentation and interpretation of the results, it might be much more informative to look at the condition means in terms of the probabilities of success in each of the four conditions. Given that we have the linear predictions for each condition, this can be easily computed by sending all posterior samples for the linear predictions through the link function. Applying the logistic function (`plogis()` in R) transforms the linear predictors to the probability scale:[1]

```
df_postSamp_pDV <- df_postSamp_pDV %>%
  mutate(p_A1_B1 = plogis(A1_B1),
         p_A1_B2 = plogis(A1_B2),
         p_A2_B1 = plogis(A2_B1),
         p_A2_B2 = plogis(A2_B2))
```

Now, we have posterior samples for each condition on the probability scale. We can take a look at the posterior means, and see that these closely correspond to the probabilities in the data that we have seen above in Table 7.3.

```
df_postSamp_pDV %>%
  select(p_A1_B1, p_A1_B2, p_A2_B1, p_A2_B2) %>%
  colMeans()
```

```
## p_A1_B1 p_A1_B2 p_A2_B1 p_A2_B2
##   0.200   0.501   0.204   0.793
```

Of course, the advantage is that we now have posterior samples for these conditions available, and can compute posterior 95% credible intervals (also see Figure 7.5). Rather than computing these manually, the function

[1]The same (with lower precision) can be achieved using `1/(1+exp(-.))`.

`conditional_effects()` can do this for us. By default, all main effects and two-way interactions estimated in the model are shown (this can be changed by including, for example, `effects = "A:B"`).

```
conditional_effects(fit_pDV_AB.sum, robust = FALSE)[]
```

```
## $A
##    A   pDV  B cond__ effect1__ estimate__     se__ lower__ upper__
## 1 A1 0.425 B1      1        A1      0.200 0.0562   0.102   0.322
## 2 A2 0.425 B1      1        A2      0.204 0.0559   0.108   0.322
##
## $B
##    B   pDV  A cond__ effect1__ estimate__     se__ lower__ upper__
## 1 B1 0.425 A1      1        B1      0.200 0.0562   0.102   0.322
## 2 B2 0.425 A1      1        B2      0.501 0.0686   0.363   0.633
##
## $`A:B`
##    A  B   pDV cond__ effect1__ effect2__ estimate__     se__ lower__
## 1 A1 B1 0.425      1        A1        B1      0.200 0.0562   0.102
## 2 A1 B2 0.425      1        A1        B2      0.501 0.0686   0.363
## 3 A2 B1 0.425      1        A2        B1      0.204 0.0559   0.108
## 4 A2 B2 0.425      1        A2        B2      0.793 0.0565   0.667
##   upper__
## 1   0.322
## 2   0.633
## 3   0.322
## 4   0.893
```

We plot the two-way interaction using `brms`, embedding the `conditional_effects()` call in `plot(.)[[1]]`. This allows us to select the first (and here the only) `ggplot2` element and to customize it.

```
plot(conditional_effects(fit_pDV_AB.sum,
                         effects = "A:B",
                         robust = FALSE),
     plot = FALSE)[[1]] +
  labs(y = "Success probability")
```

As a final remark, certain types of interactions might be present at the probability scale, but not at the logit scale (or the other way round); this can happen for any non-linear transformation (see Loftus 1978; Wagenmakers et al. 2012).

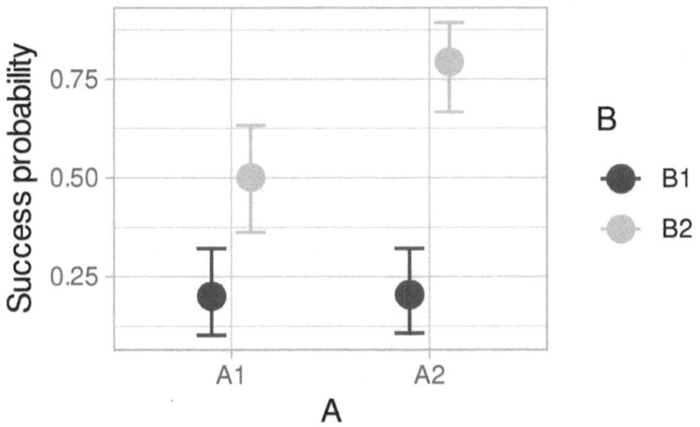

FIGURE 7.5: Means and 95% posterior credible intervals for a simulated data set of successful task performance in a 2×2 design.

7.4 Summary

To summarize, we showed that for contrasts in the context of 2×2 designs, depending on the contrast coding, the factors estimated nested effects or main effects and interactions. We also saw that it is possible to code contrasts for a 2×2 design by creating one factor comprising all design cells, and by specifying all effects of interest in a single contrast matrix. In designs with one factor and one covariate it is possible to "control" for group-differences for differences in the covariate (ANCOVA), or to estimate the extent to which regression slopes are parallel in different experimental conditions. Last, in generalized linear models with non-linear link functions it is possible to obtain posterior samples not only on the latent scale of linear predictors, but also on the scale of the response.

7.5 Further reading

Analysis of variance is discussed in detail in Maxwell, Delaney, and Kelley (2017). A practical book on ANOVA using R is Faraway (2002).

Part III

Advanced models with Stan

8

Introduction to the probabilistic programming language Stan

Introduction to Stan

Stan is a probabilistic programming language for statistical inference written in C++ that can be accessed through several interfaces (e.g., R, Python, etc.). Stan uses an advanced dynamic Hamiltonian Monte Carlo algorithm (Betancourt 2016) based on a variant of the No-U-Turn sampler (known as NUTS: Hoffman and Gelman 2014), which is, in general, more efficient than the traditional Gibbs sampler used in other probabilistic languages such as (Win)BUGS (Lunn et al. 2000) and JAGS (Plummer 2016). In this part of the book, we will focus on the package `rstan` (Guo et al. 2024) that integrates Stan (Carpenter et al. 2017) with R (R Core Team 2023).

In order to understand how to fit a model in Stan and the difficulties one might face, a minimal understanding of the Stan sampling algorithm is needed. Our descriptions of the technical details of HMC are oversimplifications that suffice for the purposes of this book. Stan takes advantage of the fact that the shape of the posterior distribution is completely determined by the priors and the likelihood; in other words, the shape of the posterior is determined by the analytical form of the unnormalized posterior, which is the upper part of the Bayes rule (abbreviated below as $q(\Theta|y)$). This is because the denominator, or marginal likelihood, "only" constitutes the normalizing constant:

$$p(\Theta|y) = \frac{p(y|\Theta) \cdot p(\Theta)}{p(y)} \tag{8.1}$$

$$q(\Theta|y) = p(y|\Theta) \cdot p(\Theta) \tag{8.2}$$

Thus, the unnormalized posterior is proportional ($\propto$) to the posterior distribution:

$$q(\Theta|y) \propto p(\Theta|y) \tag{8.3}$$

DOI: 10.1201/9780429342646-11

The Stan sampler uses Hamiltonian dynamics and treats the vector of parameters, $\boldsymbol{\Theta}$ (that could range from a vector containing a couple of parameters, e.g., $\langle \mu, \sigma \rangle$, to a vector of hundreds of parameters in hierarchical models), as the position of a frictionless particle that glides on the *negative logarithm of the unnormalized posterior*. That means that high probability places are valleys and low probability places are peaks in this space.[1] Whenever the particle comes to a halt, that location constitutes a sample from the distribution.

However, Stan doesn't just let the particle glide until it reaches the bottom of this space. If we let that happen, the particle would stop at the mode of the posterior distribution, and one would not be able to obtain samples from the rest of the distribution. Stan uses a complex algorithm to determine the weight of the particle and the momentum applied to it, as well as when to stop the particle trajectory to take a sample. Because we need to know the velocity of this particle, Stan needs to be able to calculate the derivative of the log unnormalized posterior with respect to the parameters (recall that velocity is the first derivative with respect to the position). This means that if the parameter space is differentiable and relatively smooth (i.e., does not have any big break or sharp angle), and if the tuning parameters of the Stan algorithm are well adjusted–as should happen in the warm-up period–these samples are going to represent samples of the true posterior distribution. Bear in mind that the geometry of the posterior has a big influence on whether the algorithm will converge (quickly) or not: If the space is very flat, because there isn't much data and the priors are not informative, then the particle may need to glide for a long time before it gets to a high probability area, that is a valley; if there are several valleys (multimodality) the particle may never leave the vicinity of one of them; and if the space is funnel-shaped, the particle may never explore the funnel. One of the reasons for the difficulties in exploring complicated spaces is that the continuous path of the "particle" is discretized and divided into steps, and the *step size* is optimized for the entire posterior space. In spaces that are too complex, such as a funnel, a step size might be too small to explore the wide part of the funnel, but too large to explore the narrow part; we will deal with this problem in section 9.1.2.

Although our following example assumes a vector of two parameters and thus a simple geometry, real-world examples can easily have hundreds of parameters defining an unnormalized posterior space with hundreds of dimensions.

[1]In the specific case of a model with two parameters, e.g., $\langle \mu, \sigma \rangle$, the physical analogy works quite well: The $\langle x, y \rangle$ coordinates of the particle would be determined by $\langle \mu, \sigma \rangle$, and its z coordinate would be established by the height of the negative logarithm of the unnormalized posterior.

One question that might arise here is the following: Given that we already know the shape of the posterior, why do we need samples? After all, the posterior is just the unnormalized posterior multiplied by some number, the normalizing constant.

To make this discussion concrete, let's say that we have a subject that participates in a memory test, and in each trial we get a noisy score from their true working memory score. We assume that at each trial, the score is elicited with normally distributed noise. If we want to estimate the score and how much the noise makes it vary from trial to trial, we are assuming a normal likelihood and we want to estimate its mean and standard deviation.

We will use simulate data produced by a normal distribution with a true mean of 3 and a true standard deviation of 10 (the units are arbitrary):

```
y <- rnorm(n = 100, mean = 3, sd = 10)
head(y)
```

```
## [1]  -4.41 -10.92  21.39   4.23   5.53   4.65
```

As always, given our prior knowledge, we decide on priors. In this case, we use a log-normal prior for the standard deviation, σ, since it can only be positive, but except for that, the prior distributions are quite arbitrary in this example.

$$\begin{aligned} \mu &\sim Normal(0, 20) \\ \sigma &\sim LogNormal(3, 1) \end{aligned} \tag{8.4}$$

The unnormalized posterior will be the product of the likelihood of each data point times the prior for each parameter:

$$q(\mu, \sigma | y) = \prod_{n}^{100} Normal(y_n | \mu, \sigma) \cdot Normal(\mu | 0, 20) \cdot LogNormal(\sigma | 3, 1) \tag{8.5}$$

where $y = -4.411, -10.918, \ldots$

We can also define the unnormalized posterior, $q(\cdot)$, as a function in R:

```
q <- function(mu, sigma, y) {
  dnorm(x = mu, mean = 0, sd = 20) *
    dlnorm(x = sigma, mean = 3, sd = 1) *
    prod(dnorm(x = y, mean = mu, sd = sigma))
}
```

For example, if we want to know the unnormalized posterior density for the vector of parameters $\langle \mu, \sigma \rangle = \langle 0, 5 \rangle$, we do the following:

```
q(mu = 0, sigma = 5, y = y)
```

```
## [1] 1.07e-217
```

The shape of the unnormalized posterior density is completely defined and it will look like Figure 8.1.

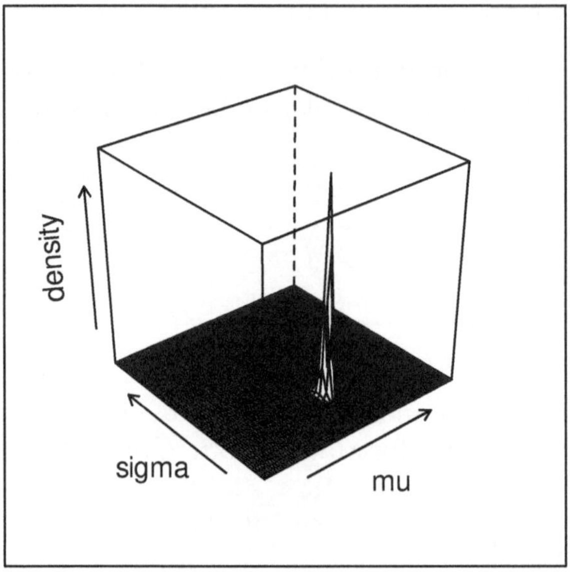

FIGURE 8.1: The unnormalized posterior defined by Equation (8.5).

Why is the shape of the unnormalized posterior density not too useful for us? The main reason is that unless we already know which probability distribution we are dealing with (e.g., normal, Bernoulli, etc.) or we can easily integrate it (which can only be done in simpler cases), we cannot do much with the analytical form of the unnormalized posterior: We cannot calculate credible intervals, or know how likely it is that the true score is above or below zero, and even the mean of the posterior is impossible to calculate. This is because the unnormalized posterior distribution represents the shape of the posterior distribution. With just the shape of an unknown distribution, we can only answer the following question: What is the most (or least) likely value of the vector of parameters? We can answer this question by searching for the highest (or lowest) place in that shape. This leads us to the maximum a posteriori (MAP) estimate (i.e., the highest point in Figure 8.1, or the lowest point of the negative log unnormalized log posterior), which is in a sense the "Bayesian counterpart" of the penalized maximum likelihood estimate (MLE); see Green

(1998) and Scheipl, Kneib, and Fahrmeir (2013) for a discussion of penalized likelihood and its connection to Bayesian inference. However, the MAP is not truly Bayesian since it doesn't take into account the uncertainty of the posterior. Even calculating the mode is not always trivial. In simple cases such as this one, one can calculate it analytically; but in more complex cases, relatively complicated algorithms are needed. (If we can recognize the shape as a distribution, we are in a different situation. In that case, we might know already the formulas for the expectation, variance, etc. This is what we did in chapter 2, but this is an unusual situation in realistic analyses.) As we mentioned before, if we want to get posterior density values, we need the denominator of the Bayes rule (or marginal likelihood), $p(y)$, which requires integrating the unnormalized posterior. Even this is not too useful if we want to communicate findings: almost every summary statistic requires us to solve more integrals, and except for a handful of cases, these integrals might not have an analytical solution.

If we want to be able to calculate summary statistics of the posterior distribution (mean, quantiles, etc.), we are going to need samples from this distribution. This is because with enough samples of a probability distribution, we can achieve very good approximations of summary statistics.

Stan will take care of returning samples from the posterior distribution, if the log unnormalized posterior distribution is differentiable and can be expressed as follows:[2]

$$\log(q(\Theta|y)) = \sum_{n} \log(p(y_n|\Theta)) + \sum_{par} \log(p(\Theta_{par})) \tag{8.6}$$

where n indicates each data point and par each parameter. In our case, this corresponds to the following:

$$\begin{aligned}\log(q(\mu, \sigma|y)) = &\sum_{n}^{100} \log(Normal(y_n|\mu, \sigma)) + \log(Normal(\mu|0, 20))\\ &+ \log(LogNormal(\sigma|3, 1))\end{aligned} \tag{8.7}$$

In the following sections, we'll see how we can implement this model and many others in Stan.

[2]Incidentally, this $\log(q(\cdot))$ is the variable `target` in a Stan model and `lp__` in its output; see the aside below.

8.1 Stan syntax

A Stan program is usually saved as a `.stan` file and accessed through R (or other interfaces) and it is organized into a sequence of optional and obligatory blocks, which must be written in order. The Stan language is different from R and it is loosely based on C++; one important aspect to pay attention to is that every statement ends in a semi-colon, `;`. Blocks (`{}`) do not end in semi-colons. Some functions in Stan are written in the same way as in R (e.g., `mean()`, `sum()`, `max()`, `min()`). But some are different; when in doubt, Stan documentation (`https://mc-stan.org/users/documentation/`) can be extremely helpful. In addition, the package `rstan` provides the function `lookup()` to look up for translations of functions. For example, in section 4.3, we saw that the R function `plogis()` is needed to convert from log-odds to probability space. If we need it in a Stan program, we can look for it in the following way:

```
lookup(plogis)
```

```
##         StanFunction                                 Arguments
## 262        inv_logit                                  (T);(int)
## 307    log_inv_logit                                  (T);(int)
## 323 logistic_ccdf_log (real, real, T);(vector, vector, vector)
## 324      logistic_cdf (real, real, T);(vector, vector, vector)
## 325  logistic_cdf_log (real, real, T);(vector, vector, vector)
## 326    logistic_lccdf (real, real, T);(vector, vector, vector)
## 327     logistic_lcdf (real, real, T);(vector, vector, vector)
##     ReturnType
## 262     T;real
## 307     T;real
## 323     T;real
## 324     T;real
## 325     T;real
## 326     T;real
## 327     T;real
```

There are three columns in the output of this call. The first one indicates Stan function names, the second one their arguments with their type, and the third one the type they return. Unlike R, Stan is strict with the type of the variables.[3] In order to decide which function to use, it is necessary to look at the Stan documentation and find the function that matches our specific needs (for `plogis()`, the corresponding function would be `inv_logit()`).

Another important difference from R is that every variable needs to be declared with its type (real, integer, vector, matrix, etc.). The next two sections exemplify these details through basic Stan programs.

8.2 A first simple example with Stan: Normal likelihood

Let's fit a Stan model to estimate the simple example given at the introduction of this chapter, where we simulate data in R from a normal distribution with a true mean of `3` and a true standard deviation of `10`:

```
y <- rnorm(n = 100, mean = 3, sd = 10)
```

As mentioned earlier, Stan code is organized in blocks. The first block indicates what constitutes data for the model:

```
data {
  int<lower = 1> N;  // Total number of trials
  vector[N] y;  // Score in each trial
}
```

The variable of type `int` (integer) represents the number of trials. In addition to the type, some constraints can be indicated with `lower` and `upper`. In this case, `N` can't be smaller than `1`. These constraints serve as a sanity check; if they are not satisfied, we get an error and the model won't run. The data are stored in a vector of length `N`, unlike R, vectors (and matrices and arrays) need to be defined with their dimensions. Comments are indicated with `//` rather than `#`.

The next block indicates the parameters of the model:

```
parameters {
  real mu;
  real<lower = 0> sigma;
}
```

The two parameters are real numbers, and `sigma` is constrained to be positive.

[3] In this kind of output, there are some types that are new to the R user (but they are also used in C++). A return type `T` indicates that the type of the output of the function is the same as type of the argument.

Finally, we indicate the prior distributions and likelihood functions in the model block:

```
model {
  // Priors:
  target += normal_lpdf(mu | 0, 20);
  target += lognormal_lpdf(sigma | 3, 1);
  // Likelihood:
  for(i in 1:N)
    target += normal_lpdf(y[i] | mu, sigma);
}
```

The variable `target` is a reserved word in Stan; every statement with `target +=` adds terms to the unnormalized *log* posterior density. We do this because adding to the unnormalized log posterior amounts to multiplying a term in the numerator of the unnormalized posterior. As explained earlier, Stan uses the shape of the unnormalized posterior to sample from the actual posterior distribution. See online section B.1 for a more detailed explanation, and see online section B.2 for alternative notations.

We didn't use curly brackets with the for-loop; this is a common practice if the for-loop has only one line, but brackets can be added and are obligatory if the for-loop spans several lines.

It's also possible to avoid the for-loop since many functions are vectorized in Stan:

```
model {
  // Priors:
  target += normal_lpdf(mu | 0, 20);
  target += lognormal_lpdf(sigma | 3, 1);
  // Likelihood:
  target += normal_lpdf(y | mu, sigma);
}
```

The for-loop and vectorized versions give us the same output: The for-loop version evaluated the log-likelihood at each value of `y` and added it to `target`. The vectorized version does not create a vector of log-likelihoods; instead, it sums up the log-likelihood evaluated at each element of `y` and then it adds that to `target`.

The complete model looks like this:

```
data {
  int<lower = 1> N;  // Total number of trials
  vector[N] y;  // Score in each trial
```

```
}
parameters {
  real mu;
  real<lower = 0> sigma;
}
model {
  // Priors:
  target += normal_lpdf(mu | 0, 20);
  target += lognormal_lpdf(sigma | 3, 1);
  // Likelihood:
  target += normal_lpdf(y | mu, sigma);
}
```

You can save the above code as `normal.stan`. Alternatively, you can use the version stored in the package `bcogsci`. (Typing `?stan-normal` in the R console provides some documentation for the model.) You can access the code of the models of this book by using `system.file("stan_models", "name_of_the_model.stan", package = "bcogsci")`.

```
normal <- system.file("stan_models",
                      "normal.stan",
                      package = "bcogsci")
```

This command just points to a text file that the package `bcogsci` stores on your computer. You can open it to read the code (with any text editor, or `readLines()` in R). You'll need to compile this code and run it with `stan()`.

Stan requires the data to be in a list object in R. Below, we fit the model with the default number of chains and iterations.

```
y <- rnorm(n = 100, mean = 3, sd = 10)
lst_score_data <- list(y = y, N = length(y))
# Fit the model with the default values of number of
# chains and iterations: chains = 4, iter = 2000
fit_score <- stan(normal, data = lst_score_data)
# alternatively:
# stan("normal.stan", data = lst_score_data)
```

Inspect how well the chains mixed in Figure 8.2. The chains for each parameter should look like a "fat hairy caterpillar" (Lunn et al. 2012); see section 3.2.1.2 for a brief discussion about convergence.

```
traceplot(fit_score, pars = c("mu", "sigma"))
```

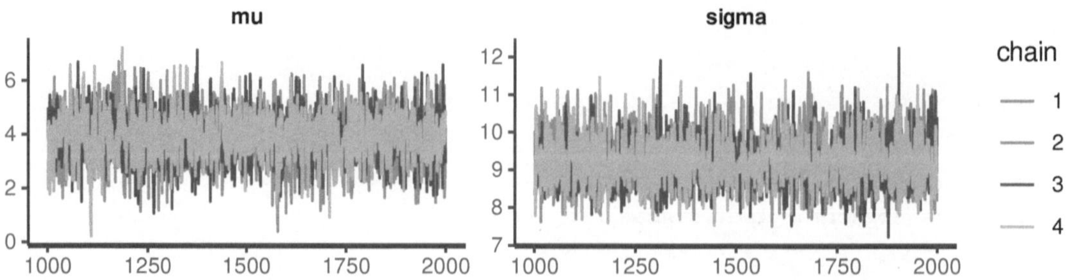

FIGURE 8.2: Traceplots of `mu` and `sigma` from the model `fit_score`.

We can see a summary of the posterior by either printing out the model fit, or by plotting it. The summary displayed by the function `print` includes means, standard deviations (`sd`), quantiles, Monte Carlo standard errors for the mean of the posterior (`se_mean`), split Rhats, and effective sample sizes (`n_eff`). The summaries are computed after removing the warmup and merging together all chains. The `se_mean` is unrelated to the `se` of an estimate in the parallel frequentist model. Similarly to a large effective sample size, small Monte Carlo standard errors indicate a successful sampling procedure: with a large value of `n_eff` and a small value for `se_mean` we can be relatively sure of the reliability of the mean of the posterior. However, what constitutes a large or small `se_mean` is harder to define (see Vehtari et al. 2021 for a more extensive discussion).[4]

```
print(fit_score, pars = c("mu", "sigma"))
```

```
## Inference for Stan model: anon_model.
## 4 chains, each with iter=2000; warmup=1000; thin=1;
## post-warmup draws per chain=1000, total post-warmup draws=4000.
##
##        mean se_mean   sd 2.5%  25%  50%  75% 97.5% n_eff Rhat
## mu     3.91    0.02 0.92 2.06 3.29 3.91 4.53   5.7  3707    1
## sigma  9.23    0.01 0.66 8.07 8.75 9.19 9.65  10.7  4026    1
##
## Samples were drawn using NUTS(diag_e) at Sat Jan 11 23:48:25 2025.
## For each parameter, n_eff is a crude measure of effective sample size,
## and Rhat is the potential scale reduction factor on split chains (at
## convergence, Rhat=1).
```

[4] We simplify the output of `print` in the text after this call, by actually calling `summary(fit, pars = pars, probs = c(0.025, 0.975))$summary`.

After transforming the `stanfit` object into a data frame, it's possible to provide summary plots as the one shown in Figure 8.3. The package `bayesplot` (Gabry and Mahr 2024) is a wrapper around `ggplot2` (Wickham, Chang, et al. 2024) and has several convenient functions to plot the samples. Bayesplot functions for posterior summaries start with `mcmc_`:

```
df_fit_score <- as.data.frame(fit_score)
mcmc_hist(df_fit_score, pars = c("mu", "sigma"))
```

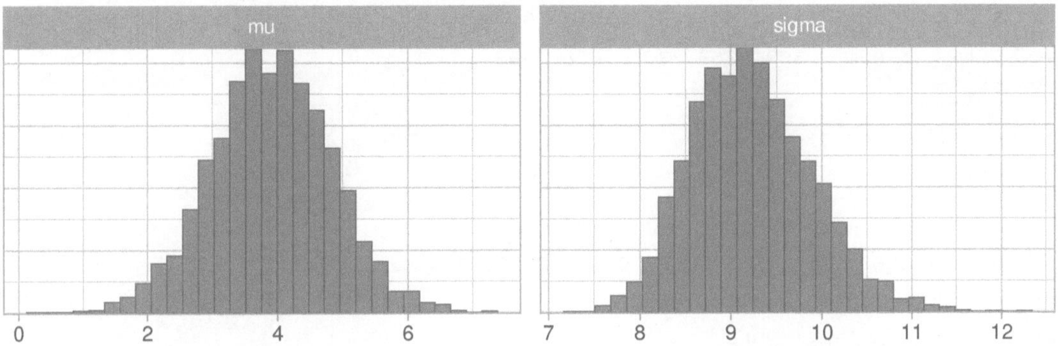

FIGURE 8.3: Histograms of the samples of the posterior distributions of `mu` and `sigma` from the model `fit_score`.

There are also several ways to get the samples for other summaries or customized plots, depending on whether we want a list, a data frame, or an array.[5]

```
# The function extract from rstan is sometimes overwritten by
# a tidyverse version with the same name, so make sure that
# you are using the right function:
rstan::extract(fit_score) %>%
  str()
```

```
## List of 3
##  $ mu   : num [1:4000(1d)] 3.82 3.54 3.81 3.66 5.15 ...
##   ..- attr(*, "dimnames")=List of 1
##   .. ..$ iterations: NULL
##  $ sigma: num [1:4000(1d)] 11.33 9.13 10.34 8.78 8.6 ...
##   ..- attr(*, "dimnames")=List of 1
##   .. ..$ iterations: NULL
```

[5]In addition, the new package `posterior` provides useful tools for working with output from Bayesian models. It allows for converting the output between many different formats, and offers consistent methods for operations commonly performed on samples, as well as summary functions.

```
## $ lp__ : num [1:4000(1d)] -372 -368 -369 -368 -369 ...
##   ..- attr(*, "dimnames")=List of 1
##   .. ..$ iterations: NULL
```

```
as.data.frame(fit_score) %>%
  str(list.len = 5)
```

```
## 'data.frame':    4000 obs. of  3 variables:
## $ mu   : num  4.92 2.6 4.91 3.07 4.02 ...
## $ sigma: num  10.04 8.51 9.97 9.39 9.04 ...
## $ lp__ : num  -369 -369 -369 -368 -368 ...
```

```
as.array(fit_score) %>%
  str()
```

```
##  num [1:1000, 1:4, 1:3] 4.92 2.6 4.91 3.07 4.02 ...
##  - attr(*, "dimnames")=List of 3
##   ..$ iterations: NULL
##   ..$ chains    : chr [1:4] "chain:1" "chain:2" "chain:3" "chain:4"
##   ..$ parameters: chr [1:3] "mu" "sigma" "lp__"
```

8.3 Another simple example: Cloze probability with Stan with the binomial likelihood

Let's fit a Stan model (`binomial_cloze.stan`) to estimate the cloze probability of a word given its context: that is, what is the probability of an upcoming word given its previous context; the model that was detailed in section 2.2 and was fit in section 3.1. We want to estimate θ, the cloze probability of producing "*umbrella*," given the following data: "*umbrella*" was produced in 80 out of 100 trials. We assume a binomial distribution as the likelihood function, and $Beta(a = 4, b = 4)$ as a prior distribution for the cloze probability.

```
data {
  // Total number of answers
  int<lower = 1> N;
  // Number of times umbrella was answered:
  int<lower = 0, upper = N> k;
}
parameters {
```

```
  // theta is a probability, must be constrained between 0 and 1
  real<lower = 0, upper = 1> theta;
}
model {
  // Prior on theta:
  target += beta_lpdf(theta | 4, 4);
  // Likelihood:
  target += binomial_lpmf(k | N, theta);
}
```

There is only one parameter in this model, cloze probability represented with the parameter `theta`, which is a real number constrained between 0 and 1. Another difference between this and the previous example is that the likelihood function ends with `_lpmf` rather than with `_lpdf`. This is because Stan differentiates between distributions of continuous variables, i.e., probability density functions (PDFs) and distributions of discrete variables, i.e., probability mass functions (PMFs).

```
lst_cloze_data <- list(k = 80, N = 100)
binomial_cloze <- system.file("stan_models",
                              "binomial_cloze.stan",
                              package = "bcogsci")
fit_cloze <- stan(binomial_cloze, data = lst_cloze_data)
```

Print the summary of the posterior distribution of θ below, and show its posterior distribution graphically (see Figure 8.4):

```
print(fit_cloze, pars = c("theta"))
```

```
##       mean 2.5% 97.5% n_eff Rhat
## theta 0.78  0.7  0.85  1371    1
```

```
df_fit_cloze <- as.data.frame(fit_cloze)
mcmc_dens(df_fit_cloze, pars = "theta") +
  geom_vline(xintercept = mean(df_fit_cloze$theta))
```

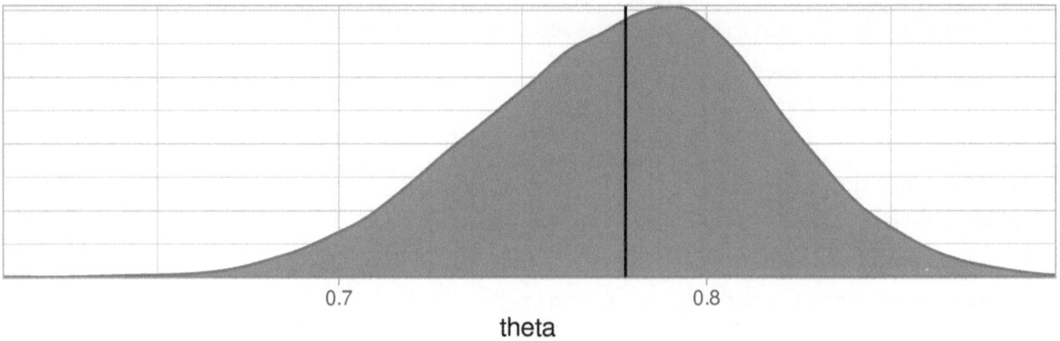

FIGURE 8.4: The posterior distribution of the cloze probability of umbrella (the parameter θ).

8.4 Regression models in Stan

In the following sections, we will revisit and expand on some of the examples that were fit with `brms` in chapter 4.

8.4.1 A first linear regression in Stan: Does attentional load affect pupil size?

As in section 4.1, we focus on the effect of cognitive load on one subject's pupil size with a subset of the data from Wahn et al. (2016). We use the following likelihood and priors. For details about our decision on priors and likelihood, see section 4.1.

$$
\begin{aligned}
p\_size_n &\sim Normal(\alpha + c\_load_n \cdot \beta, \sigma)\\
\alpha &\sim Normal(1000, 500)\\
\beta &\sim Normal(0, 100)\\
\sigma &\sim Normal_+(0, 1000)
\end{aligned}
\tag{8.8}
$$

The Stan model `pupil.stan` is as follows:

```
data {
  int<lower=1> N;
  vector[N] p_size;
  vector[N] c_load;
}
parameters {
  real alpha;
  real beta;
```

```
  real<lower = 0> sigma;
}
model {
  // priors:
  target += normal_lpdf(alpha | 1000, 500);
  target += normal_lpdf(beta | 0, 100);
  target += normal_lpdf(sigma | 0, 1000)
    - normal_lccdf(0 | 0, 1000);
  // likelihood
  target += normal_lpdf(p_size | alpha + c_load * beta, sigma);
}
```

Because we are fitting a regression, we use the location (μ) of the likelihood function to regress `p_size` with the following equation `alpha + c_load * beta`, where both `p_size` and `c_load` are vectors defined in the data block. The following line accumulates the log-likelihood of every observation:

```
target += normal_lpdf(p_size | alpha + c_load * beta, sigma);
```

This is equivalent to and slightly faster than the following lines:

```
for(n in 1:N)
    target += normal_lpdf(p_size[n] | alpha + c_load[n] * beta, sigma);
```

A statement that requires some explanation is the following:

```
target += normal_lpdf(sigma | 0, 1000)
    - normal_lccdf(0 | 0, 1000);
```

As in our original example in section 4.1, we are assuming a truncated normal distribution as a prior for σ. Not only are we setting a lower boundary to the parameter with `lower = 0`, but we are also "correcting" its prior distribution by subtracting `normal_lccdf(0 | 0, 1000)`, where `lccdf` stands for log complement of a cumulative distribution function. Once we add a lower boundary, the probability mass under half of the "regular" normal distribution should be one, that is, when we integrate from zero (rather than from minus infinity) to infinity. As discussed in online section A.2, we need to normalize the PDF by dividing it by the difference of its CDF evaluated in the new boundaries ($a = 0$ and $b = \infty$ in our case):

$$f_{[a,b]}(x) = \frac{f(x)}{F(b) - F(a)} \tag{8.9}$$

where f is the PDF and F the CDF.

This equation in log-space is:

$$log(f_{[a,b]}(x)) = log(f(x)) - log(F(b) - F(a)) \tag{8.10}$$

In Stan $\log(f(x))$ corresponds to `normal_lpdf(x |...)`, and `log(F(x))` to `normal_lcdf(x|...)`. Because in our example $b = \infty$, $F(b) = 1$, we are dealing with the complement of the log CDF evaluated at $a = 0$, $\log(1 - F(0))$, that is why we use `normal_lccdf(0 | ...)` (notice the double `c`; this symbol represents the complement of the CDF).

To be able to fit the model, Stan requires the data to be input as a list: First, load the data and center the dependent variable in a data frame, then create a list, and finally fit the model.

```
df_pupil <- df_pupil %>%
  mutate(c_load = load - mean(load))
```

```
ls_pupil <- list(p_size = df_pupil$p_size,
                 c_load = df_pupil$c_load,
                 N = nrow(df_pupil))
pupil <- system.file("stan_models",
                     "pupil.stan",
                     package = "bcogsci")
fit_pupil <- stan(pupil, data = ls_pupil)
```

Check the traceplots (Figure 8.5).

```
traceplot(fit_pupil, pars = c("alpha", "beta", "sigma"))
```

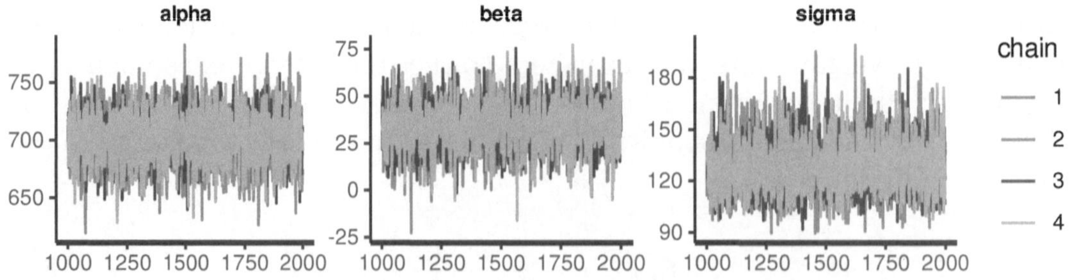

FIGURE 8.5: Traceplots of `alpha`, `beta`, and `sigma` from the model `fit_pupil`.

Examine some summaries of the marginal posterior distributions of the parameters of interest:

```
print(fit_pupil, pars = c("alpha", "beta", "sigma"))
```

```
##        mean  2.5% 97.5% n_eff Rhat
## alpha 701.6 661.5 743.1  3299    1
## beta   33.6  10.4  56.8  3207    1
## sigma 128.8 103.4 161.7  2908    1
```

Plot the posterior distributions (Figure 8.6).

```
df_fit_pupil <- as.data.frame(fit_pupil)
mcmc_hist(fit_pupil, pars = c("alpha", "beta", "sigma"))
```

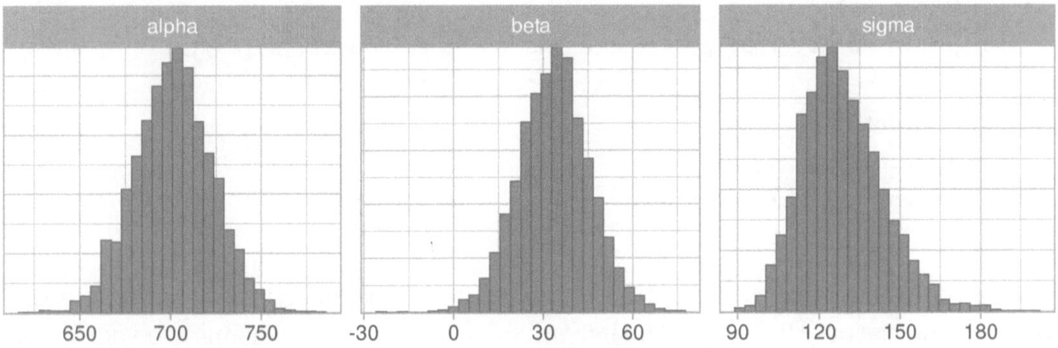

FIGURE 8.6: Histograms of the posterior samples of `alpha`, `beta`, and `sigma` from the model `fit_pupil`.

To determine the probability that the posterior for beta is larger than zero given the model and data, examine the proportion of samples above zero:

```
# We are using df_fit_pupil and not the "raw" Stanfit object.
mean(df_fit_pupil$beta > 0)
```

```
## [1] 0.997
```

To generate prior or posterior predictive distributions, we can create our own functions in R with the `purrr` function `map_dfr()` (or a for-loop) as we did in section 4.2 with the function `lognormal_model_pred()`. Alternatively, we can use the `generated quantities` block in our model:

```
data {
  int<lower = 1> N;
```

```
  vector[N] c_load;
  int<lower= 0, upper = 1> onlyprior;
  vector[N] p_size;
}
parameters {
  real alpha;
  real beta;
  real<lower = 0> sigma;
}
model {
  // priors including all constants
  target += normal_lpdf(alpha | 1000, 500);
  target += normal_lpdf(beta | 0, 100);
  target += normal_lpdf(sigma | 0, 1000)
    - normal_lccdf(0 | 0, 1000);
  if (!onlyprior)
    target += normal_lpdf(p_size | alpha + c_load * beta, sigma);
}
generated quantities {
  array[N] real p_size_pred;
  p_size_pred = normal_rng(alpha + c_load * beta, sigma);
}
```

For most of the probability functions, there is a matching pseudorandom number generator (PRNG) with the suffix `_rng`. Here we are using the vectorized function `normal_rng()`. Once `p_size_pred` is declared as an array of size `N`, the following statement generates N predictions (for each iteration of the sampler):

```
p_size_pred = normal_rng(alpha + c_load * beta, sigma);
```

At the moment not all the PRNG are vectorized, but the ones that are only allow for arrays and, confusingly enough, not vectors. We define an array by indicating `array`, between brackets, the length of each dimension, then the type, and finally the name of the variable. For example, to define an array of real numbers with three dimensions of length 6, 7, and 10, we write `array[6, 7, 10] real var`.[6] Vectors and matrices are also valid types for an array. See online section B.4 for more about the difference between arrays and vectors, and other algebra types. We also included a data variable called `onlyprior`, this is an integer that can only be set to 1 (TRUE) or 0 (FALSE). When `onlyprior == 1`, the likelihood is omitted from the model, `p_size` is ignored, and `p_size_pred` is the prior predictive distribution. When `onlyprior == 0`, the

[6] The notation for arrays has changed in recent versions, the previous notation would have been `real[6, 7, 10] var`.

likelihood is incorporated in the model (as it is in the original code `pupil.stan`) using `p_size`, and `p_size_pred` is the posterior predictive distribution.

If we want posterior predictive distributions, we fit the model to the data and set `onlyprior = 0` and if we want prior predictive distributions, we sample from the priors and set `onlyprior = 1`. Then we use `bayesplot` functions to visualize predictive checks.

For posterior predictive checks, we would do the following:

```
ls_pupil <- list(onlyprior = 0,
                 p_size = df_pupil$p_size,
                 c_load = df_pupil$c_load,
                 N = nrow(df_pupil))
pupil_gen <- system.file("stan_models",
                         "pupil_gen.stan",
                         package = "bcogsci")
fit_pupil <- stan(file = pupil_gen, data = ls_pupil)
```

Store the predicted pupil sizes in `yrep_pupil`. This variable contains an $N_{samples} \times N_{observations}$ matrix, i.e., each row of the matrix is a draw from the posterior predictive distribution, i.e., a vector with one element for each of the data points in y.

```
yrep_pupil <- extract(fit_pupil)$p_size_pred
dim(yrep_pupil)
```

```
## [1] 4000   41
```

Predictive checks functions in `bayesplot` (starting with `ppc_`) require a vector with the observations in the first argument and a matrix with the predictive distribution as its second argument. As an example, in Figure 8.7 we use an overlay of densities and we draw only 50 elements (that is 50 predicted data sets).

```
ppc_dens_overlay(df_pupil$p_size, yrep = yrep_pupil[1:50, ])
```

For prior predictive distributions, we simply set `onlyprior = 1`. The observations (`p_size`) are ignored by the model, but are required by the data block in Stan. If we haven't collected data yet, we could include a vector of zeros.

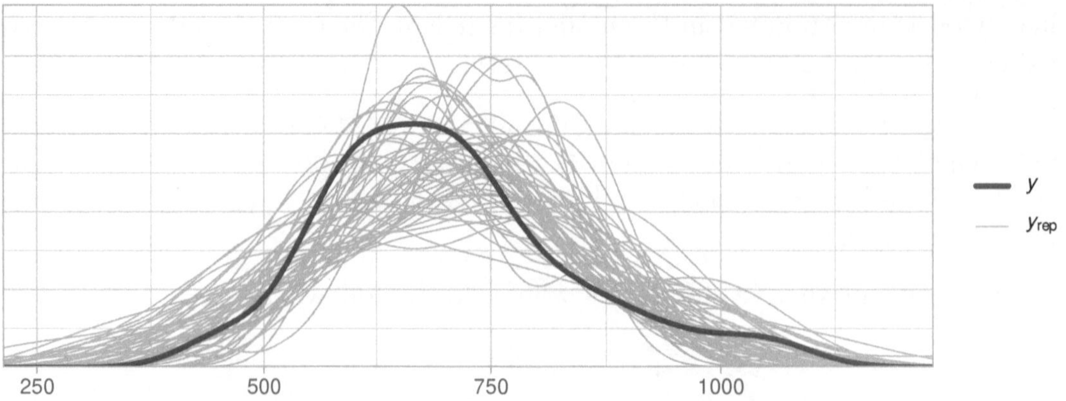

FIGURE 8.7: A posterior predictive check showing 50 predicted density plots from the model `fit_pupil` against the observed data.

```
ls_pupil_prior <- list(onlyprior = 1,
                       p_size = df_pupil$p_size,
                       # or: p_size = rep(0, nrow(df_pupil)),
                       c_load = df_pupil$c_load,
                       N = nrow(df_pupil))
prior_pupil <- stan(pupil_gen, data = ls_pupil_prior,
                    control = list(adapt_delta = 0.99))
```

To avoid divergent transitions, increase the `adapt_delta` parameter's default value from 0.8 to 0.99. It is important to highlight that we cannot safely ignore the warnings of the above model, even if we are not fitting data. This is so because in practice one is still sampling a density using Hamiltonian Monte Carlo, and thus the prior sampling process can break in the same ways as the posterior sampling process. Prior predictive distributions as the one shown in Figure 8.8 can be plot with `ppd_dens_overlay()` (and in general with functions starting with `ppd_` which don't require an argument with data).

```
yrep_prior_pupil <- extract(prior_pupil)$p_size_pred
ppd_dens_overlay(yrep_prior_pupil[1:50, ])
```

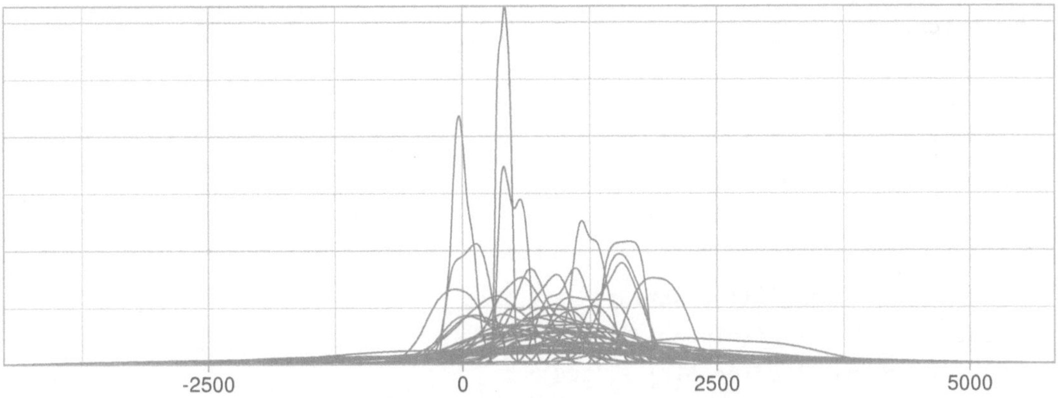

FIGURE 8.8: Prior predictive distribution showing 50 predicted density plots from the model `fit_pupil`.

8.4.2 Interactions in Stan: Does attentional load interact with trial number affecting pupil size?

We'll expand the previous model to also include the effect of (centered) trial and its interaction with cognitive load on one subject's pupil size. Our new likelihood is as follows:

$$p\_size_n \sim Normal(\alpha + c\_load_n \cdot \beta_1 + c\_trial \cdot \beta_2 + c\_load \cdot c\_trial \cdot \beta_3, \sigma) \quad (8.11)$$

Define priors for all the new βs. Since we don't have more information about the new predictors, they are sampled from identical prior distributions:

$$\begin{aligned}
\alpha &\sim Normal(1000, 500) \\
\beta_1 &\sim Normal(0, 100) \\
\beta_2 &\sim Normal(0, 100) \\
\beta_3 &\sim Normal(0, 100) \\
\sigma &\sim Normal_+(0, 1000)
\end{aligned} \quad (8.12)$$

The following Stan model, `pupil_int1.stan`, is the direct translation of the new priors and likelihood.

```
data {
  int<lower = 1> N;
  vector[N] c_load;
  vector[N] c_trial;
  vector[N] p_size;
}
parameters {
```

```
  real alpha;
  real beta1;
  real beta2;
  real beta3;
  real<lower = 0> sigma;
}
model {
  // priors including all constants
  target += normal_lpdf(alpha | 1000, 500);
  target += normal_lpdf(beta1 | 0, 100);
  target += normal_lpdf(beta2 | 0, 100);
  target += normal_lpdf(beta3 | 0, 100);
  target += normal_lpdf(sigma | 0, 1000)
    - normal_lccdf(0 | 0, 1000);
  target += normal_lpdf(p_size | alpha + c_load * beta1 +
                                 c_trial * beta2 +
                                 c_load .* c_trial * beta3, sigma);
}
```

When there are matrices or vectors involved, `*` indicates matrix multiplication whereas `.*` indicates element-wise multiplication; in R, `%*%` indicates matrix multiplication whereas `*` indicates element-wise multiplication.

There is, however, an alternative notation that can simplify our code. In the following likelihood, p_size is a vector of N observations (in this case 41), X is the model matrix with a dimension of $N \times N_{pred}$ (in this case 41×3), and β a vector of N_{pred} (in this case, 3) rows. Assuming that β is a vector, we indicate with one line that each parameter β_n is sampled from identical prior distributions.

$$\begin{aligned} p\_size &\sim Normal(\alpha + X \cdot \beta, \sigma) \\ \beta &\sim Normal(0, 100) \\ \sigma &\sim Normal_{+}(0, 1000) \end{aligned} \tag{8.13}$$

The translation into Stan code is the following:

```
data {
  int<lower = 1> N;
  int<lower = 0> K;   // number of predictors
  matrix[N, K] X;   // model matrix
  vector[N] p_size;
}
parameters {
```

```
  real alpha;
  vector[K] beta;
  real<lower = 0> sigma;
}
model {
  // priors including all constants
  target += normal_lpdf(alpha | 1000, 500);
  target += normal_lpdf(beta | 0, 100);
  target += normal_lpdf(sigma | 0, 1000)
    - normal_lccdf(0 | 0, 1000);
  target += normal_lpdf(p_size | alpha + X * beta, sigma);
}
```

For some likelihood functions, Stan provides a more efficient implementation of the linear regression than the one manually written in the previous code. It's critical to understand that, in general, a more efficient implementation should not only be faster, but should also achieve the same number of effective samples (or more) than a less efficient implementation (and should also show convergence). In this case, we can achieve that using `_glm` functions. We can replace the last line with the following statement (the order of the arguments is important):

```
target += normal_id_glm_lpdf(p_size | X, alpha, beta, sigma);
```

The most optimized model, `pupil_int.stan`, includes this last statement. We prepare the data as follows: First create a centered version of trial, `c_trial` and load `c_load`, then use the function `model.matrix()` to create the `X` matrix that contains in each column our predictors and omits the intercept with `0 +`.

```
df_pupil <- df_pupil %>%
  mutate(c_trial = trial - mean(trial),
         c_load = load - mean(load))
X <- model.matrix(~ 0 + c_load * c_trial, df_pupil)
ls_pupil_X <- list(p_size = df_pupil$p_size,
                   X = X,
                   K = ncol(X),
                   N = nrow(df_pupil))
```

```
pupil_int <- system.file("stan_models",
                         "pupil_int.stan",
                         package = "bcogsci")
fit_pupil_int <- stan(pupil_int, data = ls_pupil_X)
```

```
print(fit_pupil_int, pars = c("alpha", "beta", "sigma"))
```

```
##            mean   2.5%  97.5% n_eff Rhat
## alpha    699.63 668.03 731.54  4941    1
## beta[1]   31.26  11.88  49.85  4605    1
## beta[2]   -5.79  -8.55  -2.87  5410    1
## beta[3]   -1.82  -3.51  -0.20  4329    1
## sigma    104.48  82.88 133.55  3732    1
```

In Figure 8.9, we plot the 95% CrI of the parameters of interest. We use `regex_pars`, rather than `pars`, because we want to capture `beta[1]`, `beta[2]`, and `beta[3]`; `regex_pars` use regular expressions to select the parameters (for information about regular expressions in R, see `https://stat.ethz.ch/R-manual/R-devel/library/base/html/regex.html`).

```
df_fit_pupil_int <- as.data.frame(fit_pupil_int)
mcmc_intervals(fit_pupil_int,
               regex_pars = "beta",
               prob_outer = .95,
               prob = .8,
               point_est = "mean")
```

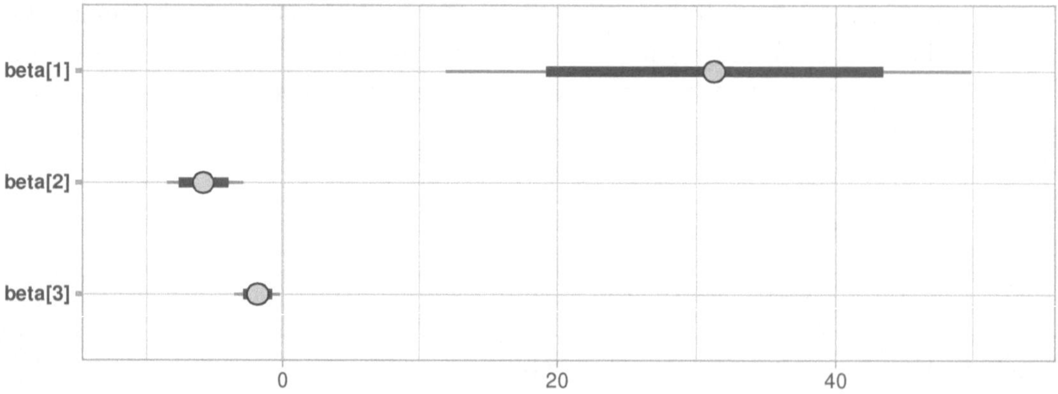

FIGURE 8.9: The means and 95% CrIs of the effect of load, `beta[1]`, the effect of trial, `beta[2]`, and their interaction, `beta[3]`.

8.4.3 Logistic regression in Stan: Does set size and trial affect free recall?

We revisit and expand on the analysis presented in section 4.3 of a subset of the data from Oberauer (2019). In this example, we will investigate whether the length of a list and trial number affect the probability of correctly recalling a word.

As in section 4.3, we assume a Bernoulli likelihood with a logit link function, and the following priors (recall that the logistic function is the inverse of the logit).

$$
\begin{aligned}
correct_n &\sim Bernoulli(logistic(\alpha + X \cdot \beta)) \\
\alpha &\sim Normal(0, 1.5) \\
\beta &\sim Normal(0, 0.1)
\end{aligned}
\tag{8.14}
$$

Where β is a vector of size $K = 2$, $\langle \beta_0, \beta_1 \rangle$. Below in `recall.stan` we present the most efficient way to code this in Stan.

```
data {
  int<lower = 1> N;
  int<lower=0> K;   // number of predictors
  matrix[N, K] X;   // model matrix
  array[N] int correct;
}
parameters {
  real alpha;
  vector[K] beta;
}
model {
  // priors including all constants
  target += normal_lpdf(alpha | 0, 1.5);
  target += normal_lpdf(beta | 0, .1);
  target += bernoulli_logit_glm_lpmf(correct | X, alpha, beta);
}
```

The dependent variable, `correct`, is an array of integers rather than a vector; this is because vectors are always composed of real numbers, but the Bernoulli likelihood only accepts the integers 1 or 0. As in the previous example, we are taking advantage of the `_glm` functions. A less efficient but more transparent option would be to replace the last statement with:

```
target += bernoulli_logit_lpmf(correct | alpha + X * beta);
```

We might want to use `bernoulli_logit_lpmf()` if we want to define a non-linear relationship between the predictors that are outside the generalized linear model framework. One example would be the following:

```
target += bernoulli_logit_lpmf(correct| alpha + exp(X * beta));
```

Another more flexible possibility when we want to indicate a Bernoulli likelihood is to use `bernoulli_lpmf()` and add the link manually. The last statement of `recall.stan` would become the following:

```
target += bernoulli_lpmf(correct| inv_logit(alpha + X * beta));
```

The function `bernoulli_lpmf()` can be useful if one wants to try other link functions; see online exercise G.8.4.

Finally, the most transparent form (but less efficient) would be the following for-loop:

```
for (n in 1:N)
  target += bernoulli_lpmf(correct[n] | inv_logit(alpha + X[n] * beta));
```

To fit the model as `recall.stan`, prepare the data by centering the trial number first:

```
df_recall <- df_recall %>%
  mutate(c_set_size = set_size - mean(set_size),
         c_trial = trial - mean(trial))
```

Next, create the model matrix, $\boldsymbol{X}$, and prepare the data as a list. As in section 8.4.2, exclude the intercept from the matrix `X` using `0 +...`. This is because in the Stan code that we are using, the intercept is kept separate from the model matrix $\boldsymbol{X}$, which only has the contrast coding for the slope parameters. The approach we take here is equivalent to defining a model matrix whose first column is a vector of ones.

```
X <- model.matrix(~ 0 + c_set_size * c_trial, df_recall)
ls_recall <- list(correct = df_recall$correct,
                  X = X,
                  K = ncol(X),
                  N = nrow(df_recall))
```

```
recall <- system.file("stan_models",
                      "recall.stan",
                      package = "bcogsci")
fit_recall <- stan(recall, data = ls_recall)
```

After fitting the model, we can print and plot summaries of the posterior distribution.

```
print(fit_recall, pars = c("alpha", "beta"))
```

```
##          mean  2.5% 97.5% n_eff Rhat
```

```
## alpha      1.99  1.40  2.62  3289    1
## beta[1] -0.19 -0.35 -0.02  4021    1
## beta[2] -0.02 -0.09  0.05  3122    1
## beta[3]  0.00 -0.03  0.04  3533    1
```

In Figure 8.10(a), we plot the 95% CrI of the parameters of interest.

```
df_fit_recall <- as.data.frame(fit_recall)
mcmc_intervals(df_fit_recall,
               regex_pars = "beta",
               prob_outer = .95,
               prob = .8,
               point_est = "mean") +
  xlab("log-odds")
```

As shown in section 4.3.4, we might want to communicate the posterior in proportions rather than in log-odds (as seen in the parameters `beta`). This can be done in R by manipulating the data frame `df_fit_recall`, or by extracting the samples with `extract(fit_recall)`. Another alternative presented here is by using the generated quantities block. To make the code more compact, we declare the type of each variable and store its content in the same line in `recall_prop.stan`.

```
generated quantities {
  real average_accuracy = inv_logit(alpha);
  vector[K] change_acc = inv_logit(alpha) - inv_logit(alpha - beta);
}
```

Recall that due to the non-linearity of the scale, the effects depend on the average accuracy, and the set size and trial that we start from (in this case we are examining the change of one unit from the average set size and the average trial).

```
recall_prop <- system.file("stan_models",
                           "recall_prop.stan",
                           package = "bcogsci")
fit_recall <- stan(recall_prop, data = ls_recall)
```

The plot in Figure 8.10(b) now shows how the average accuracy deteriorates when the subject is exposed to a set size larger than the average by one, a trial further than the middle one, and the interaction of both.

```
df_fit_recall <- as.data.frame(fit_recall) %>%
  rename(set_size = `change_acc[1]`,
         trial = `change_acc[2]`,
         interaction = `change_acc[3]`)
mcmc_intervals(df_fit_recall,
               pars = c("set_size", "trial", "interaction"),
               prob_outer = .95,
               prob = .8,
               point_est = "mean") +
  xlab("Change in accuracy")
```

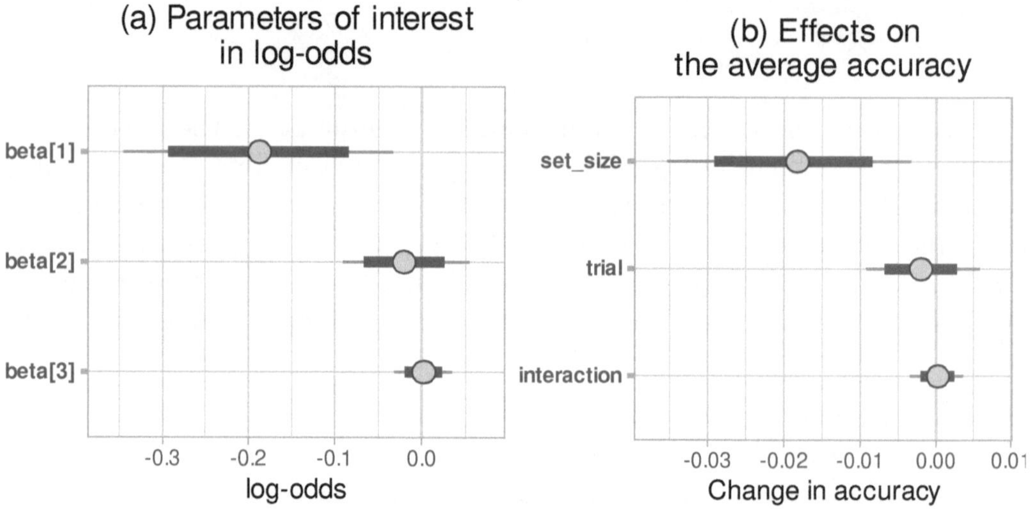

FIGURE 8.10: (a) 95% credible intervals for the `beta` parameters of the `fit_recall` model, representing the log-odds effects of set size, trial, and their interaction. (b) Impact of set size, trial, and their interaction on average recall accuracy.

The plot in Figure 8.10(b) shows that our model is estimating that by increasing the set size by one unit, the recall accuracy of the single subject deteriorates by 2%. In contrast, there is hardly any trial effect or interaction between trial and set size.

8.5 Summary

This chapter introduced basic Stan syntax for fitting some standard linear models. Example code covered the normal, binomial, Bernoulli, and log-normal likelihoods. We also saw how to express regression models using the matrix model in Stan syntax.

8.6 Further reading

For further reading on the Hamiltonian Monte Carlo algorithm, see the rather technical review of Betancourt (2017), or the more conceptual introduction provided by Monnahan, Thorson, and Branch (2017) or chapter 17 of Lambert (2018). A useful article with example R code is Neal (2011). A detailed walk-through on its implementation is also provided in Chapter 41 of MacKay (2003). The Stan documentation (Stan Development Team 2024), consisting of a User's Guide and the Language Reference Manual are important starting points for going deeper into Stan programming: `https://mc-stan.org/users/documentation/`.

9

Hierarchical models and reparameterization

Now that we know how to fit simple regression models using Stan syntax, we can turn to more complex cases, such as the hierarchical models that we fit in `brms` in chapter 5. Fitting such models in Stan allows us a great deal of flexibility. However, a price to be paid when using Stan is that we need to think about how exactly we code the model. In some cases, two pieces of computer code that are mathematically similar might behave very differently due to the computer's limitations; in this chapter, we will learn some of the more common techniques needed to optimize the model's behavior. In particular, we will learn how to deal with convergence problems using what is called the non-centered reparameterization.

9.1 Hierarchical models with Stan

In the following sections, we will revisit and expand on some of the examples from chapter 5.

9.1.1 Varying intercept model with Stan

Recall that in section 5.2, we fit models to investigate the effect of cloze probability on EEG averages in the N400 spatiotemporal time window. For our first model, we'll make the (implausible) assumption that only the average signal varies across subjects, but all subjects share the same effect of cloze probability. This means that the likelihood incorporates the assumption that the intercept, α, is adjusted with the term u_i for each subject.

$$signal_n \sim Normal(\alpha + u_{subj[n]} + c\_cloze_n \cdot \beta, \sigma) \tag{9.1}$$

DOI: 10.1201/9780429342646-12

$$
\begin{aligned}
\alpha &\sim Normal(0, 10) \\
\beta &\sim Normal(0, 10) \\
u &\sim Normal(0, \tau_u) \\
\tau_u &\sim Normal_+(0, 20) \\
\sigma &\sim Normal_+(0, 50)
\end{aligned}
\tag{9.2}
$$

Here n represents each observation, the nth row in the data frame and $subj[n]$ is the subject that corresponds to observation n. We present the mathematical notation of the likelihood with "multiple indexing" (see Stan Development Team 2024, Chapter 19 of the User's guide): the index of u is provided by the vector $subj$.

Before we discuss the Stan implementation, let's see what the vector μ, the location of the normal likelihood, looks like. There are 2863 observations; that means that $\boldsymbol{\mu} = \langle \mu_1, \mu_2, \dots, \mu_{2863} \rangle$. We have 37 subjects which means that $\boldsymbol{u} = \langle u_1, u_2, \dots, u_{37} \rangle$. The following equation shows that the use of multiple indexing allows us to have a vector of adjustments with only 37 different elements, with a total length of 2863. In the equation below, the multiplication operator $\circ$ is the Hadamard or elementwise product (Fieller 2016): when we write $\boldsymbol{X} \circ \boldsymbol{B}$, both $\boldsymbol{X}$ and $\boldsymbol{B}$ have the same dimensions $m \times n$, and each cell in location $[i, j]$ (where $i = 1, \dots, m$, and $j = 1, \dots, n$) in $\boldsymbol{X}$ and $\boldsymbol{B}$ are multiplied to give a matrix that also has dimensions $m \times n$.[1]

[1]In Equation (9.3), α and β are "broadcasted" into a 2863-vector for ease of exposition; Stan uses vector-scalar arithmetic.

$$
\begin{split}
\boldsymbol{\mu} = \begin{bmatrix} \mu_1 \\ \mu_2 \\ \ldots \\ \mu_{101} \\ \mu_{102} \\ \ldots \\ \mu_{215} \\ \mu_{216} \\ \mu_{217} \\ \ldots \\ \mu_{1000} \\ \ldots \\ \mu_{2863} \end{bmatrix} &= \begin{bmatrix} \alpha \\ \alpha \\ \ldots \\ \alpha \\ \alpha \\ \ldots \\ \alpha \\ \alpha \\ \alpha \\ \ldots \\ \alpha \\ \ldots \\ \alpha \end{bmatrix} + \begin{bmatrix} u_{subj[1]} \\ u_{subj[2]} \\ \ldots \\ u_{subj[101]} \\ u_{subj[102]} \\ \ldots \\ u_{subj[215]} \\ u_{subj[216]} \\ u_{subj[217]} \\ \ldots \\ u_{subj[1000]} \\ \ldots \\ u_{subj[2863]} \end{bmatrix} + \begin{bmatrix} ccloze_1 \\ ccloze_2 \\ \ldots \\ ccloze_{101} \\ ccloze_{102} \\ \ldots \\ ccloze_{215} \\ ccloze_{216} \\ ccloze_{217} \\ \ldots \\ ccloze_{1000} \\ \ldots \\ ccloze_{2863} \end{bmatrix} \circ \begin{bmatrix} \beta \\ \beta \\ \ldots \\ \beta \\ \beta \\ \ldots \\ \beta \\ \beta \\ \beta \\ \ldots \\ \beta \\ \ldots \\ \beta \end{bmatrix} \\
&= \begin{bmatrix} \alpha \\ \alpha \\ \ldots \\ \alpha \\ \alpha \\ \ldots \\ \alpha \\ \alpha \\ \alpha \\ \ldots \\ \alpha \\ \ldots \\ \alpha \end{bmatrix} + \begin{bmatrix} u_1 \\ u_1 \\ \ldots \\ u_2 \\ u_2 \\ \ldots \\ u_3 \\ u_3 \\ u_3 \\ \ldots \\ u_{13} \\ \ldots \\ u_{37} \end{bmatrix} + \begin{bmatrix} -0.476 \\ -0.446 \\ \ldots \\ -0.206 \\ 0.494 \\ \ldots \\ -0.136 \\ 0.094 \\ 0.294 \\ \ldots \\ 0.524 \\ \ldots \\ 0.494 \end{bmatrix} \circ \begin{bmatrix} \beta \\ \beta \\ \ldots \\ \beta \\ \beta \\ \ldots \\ \beta \\ \beta \\ \beta \\ \ldots \\ \beta \\ \ldots \\ \beta \end{bmatrix}
\end{split}
\tag{9.3}
$$

In this model, each subject has their own intercept adjustment u_i, with i indexing the subjects. If u_i is positive, the subject has a more positive EEG signal than the average over all the subjects; if u_i is negative, then the subject has a more negative EEG signal than the average; and if u_i is 0, then the subject has the same EEG signal as the average. As we discussed in section 5.2.3, since we are estimating α and u at the same time and we assume that the average of the u's is 0 (since it is assumed to be normally distributed with a mean of zero and there is an intercept to absorb any non-zero mean), whatever the subjects have in common "goes" to α, and u only "absorbs" the differences between subjects through the variance component τ_u.

This model is implemented in the file `hierarchical1.stan`, available in the `bcogsci` package:

```
data {
  int<lower=1> N;
  vector[N] signal;
  int<lower = 1> N_subj;
  vector[N] c_cloze;
  // The following line defines an array of integers;
  array[N] int<lower = 1, upper = N_subj> subj;
}
parameters {
  real<lower = 0> sigma;
  real<lower = 0>  tau_u;
  real alpha;
  real beta;
  vector[N_subj] u;
}
model {
  target += normal_lpdf(alpha| 0, 10);
  target += normal_lpdf(beta | 0, 10);
  target += normal_lpdf(sigma | 0, 50)  -
    normal_lccdf(0 | 0, 50);
  target += normal_lpdf(tau_u | 0, 20)  -
    normal_lccdf(0 | 0, 20);
  target += normal_lpdf(u | 0, tau_u);
  target += normal_lpdf(signal | alpha + u[subj] +
                        c_cloze * beta, sigma);
}
```

In the Stan code above, we use

```
array [N] int<lower = 1, upper = N_subj> subj;
```

to define a one-dimensional *array* of `N` elements that contains integers (bounded between 1 and `N_subj`). As explained in online section B.4, the difference between vectors and one-dimensional arrays is that vectors can only contain real numbers and can be used with matrix algebra functions, and arrays can contain any type but can't be used with matrix algebra functions. We use `normal_lpdf()` rather than `normal_glm_lpdf()` since at the moment there is no efficient likelihood implementation of hierarchical generalized linear models.

The following code centers the predictor cloze and stores the data required by the Stan model in a list. Because we are using `subj` as a vector of indices, we need to be careful to have integers starting from `1` and ending in `N_subj` without skipping any number (but the order of the subject ids won't matter).[2]

```
data("df_eeg")
df_eeg <- df_eeg %>%
  mutate(c_cloze = cloze - mean(cloze))
ls_eeg <- list(N = nrow(df_eeg),
               signal = df_eeg$n400,
               c_cloze = df_eeg$c_cloze,
               subj = df_eeg$subj,
               N_subj = max(df_eeg$subj))
```

Fit the model:

```
hierarchical1 <- system.file("stan_models",
                             "hierarchical1.stan",
                             package = "bcogsci")
fit_eeg1 <- stan(hierarchical1, data = ls_eeg)
```

Summary of the model:

```
print(fit_eeg1, pars = c("alpha", "beta", "sigma", "tau_u"))
```

```
##        mean  2.5% 97.5% n_eff Rhat
## alpha  3.63  2.81  4.46  1444    1
## beta   2.32  1.23  3.37  5016    1
## sigma 11.64 11.34 11.94  5940    1
## tau_u  2.18  1.55  3.02  2071    1
```

9.1.2 Uncorrelated varying intercept and slopes model with Stan

In the following model, we relax the strong assumption that every subject will be affected equally by the manipulation. For ease of exposition, we start by assuming that (as we did in section 5.2.3) the adjustments for the intercept and slope are not correlated.

[2]It often happens that the subject ids are not in sequence in a given data frame. This can happen when, for example, one has lost some subjects due to incomplete data or for some other reason. In such situations, we can re-number the subjects sequentially as follows: type `as.numeric(as.factor(df_eeg$subj))` to transform a vector where some numbers are skipped into a vector with consecutive numbers. For example, both `as.numeric(as.factor(c(1, 3, 4, 7, 9)))` and `as.numeric(as.factor(paste0("subj", c(1, 3, 4, 7, 9))))` will give as output `1 2 3 4 5`.

$$signal_n \sim Normal(\alpha + u_{subj[n],1} + c\_cloze_n \cdot (\beta + u_{subj[n],2}), \sigma) \tag{9.4}$$

$$\begin{aligned}
\alpha &\sim Normal(0, 10) \\
\beta &\sim Normal(0, 10) \\
u_1 &\sim Normal(0, \tau_{u_1}) \\
u_2 &\sim Normal(0, \tau_{u_2}) \\
\tau_{u_1} &\sim Normal_+(0, 20) \\
\tau_{u_2} &\sim Normal_+(0, 20) \\
\sigma &\sim Normal_+(0, 50)
\end{aligned} \tag{9.5}$$

We implement this in Stan in `hierarchical2.stan`:

```
data {
  int<lower=1> N;
  vector[N] signal;
  int<lower = 1> N_subj;
  vector[N] c_cloze;
  array[N] int<lower = 1, upper = N_subj> subj;
}
parameters {
  real<lower = 0> sigma;
  vector<lower = 0>[2]  tau_u;
  real alpha;
  real beta;
  matrix[N_subj, 2] u;
}
model {
  target += normal_lpdf(alpha| 0, 10);
  target += normal_lpdf(beta | 0, 10);
  target += normal_lpdf(sigma | 0, 50)  -
    normal_lccdf(0 | 0, 50);
  target += normal_lpdf(tau_u[1] | 0, 20)  -
    normal_lccdf(0 | 0, 20);
  target += normal_lpdf(tau_u[2] | 0, 20)  -
    normal_lccdf(0 | 0, 20);
  target += normal_lpdf(u[, 1]| 0, tau_u[1]);
  target += normal_lpdf(u[, 2]| 0, tau_u[2]);
  target += normal_lpdf(signal | alpha + u[subj, 1] +
                      c_cloze .* (beta + u[subj, 2]), sigma);
}
```

In the previous model, we assign the same prior distribution to both `tau_u[1]` and `tau_u[2]`, and thus in principle we could have written the two statements in one (we multiply by 2 because there are two log-PDFs that need to be corrected for the truncation):

```
target += normal_lpdf(tau_u | 0, 20)  -
  2 * normal_lccdf(0 | 0, 20);
```

Fit the model as follows:

```
hierarchical2 <- system.file("stan_models",
                             "hierarchical2.stan",
                             package = "bcogsci")
fit_eeg2 <- stan(hierarchical2, data = ls_eeg)
```

```
## Warning: There were 3 chains where the estimated
## Bayesian Fraction of Missing Information was low. See
## https://mc-stan.org/misc/warnings.html#bfmi-low

## Warning: Examine the pairs() plot to diagnose sampling
## problems

## Warning: Bulk Effective Samples Size (ESS) is too low,
## indicating posterior means and medians may be unreliable.
## Running the chains for more iterations may help. See
## https://mc-stan.org/misc/warnings.html#bulk-ess

## Warning: Tail Effective Samples Size (ESS) is too low,
## indicating posterior variances and tail quantiles may be
## unreliable. Running the chains for more iterations may
## help. See
## https://mc-stan.org/misc/warnings.html#tail-ess
```

We see that there are warnings. As we increase the complexity and the number of parameters, the sampler has a harder time exploring the parameter space.

Show the summary of the model below:

```
print(fit_eeg2, pars = c("alpha", "beta", "tau_u", "sigma"))
```

```
##            mean  2.5% 97.5% n_eff Rhat
## alpha      3.65  2.84  4.45  1418 1.00
## beta       2.32  1.08  3.60  4104 1.00
## tau_u[1]   2.18  1.54  2.96  1956 1.00
## tau_u[2]   1.61  0.31  3.42    90 1.04
```

```
## sigma     11.62 11.31 11.95  4474 1.00
```

We see that `tau_u[2]` has a low number of effective samples (`n_eff`).

The traceplots are displayed in Figure 9.1:

```
traceplot(fit_eeg2, pars = c("alpha", "beta", "tau_u", "sigma"))
```

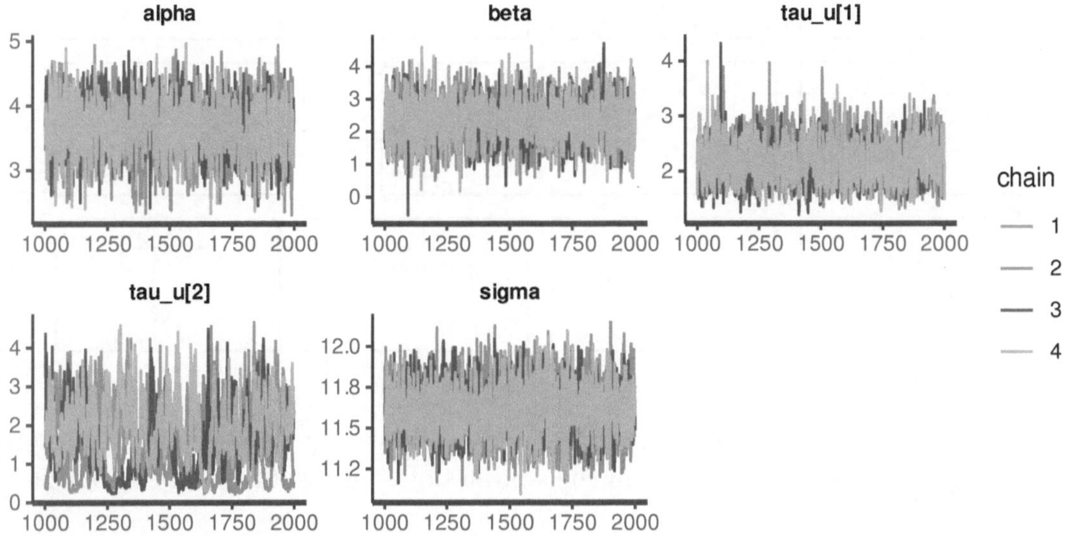

FIGURE 9.1: Traceplots of `alpha`, `beta`, `tau_u`, and `sigma` from the model `fit_eeg2`.

Figure 9.1 shows that the chains of the parameter `tau_u[2]` are not mixing properly. This parameter is especially problematic because there are not enough data from each subject to estimate this parameter accurately; its estimated mean is quite small (in comparison with `sigma`), it's bounded by zero, and there is a dependency between this parameter and `u[, 2]`. This makes the exploration by the sampler quite hard.

Pairs plots can be useful to uncover pathologies in the sampling, since we can visualize dependencies between samples, which are in general problematic.[3] The following code creates a pair plot where we see the samples of `tau_u[2]` against some of the adjustments to the slope `u`; see Figure 9.2.

```
pairs(fit_eeg2, pars = c("tau_u[2]", "u[1,2]", "u[2,2]", "u[3,2]"))
```

[3]See `https://mc-stan.org/misc/warnings.html`.

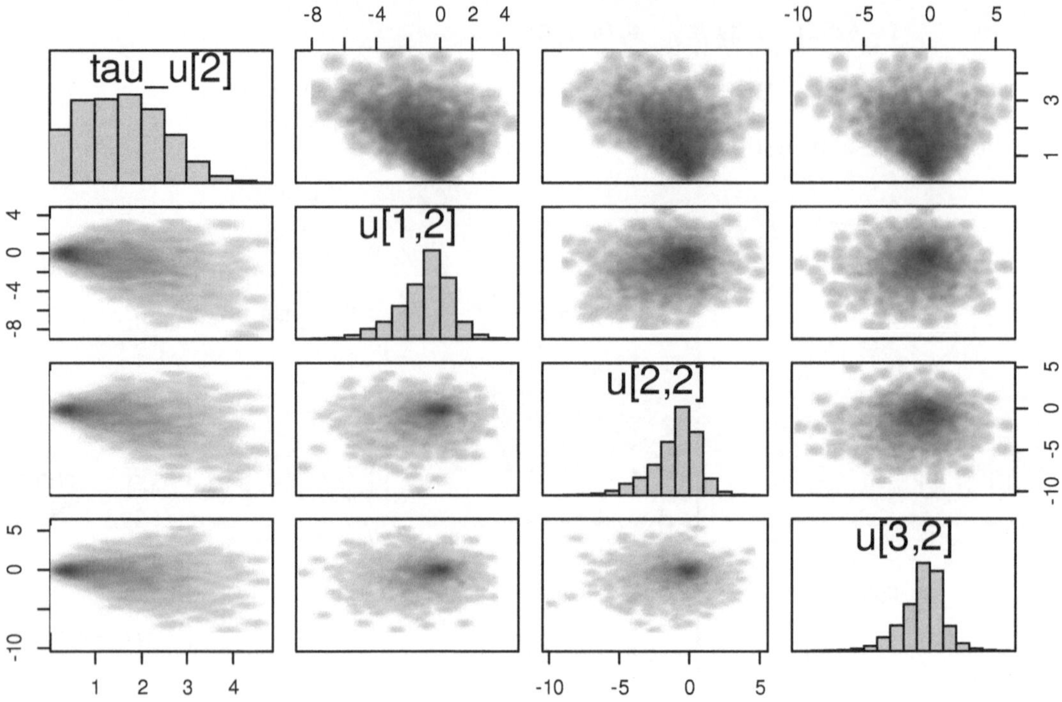

FIGURE 9.2: Pair plots showing a relatively strong dependency (funnel-shaped clouds of samples) between the samples of τ_2 and some of the by-subject adjustments to the slope.

Compare with `tau_u[1]` plotted against the by-subject adjustments to the intercept. In Figure 9.3, instead of funnels we see blobs, indicating no strong dependency between the parameters.

```
pairs(fit_eeg2, pars = c("tau_u[1]", "u[1,1]", "u[2,1]", "u[3,1]"))
```

In contrast to Figure 9.3, Figure 9.2 shows a relatively strong dependency between the samples of some of the parameters of the model, in particular τ_2 and the samples of $u_{i,2}$: If we see small values in the samples for τ_2, then the values for the samples of u_i are very close to zero (and they are larger if values of samples of τ_2 are larger). This strong dependency is hindering the exploration of the sampler leading to the warnings we saw in Stan. However, the problem that the sampler faces is, in fact, more serious than what our initial plots show. Stan samples in an *unconstrained* space where all the parameters can range from minus infinity to plus infinity, and then transforms back the parameters to the constrained space that we specified where, for example, a standard deviation parameter is restricted to be positive. This means that Stan is actually sampling from a transformed parameter equivalent to `log(tau_u[2])`

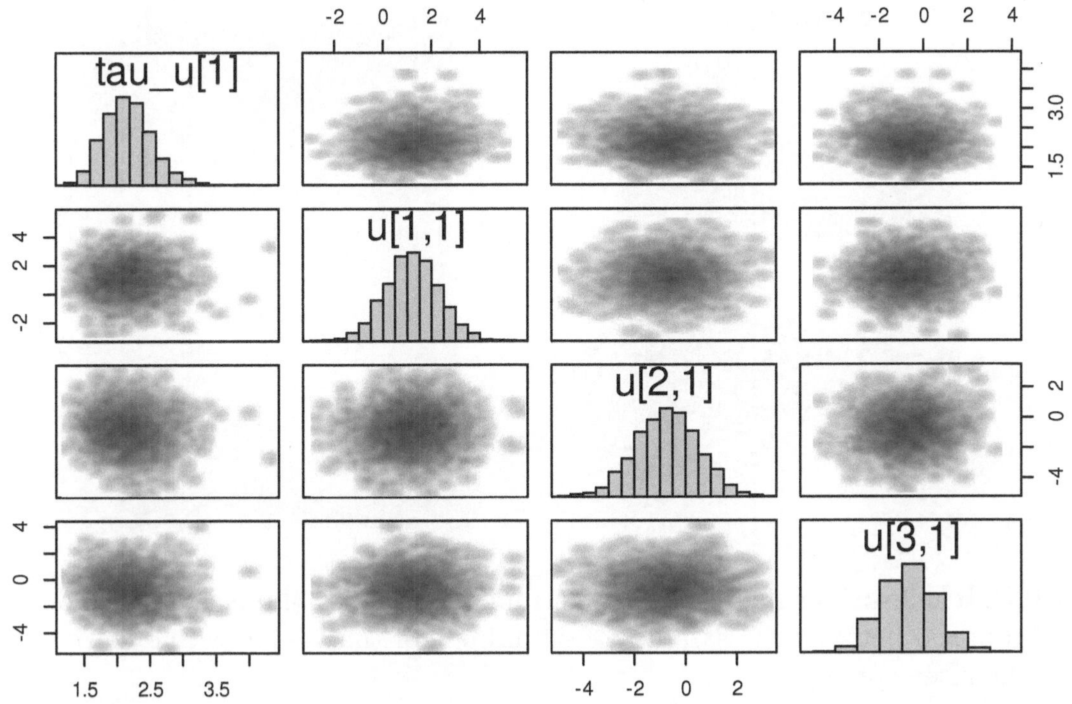

FIGURE 9.3: Pair plots showing no strong dependency (blob-shaped clouds of samples) between the samples of τ_1 and some of the by-subject adjustments to the intercept.

rather than from `tau_u[2]`. We can use `mcmc_pairs()` to see the actual funnel; see Figure 9.4.

```
mcmc_pairs(as.array(fit_eeg2),
           pars = c("tau_u[2]", "u[1,2]"),
           transform = list(`tau_u[2]` = "log"))
```

At the neck of the funnel, `tau_u[2]` is close to zero (and `log(tau_u[2])` is a negative number) and thus the adjustment `u` is constrained to be near 0. This is a problem because a step size that's optimized to work well in the broad part of the funnel will fail to work appropriately in the neck of the funnel and vice versa; see also Neal's funnel (Neal 2003) and the optimization chapter of Stan Development Team (2024) (section 26.7 of the User's guide). There are two options: We might just remove the by-subject varying slope since it's not giving us much information anyway, or we can alleviate this problem by reparameterizing the model. In general, this sort of thing is the trickiest and probably most annoying part of modeling. A model can be theoretically and mathematically sound, but still fail to converge. The best advice to solve this type of problem is to start small with simulated data where we know the true

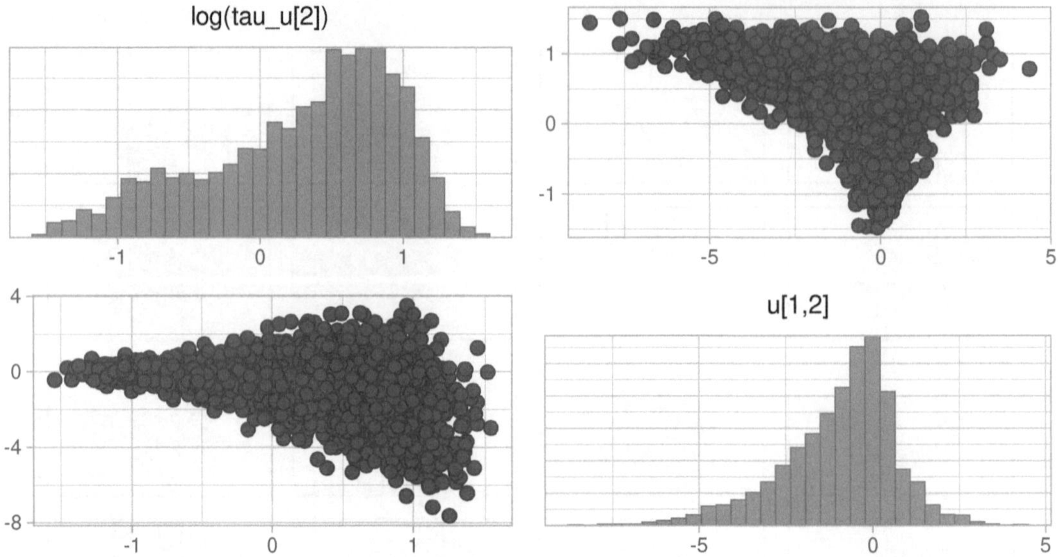

FIGURE 9.4: Pair plots showing a strong dependency (funnel-shaped clouds of samples) between the samples of τ_2 and one of the by-subject adjustments to the intercept ($u_{1,2}$).

values of the parameters, and increase the complexity of the models gradually. Although in this example the problem was clearly in the parameterization of `tau_u[2]`, in many cases the biggest hurdle is to identify where the problem lies. Fortunately, the issue with `tau_u[2]` is a common problem which is easy to solve by using a non-centered parameterization (Betancourt and Girolami 2015; Papaspiliopoulos, Roberts, and Sköld 2007). The online section B.5 explains the specific reparameterization we use for the improved version of our Stan code.

From a mathematical point of view, the following model is equivalent to the one described in Equations (9.4) and (9.5). However, as discussed previously, the computational implementation of the "new" model is more efficient. The following model includes the reparameterization of both adjustments u_1 and u_2, although the reparameterization of u_1 is not strictly necessary (we didn't see any problems in the traceplots), it won't hurt either and the Stan code will be simpler.

$$signal_n \sim Normal(\alpha + u_{subj[n],1} + c\_cloze_n \cdot (\beta + u_{subj[n],2}), \sigma) \tag{9.6}$$

$$\begin{aligned}
\alpha &\sim Normal(0, 10)\\
\beta &\sim Normal(0, 10)\\
z_{u_1} &\sim Normal(0, 1)\\
z_{u_2} &\sim Normal(0, 1)\\
\tau_{u_1} &\sim Normal_+(0, 20)\\
\tau_{u_2} &\sim Normal_+(0, 20)\\
\sigma &\sim Normal_+(0, 50)\\
u_1 &= z_{u_1} \cdot \tau_{u_1}\\
u_2 &= z_{u_2} \cdot \tau_{u_2}
\end{aligned} \tag{9.7}$$

The following Stan code (`hierarchical3.stan`) uses the previous parameterization, and introduces some new Stan functions: `to_vector()` converts a matrix into a long column vector (in column-major order, that is, concatenating the columns from left to right); and `std_normal_lpdf()` implements the log PDF of a standard normal distribution, a normal distribution with location 0 and scale 1. This function is just a more efficient version of `normal_lpdf(... | 0, 1)`. We also introduce a new optional block called `transformed parameters`. With each iteration of the sampler, the values of the parameters (i.e., `alpha`, `beta`, `sigma`, and `z`) are available at the `transformed parameters` block, and we can derive new auxiliary variables based on them. In this case, we use `z_u` and `tau_u` to obtain `u`, that then becomes available in the `model` block. Notice that both the model block and R (when we output the stan object) can access both parameters and transformed parameters.

```
data {
  int<lower=1> N;
  vector[N] signal;
  int<lower = 1> N_subj;
  vector[N] c_cloze;
  array[N] int<lower = 1, upper = N_subj> subj;
}
parameters {
  real<lower = 0> sigma;
  vector<lower = 0>[2]  tau_u;
  real alpha;
  real beta;
  matrix[N_subj, 2] z_u;
}
transformed parameters {
  matrix[N_subj, 2] u;
```

```
  u[, 1] = z_u[, 1] * tau_u[1];
  u[, 2] = z_u[, 2] * tau_u[2];
 }
model {
  target += normal_lpdf(alpha| 0, 10);
  target += normal_lpdf(beta | 0, 10);
  target += normal_lpdf(sigma | 0, 50)  -
    normal_lccdf(0 | 0, 50);
  target += normal_lpdf(tau_u[1] | 0, 20)  -
    normal_lccdf(0 | 0, 20);
 target += normal_lpdf(tau_u[2] | 0, 20)  -
    normal_lccdf(0 | 0, 20);
  target += std_normal_lpdf(to_vector(z_u));
  target += normal_lpdf(signal | alpha + u[subj, 1] +
                    c_cloze .* (beta + u[subj, 2]), sigma);
}
```

By reparameterizing the model we can also optimize it more, we can convert the matrix `z_u` into a long column vector allowing us to use a single call of `std_normal_lpdf()`. Fit the model named `hierarchical3.stan`.

```
hierarchical3 <- system.file("stan_models",
                             "hierarchical3.stan",
                             package = "bcogsci")
fit_eeg3 <- stan(hierarchical3, data = ls_eeg)
```

Verify that the model worked as expected by printing its summary and traceplots; see Figure 9.5.

```
print(fit_eeg3, pars = c("alpha", "beta", "tau_u", "sigma"))
```

```
##             mean  2.5% 97.5% n_eff Rhat
## alpha       3.64  2.79  4.44  1285 1.00
## beta        2.35  1.13  3.60  3578 1.00
## tau_u[1]    2.16  1.53  2.95  1379 1.00
## tau_u[2]    1.75  0.16  3.52   946 1.01
## sigma      11.62 11.32 11.95  4847 1.00
```

```
traceplot(fit_eeg3, pars = c("alpha", "beta", "tau_u", "sigma"))
```

Although the samples of `tau_u[2]` are still conditioned by the adjustments for the slope, `u[,2]`, these latter parameters are not the ones explored by the model,

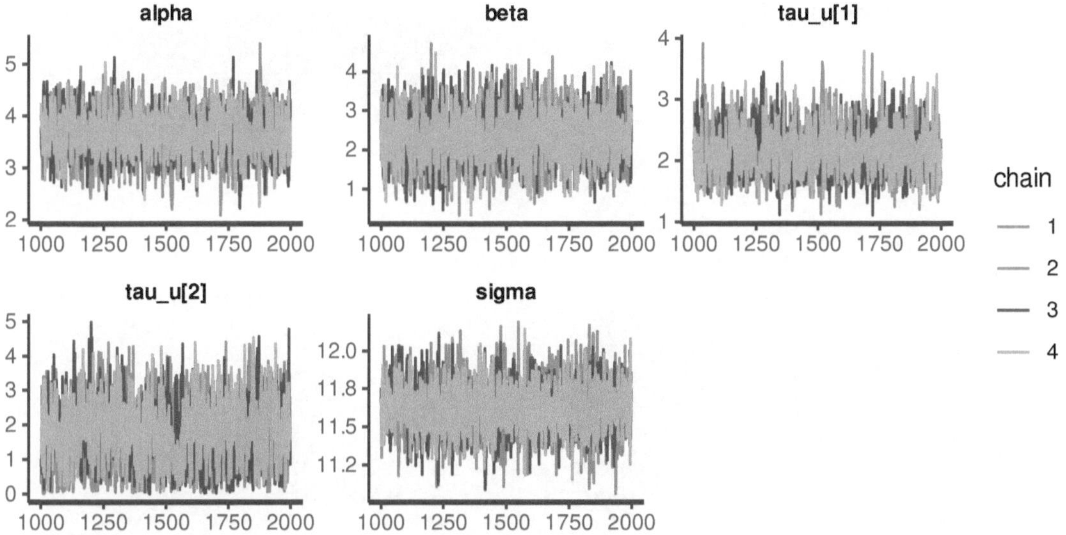

FIGURE 9.5: Traceplots of `alpha`, `beta`, `tau_u`, and `sigma` from the model `fit_eeg3`.

the auxiliary parameters, `z_u`, are the relevant ones for the sampler. The plots in Figures 9.6 and 9.7 show that although samples from `log(tau_u[2])` and `u[1,2]` depend on each other, samples from `log(tau_u[2])` and `z_u[1,2]` do not.

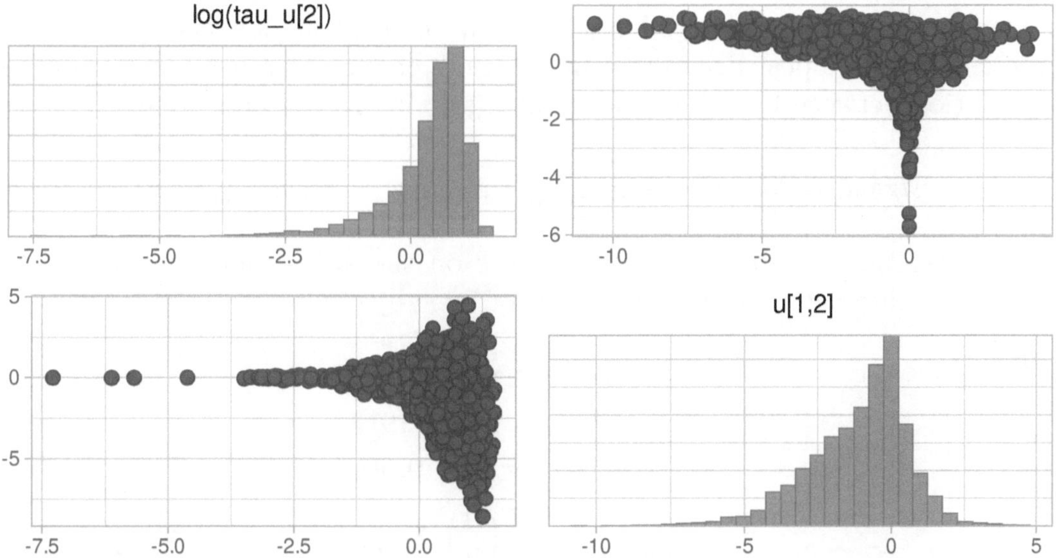

FIGURE 9.6: Pair plots showing a clear dependency (funnel-shaped clouds of samples) between the samples of τ_2 and some of the by-subject adjustments to the slope.

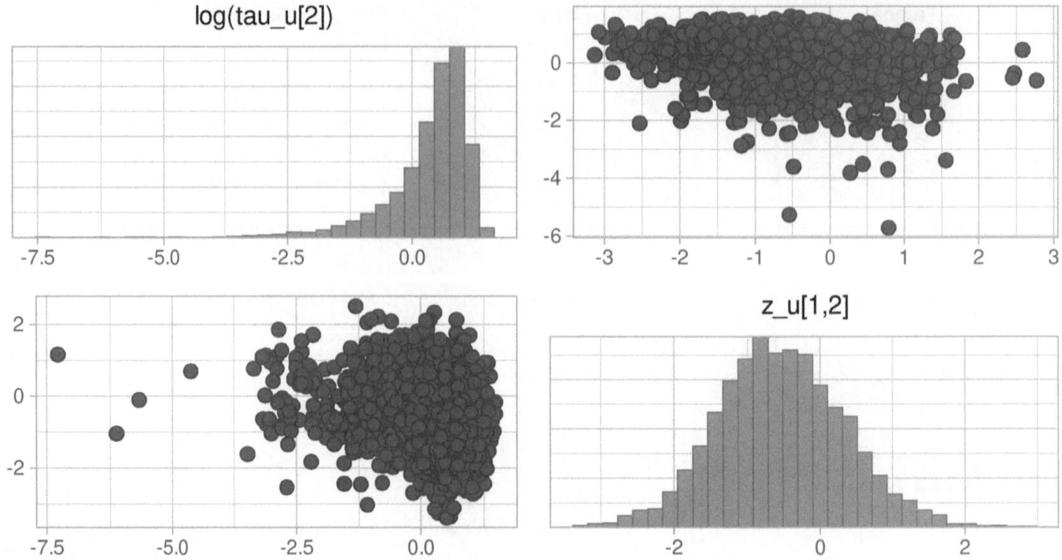

FIGURE 9.7: Pair plots showing no clear dependency between the samples of τ_2 and some of the by-subject auxiliary parameters (`z_u`) used to build the adjustments to the slope.

9.1.3 Correlated varying intercept varying slopes model

For the model with correlated varying intercepts and slopes, the likelihood remains identical to the model without a correlation between group-level intercepts and slopes. Priors and hyperpriors change to reflect the potential correlation between by-subject adjustments to intercepts and slopes:

$$signal_n \sim Normal(\alpha + u_{subj[n],1} + c\_cloze_n \cdot (\beta + u_{subj[n],2}), \sigma) \tag{9.8}$$

The correlation is indicated in the priors on the adjustments for vector of by-subject intercepts u_1 and the vector of by-subject slopes u_2.

- Priors:

$$\begin{aligned} \alpha &\sim Normal(0, 10) \\ \beta &\sim Normal(0, 10) \\ \sigma &\sim Normal_+(0, 50) \\ \begin{pmatrix} u_{i,1} \\ u_{i,2} \end{pmatrix} &\sim \mathcal{N}\left(\begin{pmatrix} 0 \\ 0 \end{pmatrix}, \boldsymbol{\Sigma}_u \right) \end{aligned} \tag{9.9}$$

where i ranges from 1 to N_{subj}

$$\boldsymbol{\Sigma}_u = \begin{pmatrix} \tau_{u_1}^2 & \rho_u \tau_{u_1} \tau_{u_2} \\ \rho_u \tau_{u_1} \tau_{u_2} & \tau_{u_2}^2 \end{pmatrix} \tag{9.10}$$

$$\begin{aligned}
\tau_{u_1} &\sim Normal_+(0, 20) \\
\tau_{u_2} &\sim Normal_+(0, 20) \\
\mathbf{R_u} &= \begin{bmatrix} 1 & \rho_u \\ \rho_u & 1 \end{bmatrix} \sim LKJcorr(2)
\end{aligned} \tag{9.11}$$

As a first attempt, we write this model following the mathematical notation as closely as possible. We'll see that this will be problematic in terms of efficient sampling and convergence. In this Stan model (`hierarchical_corr.stan`), we use some new functions and types:

- `corr_matrix[n] R;` defines a (square) matrix of n rows and n columns called $\boldsymbol{R}$, symmetrical around a diagonal of ones.
- `rep_row_vector(X, n)` creates a row vector with n columns filled with X.
- `quad_form_diag(R, v)` creates a *quadratic form* using the column vector v as a diagonal matrix (a matrix with all zeros except for its diagonal), this function corresponds in Stan to: `diag_matrix(v) * R * diag_matrix(v)` and in R to `diag(v) %*% R %*% diag(v)`. This computes a variance-covariance matrix from the vector of standard deviations, v, and the correlation matrix, $\boldsymbol{R}$ (recall the generation of multivariate data in section 1.6.3).

```
data {
  int<lower=1> N;
  vector[N] signal;
  int<lower = 1> N_subj;
  vector[N] c_cloze;
  array[N] int<lower = 1, upper = N_subj> subj;
}
parameters {
  real<lower = 0> sigma;
  vector<lower = 0>[2]  tau_u;
  real alpha;
  real beta;
  matrix[N_subj, 2] u;
  corr_matrix[2] R_u;
}
model {
  target += normal_lpdf(alpha| 0,10);
  target += normal_lpdf(beta | 0,10);
  target += normal_lpdf(sigma | 0, 50)  -
    normal_lccdf(0 | 0, 50);
  target += normal_lpdf(tau_u | 0, 20)  -
    2 * normal_lccdf(0 | 0, 20);
```

```
  target += lkj_corr_lpdf(R_u | 2);
  for(i in 1:N_subj)
    target +=  multi_normal_lpdf(u[i,] |
                                 rep_row_vector(0, 2),
                                 quad_form_diag(R_u, tau_u));
  target += normal_lpdf(signal | alpha + u[subj, 1] +
                        c_cloze .* (beta + u[subj, 2]), sigma);
}
```

Problematic aspects of the first model presented in section 9.1.2 (before the reparameterization); that is, dependencies between parameters, are also present here. Fit the model as follows:

```
hierarchical_corr <- system.file("stan_models",
                                 "hierarchical_corr.stan",
                                 package = "bcogsci")
fit_eeg_corr <- stan(hierarchical_corr, data = ls_eeg)
```

```
## Warning: There were 6 divergent transitions after
## warmup. See
## https://mc-stan.org/misc/warnings.html#divergent-transitions-after-warmup
## to find out why this is a problem and how to eliminate
## them.

## Warning: There were 3 chains where the estimated
## Bayesian Fraction of Missing Information was low. See
## https://mc-stan.org/misc/warnings.html#bfmi-low

## Warning: Examine the pairs() plot to diagnose sampling
## problems

## Warning: The largest R-hat is NA, indicating chains have
## not mixed. Running the chains for more iterations may
## help. See https://mc-stan.org/misc/warnings.html#r-hat

## Warning: Bulk Effective Samples Size (ESS) is too low,
## indicating posterior means and medians may be unreliable.
## Running the chains for more iterations may help. See
## https://mc-stan.org/misc/warnings.html#bulk-ess

## Warning: Tail Effective Samples Size (ESS) is too low,
## indicating posterior variances and tail quantiles may be
## unreliable. Running the chains for more iterations may
## help. See
## https://mc-stan.org/misc/warnings.html#tail-ess
```

As we expected, there are warnings and bad mixing of the chains for `tau_u[2]`; see also the traceplots in Figure 9.8.

```
print(fit_eeg_corr, pars = c("alpha", "beta", "tau_u", "sigma"))
```

```
##             mean  2.5% 97.5% n_eff Rhat
## alpha       3.65  2.80  4.48  1219 1.00
## beta        2.34  1.12  3.53  4014 1.00
## tau_u[1]    2.20  1.53  3.01  1642 1.00
## tau_u[2]    1.66  0.29  3.46    99 1.08
## sigma      11.62 11.31 11.93  4285 1.00
```

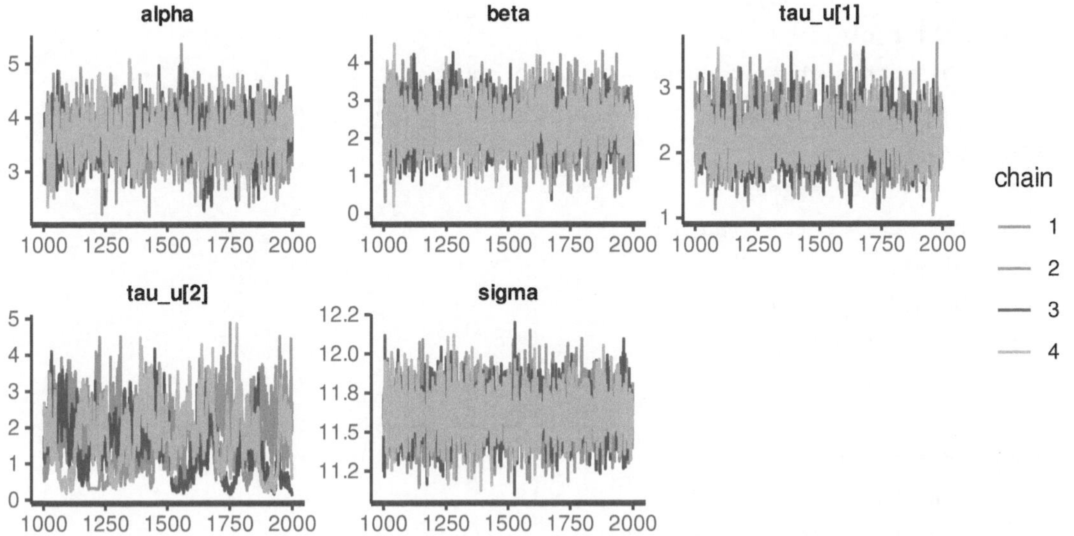

FIGURE 9.8: Traceplots of `alpha`, `beta`, `tau_u`, and `sigma` from the model `fit_eeg_corr`.

The problem (which can also be discovered in a pairs plot) is the same one that we saw before: There is a strong dependency between `tau_u[2]` (in fact, `log(tau_u[2])`, which is the parameter dimension that the sampler considers) and `u`, creating a funnel.

The solution to this problem is the reparameterization of this model. The reparameterization for this type of model requires us to use Cholesky factorization (Fieller 2016). The mathematics and the intuition behind this parameterization are explained in online section B.6.

The reparameterization of the model, which allows for a correlation between adjustments for the intercepts and slopes, in `hierarchical_corr2.stan` is shown below. The code implements the following new types and functions:

- `cholesky_factor_corr[2] L_u`, which defines `L_u` as a lower triangular (2×2) matrix which has to be the Cholesky factor of a correlation.
- `diag_pre_multiply(tau_u,L_u)` which makes a diagonal matrix out of the vector `tau_u` and multiplies it by `L_u`.
- `to_vector(z_u)` makes a long vector out the matrix `z_u`.
- `lkj_corr_cholesky_lpdf(L_u | 2)` is the Cholesky factor associated with the LKJ correlation distribution. It implies that `lkj_corr_lpdf(L_u * L_u'| 2)`. The symbol `'` indicates transposition (in R, this is the function `t()`).

```
data {
  int<lower=1> N;
  vector[N] signal;
  int<lower = 1> N_subj;
  vector[N] c_cloze;
  array[N] int<lower = 1, upper = N_subj> subj;
}
parameters {
  real<lower = 0> sigma;
  vector<lower = 0>[2]  tau_u;
  real alpha;
  real beta;
  matrix[2, N_subj] z_u;
  cholesky_factor_corr[2] L_u;
}
transformed parameters {
  matrix[N_subj, 2] u;
  u = (diag_pre_multiply(tau_u, L_u) * z_u)';
}
model {
  target += normal_lpdf(alpha| 0,10);
  target += normal_lpdf(beta | 0,10);
  target += normal_lpdf(sigma | 0, 50)  -
    normal_lccdf(0 | 0, 50);
  target += normal_lpdf(tau_u | 0, 20)  -
    2 * normal_lccdf(0 | 0, 20);
  target += lkj_corr_cholesky_lpdf(L_u | 2);
  target += std_normal_lpdf(to_vector(z_u));
  target += normal_lpdf(signal | alpha + u[subj, 1] +
                      c_cloze .* (beta + u[subj, 2]), sigma);
}
generated quantities {
  matrix[2, 2] R_u = (L_u * L_u');
```

```
  real rho_u = R_u[1, 2];
  vector[N_subj] effect_by_subj = beta + u[ , 2];
}
```

In this Stan model, we also defined an `effect_by_subject` in the generated quantities. This would allow us to plot or to summarize by-subject effects of cloze probability.

One can recover the correlation parameter by adding in the `generated quantities` section code to extract one element of the 2×2 correlation matrix. The correlation matrix is recovered with `L_u * L_u'`, and then `[1, 2]` extracts the element in the first row and second column (recall that the diagonal is filled with ones).

Fit the new model:

```
hierarchical_corr2 <- system.file("stan_models",
                                  "hierarchical_corr2.stan",
                                  package = "bcogsci")
fit_eeg_corr2 <- stan(hierarchical_corr2, data = ls_eeg)
```

The Cholesky matrix has some elements which are always zero or one, and thus the variance within and between chains (and therefore Rhat) are not defined. However, the rest of the parameters of the model have an appropriate number of effective sample size (more than 10% of the total number of post-warmup samples), Rhats are close to one, and the chains are mixing well; see also the traceplots in Figure 9.9.

```
print(fit_eeg_corr2,
      pars =
        c("alpha", "beta", "tau_u", "rho_u", "sigma", "L_u"))
```

```
##           mean  2.5% 97.5% n_eff Rhat
## alpha     3.64  2.82  4.45  1432    1
## beta      2.31  1.06  3.53  3868    1
## tau_u[1]  2.20  1.56  3.02  1498    1
## tau_u[2]  1.61  0.13  3.48   990    1
## rho_u     0.17 -0.56  0.77  3721    1
## sigma    11.61 11.33 11.91  5204    1
## L_u[1,1]  1.00  1.00  1.00   NaN  NaN
## L_u[1,2]  0.00  0.00  0.00   NaN  NaN
## L_u[2,1]  0.17 -0.56  0.77  3721    1
## L_u[2,2]  0.92  0.61  1.00  2136    1
```

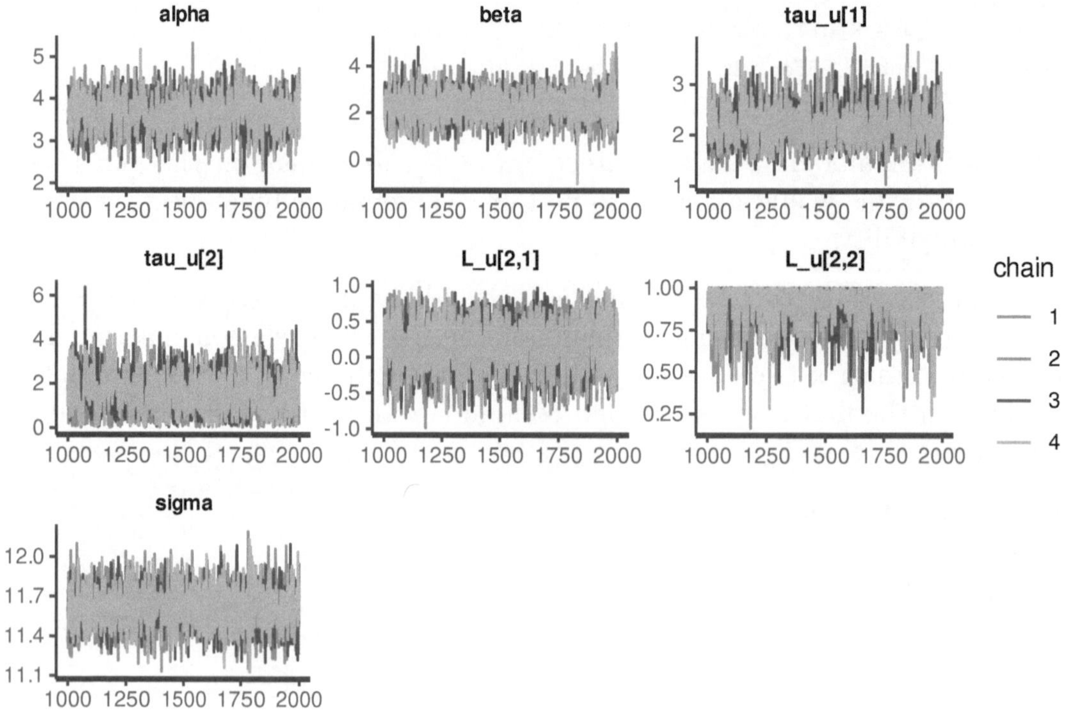

FIGURE 9.9: Traceplots of `alpha`, `beta`, `tau_u`, `L_u`, and `sigma` from the model `fit_eeg_corr`.

Is there a correlation between the by-subject intercept and slope?

Let's visualize some of the posteriors with the following code (see Figure 9.10):

```
mcmc_hist(as.data.frame(fit_eeg_corr2),
          pars = c("beta", "rho_u"))
```

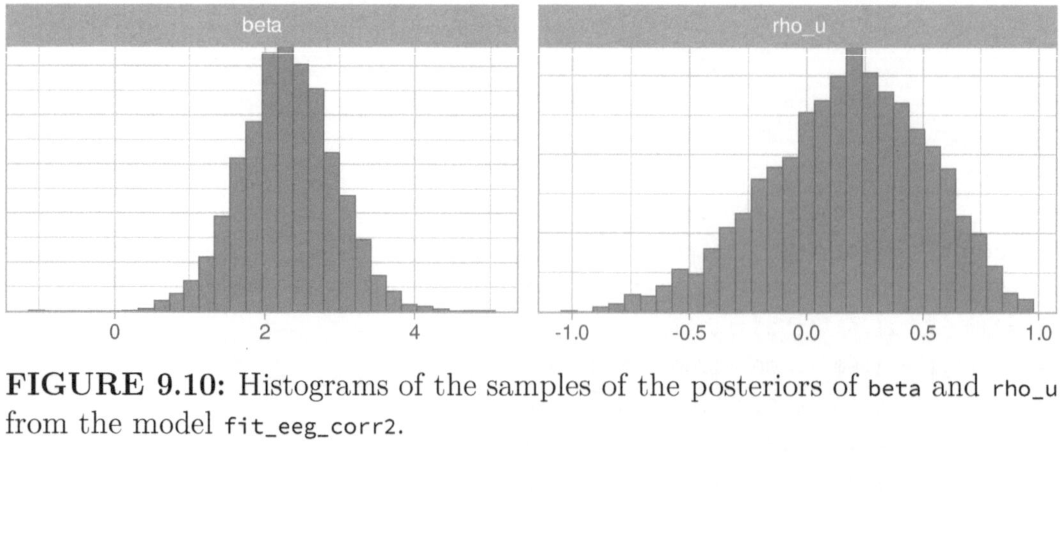

FIGURE 9.10: Histograms of the samples of the posteriors of `beta` and `rho_u` from the model `fit_eeg_corr2`.

Figure 9.10 shows that the posterior distribution is widely spread out between −1 and +1. One can't really learn from these data whether the by-subject intercepts and slopes are correlated. The broad spread of the posterior indicates that we don't have enough data to estimate this parameter with high enough precision: the posterior is basically just reflecting the prior specification (the LKJcorr prior with parameter $\eta = 2$).

9.1.4 By-subject and by-items correlated varying intercept varying slopes model

We extend the previous model by adding by-items intercepts and slopes, and priors and hyperpriors that reflect the potential correlation between by-items adjustments to intercepts and slopes:

$$\begin{aligned} signal_n \sim Normal(&\alpha + u_{subj[n],1} + w_{item[n],1} + \\ &c\_cloze_n \cdot (\beta + u_{subj[n],2} + w_{item[n],2}), \sigma) \end{aligned} \quad (9.12)$$

The correlation is indicated in the priors on the adjustments for the vectors representing the varying intercepts u_1 and varying slopes u_2 for subjects, and the varying intercepts w_1 and varying slopes w_2 for items.

- Priors:

$$\begin{aligned} \alpha &\sim Normal(0, 10) \\ \beta &\sim Normal(0, 10) \\ \sigma &\sim Normal_+(0, 50) \\ \begin{pmatrix} u_{i,1} \\ u_{i,2} \end{pmatrix} &\sim \mathcal{N}\left(\begin{pmatrix} 0 \\ 0 \end{pmatrix}, \boldsymbol{\Sigma_u}\right) \\ \begin{pmatrix} w_{i,1} \\ w_{i,2} \end{pmatrix} &\sim \mathcal{N}\left(\begin{pmatrix} 0 \\ 0 \end{pmatrix}, \boldsymbol{\Sigma_w}\right) \end{aligned} \quad (9.13)$$

$$\boldsymbol{\Sigma_u} = \begin{pmatrix} \tau_{u_1}^2 & \rho_u \tau_{u_1} \tau_{u_2} \\ \rho_u \tau_{u_1} \tau_{u_1} & \tau_{u_2}^2 \end{pmatrix} \quad (9.14)$$

$$\boldsymbol{\Sigma_w} = \begin{pmatrix} \tau_{w_1}^2 & \rho_w \tau_{w_1} \tau_{w_2} \\ \rho_w \tau_{w_1} \tau_{w_1} & \tau_{w_2}^2 \end{pmatrix} \quad (9.15)$$

$$\begin{aligned} \tau_{u_1} &\sim Normal_+(0, 20) \\ \tau_{u_2} &\sim Normal_+(0, 20) \\ \mathbf{R_u} &= \begin{bmatrix} 1 & \rho_u \\ \rho_u & 1 \end{bmatrix} \sim LKJcorr(2) \end{aligned} \quad (9.16)$$

$$
\begin{aligned}
\tau_{w_1} &\sim Normal_+(0,20)\\
\tau_{w_2} &\sim Normal_+(0,20)\\
\mathbf{R_w} &= \begin{bmatrix} 1 & \rho_w \\ \rho_w & 1 \end{bmatrix} \sim LKJcorr(2)
\end{aligned}
\tag{9.17}
$$

The translation to Stan (`hierarchical_corr_by.stan`) looks as follows:

```
data {
  int<lower=1> N;
  vector[N] signal;
  int<lower = 1> N_subj;
  int<lower = 1> N_item;
  vector[N] c_cloze;
  array[N] int<lower = 1, upper = N_subj> subj;
  array[N] int<lower = 1, upper = N_item> item;
}
parameters {
  real<lower = 0> sigma;
  vector<lower = 0>[2]  tau_u;
  vector<lower = 0>[2]  tau_w;
  real alpha;
  real beta;
  matrix[2, N_subj] z_u;
  matrix[2, N_item] z_w;
  cholesky_factor_corr[2] L_u;
  cholesky_factor_corr[2] L_w;
}
transformed parameters {
  matrix[N_subj, 2] u;
  matrix[N_item, 2] w;
  u = (diag_pre_multiply(tau_u, L_u) * z_u)';
  w = (diag_pre_multiply(tau_w, L_w) * z_w)';
}
model {
  target += normal_lpdf(alpha| 0,10);
  target += normal_lpdf(beta | 0,10);
  target += normal_lpdf(sigma | 0, 50)  -
    normal_lccdf(0 | 0, 50);
  target += normal_lpdf(tau_u | 0, 20) -
    2 * normal_lccdf(0 | 0, 20);
  target += normal_lpdf(tau_w | 0, 20) -
    2* normal_lccdf(0 | 0, 20);
```

```
  target += lkj_corr_cholesky_lpdf(L_u | 2);
  target += lkj_corr_cholesky_lpdf(L_w | 2);
  target += std_normal_lpdf(to_vector(z_u));
  target += std_normal_lpdf(to_vector(z_w));
  target += normal_lpdf(signal | alpha + u[subj, 1] + w[item, 1] +
                        c_cloze .* (beta + u[subj, 2] + w[item, 2]),
                        sigma);
}
generated quantities {
  real rho_u = (L_u * L_u')[1, 2];
  real rho_w = (L_w * L_w')[1, 2];
}
```

Add item as a number to the data and store it in a list:

```
df_eeg <- df_eeg %>%
  mutate(item = as.numeric(as.factor(item)))
ls_eeg <- list(N = nrow(df_eeg),
               signal = df_eeg$n400,
               c_cloze = df_eeg$c_cloze,
               subj = df_eeg$subj,
               item = df_eeg$item,
               N_subj = max(df_eeg$subj),
               N_item = max(df_eeg$item))
```

Fit the model:

```
hierarchical_corr_by <- system.file("stan_models",
                                    "hierarchical_corr_by.stan",
                                    package = "bcogsci")
fit_eeg_corr_by <- stan(hierarchical_corr_by, data = ls_eeg)
```

Print the summary:

```
print(fit_eeg_corr_by,
      pars = c("alpha", "beta", "sigma", "tau_u", "tau_w",
               "rho_u", "rho_w"))
```

```
##            mean  2.5% 97.5% n_eff Rhat
## alpha      3.65  2.72  4.57  1851    1
## beta       2.32  0.94  3.63  3076    1
## sigma     11.48 11.19 11.80  4485    1
```

```
## tau_u[1]  2.22  1.56  3.06  1510     1
## tau_u[2]  1.50  0.09  3.38  1229     1
## tau_w[1]  1.52  0.88  2.20  1261     1
## tau_w[2]  2.30  0.28  4.24   982     1
## rho_u     0.14 -0.61  0.77  3465     1
## rho_w    -0.40 -0.89  0.33  2131     1
```

The summary above shows that the data are far too sparse to get tight estimates of the correlation parameters `rho_u` and `rho_w`. Both posteriors are widely spread out.

This completes our review of hierarchical models and their implementation in Stan. The importance of coding a hierarchical model directly in Stan rather than using `brms` is that this increases the flexibility of the type of models that we can fit. In fact, we will see in chapters 15–18 that the same "machinery" can be used to have hierarchical parameters in cognitive models.

9.2 Summary

In this chapter, we learned to fit the four standard types of hierarchical models that we encountered in earlier chapters:

- The by-subjects varying intercepts model.
- The by-subjects varying intercepts and varying slopes model without any correlation.
- The by-subjects varying intercepts and varying slopes model with correlation.
- The hierarchical model, with a full variance covariance matrix for both subjects and items.

We also learned about some important and powerful tools for making the Stan models more efficient at sampling: the non-centered parameterization and the Cholesky factorization. One important takeaway was that if data are sparse, the posteriors will just reflect the priors. We saw examples of this situation when investigating the posteriors of the correlation parameters.

9.3 Further reading

Gelman and Hill (2007) provides a comprehensive introduction to Bayesian hierarchical models, although that edition does not use Stan but rather WinBUGS.

Sorensen, Hohenstein, and Vasishth (2016) is a short tutorial on hierarchical modeling using Stan, especially tailored for psychologists and linguists.

An additional example of reparameterization (suggested by Martin Modrák, personal communication) involves reparameterizing a varying intercept/effects model in terms of total variance and its distribution across individual components. This approach has both theoretical advantages, such as making it easier to elicit priors on total variance and its split, and practical benefits, such as allowing us to work with data sets where only a small number of subjects have repeated measurements, which might otherwise pose computational challenges. This concept is explored in more depth in Hem, Fuglstad, and Riebler (2022). Additionally, a short gist provided by Martin Modrák demonstrates how this reparameterization resolves divergences and increases the effective sample size in a model adapted from this chapter. At the time of writing, Modrák's gist is available at `https://gist.github.com/martinmodrak/61a68f22ed954ae68abc76faeef860b6`.

10

Custom distributions in Stan

Stan includes a large number of distributions, but what happens if we need a distribution that is not provided? In many cases, we can simply build a custom distribution by combining the ever-growing number of functions available in the Stan language.

10.1 A change of variables with the reciprocal normal distribution

In previous chapters, when faced with response times, we assumed a log-normal distribution. The log-normal distribution moves the inferences relating to the location parameter into a multiplicative frame (see online section A.4). Another alternative, however, is to assume a "reciprocal"-normal distribution of response times. This is referred to hereafter as rec-normal in the text, and as the *RecNormal* distribution in equations. That is, we may want to assume that the reciprocal of the response times are normally distributed. This assumption may be theoretically motivated (Harris et al. 2014; Harris and Waddington 2012) or it may arise from the application of the Box-Cox variance stabilizing transform procedure (Box and Cox 1964). An example from psycholinguistics of a data analysis with a reciprocal transform of reading time data appears in Wu, Kaiser, and Vasishth (2017).

$$\begin{aligned} 1/y &\sim \mathit{Normal}(\mu, \sigma) \\ y &\sim \mathit{RecNormal}(\mu, \sigma) \end{aligned} \tag{10.1}$$

An interesting aspect of the rec-normal is that it affords an interpretation of the location parameter in terms of rate or speed rather than time. Analogously to the case of the log-normal, neither the location μ nor the scale σ are in the same scale as the dependent variable y. These parameters are in the same scale as the transformed dependent variable (here, $1/y$) that is assumed to be normally distributed.

By setting $\mu = 0.002$ and $\sigma = 0.0004$, data can be generated from the rec-normal that looks right-skewed and not unlike a distribution of response times. The code below shows some summary statistics of the generated data.

DOI: 10.1201/9780429342646-13

```
N <- 100
mu <- 0.002
sigma <- 0.0004
rt <- 1 / rnorm(N, mu, sigma)
c(mean = mean(rt), sd = sd(rt), min = min(rt), max = max(rt))
```

```
## mean   sd  min  max
##  523  104  303  848
```

Figure 10.1 shows the distribution generated with the code shown above.

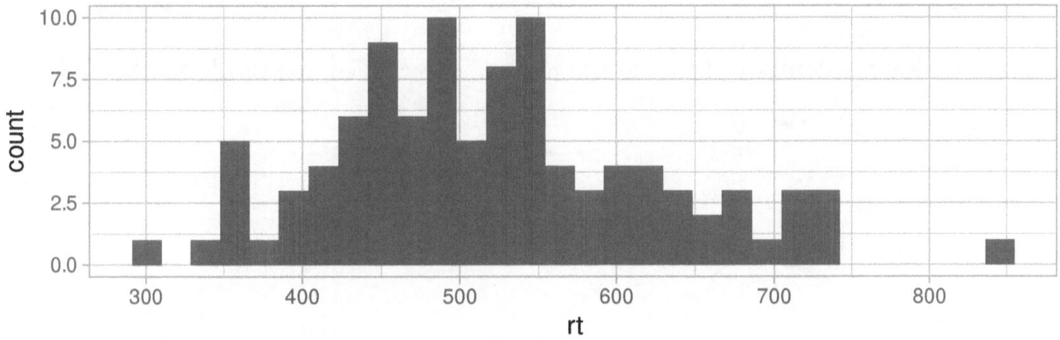

FIGURE 10.1: Distribution of synthetic data with a rec-normal distribution with parameters $\mu = 0.002$ and $\sigma = 0.0004$.

We can fit the rec-normal distribution to the response times in a simple model with a normal likelihood, by storing the reciprocal of the response times (`1/RT`) in the vector variable `recRT` (rather than storing the "raw" response times):

```
model {
\\ priors go here
target += normal_lpdf(recRT | mu, sigma);
}
```

One issue here is that the parameters of the likelihood, `mu` and `sigma` are going to be very far away from the unit scale (0.002 and 0.0004, respectively). Due to the way Stan's sampler is built, parameters that are too small (much smaller than 1) or too large (much larger than 1) can cause convergence problems. A straightforward solution is to fit the normal distribution to parameters in reciprocal of seconds rather than milliseconds; this would make the parameters have values that are 1000 times larger (2 and 0.4, respectively). We can do this using the `transformed parameters` block.[1] Although we use `mu` and `sigma` to

[1]In this specific case, we could also transform the dependent variable first, but as will become clear later, we want to avoid changing our dependent variable.

fit the data, the priors are defined on the parameters `mu_s` and `sigma_s` (μ_s and σ_s).

Unless one can rely on previous estimates, finding good priors for μ_s and σ_s is not trivial. To define appropriate priors, we would need to start with relatively arbitrary priors, inspect the prior predictive distributions, adjust the priors, and repeat the inspection of prior predictive distributions until these distributions start to look realistic. In the interest of conserving space, we skip this iterative process here, and assign the following priors:

$$\begin{aligned} \mu_s &\sim Normal(2, 1) \\ \sigma_s &\sim Normal_+(0.4, 0.2) \end{aligned} \tag{10.2}$$

This model is implemented in the file `normal_recrt.stan`, available in the `bcogsci` package:

```
data {
  int<lower = 1> N;
  vector[N] recRT;
}
parameters {
  real mu_s;
  real<lower = 0> sigma_s;
}
transformed parameters {
  real mu = mu_s / 1000;
  real sigma = sigma_s / 1000;
}
model {
  target += normal_lpdf(mu_s | 2, 1);
  target += normal_lpdf(sigma_s | 0.4, 0.2) +
            - normal_lccdf(0 | 0.4, 0.2);
  target += normal_lpdf(recRT | mu, sigma);
}
```

Fit and display the summary of the previous model:

```
normal_recrt <- system.file("stan_models",
                            "normal_recrt.stan",
                            package = "bcogsci")
fit_rec <- stan(normal_recrt, data = list(N = N, recRT = 1 / rt))
```

```
print(fit_rec, pars = c("mu", "sigma"), digits = 4)
```

```
##          mean   2.5%  97.5% n_eff Rhat
## mu     0.0020 0.0019 0.0021  3197    1
## sigma  0.0004 0.0004 0.0005  3172    1
```

Is a rec-normal likelihood more appropriate than a log-normal likelihood? As things stand, we cannot compare the models with these two likelihoods. This is because the dependent variables are different: we cannot compare reciprocal response times with untransformed response times on the millisecond scale. Model comparison with the Bayes factor or cross-validation can only compare models with the same dependent variables; see chapters 12–14.

If we do want to compare the reciprocal-normal likelihood and the log-normal likelihood, we have to set up the models with the two likelihoods in such a way that the dependent measure is on the raw millisecond scale in each model. This means that for the reciprocal normal likelihood, the model will receive as data raw reading times in milliseconds, and these will be treated as a transformed random variable from reciprocal reading times. This approach is discussed next, but requires knowledge of the Jacobian adjustment (see Ross 2002).

10.1.1 Scaling a probability density with the Jacobian adjustment

To work with the original dependent variable (`RT` rather than `1/RT`) we need to perform a *change of variables*: If the random variable X represents the reciprocal reading times ($1/RT$), we can transform this random variable to a new one, $Y = 1/X = 1/(1/RT) = RT$, yielding a transformed random variable Y which represents reading time.

This change of variables requires an adjustment to the (unnormalized log) posterior probability to account for the distortion caused by the transform.[2] The probability must be scaled by a *Jacobian* adjustment, which, in the univariate case as this one, is the absolute value of the derivative of the transform.[3] The absolute value of any number is represented by enclosing it in two vertical bars: e.g., $|-2| = 2$.

[2]Not every transformation is valid, univariate changes of variables must be monotonic and differentiable, multivariate changes of variables must be surjective and differentiable.

[3]In the multivariate case, it is equal to the absolute determinant of the Jacobian, the matrix of all its first-order partial derivatives, of the transform; see section 6.7 of Ross (2002), and p. 707 of Adams and Essex (2018).

$$
\begin{aligned}
p(RT_n|\mu,\sigma) = Normal(1/RT_n|\mu,\sigma)\left|\frac{\mathrm{d}}{\mathrm{d}RT_n}1/RT_n\right| = \\
Normal(1/RT_n|\mu,\sigma)\cdot\left|-1/RT_n^2\right| = \\
Normal(1/RT_n|\mu,\sigma)\cdot 1/RT_n^2
\end{aligned}
\tag{10.3}
$$

If we omit the Jacobian adjustment, we are essentially fitting a different and unnormalized probability density function (this will be shown later in this chapter). The discrepancy between the correct and incorrect posterior would depend on how large the Jacobian adjustment for our model is.

Because Stan works in log space, rather than multiplying the Jacobian adjustment, we add its logarithm to the log-probability density ($\log(1/RT_n^2) = -2\cdot\log(RT_n)$). The contribution to the log likelihood (of μ and σ) with its adjustment based from an individual observation, RT_n, would be as follows.

$$
\log(Normal(1/RT_n|\mu,\sigma)) - 2\cdot\log(RT_n) \tag{10.4}
$$

We obtain the log likelihood based on all the N observations by summing the log likelihood of individual observations.

$$
\log\mathcal{L} = \sum_{n=1}^{N}\log(Normal(1/RT_n|\mu,\sigma)) - \sum_{n=1}^{N}2\cdot\log(RT_n) \tag{10.5}
$$

In Stan, this summing up is done as follows. The function `normal_lpdf()` applied to a vector already returns the sum of the individual log-probability densities. The Jacobian adjustment `2*log(RT)`, which returns a vector of values, has to be summed up and added in manually (the term therefore becomes `-sum(2*log(RT))`).

```
target += normal_lpdf(1 ./ RT | mu, sigma)
          - sum(2 * log(RT));
```

Before fitting the model with the change of variables, we are going to truncate the distribution. We didn't encounter negative values in our synthetic data, but this was because the distribution was not too spread out; that is, the scale was much smaller than the location ($\sigma << \mu$). However, in principle we could end up generating negative values. For this reason, we truncate the underlying normal distribution (for more details, see online section A.2). The reciprocal truncated normal distribution has been argued to be an appropriate model of response times, neural inter-spike intervals, and latency distributions of saccades in a simple optimality model in which reward is maximized to yield an optimal response rate (Harris et al. 2014; Harris and Waddington 2012).

Because we have N observations, the truncation consists of adding `- N * normal_lccdf(0 | mu, sigma)` to `target`; also see section 8.4.1. The complete model, `recnormal_rt.stan`, including the truncation is as follows:

```
data {
  int<lower = 1> N;
  vector[N] RT;
}
parameters {
  real mu_s;
  real<lower = 0> sigma_s;
}
transformed parameters {
  real mu = mu_s / 1000;
  real sigma = sigma_s / 1000;
}
model {
  target += normal_lpdf(mu_s | 2, 1);
  target += normal_lpdf(sigma_s | 0.4, 0.2)
            - normal_lccdf(0 | 0.4, 0.2);
  target += normal_lpdf(1 ./ RT | mu, sigma)
      - N * normal_lccdf(0 | mu, sigma)
     - sum(2 * log(RT));
}
```

Next, generate data from a reciprocal truncated normal distribution:

```
N <- 500
mu <- .002
sigma <- .0004
rt <- 1 / rtnorm(N, mu, sigma, a = 0)
```

Fit the model to the data:

```
recnormal_rt <- system.file("stan_models",
  "recnormal_rt.stan",
  package = "bcogsci")
fit_rec <- stan(recnormal_rt, data = list(N = N, RT = rt))
```

Print the posterior summary:

```r
print(fit_rec, pars = c("mu", "sigma"), digits = 4)
```

```
##         mean   2.5%  97.5% n_eff Rhat
## mu    0.0020 0.0020 0.0020  3779    1
## sigma 0.0004 0.0004 0.0004  3522    1
```

We get the same results as before, but now we could potentially compare this model to one that assumes a log-normal likelihood, for example.

An important question is the following: Does every transformation of a random variable require a Jacobian adjustment? Essentially, if we assign a distribution to a random variable and then transform it, this does not require a Jacobian adjustment.

Alternatively, if we apply a transformation on a random variable first, and then assign a distribution to the transformed random variable afterward, we have a change of variables. This requires a Jacobian adjustment. This is the reason why `1 ./ RT` requires a Jacobian adjustment, but not `mu_s`, `mu`, `sigma_s`, or `sigma_s` in the previous model.

As a last step, we can encapsulate our new distribution in a function, and also create a random number generator function. This is done in a block called `functions`. To create a function, we need to specify the type of every argument and the type that the function returns.

As a simple example, if we would like a function to center a vector, we would do it as follows:

```
functions {
  vector center(vector x) {
    vector[num_elements(x)] centered;
    centered = x - mean(x);
    return centered;
  }
}
data {
...
}
parameters {
...
}
model {
...
}
```

We want to create a log(PDF) function, similar to the native Stan functions that end in `_lpdf`. Our function will take as arguments a vector `RT`, and real numbers `mu` and `sigma`. In `_lpdf` functions, (some of) the arguments (e.g., `RT`, `mu`, and `sigma`) can be vectorized, but the output of the function is always a real number: the sum of the log(PDF) evaluated at every value of the random variable at the left of the pipe. As we show below, to do that we just move the right hand side of `target` in our original model inside our new function. See the section entitled Further Reading at the end of this chapter for more about Stan functions.

For our custom random number generator function (which should end with `_rng`), we have the added complication that the function is truncated. We simply generate random values from a normal distribution, do the reciprocal transformation and if the number is above zero, this number is returned; otherwise, a new number is drawn. This implementation may be inefficient when rejections occur frequently. Another, more complex implementation using the inverse CDF approach can be found in Stan Development Team (2024) (section 21.10 of the User's guide).

The complete code can be found in `recnormal_rt_f.stan`, and is also shown below:

```
functions {
  real recnormal_lpdf(vector y, real mu, real sigma){
    real lpdf;
    lpdf = normal_lpdf(1 ./ y | mu, sigma)
    - num_elements(y) * normal_lccdf(0 | mu, sigma)
    - sum(2 * log(y));
    return lpdf;
  }
  real recnormal_rng(real mu, real sigma){
    real y = 0;
    while (y <= 0)
      y = 1 / normal_rng(mu, sigma);
    return y;
  }
}
data {
  int<lower = 1> N;
  vector<lower = 0>[N] RT;
}
parameters {
  real mu_s;
```

```
  real<lower = 0> sigma_s;
}
transformed parameters {
  real mu = mu_s / 1000;
  real sigma = sigma_s / 1000;
}
model {
  target += normal_lpdf(mu_s | 2, 1);
  target += normal_lpdf(sigma_s | 0.4, 0.2)
            - normal_lccdf(0 | 0.4, 0.2);
  target += recnormal_lpdf(RT | mu, sigma);
}
generated quantities {
  array[N] real rt_pred;
  for (n in 1:N)
    rt_pred[n] = recnormal_rng(mu, sigma);
}
```

Fit the model to the simulated data:

```
recnormal_rt_f <- system.file("stan_models",
                              "recnormal_rt_f.stan",
                              package = "bcogsci")
fit_rec_f <- stan(recnormal_rt_f, data = list(N = N, RT = rt))
```

Print the summary:

```
print(fit_rec_f, pars = c("mu", "sigma"), digits = 4)
```

```
##         mean   2.5%  97.5% n_eff  Rhat
## mu    0.0020 0.0020 0.0020  3643 0.999
## sigma 0.0004 0.0004 0.0004  3476 1.000
```

10.2 Validation of a computed posterior distribution

The model converged, but did it work? At the very least we expect that the simulated data that we used to test the model shows a very similar distribution to the posterior predictive distributions. This is because we are fitting the data with the same function that we use to generate the data. Figure 10.2 shows that the simulated data and the posterior predictive distributions are similar.

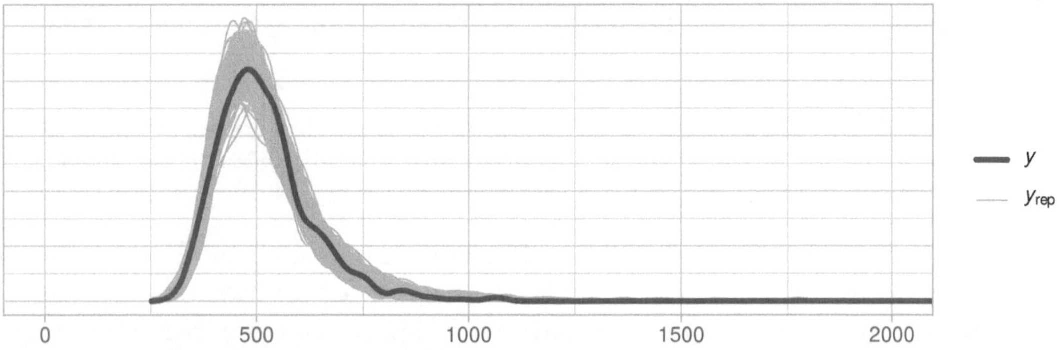

FIGURE 10.2: A posterior predictive check of `fit_rec` in comparison with the simulated data.

Our synthetic data set was generated by first defining a *ground truth* vector of parameters $\langle \mu; \sigma \rangle$. We would expect that the true values of the parameters μ and σ should be well inside the bulk of the posterior distribution of our model. We investigate this by plotting the true values of the parameters together with their posterior distributions using the function `mcmc_recover_hist()` of `bayesplot` as shown below in Figure 10.3.

```
post_rec <- as.data.frame(fit_rec_f) %>%
  select(mu, sigma)
mcmc_recover_hist(post_rec, true = c(mu, sigma))
```

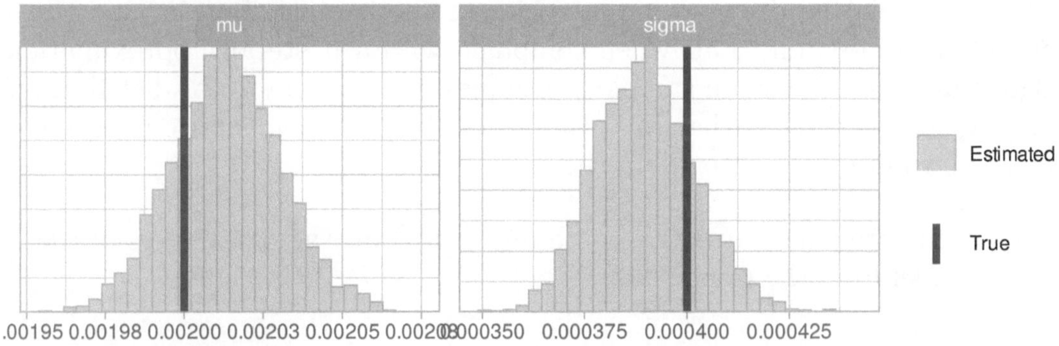

FIGURE 10.3: Posterior distributions of the parameters of `fit_rec` together with their true values.

Even though this approach can capture serious misspecifications in our models, it has three important shortcomings.

First, it's unclear what we should conclude if the true value of a parameter is not in the bulk of its posterior distribution. Even in a well-specified model, if we inspect enough parameters we will find that for some parameters, the bulks of their respective marginal posterior distributions can be relatively far away from their true values.

The second shortcoming is that if the posteriors that we obtain are very wide, it might not be clear what constitutes the bulk of the posterior distribution that needs to be inspected.

The third shortcoming is that regardless of how a successful recovery of parameters is defined, we might be able to recover a posterior based on data generated from some parts of the parameter space, but not based on data generated from other parts of parameter space (Talts et al. 2018). However, inspecting the entire parameter space is infeasible.

An alternative to our approach of plotting the true values of the parameters together with their posterior distributions is called simulation-based calibration (Talts et al. 2018; Modrák et al. 2023), which extends ideas of Cook, Gelman, and Rubin (2006). This is discussed next.

10.2.1 The simulation-based calibration procedure

Talts et al. (2018) suggest the simulation-based calibration as the procedure for validating a model. We check that the Bayesian computation is *faithful* in the sense that it is not biased. Our goal is to ensure that our model neither overestimates nor underestimates, nor does it possess incorrect precision in the parameters, given the data and priors. As initial steps, we employ two different implementations of the same statistical model: (1–2) a generator capable of directly simulating draws from the prior predictive distribution, and (3) a probabilistic program that, when combined with a posterior approximation algorithm (such as a Stan sampler), samples from the posterior distribution. Finally, (4–5) we verify that the results from both implementations have the same distribution conditional on the data.

1. Generate true values (i.e., a ground truth) for the parameters of the model from the priors:

$$\tilde{\Theta}_n \sim p(\boldsymbol{\Theta}) \tag{10.6}$$

Here, $p(\boldsymbol{\Theta})$ represents a joint prior distributions of vector of parameters. In our previous example, the vector of parameters is $\langle\mu;\sigma\rangle$, and the joint prior distribution is two independent distributions, a normal and a truncated normal. Crucially, the prior distributions should be meaningful; that is, one should use priors that are at least regularizing or weakly informative. The prior space plays a crucial role because it determines the parameter space where we will verify the correctness or faithfulness of the model.

Assuming 150 samples (from the priors), this step for our model would be as follows:

```
N_sim <- 150
mu_tilde <- rnorm(N_sim, 2, 1) / 1000
sigma_tilde <- rtnorm(N_sim, 0.4, 0.2, a = 0) / 1000
```

2. Generate multiple (`N_sim`) data sets based on the probability density function used as the likelihood:

$$\tilde{\boldsymbol{D}}_n \sim p(\boldsymbol{D}|\tilde{\Theta}_n) \tag{10.7}$$

The previous equation indicates that each data set $\tilde{\boldsymbol{D}}_n$ is sampled from each of the generated parameters, $\tilde{\Theta}_n$. Following our previous example, use the reciprocal truncated normal distribution to generate data sets of response times:

```
## Create a place holder for all the simulated datasets
rt_tilde <- vector(mode = "list", length = N_sim)
# Number of observations
N <- 500
for (n in 1:N_sim) {
  rt_tilde[[n]] <- 1 / rtnorm(N, mu_tilde[n], sigma_tilde[n], a = 0)
}
```

Now, `rt_tilde` consists of a list with 150 vectors of 500 observations each. Each list represents a simulated data, which corresponds to $\tilde{\boldsymbol{D}}_n$.

3. Fit the same Bayesian model to each of the generated data sets.

After fitting the same model to each data set, we should compare the recovered posterior distribution for each parameter of each simulated data set; that is, $p(\Theta_n|\tilde{\boldsymbol{D}}_n)$, with the parameters that were used to generate the data, $\tilde{\Theta}_n$. This comparison is represented by ... in the code below (that is, it is not yet implemented in the code shown below). This raises the question of how to compare the posterior distribution with the ground truth values of the parameters.

```
for (i in 1:N_sim) {
  fit <- stan(recnormal_rt,
```

```
    data = list(N = N, rt = rt_tilde[[n]])
  )
  ...
}
```

The procedure shown before (1–3) defines a natural condition for assessing whether the computed posterior distributions match the exact posterior distributions (Talts et al. 2018; Cook, Gelman, and Rubin 2006). This condition is met if the generator, probabilistic program, and posterior approximation algorithm function match: The generator and the probabilistic program must accurately represent the same data-generating process, while the posterior approximation algorithm should yield samples that closely match the correct posterior for the probabilistic program, based on data simulated from the prior. Any discrepancy suggests a mismatch among the components (Modrák et al. 2023).[4]

The next two steps describe a well-defined comparison of the exact posterior and prior distribution (different calibration checking methods differ in how exactly they test for mismatches between the distributions; Modrák et al. 2023).

4. Generate an ensemble of rank statistics of prior samples relative to corresponding posterior samples.

Talts et al. (2018) demonstrate that the match between the exact posterior and prior can be evaluated by examining whether, for each element of $\boldsymbol{\Theta}$ (e.g., $\langle \mu; \sigma \rangle$), the *rank statistics* of the prior sample relative to S posterior samples will be uniformly distributed across $[0, S]$. In other words, for each sample of the prior (e.g., $\tilde{\mu}_n$), we calculate the number of samples of the posterior that is smaller than the corresponding value of the parameter sampled, and we examine its distribution. A mismatch between the exact posterior and our posterior would show up as a non-uniform distribution.

As a next step, we examine this new distribution visually using a histogram with $S + 1$ possible ranks (from 0 to S). There are two issues that we need to take into account (Talts et al. 2018):

[4]We are working under the assumption that Stan, which yields the posterior approximation, works correctly. In principle, if we assume that our model is correct, we can also use simulation-based calibration to examine whether our approximation to the posterior distribution is correct (that is, whether Stan's sampler, or any posterior approximation method works correctly); for an example, see Yao et al. (2018).

(1) Regardless of our model, histograms will deviate from uniformity if the posterior samples are dependent. This can be solved by *thinning* the samples of the posterior, that is removing a number of intermediate samples (Talts et al. 2018 recommend between 6 and 10).

(2) To reduce the noise in the histogram, we should have bins of equal size. If $S+1$ are divisible by a large power of 2, e.g., 1024, we will be able to re-bin the histogram easily with bins of equal size. We complete the code shown in step 3 by generating an ensemble of ranks statistics as well.

Here we encapsulate steps 1 to 4 into a function.

```
gen_ranks <- function(...,
                      N_sim = 150,
                      N_s = 1024 * 6, # Total samples from the posterior
                      N = 500 # Number of observations
                      ){
  mu_tilde <- rnorm(N_sim, 2, 1) / 1000
  sigma_tilde <- rtnorm(N_sim, 0.4, 0.2, a = 0) / 1000
  # Thin by 6, and remove one extra sample so that S + 1 = 1024
  thinner <- seq(from = 1, to = N_s, by = 6)[-1]
  # Placeholders
  rank_mu <- rep(NA, N_sim)
  rank_sigma <- rep(NA, N_sim)
  rt_tilde <- vector(mode = "list", length = N_sim)
  # Loop over the simulations
  for (n in 1:N_sim) {
    message("Fit number ", n)
    # it will repeat the sampling unless the model converged
    repeat{
      rt_tilde[[n]] <- 1 / rtnorm(N,
                                  mu_tilde[n],
                                  sigma_tilde[n],
                                  a = 0
                                  )
      # Fit is only stored temporarily
      fit <- stan(...,
                  data = list(N = N,
                              RT = rt_tilde[[n]]),
                  warmup = 2500,
                  iter = 2500 + N_s / 4,
```

```
                   refresh = -1,
                   control = list(adapt_delta = .999,
                                  max_treedepth = 14))
      # continue if there are no divergent transitions
      # and all Rhat are smaller than 1.05
      if (!get_num_divergent(fit) &
          all(monitor(fit)[,"Rhat"] <= 1.05)) break
      message("Repeating fit ",n)
    }
    post <- extract(fit)
    # Number of samples of the posterior that is smaller than
    # the corresponding value of the parameter sampled
    rank_mu[n] <- sum(post$mu[thinner] < mu_tilde[n])
    rank_sigma[n] <- sum(post$sigma[thinner] < sigma_tilde[n])
  }
  #data frame with ranks:
  tibble(sim = rep(1:N_sim, 2),
         variable = rep(c("mu", "sigma"),
                        each = N_sim),
         rank = c(rank_mu, rank_sigma))
}

set.seed(123)
# generate ranks with the default values:
df_rank <- gen_ranks(recnormal_rt)
```

5. Graphically assess the correctness of the model using rank histograms

As a last step, we use a histogram for each parameter to identify deviations of uniformity. If the posterior estimates are correct, then each of the B bins has a probability of $1/B$ that a simulation (i.e., an individual rank) falls into it: $Binomial(N_{sim}, 1/B)$. This allows us to complement the histograms with confidence intervals, indicating where the variation expected from a uniform histogram should be.

Use the code below to build the plot shown in Figure 10.4. We can conclude that the implementation of our model (in the parameter space determined by the priors) is correct.

```
B <- 16
bin_size <- 1024 / 16
ci_l <- qbinom(0.005, size = N_sim, prob = 1 / B)
ci_u <- qbinom(0.995, size = N_sim, prob = 1 / B)
ggplot(df_rank, aes(x = rank)) +
  geom_histogram(breaks = seq(0, to = 1023, length.out = B + 1),
                 closed = "left", colour = "black") +
  scale_y_continuous("count") +
  geom_hline(yintercept = c(ci_l, ci_u), linetype = "dashed") +
  facet_wrap(~variable, scales = "free_y")
```

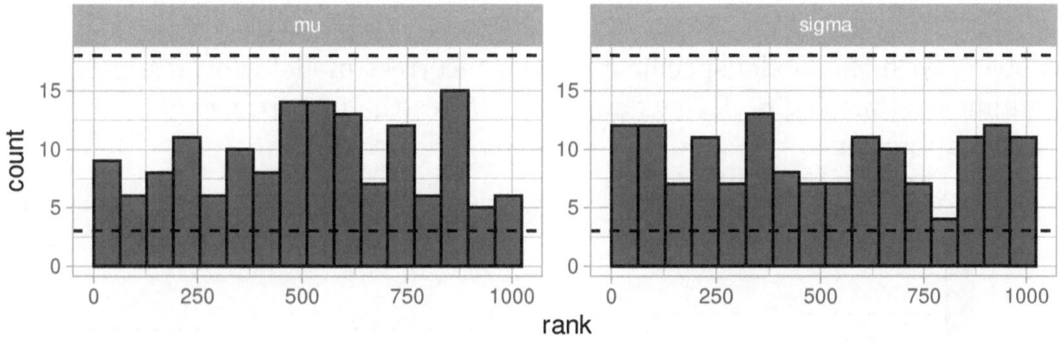

FIGURE 10.4: Rank histograms of μ and σ from the reciprocal truncated normal distribution. The dashed lines represent the 99% confidence interval.

Next, we consider an example where simulation-based calibration can reveal a problem.

10.2.2 An example where simulation-based calibration reveals a problem

Let's assume that we made an error in the model implementation. We'll fit an **incorrect** model: rather than the complete likelihood including the normal PDF and the truncation and Jacobian adjustments, we will fit only the normal PDF evaluated at the reciprocal of the response times, that is `target += normal_lpdf(1 ./ RT | mu, sigma);`.

Assessing the correctness of the model by looking at the recovery of the parameters is misleading in this case; this is evident from Figure 10.5. One could conclude that the model is fine based on this plot, since the true values of the parameters are well inside the bulk of their posterior distributions.

However, the rank histograms produced by the simulation-based calibration procedure in Figure 10.6 show very tall bins at the left and at the right of each histogram, exceeding the 99% CI. This is a very clear indication that there is a

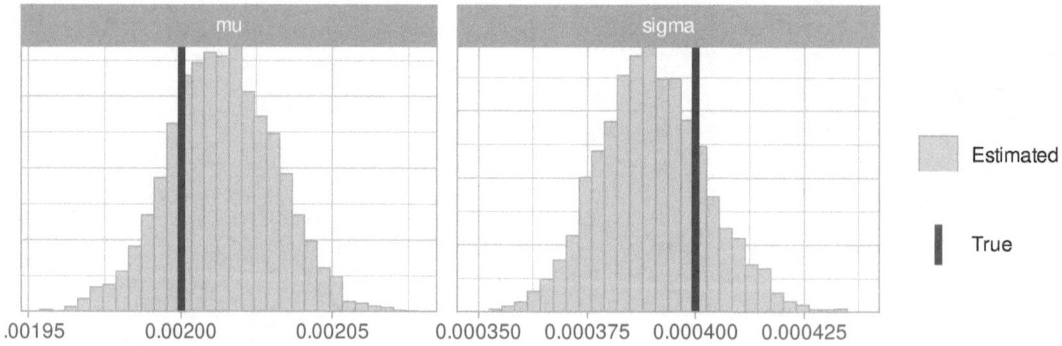

FIGURE 10.5: Posterior distributions of the parameters of an incorrect model (truncation and Jacobian adjustments missing) together with their true values.

problem with the model specification: our incorrect model is overestimating μ and underestimating σ. This example illustrates the importance of simulation-based calibration.

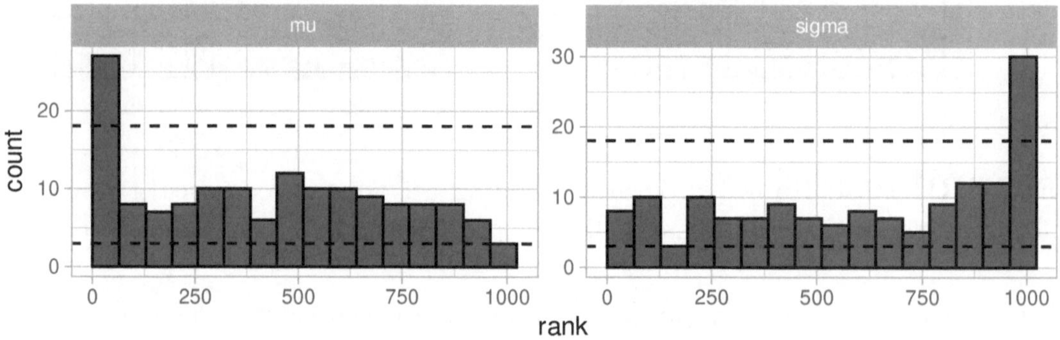

FIGURE 10.6: Rank histograms of μ and σ from the **incorrect** implementation of the reciprocal truncated normal distribution. The dashed lines represent the 99% confidence interval.

10.2.3 Issues with and limitations of simulation-based calibration

In the previous sections, we have used only rank histograms. Even though histograms are a very intuitive means of visualization to assess model correctness, they might not be sensitive enough for small deviations (Talts et al. 2018). Other visualizations might be better suited. Another limitation of histograms is that they are sensitive to the number of bins (Säilynoja, Bürkner, and Vehtari 2022). One could bin the histograms multiple times, but this approach is difficult to interpret when there are many parameters and can be vulnerable to multiple testing biases (Talts et al. 2018). One alternative to rank histograms is visualizations based on the empirical cumulative density function of the ranks. This is briefly presented in online section B.7.

Even though simulation-based calibration is a comprehensive approach to test model correctness, it has several drawbacks, regardless of the means of visualization:

(i) This approach requires fitting a model many times, which can be too time intensive for complex models. It's worth bearing in mind that code with errors can sometimes show clear "catastrophic" failures in the recovery of the parameters. For this reason, one could start the verification of model correctness with a simple recovery of parameters (see Talts et al. 2018; Gelman et al. 2020), and then proceed with the simulation-based calibration procedure for at least some suspect aspects of the model (e.g., when working with a custom likelihood).

(ii) Another major limitation of simulation-based calibration is that it is concerned exclusively with the computational aspects of the analysis and offers no guarantee for any single observation. A complete Bayesian workflow should include prior and posterior predictive checks, see the online chapter F and Schad, Betancourt, and Vasishth (2020).

(iii) Finally, it's important to apply simulation-based calibration only after appropriate priors are set. This is important because only the priors determine the parameter space that will be inspected. Weak priors that may be good enough for estimation may nevertheless lead to a parameter space that is unrealistically large since, during the calibration stage, there are no independent data to constrain the space. This means that it is important to conduct prior predictive checks before assessing the correctness of the model with simulation-based calibration.

10.3 Another custom distribution: The exponential distribution implemented manually

There are cases when one needs a random variable with a distribution that is not included in Stan. Often one can find the PDF (or a CDF) derived in a paper, and one needs to implement it manually by writing the log PDF in the Stan language. Even though the exponential distribution is included in Stan, we demonstrate how we would include it step-by-step as if it weren't available. This example extends the demonstration in section 22.1 of the User's guide (Stan Development Team 2024).

The exponential distribution is used to model waiting times until a certain event occurs. Its PDF is the following:

$$f(x|\lambda) = \begin{cases} \lambda e^{-\lambda x} & x \geq 0, \\ 0 & x < 0. \end{cases} \tag{10.8}$$

The parameter λ is often called the rate parameter and must be positive. A higher value for the rate parameter leads to shorter waiting times on average. The mean of the distribution is $1/\lambda$. The exponential distribution has the key property of being *memoryless*. What this means is explained in Ross (2002) as follows: Suppose that X is a random variable with an exponential distribution as a PDF; the random variable represents the lifetime of an item (e.g., X could represent the time it takes for a radioactive particle to completely decay). If the item is t time-units old, then the remaining life s of that item has the same probability distribution as the life of a new item. Mathematically, this amounts to stating that

$$P(X > s + t|X > t) = P(X > s) \tag{10.9}$$

The implication of the memoryless property is that we do not need to know the age of an item to know what the distribution of its remaining life is.

To give another example of memorylessness, the conditional probability that a certain event will happen in the next 100 ms is the same regardless of whether we have been already waiting 1000 ms, 10 ms, or 0 ms. Although the exponential distribution is not commonly used for modeling response times in cognitive science, it has been used in the past (Ashby and Townsend 1980; Ashby 1982). We focus on this distribution because of its simple analytical form.

As a first step, we make our own version of the PDF in R, and then check that it integrates to one to verify that this is a proper distribution (and that we haven't introduced a typo in the formula).

To avoid underflow, that is getting a zero instead of a very small number, we'll work in the log-scale. This will also be useful when we implement this function in Stan, thus the log PDF is

$$\log(f(x|\lambda)) = \log(\lambda) - \lambda x \tag{10.10}$$

where $x > 0$.

Implement this function in R with the same arguments as the `d*` family of functions: if `log = TRUE` the output is a log density; call this new function `dexp2`.

```
dexp2 <- function(x, lambda = 1, log = FALSE) {
  log_density <- log(lambda) - lambda * x
  if (log == FALSE) {
    exp(log_density)
  } else {
    log_density
  }
}
```

Verify that this function integrates to one for some point values of its parameter λ (here, $\lambda = 1$ and $\lambda = 20$):

```
dexp2_l1 <- function(x) dexp2(x, 1)
integrate(dexp2_l1, lower = 0, upper = Inf)$value
```

```
## [1] 1
```

```
dexp2_l20 <- function(x) dexp2(x, 20)
integrate(dexp2_l20, lower = 0, upper = Inf)$value
```

```
## [1] 1
```

To test our function, we'll also need to generate random values from the exponential distribution. If the quantile function of the distribution exists, the inverse transform sampling is a relatively straightforward way to get pseudo random numbers sampled from a target distribution (for an accessible introduction to inverse sampling, see Lynch 2007). Given a target distribution with a PDF f, and a quantile function F^{-1} (the inverse of the CDF), the inverse transform sampling method consists of the following:

1. Sample one number u from $Uniform(0, 1)$.
2. Then $z = F^{-1}(u)$ is a draw from $f(x)$.

In this case, the quantile function (the inverse of the CDF) is the following:

$$-\log(1 - p)/\lambda \tag{10.11}$$

Here is how one can derive this CDF and its inverse if they were not available to us. First, consider the fact that the CDF of the exponential distribution is as follows. The term q is some quantile of the distribution:

$$F(q) = \int_0^q \lambda \exp(-\lambda x)\, \mathrm{d}x \tag{10.12}$$

We can solve this integral by using the u-substitution method (Salas, Etgen, and Hille 2003). First, define

$$u = g(x) = -\lambda x \tag{10.13}$$

Then, the derivative du/dx is:

$$\frac{\mathrm{d}u}{\mathrm{d}x} = -\lambda \tag{10.14}$$

This implies that $\mathrm{d}u = -\lambda \mathrm{d}x$, or that $-\mathrm{d}u = \lambda \mathrm{d}x$.

In the CDF, replace the term $\lambda \mathrm{d}x$ with $-\mathrm{d}u$:

$$F(q) = \int_0^q \lambda \exp(-\lambda x)\, \mathrm{d}x = \int_0^{-\lambda q} (-\exp(-\lambda x)\, \mathrm{d}u) \tag{10.15}$$

Rewriting $-\lambda x$ as u, the CDF simplifies to:

$$F(q) = \int_0^q \lambda \exp(-\lambda x)\, \mathrm{d}x = \int_0^q (-\exp(u)\, \mathrm{d}u) \tag{10.16}$$

We know from calculus that the integral of $-\exp(u)$ is $-\exp(u)$. So, the integral becomes:

$$F(q) = [-\exp(u)]_0^q \tag{10.17}$$

Replacing u with $-\lambda x$, we get:

$$F(q) = [-\exp(-\lambda x)]_0^q = 1 - \exp(-\lambda q) \tag{10.18}$$

Thus, we know that the CDF is $F(t) = 1 - \exp(-\lambda q) = p$, where p is the probability of observing the quantile q or some value smaller than q. To derive the inverse of the CDF, solve the equation below for q:

$$\begin{aligned}
p &= 1 - e^{-\lambda q} \\
p - 1 &= -e^{-\lambda q} \\
-p + 1 &= e^{-\lambda q} \\
\log(1 - p) &= -\lambda q \\
-\log(1 - p)/\lambda &= q
\end{aligned} \tag{10.19}$$

Write the quantile function (the inverse of the CDF) and the random number generator for the exponential distribution in R. To differentiate it from the built-in function in R, we will call it `qexp2()`:

```
qexp2 <- function(p, lambda = 1) {
  -log(1 - p) / lambda
}
rexp2 <- function(n, lambda = 1) {
  u <- runif(n, 0, 1)
  qexp2(u, lambda)
}
```

The functions that we would use in a Stan model are relatively faithful to the R code, but follow Stan conventions: The expression `log1m(p)` is more arithmetically stable than `log(1 - .)` for values of `p` close to one, the function `exp_lpdf()` stores the sum of the log(PDF) evaluated at each value of x, this would be analogous to doing `sum(dexp2(x, lambda, log = TRUE))`; the function `exp_rng()` implements a non-vectorized version of `rexp2()` and uses the auxiliary function `exp_icdf()` (icdf stands for inverse CDF), which is similar to `qexp2()`.

```
functions {
  real exp_lpdf(vector x, real lambda){
    vector[num_elements(x)] lpdf = log(lambda) - lambda * x;
    return sum(lpdf);
  }
  real exp_icdf(real p, real lambda){
    return -log1m(p) / lambda;
  }
  real exp_rng(real lambda){
    real u = uniform_rng(0, 1);
    return exp_icdf(u, lambda);
  }
}
```

We are now ready to generate synthetic data and fit the distribution in Stan. Generate 1000 observations.

```
N <- 1000
lambda <- 1 / 200
rt <- rexp2(N, lambda)
```

Use `exponential.stan`, which includes the function block shown earlier, and the following obligatory blocks:

```
data {
  int<lower = 1> N;
  vector<lower = 0>[N] RT;
}
parameters {
  real<lower = 0> lambda;
}
model {
  target += normal_lpdf(lambda | 0, .1) -
    normal_lccdf(0 | 0, .1);
  target += exp_lpdf(RT | lambda);
}
generated quantities {
  array[N] real rt_pred;
  for (n in 1:N)
    rt_pred[n] = exp_rng(lambda);
}
```

Fit the data with Stan.

```
exponential <- system.file("stan_models",
                           "exponential.stan",
                           package = "bcogsci")
fit_exp <- stan(exponential, data = list(N = N, RT = rt))
```

Print the summary:

```
print(fit_exp, pars = c("lambda"))
```

```
##        mean 2.5% 97.5% n_eff Rhat
## lambda 0.01    0  0.01  1471    1
```

Carry out a quick check first, verifying that the true value of the parameter λ is reasonably inside the bulk of the posterior distribution. This is shown in Figure 10.7(a). Figure 10.7(b) shows the results of the simulation-based calibration procedure and suggests that our implementation was correct.

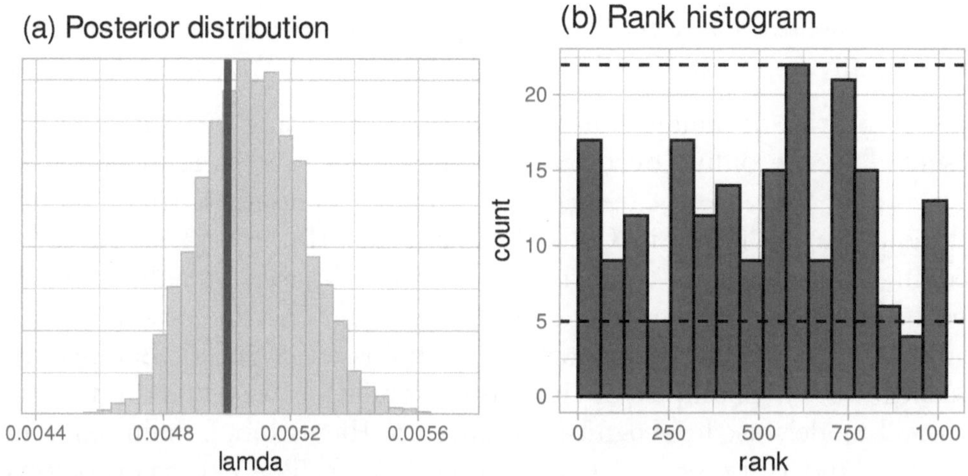

FIGURE 10.7: (a) Posterior distribution of the parameter λ of `fit_exp` together with its true values as a black line. (b) Rank histogram of λ and from our hand-made implementation of the exponential distribution. The dashed lines represent the 99% confidence interval.

10.4 Summary

In this chapter, we learned how to create a PDF that is not provided by Stan. We learned how to do this by doing a change of variables (using the Jacobian adjustment) and by building the distribution from scratch in a function ending in `_lpdf`. We also learned how to verify the correctness of the new functions by recovering the true values of the parameters and by using simulation-based calibration.

10.5 Further reading

Jacobian adjustments in Bayesian models are a common source of confusion and there are several posts in blogs and study cases that try to shed light on them:

- `https://rstudio-pubs-static.s3.amazonaws.com/486816_440106f76c944734a7d4c84761e37388.html`
- `https://betanalpha.github.io/assets/case_studies/probability_theory.html#42_probability_density_functions`
- `https://jsocolar.github.io/jacobians/`
- `https://mc-stan.org/learn-stan/case-studies/mle-params.html`

Jacobian adjustments are also relevant for adjusting priors for order constraints, which is especially important for Bayes factors (Heck and Wagenmakers 2016).

A complete tutorial of simulation-based calibration using the `SBC` package was presented in the online event Stanconnect 2021, and it is available in `https://www.martinmodrak.cz/post/2021-sbc_tutorial/`. The ideas that predate this technique can be found in Cook, Gelman, and Rubin (2006) and were extended in Talts et al. (2018). The use of ECDF-based visualizations is discussed in Säilynoja, Bürkner, and Vehtari (2022). The role of simulation-based calibration in the Bayesian workflow is discussed in Schad, Betancourt, and Vasishth (2020). Examples of the use of simulation-based calibration to validate novel models used in cognitive science are Hartmann, Johannsen, and Klauer (2020) and Bürkner and Charpentier (2020). The extension of this procedure to validate Bayes factors is discussed in Schad et al. (2021).

Custom functions in general and custom probability functions are treated chapter 21 of the User's guide (Stan Development Team 2024).

Part IV

Evidence synthesis and measurement error

11

Meta-analysis and measurement error models

In this chapter, we introduce two relatively underutilized modeling approaches that are potentially very important for cognitive science: meta-analysis and measurement-error models.

Meta-analysis can be very informative when carrying out systematic reviews, and measurement-error models are able to take into account uncertainty in one's dependent or independent variable (or both). What's common to these two classes of model is that they both assume that the n-th measured data point y_n has a location parameter, say ζ_n (pronounced *zeta en*), that is measured with some uncertainty that can be represented by the standard error SE_n of the measurement y_n:

$$y_n \sim Normal(\zeta_n, SE_n)$$

In both classes of model, the goal is to obtain a posterior distribution of a latent parameter ζ which is assumed to generate the ζ_n, with some standard deviation τ. The parameter τ quantifies the noise in the measurement process or the between-study variability in a meta-analysis.

$$\zeta_n \sim Normal(\zeta, \tau)$$

The main parameter of interest is usually ζ, but the posterior distributions of τ and ζ_n can also be informative. The above model specification should remind you of the hierarchical models we saw in earlier chapters.

11.1 Meta-analysis

Once a number of studies have accumulated on a particular topic, it can be very informative to synthesize the data. Here is a commonly used approach–a random-effects meta-analysis.

DOI: 10.1201/9780429342646-15

11.1.1 A meta-analysis of similarity-based interference in sentence comprehension

The model is set up as follows. For each study n, let effect_n be the effect of interest, and let SE_n be the standard error of the effect. A concrete example of a recent meta-analysis is the effect of similarity-based interference in sentence comprehension (Jäger, Engelmann, and Vasishth 2017); when two nouns are more similar to each other, there is greater processing difficulty (i.e., longer reading times in milliseconds) when an attempt is made to retrieve one of the nouns to complete a linguistic dependency (such as a subject-verb dependency). The estimate of the effect and its standard error is the information we have from each study n.

First, load the data, and add an id variable that identifies each experiment.

```
data("df_sbi")
(df_sbi <- df_sbi %>%
  mutate(study_id = 1:n()))
```

```
## # A tibble: 12 x 4
##   publication        effect    SE study_id
##   <chr>               <int> <int>    <int>
## 1 VanDyke07E1LoSem      13    30        1
## 2 VanDyke07E2LoSem      37    21        2
## 3 VanDyke07E3LoSem      20    11        3
## # i 9 more rows
```

The effect size and standard errors were estimated from published summary statistics in the respective article. In some cases, this involved a certain amount of guesswork; the details are documented in the online material accompanying Jäger, Engelmann, and Vasishth (2017).

We begin with the assumption that there is a true (unknown) effect ζ_n that lies behind each of these studies. Each of the observed effects has an uncertainty associated with it, SE_n. We can therefore assume that each observed effect, effect_n, is generated as follows:

$$\text{effect}_n \sim Normal(\zeta_n, SE_n) \tag{11.1}$$

Each study is assumed to have a different true effect ζ_n because each study will have been carried out under different conditions: in a different lab with different protocols and workflows, with different subjects, possibly with different languages, with slightly different experimental designs, etc.[1]

Further, each of the true underlying effects ζ_n has behind it some true unknown value ζ. The parameter ζ represents the underlying effect of similarity-based interference across experiments. Our goal is to obtain the posterior distribution of this overall effect.

We can write the above statement as follows:

$$\zeta_n \sim Normal(\zeta, \tau) \tag{11.2}$$

τ is the between-study standard deviation; this expresses the assumption that there will be some variability between the true effects ζ_n.

To summarize the model:

- effect_n is the observed effect (in this example, in milliseconds) in the n-th study.
- ζ_n is the true (unknown) effect in each study.
- ζ is the true (unknown) effect of the experimental manipulation, namely, the similarity-based interference effect.
- Each SE_n is estimated from the standard error available from study n.
- The parameter τ represents between-study standard deviation.

We can construct a hierarchical model as follows:

$$\begin{aligned}
\text{effect}_n &\sim Normal(\zeta_n, SE_n) \quad n = 1, \ldots, N_{studies}\\
\zeta_n &\sim Normal(\zeta, \tau)\\
\zeta &\sim Normal(0, 100)\\
\tau &\sim Normal_+(0, 100)
\end{aligned} \tag{11.3}$$

[1]In the current example, the dependent variable is either self-paced reading time or first-pass reading time from eyetracking. A possible criticism here is that these two different dependent measures should not appear in the same meta-analysis. Despite these being quite different dependent variables, the simplifying assumption we made was that the dependent measure, which is in milliseconds in both the self-paced reading studies and eyetracking studies, is a difference of means between two (sets of) conditions, and therefore gives an estimate of the effect of interest. An alternative approach would have been to standardize the effect size (Cohen's d) in each study; but this would be difficult to carry out in this example as much of the published work did not provide the original data; in many of the cases, only summary statistics from the publication were available. Such standardized effect sizes would also be open to criticism, because the effect size in self-paced reading and eyetracking may also not be comparable.

The priors are based on domain knowledge; it seems reasonable to allow the effect to range a priori from −200 to +200 ms with probability 95%. Of course, a sensitivity analysis is necessary (but skipped here).

This model can be implemented in `brms` in a relatively straightforward way as shown below. We show the Stan version later in the chapter (section 11.1.1.2); the Stan version presents some interesting challenges that can be useful for the reader interested in deepening their Stan modeling knowledge.

11.1.1.1 `brms` version of the meta-analysis model

First, define the priors:

```
priors <- c(prior(normal(0, 100), class = Intercept),
            prior(normal(0, 100), class = sd))
```

Fit the model as follows. Because of our relatively uninformative priors and the few data points, the models of this chapter require us to tune the `control` parameter, increasing `adapt_delta` and `max_treedepth`.

```
fit_sbi <- brm(effect | resp_se(`SE`, sigma = FALSE) ~
                 1 + (1 | study_id),
               data = df_sbi,
               prior = priors,
               control = list(adapt_delta = 0.99, max_treedepth = 10))
```

The posterior of ζ and τ are summarized below as `Intercept` and `sd(Intercept)`.

```
fit_sbi
```

```
## ...
## Group-Level Effects:
## ~study_id (Number of levels: 12)
##               Estimate Est.Error l-95% CI u-95% CI Rhat Bulk_ESS
## sd(Intercept)    11.51      7.68     0.52    29.52 1.00      860
##               Tail_ESS
## sd(Intercept)     1474
##
## Population-Level Effects:
##           Estimate Est.Error l-95% CI u-95% CI Rhat Bulk_ESS Tail_ESS
## Intercept    13.27      6.23     2.62    27.09 1.00     1105      966
##
## Family Specific Parameters:
```

```
##       Estimate Est.Error l-95% CI u-95% CI Rhat Bulk_ESS Tail_ESS
## sigma     0.00      0.00     0.00     0.00   NA       NA       NA
##
## ...
```

The `sigma` parameter does not play any role in this model, but appears in the `brms` output anyway. In the model specification, `sigma` was explicitly removed by writing `sigma = FALSE`. For this reason, we can ignore that parameter in the model summary output above. Online section C.1 explains what happens if we set `sigma = TRUE`.

As theory predicts, the overall effect from these studies has a positive sign.

One advantage of such a meta-analysis is that the posterior can now be used as an informative prior for a future study. This is especially important when doing an analysis using Bayes factors. But this meta-analysis posterior could also be used as an informative prior in a future experiment; that would allow the researcher to build on what is known so far from published studies.

Another interesting by-product of a random-effects meta-analysis is the possibility of displaying a forest plot (Figure 11.1). A forest plot shows the meta-analytic estimate (the parameter `b_Intercept` in `brms`) alongside the original estimates effect_n (and their SE_n) and the posterior distributions of the ζ_n for each study (we reconstruct these estimates by adding `b_Intercept` to the parameters starting with `r_` in `brms`). The original estimates are the ones fed to the model as data and the posterior distributions of the ζ_n are calculated, as in previous hierarchical models, after the information from all studies is pooled together. The ζ_n estimates are shrunken estimates of each study's (unknown) true effect, shrunken toward the grand mean ζ, and weighted by the standard error observed in each study n. The ζ_n for a particular study is shrunk more toward the grand mean ζ when the study's standard error is large (i.e., when the estimate is very imprecise). The code below shows how to build a forest plot step by step.

First, change the format of the data so that it looks like the output of `brms`:

```
df_sbi <- df_sbi %>%
  mutate(Q2.5 = effect - 2 * SE,
         Q97.5 = effect + 2 * SE,
         Estimate = effect,
         type = "original")
```

Extract the meta-analytical estimate:

```
df_Intercept <- posterior_summary(fit_sbi,
                                  variable = c("b_Intercept")) %>%
  as.data.frame() %>%
  mutate(publication = "M.A. estimate", type = "")
```

For the pooled estimated effect (or fitted value) of the individual studies, we need the sum of the meta-analytical estimate (intercept) and each of the by-study adjustment. Obtain this with the `fitted()` function:

```
df_model <- fitted(fit_sbi) %>%
  # Convert matrix to data frame:
  as.data.frame() %>%
  # Add a column to identify the estimates,
  # and another column to identify the publication:
  mutate(type = "adjusted",
         publication = df_sbi$publication)
```

Bind the observed effects, the meta-analytical estimate, and the fitted values of the studies together, and plot the data:

```
# the adjusted estimates and the meta-analysis estimate:
bind_rows(df_sbi, df_model, df_Intercept) %>%
  # Plot:
  ggplot(aes(x = Estimate,
             y = publication,
             xmin = Q2.5,
             xmax = Q97.5,
             color = type)) +
  geom_point(position = position_dodge(.5)) +
  geom_errorbarh(position = position_dodge(.5)) +
  # Add the meta-analytic estimate and Credible Interval:
  geom_vline(xintercept = df_Intercept$Q2.5,
             linetype = "dashed",
             alpha = .3) +
  geom_vline(xintercept = df_Intercept$Q97.5,
             linetype = "dashed",
             alpha = .3) +
  geom_vline(xintercept = df_Intercept$Estimate,
             linetype = "dashed",
```

```
            alpha = .5) +
  scale_color_discrete(breaks = c("adjusted", "original"))
```

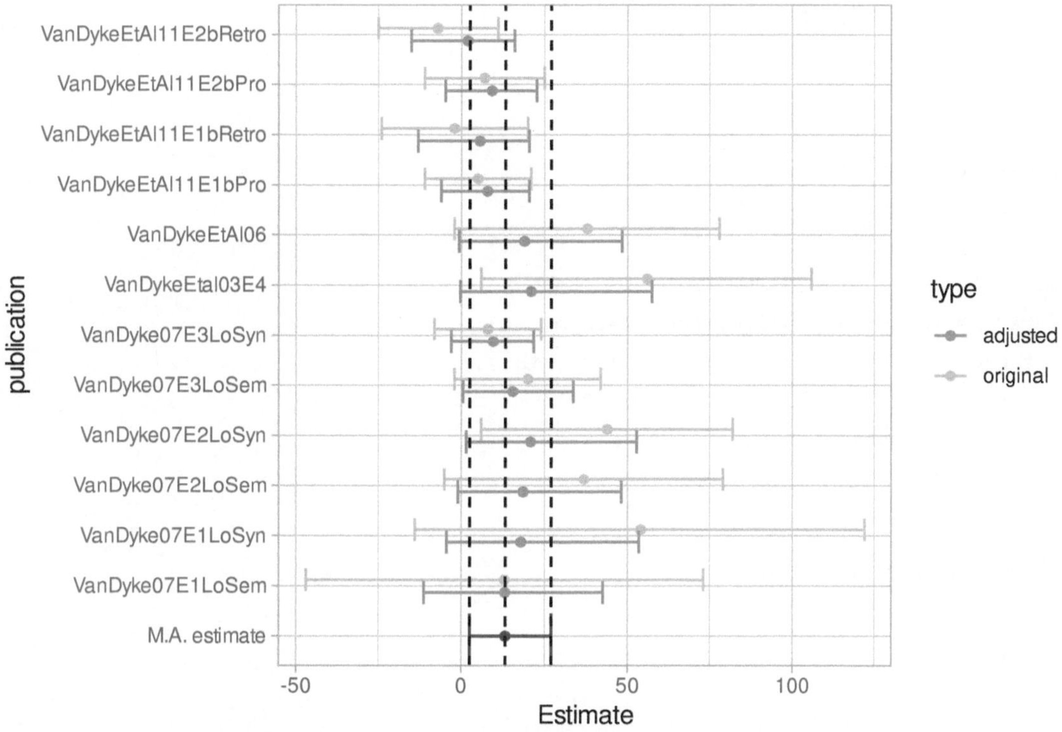

FIGURE 11.1: Forest plot showing the original and the adjusted estimates computed from each study from the random-effects meta-analysis. The error bars on the original estimates show 95% confidence intervals, and those on the adjusted estimates show 95% credible intervals.

It is important to keep in mind that a meta-analysis is always going to yield biased estimates as long as we have publication bias: if a field has a tendency to allow only "big news" studies to be published, then the literature that will appear in the public domain will be biased, and any meta-analysis based on such information will be biased. Despite this limitation, a meta-analysis is still a useful way to synthesize the known evidence; one just has to remember that the estimate from the meta-analysis is almost certain to be biased.

11.1.1.2 Stan version of the meta-analysis model

Even though `brms` can handle meta-analyses, fitting them in Stan allows us for more flexibility, which might be necessary in some cases. As a first attempt we could build a model that closely follows the formal specification given in Equation (11.3).

```
data {
  int<lower=1> N;
  vector[N] effect;
  vector[N] SE;
  vector[N] study_id;
}
parameters {
  real zeta;
  real<lower = 0> tau;
  vector[N] zeta_n;
}
model {
  target += normal_lpdf(effect| zeta_n, SE);
  target += normal_lpdf(zeta_n | zeta, tau);
  target += normal_lpdf(zeta | 0, 100);
  target += normal_lpdf(tau | 0, 100)
    - normal_lccdf(0 | 0, 100);
}
```

Fit the model as follows:

```
ma0 <- system.file("stan_models",
                   "meta-analysis0.stan",
                   package = "bcogsci")
ls_sbi <- list(N = nrow(df_sbi),
               effect = df_sbi$effect,
               SE = df_sbi$SE,
               study_id = df_sbi$study_id)
fit_sbi0 <- stan(ma0,
                 data = ls_sbi,
                 control = list(adapt_delta = 0.999, max_treedepth = 12))
```

```
## Warning: There were 3 divergent transitions after
## warmup. See
## https://mc-stan.org/misc/warnings.html#divergent-transitions-after-warmup
## to find out why this is a problem and how to eliminate
## them.
```

```
## Warning: There were 1 chains where the estimated
## Bayesian Fraction of Missing Information was low. See
## https://mc-stan.org/misc/warnings.html#bfmi-low
```

```
## Warning: Examine the pairs() plot to diagnose sampling
```

```
## problems

## Warning: Bulk Effective Samples Size (ESS) is too low,
## indicating posterior means and medians may be unreliable.
## Running the chains for more iterations may help. See
## https://mc-stan.org/misc/warnings.html#bulk-ess

## Warning: Tail Effective Samples Size (ESS) is too low,
## indicating posterior variances and tail quantiles may be
## unreliable. Running the chains for more iterations may
## help. See
## https://mc-stan.org/misc/warnings.html#tail-ess
```

We see that there are warnings. As discussed in section 9.1.2, we can use pairs plots to uncover pathologies in the sampling. Here we see the samples of `zeta` and `tau` are highly correlated:

```
pairs(fit_sbi0, pars = c("zeta", "tau"))
```

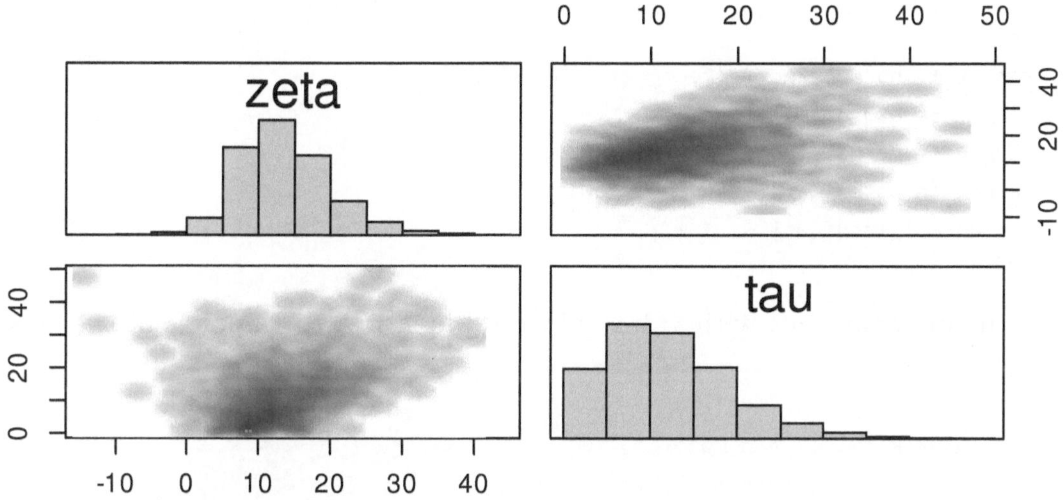

We face a similar problem as we faced in section 9.1.2, namely, the sampler cannot properly explore the neck of the funnel-shaped space, because of the strong correlation between the parameters. The solution is, as in section 9.1.2, a non-centered parameterization. Re-write Equation (11.3) as follows:

$$
\begin{aligned}
z_n &\sim Normal(0, 1) \\
\zeta_n &= z_n \cdot \tau + \zeta \\
\text{effect}_n &\sim Normal(\zeta_n, SE_n) \\
\zeta &\sim Normal(0, 100) \\
\tau &\sim Normal_+(0, 100)
\end{aligned}
\tag{11.4}
$$

This works because if $X \sim Normal(a, b)$ and $Y \sim Normal(0, 1)$, then $X = a + Y \cdot b$. You can re-visit section 9.1.2 for more details.

Translate Equation (11.4) into Stan code as follows in `meta-analysis1.stan`:

```
data {
  int<lower=1> N;
  vector[N] effect;
  vector[N] SE;
  vector[N] study_id;
}
parameters {
  real zeta;
  real<lower = 0> tau;
  vector[N] z;
}
transformed parameters {
  vector[N] zeta_n = z * tau + zeta;
}
model {
  target += normal_lpdf(effect| zeta_n, SE);
  target += std_normal_lpdf(z);
  target += normal_lpdf(zeta | 0, 100);
  target += normal_lpdf(tau | 0, 100)
    - normal_lccdf(0 | 0, 100);
}
```

The model converges with values virtually identical to the ones of the `brms` model.

```
ma1 <- system.file("stan_models",
                   "meta-analysis1.stan",
                   package = "bcogsci")
fit_sbi1 <- stan(ma1,
                 data = ls_sbi,
                 control = list(adapt_delta = 0.999,
                                max_treedepth = 12))
```

```
print(fit_sbi1, pars = c("zeta", "tau"))
```

```
##      mean 2.5% 97.5% n_eff Rhat
## zeta 13.4 2.71  28.3   739    1
```

```
## tau   11.6 0.55  30.2   729    1
```

We can also reparameterize the model slightly differently, if we set $U_n \sim Normal(0, SE_n)$ then,

$$\text{effect}_n = U_n + \zeta_n \tag{11.5}$$

Then, given that $\zeta_n \sim Normal(\zeta, \tau)$,

$$\text{effect}_n \sim Normal(\zeta, \sqrt{SE^2 + \tau^2}) \tag{11.6}$$

See online section C.1 if it's not clear why this reparameterization works.

This is equivalent to the `brms` model where `sigma = TRUE`. Parameterizing the model in this way causes us to lose the possibility of estimating the posterior of the true effect of the individual studies.

Write this in Stan as follows; this code is available in the file `meta-analysis2.stan` within the `bcogsci` package:

```
data {
  int<lower=1> N;
  vector[N] effect;
  vector[N] SE;
  vector[N] study_id;
}
parameters {
  real zeta;
  real<lower = 0> tau;
}
model {
  target += normal_lpdf(effect| zeta,
                        sqrt(square(SE) + square(tau)));
  target += normal_lpdf(zeta | 0, 100);
  target += normal_lpdf(tau | 0, 100)
    - normal_lccdf(0 | 0, 100);
}
```

Fit the model:

```
ma2 <- system.file("stan_models",
                   "meta-analysis2.stan",
                   package = "bcogsci")
fit_sbi2 <- stan(ma2,
```

```
              data = ls_sbi,
              control = list(adapt_delta = 0.9))
```

```
print(fit_sbi2, pars = c("zeta", "tau"))
```

```
##       mean 2.5% 97.5% n_eff Rhat
## zeta 13.7 3.23  29.6   822    1
## tau  11.7 0.70  29.3  1000    1
```

This summary could be reported in an article by displaying the posterior means and 95% credible intervals of the parameters.

11.2 Measurement-error models

Measurement error models deal with the situation where some predictor or the dependent variable, or both, are observed with measurement error. This measurement error could arise because a variable is an average (i.e., its standard error can also be estimated), or because we know that our measurement is noisy due to limitations of our equipment (e.g., delays in the signal from the keyboard to the motherboard, impedance in the electrodes in an EEG system, etc.).

11.2.1 Accounting for measurement error in individual differences in working memory capacity and reading fluency

As a motivating example, consider the following data from Nicenboim, Vasishth, et al. (2018). For each subject, we have the partial-credit unit (PCU) scores of an operation span task as a measure of their working memory capacity (Conway et al. 2005) along with their standard error. In addition, the reading fluency of each subject is calculated from a separate set of data based on the mean reading speeds (character/second) in a rapid automatized naming task (RAN, Denckla and Rudel 1976); the standard error of the reading speed is also available.

Of interest here is the extent of the association between working memory capacity (measured as PCU) and reading fluency (measured as reading speed in 50 characters per second). We avoid making any causal claims: It could be that our measure of working memory capacity really affects reading fluency or it could be the other way around. A third possibility is that there is a third variable (or several) that affects both reading fluency and working memory capacity. A treatment of causal inference in Bayesian models can be found in chapters 5 and 6 of McElreath (2020).

```
data("df_indiv")
df_indiv
```

```
## # A tibble: 100 x 5
##     subj mean_rspeed se_rspeed mean_pcu se_pcu
##    <dbl>       <dbl>     <dbl>    <dbl>  <dbl>
## 1      1      0.0521   0.00113    0.738 0.0648
## 2      2      0.0479   0.00121    0.292 0.0315
## 3      3      0.0601   0.00117    0.408 0.0900
## # i 97 more rows
```

At first glance, we see a relationship between mean PCU scores and mean reading speed; see Figure 11.2. However, this relationship seems to be driven by two extreme data points on the top left corner of the plot.

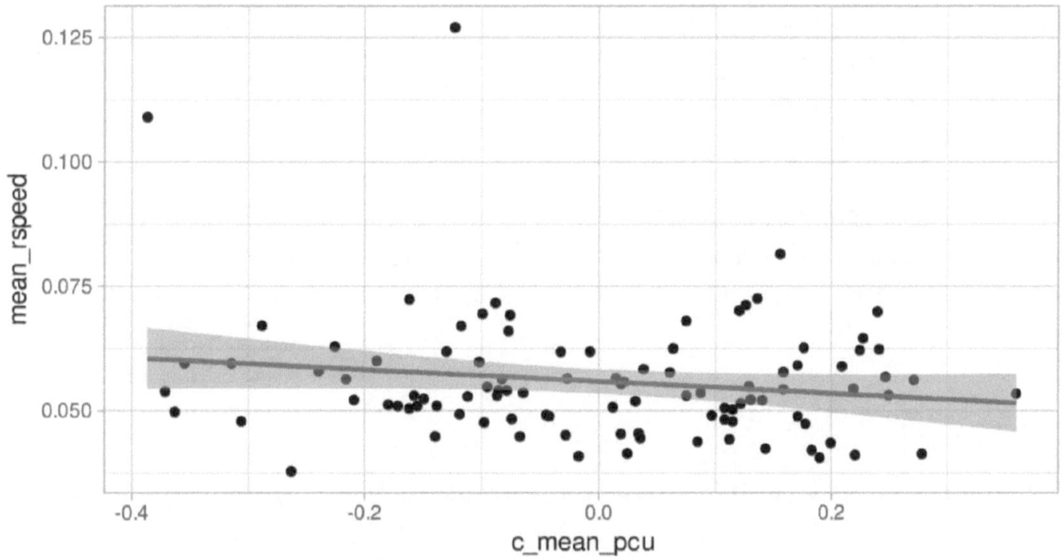

FIGURE 11.2: The relationship between (centered) mean PCU scores and mean reading speed.

A simple linear model shows a somewhat weak association between mean reading speed and centered mean PCU. The priors are relatively arbitrary but they are in the right order of magnitude given that reading speeds are quite short and well below 1.

```
df_indiv <- df_indiv %>%
  mutate(c_mean_pcu = mean_pcu - mean(mean_pcu))
priors <- c(prior(normal(0, 0.5), class = Intercept),
  prior(normal(0, 0.5), class = b),
  prior(normal(0, 0.5), class = sigma))
fit_indiv <- brm(mean_rspeed ~ c_mean_pcu,
                 data = df_indiv,
                 family = gaussian(),
                 prior = priors)
```

```
fit_indiv
```

```
## ...
## Population-Level Effects:
##            Estimate Est.Error l-95% CI u-95% CI Rhat Bulk_ESS
## Intercept      0.06      0.00     0.05     0.06 1.00     5142
## c_mean_pcu    -0.01      0.01    -0.03     0.00 1.00     2640
##            Tail_ESS
## Intercept      2744
## c_mean_pcu     2361
##
## Family Specific Parameters:
##       Estimate Est.Error l-95% CI u-95% CI Rhat Bulk_ESS Tail_ESS
## sigma     0.01      0.00     0.01     0.01 1.00     2014     2209
##
## ...
```

```
# Proportion of samples below zero
(Pb <- mean(as_draws_df(fit_indiv)$b_c_mean_pcu < 0))
```

```
## [1] 0.942
```

Figure 11.3(a) shows the posterior distribution of the slope in this model. Most of the probability mass is negative (94.2%), suggesting that a better PCU score is associated with slower reading speed rather than faster; that is, that a larger working memory capacity is associated with less reading fluency. This is not a very intuitive result and it could be the case that is driven by the two extreme data points. Rather than removing these data points, we'll examine what happens when the uncertainty of the measurements is taken into account.

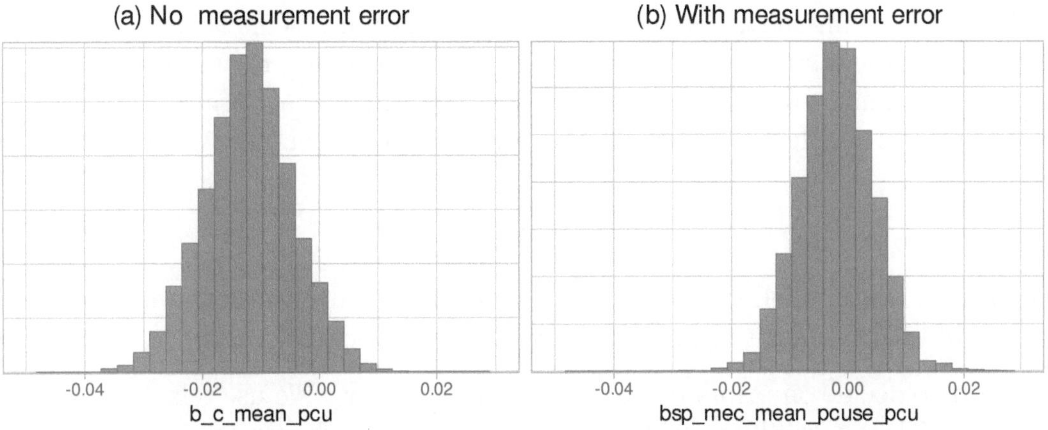

FIGURE 11.3: (a) Posterior distribution of the slope for the effect of centered mean PCU on mean reading speed (50 characters per second) in a model without measurement error (`fit_indiv`). (b) Posterior distribution of the slope for the same dependent variable (mean reading speed) and predictor (centered mean PCU) in a model that accounts for measurement error (`fit_indiv_me`).

Taking this uncertainty of the measurement is important; in many practical research problems, researchers will often take average measurements like these and examine the correlation between them. However, each of those data points is being measured with some error (uncertainty), but this error is being ignored when we take the averaged values. Ignoring this uncertainty leads to over-enthusiastic inferences. A measurement-error model solves this issue.

The measurement error model is stated as follows. There is assumed to be a true unobserved value $y_{n,TRUE}$ for the dependent variable, and a true unobserved value $x_{n,TRUE}$ for the predictor, where n is indexing the observation number. The observed values y_n and the predictor x_n are assumed to be generated with some error:

$$\begin{aligned} y_n &\sim Normal(y_{n,TRUE}, SE_{y_n}) \\ x_n &\sim Normal(x_{n,TRUE}, SE_{x_n}) \end{aligned} \tag{11.7}$$

The regression is fit to the (unknown) *true* values of the dependent and independent variables:

$$y_{n,TRUE} \sim Normal(\alpha + \beta x_{n,TRUE}, \sigma) \tag{11.8}$$

In addition, there is also an unknown standard deviation (standard error) of the latent unknown means that generate the underlying PCU means. I.e., we assume that each of the observed centered PCU scores is normally distributed with an underlying mean, χ, and a standard deviation τ. This is very similar to the meta-analysis situation we saw earlier: $\zeta_n \sim Normal(\zeta, \tau)$, where ζ_n was the location parameter of each study, and ζ was the (unknown) location parameter representing the effect of interest, and τ was the between-study variability.

$$x_{n,TRUE} \sim Normal(\chi, \tau) \tag{11.9}$$

The goal of the modeling is to obtain posterior distributions for the intercept and slope α and β (and the residual error standard deviation σ).

We need to decide on priors for all the parameters now. We use relatively vague priors, which can still be considered regularizing priors based on our knowledge of the order of magnitude of the measurements. In situations where not much is known about a research question, one could use such vague priors.

$$\begin{aligned}
\alpha &\sim Normal(0, 0.5)\\
\beta &\sim Normal(0, 0.5)\\
\chi &\sim Normal(0, 0.5)\\
\sigma &\sim Normal_{+}(0, 0.5)\\
\tau &\sim Normal_{+}(0, 0.5)
\end{aligned} \tag{11.10}$$

11.2.1.1 The `brms` version of the measurement error model

In `brms`, the model specification would be as follows:

```
priors_me <- c(prior(normal(0, 0.5), class = Intercept),
               prior(normal(0, 0.5), class = b),
               prior(normal(0, 0.5), class = meanme),
               prior(normal(0, 0.5), class = sdme),
               prior(normal(0, 0.5), class = sigma))
```

Here, the parameter with class `meanme` and `sdme` refer to the unknown mean and standard deviation (standard error) of the latent unknown means that generate the underlying PCU means, χ and τ in Equation (11.10). Once we decide on the priors, we use `resp_se(.)` with `sigma = TRUE` (i.e, we don't estimate $y_{n,TRUE}$ explicitly) and we use `me(c_meanpcu, se_pcu)` to indicate that the dependent variable `c_mean_pcu` is measured with error and `se_pcu` is its SE.

```
fit_indiv_me <- brm(mean_rspeed | resp_se(se_rspeed, sigma = TRUE) ~
                      me(c_mean_pcu, se_pcu),
                    data = df_indiv,
```

```
                family = gaussian(),
                prior = priors_me)
```

```
fit_indiv_me
```

```
## ...
## Population-Level Effects:
##                      Estimate Est.Error l-95% CI u-95% CI Rhat Bulk_ESS
## Intercept                0.05      0.00     0.05     0.06 1.00     3909
## mec_mean_pcuse_pcu      -0.00      0.01    -0.01     0.01 1.00     6498
##                      Tail_ESS
## Intercept                2571
## mec_mean_pcuse_pcu       2990
##
## Family Specific Parameters:
##       Estimate Est.Error l-95% CI u-95% CI Rhat Bulk_ESS Tail_ESS
## sigma     0.01      0.00     0.01     0.01 1.00     6428     2811
##
## ...
```

```
# Proportion of samples below zero
# Parameter names can be found out with `variables(fit_indiv_me)`
(Pb_me <- mean(as_draws_df(fit_indiv_me)$bsp_mec_mean_pcuse_pcu < 0))
```

```
## [1] 0.616
```

The posterior for the slope is plotted in Figure 11.3(b); this figure shows that the association between PCU scores and reading speed is much weaker once measurement error is taken into account: The posterior is much more uncertain (much more widely distributed) than in the simple linear model we fit above (compare Figure 11.3(b) with Figure 11.3(a)), and the direction of the association is now unclear, with 62% of the probability mass below zero, rather than 94%.

Figure 11.4 visualizes the main reason why we have no clear association in the measurement error analysis: the two points at the top left part of the plot that were driving the effect have very large SE for the measurement of reading speed. The code to produce Figure 11.4 appears below and overlays several (250) regression lines that correspond to different samples of the posterior distribution with the measurements of reading speed and PCU.

```
df_reg <- as_draws_df(fit_indiv_me) %>%
  select(alpha = b_Intercept, beta = bsp_mec_mean_pcuse_pcu) %>%
  slice(1:250)
ggplot(df_indiv, aes(x = c_mean_pcu, y = mean_rspeed)) +
  geom_point() +
  geom_errorbarh(aes(xmin = c_mean_pcu - 2 * se_pcu,
                     xmax = c_mean_pcu + 2 * se_pcu),
                 alpha = .5, linetype = "dotted") +
  geom_errorbar(aes(ymin = mean_rspeed - 2 * se_rspeed,
                    ymax = mean_rspeed + 2 * se_rspeed),
                alpha = .5, linetype = "dotted") +
  geom_abline(aes(intercept = alpha, slope = beta),
              data = df_reg,
              alpha = .02)
```

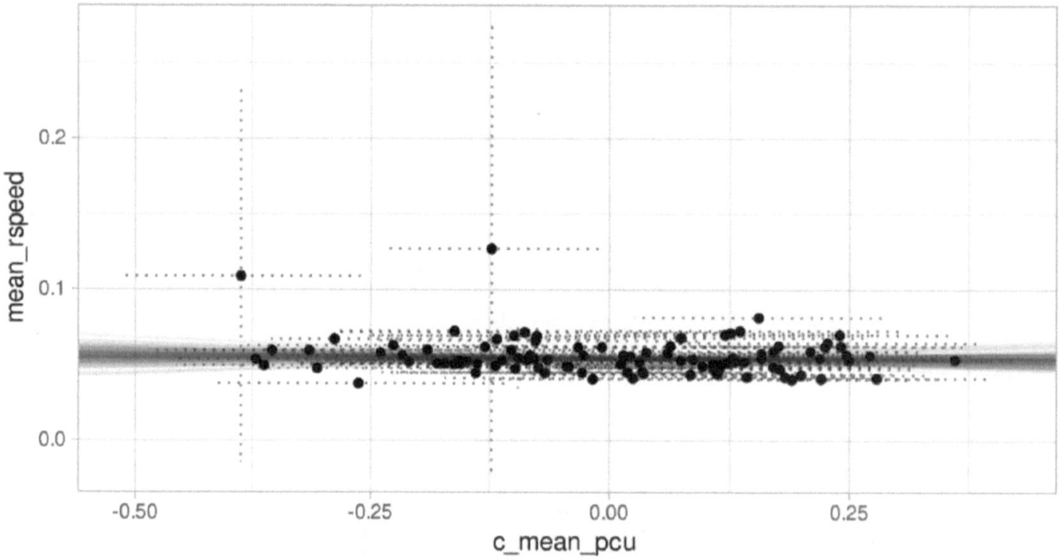

FIGURE 11.4: The relationship between centered mean PCU scores and mean reading speed accounting for measurement error. The error bars represent two standard errors. The regression lines are produced with 250 samples of the intercept and slope from the posterior distribution.

Of course, the conclusion here cannot be that there is no association between PCU scores and reading speed. In order to claim an absence of an effect, we would need to use Bayes factors (see chapter 13) or cross-validation (see chapter 14).

11.2.1.2 The Stan version of the measurement error model

As it happened when we carried out the meta-analysis, the main difficulty for modeling measurement error models directly in Stan is that we need to reparameterize the models to avoid dependencies between samples of different parameters. The two changes that we need to do to the parameterization of our model presented in Equation (11.10) are the following.

1. Sample from an auxiliary parameter z_n rather than directly from $x_{n,TRUE}$, as we did in Equation (11.4):

$$\begin{aligned} z_n &\sim Normal(0, 1) \\ x_{n,TRUE} &= z_n \cdot \tau + \chi \\ x_n &\sim Normal(x_{n,TRUE}, SE_{x_n}) \end{aligned} \tag{11.11}$$

2. Don't model $y_{n,TRUE}$ explicitly as in Equation (11.8); rather take into account the SE and the variation on $y_{n,TRUE}$ in the following way:

$$y_n \sim Normal\left(\alpha + \beta x_{n,TRUE}, \sqrt{SE_{y_n}^2 + \sigma^2}\right) \tag{11.12}$$

We are now ready to write this in Stan; the code is in the model called `me.stan`:

```
data {
  int<lower=1> N;
  vector[N] x;
  vector[N] SE_x;
  vector[N] y;
  vector[N] SE_y;
}
parameters {
  real alpha;
  real beta;
  real chi;
  real<lower = 0> sigma;
  real<lower = 0> tau;
  vector[N] z;
}
transformed parameters {
  vector[N] x_true = z * tau + chi;
}
model {
```

```
  target += normal_lpdf(x | x_true, SE_x);
  target += normal_lpdf(y | alpha + beta * x_true,
                        sqrt(square(SE_y) + square(sigma)));
  target += std_normal_lpdf(z);
  target += normal_lpdf(alpha | 0, 0.5);
  target += normal_lpdf(beta | 0, 0.5);
  target += normal_lpdf(chi | 0, 0.5);
  target += normal_lpdf(sigma| 0, 0.5)
    - normal_lccdf(0 | 0, 0.5);
  target += normal_lpdf(tau | 0, 0.5)
    - normal_lccdf(0 | 0, 0.5);
}
```

Fit the model:

```
me <- system.file("stan_models",
                  "me.stan",
                  package = "bcogsci")
ls_me <- list(N = nrow(df_indiv),
              y = df_indiv$mean_rspeed,
              SE_y = df_indiv$se_rspeed,
              x = df_indiv$c_mean_pcu,
              SE_x = df_indiv$se_pcu)
fit_indiv_me_stan <- stan(me, data = ls_me)
```

```
print(fit_indiv_me_stan, pars = c("alpha", "beta", "sigma"))
```

```
##       mean  2.5% 97.5% n_eff Rhat
## alpha 0.05  0.05  0.06  3610    1
## beta  0.00 -0.01  0.01  5216    1
## sigma 0.01  0.01  0.01  5592    1
```

The posterior distributions are similar to those that we obtained with `brms`.

11.3 Summary

This chapter introduced two statistical tools that are potentially of great relevance to cognitive science: random-effects meta-analysis and measurement error models. Despite the inherent limitations of meta-analysis, these should be used routinely to accumulate knowledge through systematic evidence synthesis.

Measurement errors can also prevent over-enthusiastic conclusions that are often made based on noisy data.

11.4 Further reading

For some examples of Bayesian meta-analyses in psycholinguistics, see Vasishth et al. (2013), Jäger, Engelmann, and Vasishth (2017), Nicenboim, Roettger, and Vasishth (2018), Nicenboim, Vasishth, and Rösler (2020), Bürki et al. (2020), Cox et al. (2022), and Bürki, Alario, and Vasishth (2023). A frequentist meta-analysis of priming effects in psycholinguistics appears in Mahowald et al. (2016). Sutton et al. (2012) and Higgins and Green (2008) are two useful general introductions that discuss systematic reviews, meta-analysis, and evidence synthesis; these two references are from medicine, where meta-analysis is more widely used than in cognitive science. A potentially important article for meta-analysis introduces a methodology for modeling bias, to adjust for different kinds of bias in the data (Turner et al. 2008). Meta-analyses have important limitations and should not be taken at face value; this point is discussed in, e.g., Spector and Thompson (1991). A practical book-length introduction to Bayesian meta-analysis is Grant and Di Tanna (2025).

Further reading

Part V

Model comparison

12

Introduction to model comparison

A key goal of cognitive science is to decide which theory, among multiple candidates under consideration, accounts for the experimental data better. This can be accomplished by implementing the theories (or simplified versions of them) as Bayesian models and comparing their predicting power. This kind of model comparison is closely related to hypothesis testing. There are two Bayesian perspectives on model comparison: a *prior* predictive perspective based on the Bayes factor using marginal likelihoods, and a *posterior* predictive perspective based on cross-validation. The main characteristic difference between the prior predictive approach (Bayes factor) versus the posterior predictive approach (cross-validation) is the following: The Bayes factor examines how well the model (prior and likelihood) explains the experimental data. By contrast, the posterior predictive approach assesses model predictions for held-out data after seeing most of the data.

12.1 Prior predictive vs. posterior predictive model comparison

The predictive accuracy of the Bayes factor is only based on its prior predictive distribution. In Bayes factor analyses, the prior model predictions are used to evaluate the support that the data give to the model. By contrast, in cross-validation, the model is fit to a large subset of the data (i.e., the training data). The posterior distributions of the parameters of this fitted model are then used to make predictions for held-out or validation data, and model fit is assessed on this subset of the data. Typically, this process is repeated several times, until the subsets of the entire data set are assessed as held-out data. This approach attempts to assess whether the model will generalize to truly new, unobserved data. Of course, the held-out data is usually not "truly new" because it is part of the data that was collected, but at least it is data that the model has not been exposed to. That is, the predictive accuracy of cross-validation methods is based on how well the posterior predictive distribution that is fit to most of the data (i.e., the training data) characterizes out-of-sample data (i.e., the test or held-out data). Notice that one could in principle use the posterior

DOI: 10.1201/9780429342646-17

predictive approach using truly new data, by just repeating the experiment with new subjects and then treating that new data as held-out data.

Coming back to Bayes factors, the prior predictive distribution is obviously highly sensitive to the priors: it evaluates the probability of the observed data under prior assumptions. By contrast, the posterior predictive distribution is less dependent on the priors because the priors are combined with the likelihood (and are thus less influential, given sufficient data) before making predictions for held-out validation data.

Jaynes (2003, Chapter 20) compares these two perspectives to "a cruel realist" and "a fair judge." According to Jaynes, Bayes factor adopts the posture of a cruel realist, who "judge[s] each model taking into account the prior information we actually have pertaining to it; that is, we penalize a model if we do not have the best possible prior information about its parameters, although that is not really a fault of the model itself." By contrast, cross-validation adopts the posture of a scrupulously fair judge, "who insists that fairness in comparing models requires that each is delivering the best performance of which it is capable, by giving each the best possible prior probability for its parameters (similarly, in Olympic games we might consider it unfair to judge two athletes by their performance when one of them is sick or injured; the fair judge might prefer to compare them when both are doing their absolute best)."

12.2 Some important points to consider when comparing models

Regardless of whether we use Bayes factor or cross-validation or any other method for model comparison, there are several important points that one should keep in mind:

1. Although the objective of model comparison might ultimately be to find out which of the models under consideration generalizes better, this generalization can only be done well within the range of the observed data (see Vehtari and Lampinen 2002; Vehtari and Ojanen 2012). That is, if one hypothesis implemented as the model $\mathcal{M}_1$ is shown to be superior to a second hypothesis, implemented as the model $\mathcal{M}_2$, according to Bayes factor and/or cross-validation, and evaluated with a young Western university student population, this doesn't mean that $\mathcal{M}_1$ will be superior to $\mathcal{M}_2$ when it is evaluated with a broader population (and in fact it seems that many times it won't, see Henrich, Heine, and Norenzayan 2010). However, if we can't generalize even within the range of the observed data (e.g., university students

in the northern part of the western hemisphere), there is no hope of generalizing outside of that range (e.g., non-University students). Navarro (2019) argues that one of the most important functions of a model is to encourage directed exploration of new territory; our view is that this makes sense only if historical data can also be accounted for. In practice, what that means for us is that evaluating a model's performance should be carried out using historical benchmark data in addition to any new data one has; just using isolated pockets of new data to evaluate a model is not convincing. For examples from psycholinguistics of model evaluation using historical benchmark data, see Engelmann, Jäger, and Vasishth (2020), Nicenboim, Vasishth, and Rösler (2020), and Yadav et al. (2023).

2. Model comparison can provide a quantitative way to evaluate models, but this cannot replace understanding the qualitative patterns in the data (see, e.g., Navarro 2019). A model can provide a good fit by behaving in a way that contradicts our substantive knowledge. For example, Lissón et al. (2021) examine two computational models of sentence comprehension. One of the models yielded higher predictive accuracy when the parameter that is related to the probability of correctly comprehending a sentence was higher for impaired subjects (individuals with aphasia) than for the control population. This contradicts domain knowledge—impaired subjects are generally observed to show worse performance than unimpaired control subjects—and led to a re-evaluation of the model.

3. Model comparison is based on finding the most "useful model" for characterizing our data, but neither the Bayes factor nor cross-validation (nor any other method that we are aware of) guarantees selecting the model closest to the truth (even with enough data). This is related to our previous point: A model that's closest to the true generating data process is not guaranteed to produce the best (prior or posterior) predictions, and a model with a clearly wrong generating data process is not guaranteed to produce poor (prior or posterior) predictions. See Wang and Gelman (2014), for an example with cross-validation; and Navarro (2019) for a toy example with Bayes factors.

4. One should also check that the precision of the data being modeled is high; if an effect is being modeled that has high uncertainty (the posterior distribution of the target parameter is widely spread out), then any measure of model fit can be uninformative because we don't have accurate estimates of the effect of interest. In the Bayesian context, this implies that the posterior predictive distributions of the

effects generated by the model should be theoretically plausible and reasonably constrained, and the target parameter of interest should have as high precision as possible; this implies that we need to have sufficient data if we want to obtain precise estimates of the parameter of interest. What counts as sufficient will depend on the topic being studied.[1] Later in this part of the book, we will discuss the adverse impact of imprecision in the data on model comparison (see section 13.5.2). We will show that, in the face of low precision, we generally won't learn much from model comparison.

5. When comparing a null model with an alternative model, it is important to be clear about what the null model specification is. For example, in section 5.2.4, we encountered the correlated varying intercepts and varying slopes model for the Stroop effect. The `brms` formula for the full model was:

```
n400 ~ 1 + c_cloze + (1 + c_cloze | subj)
```

The formal statement of this model is:

$$signal_n \sim Normal(\alpha + u_{subj[n],1} + c\_cloze_n \cdot (\beta + u_{subj[n],2}), \sigma) \tag{12.1}$$

If we want to test the null hypothesis that centered cloze has no effect on the dependent variable, one null model is:

```
n400 ~ 1 + (1 + c_cloze | subj) (Model M0a)
```

Formally, this would be stated as follows (the β term is removed as it is assumed to be 0):

$$signal_n \sim Normal(\alpha + u_{subj[n],1} + c\_cloze_n \cdot u_{subj[n],2}, \sigma) \tag{12.2}$$

In model `M0a`, by-subject variability is allowed; just the population-level (or fixed) effect of centered cloze is assumed to be zero. This is called a nested model comparison, because the null model is subsumed in the full model.

An alternative null model could remove only the varying slopes:

```
n400 ~ 1 + c_cloze + (1 | subj) (Model M0b)
```

[1] As an example from psycholinguistics, strong garden-path effects like those elicited by "The horse (that was) raced past the barn fell" (Bever 1970) may be easy to detect with high precision with a relatively small number of subjects, but subtle effects such as local coherence (Tabor, Galantucci, and Richardson 2004) will probably require a much larger sample size to detect the effect with high precision (Paape, Smith, and Vasishth 2024).

Formally:

$$signal_n \sim Normal(\alpha + u_{subj[n],1} + c\_cloze_n \cdot \beta, \sigma) \tag{12.3}$$

Model `M0b`, which is also nested inside the full model, can be used to test a different null hypothesis than `M0a` above: is the between-subject variability in the centered cloze effect zero?

Yet another possibility is to remove both the population-level and group-level (or random) effects of centered cloze:

```
n400 ~ 1 + (1 | subj) (Model M0c)
```

Formally:

$$signal_n \sim Normal(\alpha + u_{subj[n],1}, \sigma) \tag{12.4}$$

Model `M0c` is also nested inside the full model, but it now has two parameters missing instead of one: β and $u_{subj[n],1}$. Usually, it is best to compare models by removing one parameter; otherwise one cannot be sure which parameter was responsible for our rejecting or accepting the null hypothesis.

12.3 Further reading

Roberts and Pashler (2000) and Pitt and Myung (2002) argue for the need of going beyond "a good fit" (this is a good posterior predictive check in the context of Bayesian data analysis) and argue for the need of model comparison and a focus on measuring the generalizability of a model. Navarro (2019) deals with the problematic aspects of model selection in the context of psychological literature and cognitive modeling. Fabian Dablander's blog post, `https://fabiandablander.com/r/Law-of-Practice.html`, shows a very clear comparison between Bayes factor and PSIS-LOO-CV. Rodriguez, Williams, and Rast (2022) provides JAGS code for fitting models with spike-and-slab priors. Fabian Dablander has a comprehensive blog post on how to implement a Gibbs sampler in R when using such a prior: `https://fabiandablander.com/r/Spike-and-Slab.html`. Yadav et al. (2023) uses 17 different data sets for model comparison using cross-validation, holding out each data set successively; this is an example of evaluating the predictive performance of a model on truly new data.

13

Bayes factors

This chapter is based on a longer manuscript available on arXiv: Schad et al. (2021); our terminology used here is based on the conventions used in that paper. A published version of the arXiv article appears in Schad, D. J., Nicenboim, B., Bürkner, P. C., Betancourt, M., and Vasishth, S. (2022). Workflow techniques for the robust use of Bayes factors. Psychological Methods. In this chapter, whenever we refer to the published version of the arXiv paper, we mean this article.

Bayesian approaches provide tools for different aspects of data analysis. A key contribution of Bayesian data analysis to cognitive science is that it furnishes probabilistic ways to quantify the evidence that data provide in support of one model or another. Models provide ways to implement scientific hypotheses; as a consequence, model comparison and hypothesis testing are closely related. Bayesian hypothesis testing comparing any kind of hypotheses is implemented using Bayes factors (Rouder, Haaf, and Vandekerckhove 2018; Schönbrodt and Wagenmakers 2018; Wagenmakers et al. 2010; Kass and Raftery 1995; Gronau et al. 2017; Jeffreys 1939), which quantify evidence in favor of one statistical (or computational) model over another. This chapter will focus on Bayes factors as the way to compare models and to obtain evidence for (general) hypotheses.

There are subtleties associated with Bayes factors that are not widely appreciated. For example, the results of Bayes factors analyses are highly sensitive to and crucially depend on prior assumptions about model parameters (we will illustrate this below); these priors can vary depending on the experiment/research problem and even differ subjectively among researchers. Many authors use or recommend so-called default prior distributions, where the prior parameters are fixed, and are independent of the scientific problem in question (e.g., Navarro 2015). However, default priors can result in an overly simplistic perspective on Bayesian hypothesis testing, and can be misleading. For this reason, even though leading experts in the use of the Bayes factor, such as Rouder et al. (2009), often provide default priors for computing Bayes factors, they also make it clear that: "simply put, principled inference is a thoughtful process that cannot be performed by rigid adherence to defaults" (Rouder et al. 2009, 235). However, this observation does not seem to have had much impact

DOI: 10.1201/9780429342646-18

on how Bayes factors are used in fields like psychology and psycholinguistics; the use of default priors when computing Bayes factors seems to be widespread.

Given the key influence of priors on Bayes factors, defining priors becomes a central issue when using Bayes factors. The priors determine which models will be compared.

In this chapter, we demonstrate how Bayes factors should be used in practical settings in cognitive science. In doing so, we demonstrate the strength of this approach and some important pitfalls that researchers should be aware of.

13.1 Hypothesis testing using the Bayes factor

13.1.1 Marginal likelihood

Bayes' rule can be written with reference to a specific statistical model $\mathcal{M}_1$.

$$p(\boldsymbol{\Theta} \mid \boldsymbol{y}, \mathcal{M}_1) = \frac{p(\boldsymbol{y} \mid \boldsymbol{\Theta}, \mathcal{M}_1) p(\boldsymbol{\Theta} \mid \mathcal{M}_1)}{p(\boldsymbol{y} \mid \mathcal{M}_1)} \tag{13.1}$$

Here, $\boldsymbol{y}$ refers to the data and $\boldsymbol{\Theta}$ is a vector of parameters; for example, this vector could include the intercept, slope, and variance component in a linear regression model.

The denominator $p(\boldsymbol{y} \mid \mathcal{M}_1)$ is the marginal likelihood, and is a single number that gives us the likelihood of the observed data $\boldsymbol{y}$ given the model $\mathcal{M}_1$ (and only in the discrete case, it gives us the probability of the observed data $\boldsymbol{y}$ given the model; see section 1.7). Because in general it's not a probability, it should be interpreted relative to another marginal likelihood (evaluated with the same $\boldsymbol{y}$).

In frequentist statistics, it's also common to quantify evidence for the model by determining the maximum likelihood, that is, the likelihood of the data given the best-fitting model parameter. Thus, the data is used twice: once for fitting the parameter, and then for evaluating the likelihood. Importantly, this inference completely hinges upon this best-fitting parameter to be a meaningful value that represents well what we know about the parameter, and doesn't take the uncertainty of the estimates into account. Bayesian inference quantifies the uncertainty that is associated with a parameter; that is, one accepts that the knowledge about the parameter value is uncertain. Computing the marginal likelihood entails computing the likelihood given all plausible values for the model parameter.

One difficulty in the above equation showing Bayes' rule is that the marginal likelihood $p(\boldsymbol{y} \mid \mathcal{M}_1)$ in the denominator cannot in general be easily computed:

$$p(\boldsymbol{\Theta} \mid \boldsymbol{y}, \mathcal{M}_1) = \frac{p(\boldsymbol{y} \mid \boldsymbol{\Theta}, \mathcal{M}_1) p(\boldsymbol{\Theta} \mid \mathcal{M}_1)}{p(\boldsymbol{y} \mid \mathcal{M}_1)} \tag{13.2}$$

In the marginal likelihood, the parameters are being integrated out:

$$p(\boldsymbol{y} \mid \mathcal{M}_1) = \int p(\boldsymbol{y} \mid \boldsymbol{\Theta}, \mathcal{M}_1) p(\boldsymbol{\Theta} \mid \mathcal{M}_1) \, \mathrm{d}\boldsymbol{\Theta} \tag{13.3}$$

The likelihood is evaluated for every possible parameter value, weighted by the prior plausibility of the parameter values. The product $p(\boldsymbol{y} \mid \boldsymbol{\Theta}, \mathcal{M}_1) p(\boldsymbol{\Theta} \mid \mathcal{M}_1)$ is then summed up (that is what the integral does).

For this reason, the prior is as important as the likelihood. Equation (13.3) also looks almost identical to the prior predictive distribution from section 3.3 (that is, the predictions that the model makes before seeing any data). The prior predictive distribution is repeated below for convenience:

$$\begin{aligned} p(\boldsymbol{y_{pred}}) &= p(y_{pred_1}, \dots, y_{pred_n}) \\ &= \int_{\boldsymbol{\Theta}} p(y_{pred_1} | \boldsymbol{\Theta}) \cdot p(y_{pred_2} | \boldsymbol{\Theta}) \cdots p(y_{pred_N} | \boldsymbol{\Theta}) p(\boldsymbol{\Theta}) \, d\boldsymbol{\Theta} \end{aligned} \tag{13.4}$$

However, while the prior predictive distribution describes possible observations, the marginal likelihood is evaluated on the actually observed data.

Let's compute the Bayes factor for a very simple example case. We assume a study where we assess the number of "successes" observed in a fixed number of trials. For example, suppose that we have 80 "successes" out of 100 trials. A simple model of this data can be built by assuming, as we did in section 1.4, that the data are distributed according to a binomial distribution. In a binomial distribution, n independent trials are performed, where the result of each experiment is either a "success" or "no success," with the probability of "success" being θ. The binomial distribution is the probability distribution of the number of successes k (number of "success" responses) in this situation for a given sample of experiments X.

Suppose now that we have prior information about the probability parameter θ. As we explained in section 2.2, a typical prior distribution for θ is a beta distribution. The beta distribution defines a probability distribution on the interval $[0, 1]$, which is the interval on which the probability θ is defined. It has two parameters a and b, which determine the shape of the distribution. The prior parameters a and b can be interpreted as the a priori number of "successes" versus "failures." These could be based on previous evidence, or on the researcher's beliefs, drawing on their domain knowledge (O'Hagan et al. 2006).

Here, to illustrate the calculation of the Bayes factor, we assume that the parameters of the beta distribution are $a = 4$ and $b = 2$. As mentioned above, these parameters can be interpreted as representing "success" (4 prior observations representing success), and "no success" (2 prior observations representing "no success"). The resulting prior distribution is visualized in Figure 13.1. A $Beta(a = 4, b = 2)$ prior on θ amounts to a regularizing prior with some, but no clear prior evidence for more than 50% of success.

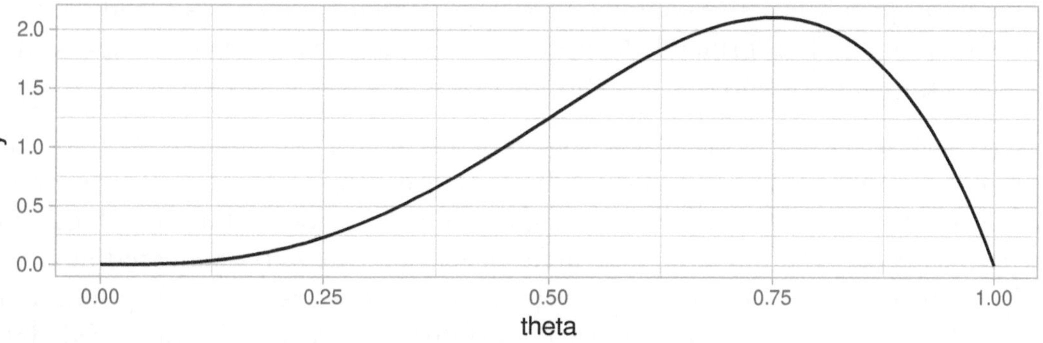

FIGURE 13.1: The beta distribution with parameters a = 4 and b = 2.

To compute the marginal likelihood, Equation (13.3) shows that we need to multiply the likelihood with the prior. The marginal likelihood is then the area under the curve, that is, the likelihood averaged across all possible values for the model parameter (the probability of success).

Based on this data, likelihood, and prior we can calculate the marginal likelihood, that is, the area under the curve, in the following way using R:

```
# First we multiply the likelihood with the prior
plik1 <- function(theta) {
  dbinom(x = 80, size = 100, prob = theta) *
    dbeta(x = theta, shape1 = 4, shape2 = 2)
}
# Then we integrate (compute the area under the curve):
(MargLik1 <- integrate(f = plik1, lower = 0, upper = 1)$value)
```

```
## [1] 0.02
```

One would prefer a model that gives a higher marginal likelihood, i.e., a higher likelihood of observing the data after integrating out the influence of the model parameter(s). A model will yield a high marginal likelihood if it makes a high proportion of good predictions (i.e., model 2 in Figure 13.2; the figure is adapted from Bishop 2006). Model predictions are normalized, that is, the total probability that models assign to different expected data patterns is the same for all models. Models that are too flexible (model 3 in Figure 13.2) will divide their prior predictive probability density across all of their predictions. Such models can predict many different outcomes. Thus, they likely can also predict the actually observed outcome. However, due to the normalization, they cannot predict it with high probability, because they also predict all kinds of other outcomes. This is true for both models with priors that are too wide or for models with too many parameters. Bayesian model comparison inherently penalizes more complex models, a principle known as *Occam's razor* (MacKay 2003). This principle favors simpler explanations when multiple viable explanations are possible.

By contrast, good models (Figure 13.2, model 2) will make very specific predictions, where the specific predictions are consistent with the observed data. Here, all the predictive probability density is located at the "location" where the observed data fall and little probability density is located at other places, providing good support for the model. Of course, specific predictions can also be wrong, when expectations differ from what the observed data actually look like (Figure 13.2, model 1).

Having a natural Occam's razor is good for posterior inference, i.e., for assessing how much (continuous) evidence there is for one model or another. However, it doesn't necessarily imply good decision making or hypothesis testing, i.e., to make discrete decisions about which model explains the data best, or on which model to base further actions.

Here, we provide two examples of more flexible models. First, the following model assumes the same likelihood and the same distribution function for the prior. However, we assume a flat, uninformative prior, with prior parameters $a = 1$ and $b = 1$ (i.e., only one prior "success" and one prior "failure"), which provides more prior spread than the first model. Again, we can formulate our model as multiplying the likelihood with the prior, and integrating out the influence of the parameter θ:

```
plik2 <- function(theta) {
  dbinom(x = 80, size = 100, prob = theta) *
    dbeta(x = theta, shape1 = 1, shape2 = 1)
}
(MargLik2 <- integrate(f = plik2, lower = 0, upper = 1)$value)
```

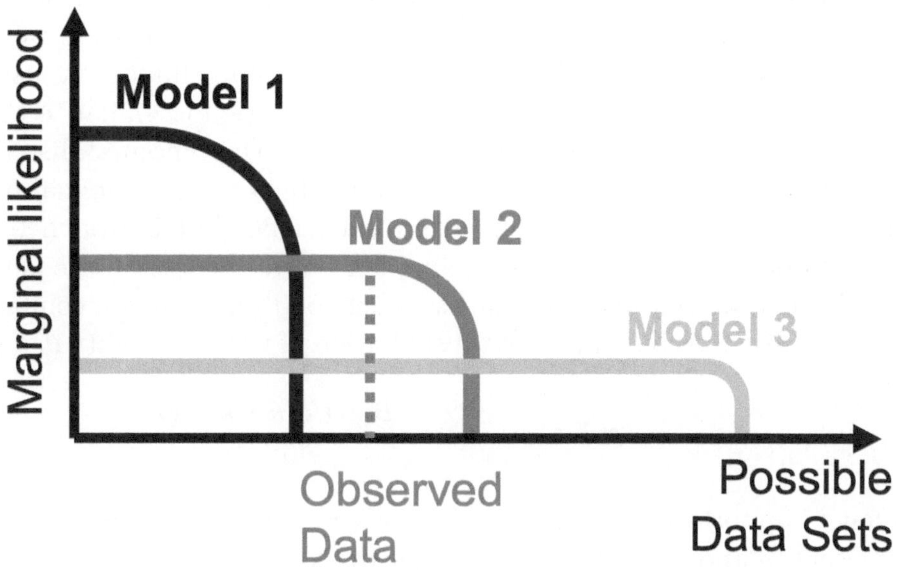

FIGURE 13.2: Shown are the schematic marginal likelihoods that each of three models assigns to different possible data sets. The total probability each model assigns to the data is equal to one, i.e., the areas under the curves of all three models are the same. Model 1 (black), the low complexity model, assigns all the probability to a narrow range of possible data, and can predict these possible data sets with high likelihood. Model 3 (light gray) assigns its probability to a large range of different possible outcomes, but predicts each individual observed data set with low likelihood (high complexity model). Model 2 (dark gray) takes an intermediate position (intermediate complexity). The vertical dashed line (dark gray) illustrates where the actual empirically observed data fall. The data most support model 2, since this model predicts the data with highest likelihood. The figure is closely based on Figure 3.13 in Bishop (2006).

```
## [1] 0.0099
```

We can see that this second model is more flexible: due to the more spread-out prior, it is compatible with a larger range of possible observed data patterns. However, when we integrate out the θ parameter to obtain the marginal likelihood, we can see that this flexibility also comes with a cost: the model has a smaller marginal likelihood (0.0099) than the first model (0.02). Thus, on average (averaged across all possible values of θ) the second model performs worse in explaining the specific data that we observed compared to the first model, and has less support from the data.

A model might be more "complex" because it has a more spread-out prior, or alternatively because it has a more complex likelihood function, which uses a larger number of parameters to explain the same data. Here we implement a third model, which assumes a more complex likelihood by using a beta-binomial distribution. The beta-binomial distribution is similar to the binomial distribution, with one important difference: In the binomial distribution the probability of success θ is fixed across trials. In the beta-binomial distribution, the probability of success is fixed for each trial, but is drawn from a beta distribution across trials. Thus, θ can differ between trials. In the beta-binomial distribution, we thus assume that the likelihood function is a combination of a binomial distribution and a beta distribution of the probability θ, which yields:

$$p(X = k \mid a, b) = \frac{B(k + a, n - k + b)}{B(a, b)} \tag{13.5}$$

What is important here is that this more complex distribution has two parameters (a and b; rather than one, θ) to explain the same data. We assume log-normally distributed priors for the a and b parameters, with location zero and scale 100. The likelihood of this combined beta-binomial distribution is given by the R function `dbbinom()` in the package `extraDistr`. We can now write down the likelihood times the priors (given as log-normal densities, `dlnorm()`), and integrate out the influence of the two free model parameters a and b using numerical integration (applying `integrate()` twice):

```
plik3 <- function(a, b) {
  dbbinom(x = 80, size = 100, alpha = a, beta = b) *
    dlnorm(x = a, meanlog = 0, sdlog = 100) *
    dlnorm(x = b, meanlog = 0, sdlog = 100)
}
# Compute marginal likelihood by applying integrate twice
f <- function(b) {
  integrate(function(a) plik3(a, b), lower = 0, upper = Inf)$value
  }
# integrate requires a vectorized function:
(MargLik3 <- integrate(Vectorize(f), lower = 0, upper = Inf)$value)
```

```
## [1] 0.00000707
```

The results show that this third model has an even lower marginal likelihood compared to the first two (0.00000707). With its two parameters a and b, this third model has a lot of flexibility to explain a lot of different patterns of observed empirical results. However, again, this increased flexibility comes at a cost, and the simple pattern of observed data does not seem to require such complex model assumptions. The small value for the marginal likelihood indicates that this complex model has less support from the data.

That is, for this present simple example case, we would prefer model 1 over the other two, since it has the largest marginal likelihood (0.02), and we would prefer model 2 over model 3, since the marginal likelihood of model 2 (0.0099) is larger than that of model 3 (0.00000707). The decision about which model is preferred is based on comparing the marginal likelihoods.

13.1.2 The Bayes factor

The Bayes factor is a measure of relative evidence, the comparison of the predictive performance of one model against another one. This comparison is a ratio of marginal likelihoods:

$$BF_{12} = \frac{P(\boldsymbol{y} \mid \mathcal{M}_1)}{P(\boldsymbol{y} \mid \mathcal{M}_2)} \tag{13.6}$$

BF_{12} indicates the extent to which the data are more likely under $\mathcal{M}_1$ over $\mathcal{M}_2$, or in other words, the relative evidence that we have for $\mathcal{M}_1$ over $\mathcal{M}_2$. Values larger than one indicate evidence in favor of $\mathcal{M}_1$, smaller than one indicate evidence in favor of $\mathcal{M}_2$, and values close to one indicate that the evidence is inconclusive. This model comparison does not depend on a specific parameter value. Instead, all possible prior parameter values are taken into account simultaneously. This is in contrast with the likelihood ratio test, as it is explained in the online section D.2.

A scale (see Table 13.1) has been proposed to interpret Bayes factors according to the strength of evidence in favor of one model (corresponding to some hypothesis) over another (Jeffreys 1939); but this scale should not be regarded as a hard and fast rule with clear boundaries.

TABLE 13.1: The Bayes factor scale as proposed by Jeffreys (1939). This scale should not be regarded as a hard and fast rule.

BF_{12}	Interpretation
> 100	Extreme evidence for $\mathcal{M}_1$.
$30 - 100$	Very strong evidence for $\mathcal{M}_1$.
$10 - 30$	Strong evidence for $\mathcal{M}_1$.
$3 - 10$	Moderate evidence for $\mathcal{M}_1$.
$1 - 3$	Anecdotal evidence for $\mathcal{M}_1$.
1	No evidence.
$\frac{1}{1} - \frac{1}{3}$	Anecdotal evidence for $\mathcal{M}_2$.

BF_{12}	Interpretation
$\frac{1}{3} - \frac{1}{10}$	Moderate evidence for $\mathcal{M}_2$.
$\frac{1}{10} - \frac{1}{30}$	Strong evidence for $\mathcal{M}_2$.
$\frac{1}{30} - \frac{1}{100}$	Very strong evidence for $\mathcal{M}_2$.
$< \frac{1}{100}$	Extreme evidence for $\mathcal{M}_2$.

So if we go back to our previous example, we can calculate BF_{12}, BF_{13}, and BF_{23}. The subscript represents the order in which the models are compared; for example, BF_{21} is simply $\frac{1}{BF_{12}}$.

$$BF_{12} = \frac{marginal\ likelihood\ model\ 1}{marginal\ likelihood\ model\ 2} = \frac{MargLik1}{MargLik2} = 2 \tag{13.7}$$

$$BF_{13} = \frac{MargLik1}{MargLik3} = 2825.4 \tag{13.8}$$

$$BF_{32} = \frac{MargLik3}{MargLik2} = 0.001 = \frac{1}{BF_{23}} = \frac{1}{1399.9} \tag{13.9}$$

However, if we want to know, given the data y, what the probability for model $\mathcal{M}_1$ is, or how much more probable model $\mathcal{M}_1$ is than model $\mathcal{M}_2$, then we need the prior odds, that is, we need to specify how probable $\mathcal{M}_1$ is compared to $\mathcal{M}_2$ *a priori.*

$$\frac{p(\mathcal{M}_1 \mid y)}{p(\mathcal{M}_2 \mid y)} = \frac{p(\mathcal{M}_1)}{p(\mathcal{M}_2)} \times \frac{P(y \mid \mathcal{M}_1)}{P(y \mid \mathcal{M}_2)} \tag{13.10}$$

$$\text{Posterior odds}_{12} = \text{Prior odds}_{12} \times BF_{12} \tag{13.11}$$

The Bayes factor indicates the amount by which we should update our relative belief between the two models in light of the data and priors. However, **the Bayes factor alone cannot tell us which one of the models is the most probable**. Given our priors for the models and the Bayes factor, we can calculate the odds between the models.

Here we compute posterior model probabilities for the case where we compare two models against each other. However, posterior model probabilities can also be computed for the more general case, where more than two models are considered:

$$p(\mathcal{M}_1 \mid \mathbf{y}) = \frac{p(\mathbf{y} \mid \mathcal{M}_1)p(\mathcal{M}_1)}{\sum_n p(\mathbf{y} \mid \mathcal{M}_n)p(\mathcal{M}_n)} \tag{13.12}$$

For simplicity, we mostly constrain ourselves to two models. (However, the sensitivity analyses we carry out below compare more than two models.)

Bayes factors (and posterior model probabilities) tell us how much evidence the data (and priors) provide in favor of one model or another. Put differently, they enable us to draw conclusions about the model space, that is, to determine the degree to which each hypothesis agrees with the available data.

A completely different issue, however, is the question of how to perform (discrete) decisions based on continuous evidence. The question here is: which hypothesis should one choose to maximize utility? While Bayes factors have a clear rationale and justification in terms of the (continuous) evidence they provide, there is not a clear and direct mapping from inferences to how to perform decisions based on them. To derive decisions based on posterior model probabilities, utility functions are needed. Indeed, the utility of different possible actions (i.e., to accept and act based on one hypothesis or another) can differ quite dramatically in different situations. Erroneously rejecting a novel therapy could have significant negative consequences for a researcher attempting to implement life-saving treatment, but adopting the therapy incorrectly might have less of an adverse impact. By contrast, erroneously claiming a new discovery in fundamental research may have bad consequences (low utility), whereas erroneously missing a new discovery claim may be less problematic if further evidence can be accumulated. Thus, Bayesian evidence (in the form of Bayes factors or posterior model probabilities) must be combined with utility functions in order to perform decisions based on them. For example, this could imply specifying the utility of a true discovery and the utility of a false discovery. Calibration (i.e., simulations) can then be used to derive decisions that maximize overall utility.[1]

The question now is: how do we extend the Bayes factor calculation to models that we care about, i.e., to models that represent more realistic data analysis situations. In cognitive science, we typically fit fairly complex hierarchical models with many variance components. The major problem is that we won't be able to calculate the marginal likelihood for hierarchical models (or any other complex model) either analytically or just using the R functions shown above. There are two very useful methods for calculating the Bayes factor for complex models: the Savage–Dickey density ratio method (Dickey and Lientz 1970; Wagenmakers et al. 2010), and bridge sampling (Bennett 1976; Meng and Wong 1996).

[1]See Schad, D. J., Nicenboim, B., Bürkner, P. C., Betancourt, M., and Vasishth, S. (2022). Workflow techniques for the robust use of Bayes factors. Psychological Methods.

The Savage–Dickey density ratio method has the advantage that is a straightforward way to compute the Bayes factor, but it is limited to nested models, i.e., models with point null hypotheses (e.g., $\mu = 0$). At the time of writing, the implementation of the Savage–Dickey method in `brms` can be unstable, especially in cases where the posterior is far away from zero. Despite this instability in some special cases, in many situations it can make sense to use this approach instead of bridge sampling (Oberauer, Musfeld, and Aust 2024).

Bridge sampling is in general a much more powerful method, but it requires many more effective samples than what is normally required for parameter estimation, and it may be difficult to obtain stable Bayes factors with very large data sets. In this chapter, we will primarily use bridge sampling from the `bridgesampling` package (Gronau et al. 2017; Gronau, Singmann, and Wagenmakers 2020) with the function `bayes_factor()` to calculate the Bayes factor in the first examples.

13.2 Examining the N400 effect with the Bayes factor

In section 5.2 we estimated the effect of cloze probability on the N400 average signal. This yielded a posterior credible interval for the effect of cloze probability. It is certainly possible to check whether, e.g., the 95% posterior credible interval overlaps with zero or not. However, such estimation cannot really answer the following question: How much evidence do we have in support for an effect? A 95% credible interval that doesn't overlap with zero, or a high probability mass away from zero may hint that the predictor may be needed to explain the data, but it is not really answering how much evidence we have in favor of an effect (see the online section D.1, and for discussion, see Royall 1997; Wagenmakers et al. 2020; Rouder, Haaf, and Vandekerckhove 2018). The Bayes factor answers this question about the evidence in favor of an effect by explicitly conducting a model comparison. We will compare a model that assumes the presence of an effect, with a null model that assumes no effect.

As we saw before, the Bayes factor is highly sensitive to the priors. In the example presented above, both models are identical except for the effect of interest, β, and so the prior on this parameter will play a major role in the calculation of the Bayes factor.

Next, we will run a hierarchical model which includes group-level (or random) intercepts and slopes by items and by subjects. We will use regularizing priors on all the parameters–this speeds up computation and implies realistic expectations about the parameters. However, the prior on β will be crucial for the calculation of the Bayes factor.

One possible way to build a good prior for the parameter β, which estimates the influence of cloze probability, is as follows (see the online chapter E for a more detailed discussion on prior selection). The reasoning below is based on domain knowledge; but there is room for differences of opinion here. In a realistic data analysis situation, we would carry out a sensitivity analysis using a range of priors to determine the extent of influence of the priors.

1. One may want to be agnostic regarding the direction of the effect; that means that we will center the prior of β on zero by specifying that the mean of the prior distribution is zero. However, we are still not sure about the variance of the prior on β.
2. One would need to know a bit about the variation on the dependent variable that we are analyzing. After re-analyzing the data from a couple of EEG experiments available from osf.io, we can say that for N400 averages, the standard deviation of the signal is between 8 and 15 microvolts (Nicenboim, Vasishth, and Rösler 2020).
3. Based on published estimates of effects in psycholinguistics, we can conclude that they are generally rather small, often representing between 5% and 30% of the standard deviation of the dependent variable.
4. The effect of noun predictability on the N400 is one of the most reliable and strongest effects in neurolinguistics (together with the P600 that might even be stronger), and the slope β represents the average change in voltage when moving from a cloze probability of zero to one–the strongest prediction effect.

An additional way to obtain good priors is to perform prior predictive checks (Schad, Betancourt, and Vasishth 2020, also see the online chapter F, which presents a principled Bayesian workflow). Here, the idea is to simulate data from the model and the priors, and then to analyze the simulated data using summary statistics. For example, it would be possible to compute the summary statistic of the difference in the N400 between high versus low cloze probability. The simulations would yield a distribution of differences. Arguably, this distribution of differences, that is, the data analyses of the simulated data, are much easier to judge for plausibility than the prior parameters specifying prior distributions. That is, we might find it easier to judge whether a difference in voltage between high and low cloze probability is plausible rather than judging the parameters of the model. For reasons of brevity, we skip this step here.

Instead, we will start with the prior $\beta \sim Normal(0, 5)$ (since 5 microvolts is roughly 30% of 15, which is the upper bound of the expected standard deviation of the EEG signal).

```
priors1 <- c(prior(normal(2, 5), class = Intercept),
             prior(normal(0, 5), class = b),
             prior(normal(10, 5), class = sigma),
             prior(normal(0, 2), class = sd),
             prior(lkj(4), class = cor))
```

We load the data set on N400 amplitudes, which has data on cloze probabilities (Nieuwland et al. 2018). The cloze probability measure is mean-centered to make the intercept and the random intercepts easier to interpret (i.e., after centering, they represent the grand mean and the average variability around the grand mean across subjects or items).

```
data(df_eeg)
df_eeg <- df_eeg %>%
  mutate(c_cloze = cloze - mean(cloze))
```

A large number of effective samples will be needed to be able to get stable estimates of the Bayes factor with bridge sampling. For this reason a large number of sampling iterations (`n = 20000`) is specified. The parameter `adapt_delta` is set to 0.9 to ensure that the posterior sampler is working correctly. For Bayes factors analyses, it's necessary to set the argument `save_pars = save_pars(all = TRUE)`. This setting is a precondition for performing bridge sampling to compute the Bayes factor.

```
fit_N400_h_linear <- brm(n400 ~ c_cloze +
                           (c_cloze | subj) + (c_cloze | item),
                         prior = priors1,
                         warmup = 2000,
                         iter = 20000,
                         control = list(adapt_delta = 0.9),
                         save_pars = save_pars(all = TRUE),
                         data = df_eeg)
```

Next, take a look at the population-level (or fixed) effects from the Bayesian modeling.

```
fixef(fit_N400_h_linear)
```

```
##           Estimate Est.Error Q2.5 Q97.5
## Intercept     3.65      0.45 2.77  4.54
```

```
## c_cloze        2.33      0.64 1.05  3.58
```

We can now take a look at the estimates and at the credible intervals. The effect of cloze probability (`c_cloze`) is 2.33 microvolts with a 95% credible interval ranging from 1.05 to 3.58 microvolts. While this provides an initial hint that highly probable words may elicit a stronger N400 compared to low probable words, by just looking at the posterior there is no way to quantify evidence for the question whether this effect is different from zero. Model comparison is needed to answer this question.

To this end, we run the model again, now without the parameter of interest, i.e., the null model. This is a model where our prior for β is that it is exactly zero.

```
fit_N400_h_null <- brm(n400 ~ 1 +
                         (c_cloze | subj) + (c_cloze | item),
                       prior = priors1[priors1$class != "b", ],
                       warmup = 2000,
                       iter = 20000,
                       control = list(adapt_delta = 0.9),
                       save_pars = save_pars(all = TRUE),
                       data = df_eeg)
```

Now everything is ready for computing the log marginal likelihood, that is, the likelihood of the data given the model, after integrating out the model parameters. In the toy examples shown above, we had used the R function `integrate()` to perform this integration. This is not possible for the more realistic and more complex models that are considered here because the integrals that have to be solved are too high-dimensional and complex for these simple functions to do the job. Instead, a standard approach to deal with realistic complex models is to use bridge sampling (Gronau et al. 2017; Gronau, Singmann, and Wagenmakers 2020). We perform this integration using the function `bridge_sampler()` for each of the two models:

```
margLogLik_linear <- bridge_sampler(fit_N400_h_linear, silent = TRUE)
margLogLik_null <- bridge_sampler(fit_N400_h_null, silent = TRUE)
```

This gives us the marginal log likelihoods for each of the models. From these, we can compute the Bayes factors. The `bayes_factor()` function is a convenient tool for calculating the Bayes factor.

```
(BF_ln <- bayes_factor(margLogLik_linear, margLogLik_null))
```

```
## Estimated Bayes factor in favor of x1 over x2: 52.26813
```

Alternatively, the Bayes factor can be computed manually as well. First, compute the difference in marginal log likelihoods, then transform this difference in log likelihoods to the likelihood scale (using `exp()`). A difference in the exponential scale is a ratio: $exp(a - b) = exp(a)/exp(b)$. This computation yields the Bayes factor. However, the values `exp(ml1)` and `exp(ml2)` are too small to be represented accurately by R. Therefore, for numerical reasons, it is important to take the difference first and only then compute the exponential $exp(a - b)$, i.e., `exp(margLogLik_linear$logml - margLogLik_null$logml)`, which yields the same result as the `bayes_factor()` command.

```
exp(margLogLik_linear$logml - margLogLik_null$logml)
```

```
## [1] 52.3
```

The Bayes factor is quite large in this example, and furnishes strong evidence for the alternative model, which includes a coefficient representing the effect of cloze probability. That is, under the criteria shown in Table 13.1, the Bayes factor furnishes strong evidence for an effect of cloze probability.

In this example, there was good prior information about the model parameter β. What transpires, though, if we are unsure of the prior for the model parameter? Because our prior for β is inappropriate, it is possible that we end up comparing the null model with an extremely "bad" alternative model.

For example, assuming that we do not know much about N400 effects, or that we do not want to make strong assumptions, we might be inclined to use an uninformative prior. The new prior could look as follows (all the priors except for `b` remain unchanged):

```
priors_vague <- c(prior(normal(2, 5), class = Intercept),
                  prior(normal(0, 500), class = b),
                  prior(normal(10, 5), class = sigma),
                  prior(normal(0, 2), class = sd),
                  prior(lkj(4), class = cor))
```

We can use this uninformative prior in the Bayesian model:

```
fit_N400_h_linear_vague <- brm(n400 ~ c_cloze +
                                 (c_cloze | subj) + (c_cloze | item),
                               prior = priors_vague,
                               warmup = 2000,
                               iter = 20000,
                               control = list(adapt_delta = 0.9),
                               save_pars = save_pars(all = TRUE),
                               data = df_eeg)
```

Interestingly, we can still estimate the effect of cloze probability fairly well:

```
posterior_summary(fit_N400_h_linear_vague, variable = "b_c_cloze")
```

```
##             Estimate Est.Error Q2.5 Q97.5
## b_c_cloze       2.36     0.649 1.08  3.63
```

Next, we again perform the bridge sampling for the alternative model.

```
margLogLik_linear_vague <- bridge_sampler(fit_N400_h_linear_vague,
                                          silent = TRUE)
```

We compute the Bayes factor for the alternative over the null model, BF_{10}:

```
(BF_lnVague <- bayes_factor(margLogLik_linear_vague,
                            margLogLik_null))
```

```
## Estimated Bayes factor in favor of x1 over x2: 0.58436
```

This is easier to read as the evidence for null model over the alternative:

```
1 / BF_lnVague[[1]]
```

```
## [1] 1.71
```

The result is inconclusive: there is no evidence in favor of or against the effect of cloze probability. The reason for that is that priors are never uninformative when it comes to Bayes factors. The wide prior specifies that both very small and very large effect sizes are possible (with some considerable probability), but there is relatively little evidence in the data for such large effect sizes.

The above example is related to a criticism of Bayes factors by Uri Simonsohn that Bayes factors can provide evidence in favor of the null and against a very specific alternative model, when the researchers only know the direction of the effect (see `https://datacolada.org/78a`). This can happen when an uninformative prior is used.

One way to overcome this problem is to actually try to learn about the effect size that we are investigating. This can be done by first running an exploratory experiment and analysis without computing any Bayes factor, and then use the posterior distribution derived from this first experiment to calibrate the priors for the next confirmatory experiment where we do use the Bayes factor (see Verhagen and Wagenmakers 2014 for a Bayes Factor test calibrated to investigate replication success). For examples of this approach used in psycholinguistics, see Paape, Smith, and Vasishth (2024) and Schoknecht, Yadav, and Vasishth (2025).

Another possibility is to examine a lot of different alternative models, where each model uses different prior assumptions. In this manner, the degree to which the Bayes factor results rely on or are sensitive to the prior assumptions can be examined. This is an instance of a sensitivity analysis. Recall that the model is the likelihood *and* the priors. We can therefore compare models that only differ in the prior (for an example involving EEG and predictability effects, see Nicenboim, Vasishth, and Rösler 2020).

13.2.1 Sensitivity analysis

Here, we perform a sensitivity analysis by examining Bayes factors for several models. Each model has the same likelihood but a different prior for β. For all of the priors we assume a normal distribution with a mean of zero. Assuming a mean of zero asserts that we do not make any assumption a priori about the direction of the effect. The standard deviations of the various priors differ. That is, what differs is the amount of uncertainty about the effect size that we allow for in the prior. While a small standard deviation indicates that we expect the effect to be not very large, a large standard deviation permits very large effect sizes. Although a model with a wide prior (i.e., large standard deviation) also allocates prior probability to small effect sizes, it allocates much less probability to small effect sizes compared to a model with a narrow prior. Thus, if the effect size is in reality small, then a model with a narrow prior (small standard deviation) will have a better chance of detecting the effect.

Next, we try out a range of standard deviations, ranging from 1 to a much wider prior that has a standard deviation of 100. In practice, for the experiment method we are discussing here, it would not be a good idea to define very large standard deviations such as 100 microvolts, since they imply unrealistically large effect sizes. However, we include such a large value here just for illustration. Such a sensitivity analysis takes a very long time: here, we are running 11 models, where each model involves a lot of iterations to obtain stable Bayes factor estimates.

```
prior_sd <- c(1, 1.5, 2, 2.5, 5, 8, 10, 20, 40, 50, 100)
BF <- c()
for (i in 1:length(prior_sd)) {
  psd <- prior_sd[i]
  # for each prior we fit the model
  fit <- brm(n400 ~ c_cloze + (c_cloze | subj) + (c_cloze | item),
    prior =
      c(prior(normal(2, 5), class = Intercept),
        set_prior(paste0("normal(0,", psd, ")"), class = "b"),
        prior(normal(10, 5), class = sigma),
        prior(normal(0, 2), class = sd),
        prior(lkj(4), class = cor)),
    warmup = 2000,
    iter = 20000,
    control = list(adapt_delta = 0.9),
    save_pars = save_pars(all = TRUE),
    data = df_eeg)
  # for each model we run a brigde sampler
  margLogLik_linear_beta <- bridge_sampler(fit, silent = TRUE)
  # we store the Bayes factor compared to the null model
  BF <- c(BF,
        bayes_factor(margLogLik_linear_beta, margLogLik_null)$bf)
}
BFs <- tibble(beta_sd = prior_sd, BF)
```

For each model, we run bridge sampling and we compute the Bayes factor of the model against our baseline or null model, which does not contain a population-level effect of cloze probability ($\boldsymbol{BF}_{10}$). Next, we need a way to visualize all the Bayes factors. We plot them in Figure 13.3 as a function of the prior width.

This figure clearly shows that the Bayes factor provides evidence for the alternative model; that is, it provides evidence that the population-level (or fixed) effect cloze probability is needed to explain the data. This can be seen as the Bayes factor is quite large for a range of different values for the prior standard deviation. The Bayes factor is largest for a prior standard deviation of 2.5, suggesting a rather small size of the effect of cloze probability. If we assume gigantic effect sizes a priori (e.g., standard deviations of 50 or 100), then the evidence for the alternative model is weaker. Conceptually, the data

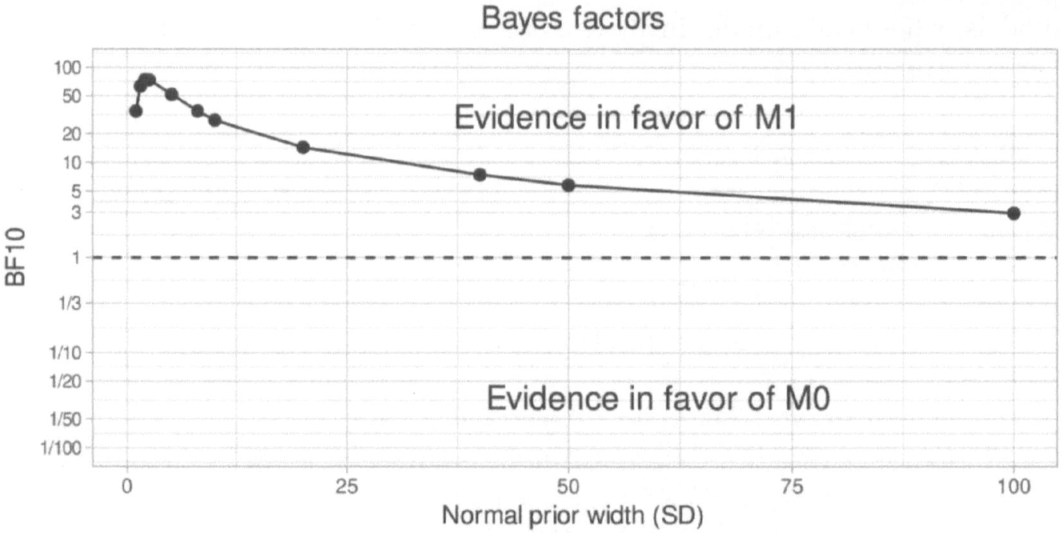

FIGURE 13.3: Prior sensitivity analysis for the Bayes factor.

do not fully support such big effect sizes, but start to favor the null model relatively more, when such big effect sizes are tested against the null. Overall, we can conclude that the data provide evidence for a not too large but robust influence of cloze probability on the N400 amplitude.

13.2.2 Non-nested models

One important advantage of computing Bayes factors using bridge sampling is that they can be used to compare models that are not nested. In nested models, the simpler model is a special case of the more complex and general model. For example, our previous model of cloze probability was a general model, allowing different influences of cloze probability on the N400. We compared this to a simpler, more specific null model, where the influence of cloze probability was not included, which means that the regression coefficient (population level or fixed effect) for cloze probability was assumed to be set to zero. Such nested models can be also compared using frequentist methods such as the likelihood ratio test (ANOVA).

By contrast, the Bayes factor also makes it possible to compare non-nested models. An example of a non-nested model would be a case where we log-transform the cloze probability variable before using it as a predictor. A model with log cloze probability as a predictor is not a special case of a model with linear cloze probability as predictor. These are just different, alternative models. With Bayes factors, we can compare these non-nested models with each other to determine which receives more evidence from the data. (For clarity: one can only apply model comparison if the same dependent variables is modeled. Hence, we can use the Bayes factor to compare a model containing log-transformed predictors with a model containing untransformed predictors, but we cannot compare a model using an untransformed dependent variable with a model using a log-transformed dependent variable.)

To do so, we first log-transform the cloze probability variable. Some cloze probabilities in the data set are equal to zero. This creates a problem when taking logs, since the log of zero is minus infinity, a value that we cannot use. We are going to overcome this problem by "smoothing" the cloze probability in this example. We use additive smoothing (also called Laplace or Lidstone smoothing; Lidstone 1920; Chen and Goodman 1999) with pseudocounts set to one, this means that the smoothed probability is calculated as the number of responses with a given gender plus one divided by the total number of responses plus two.

```
df_eeg <- df_eeg %>%
  mutate(scloze = (cloze_ans + 1) / (N + 2),
         c_logscloze = log(scloze) - mean(log(scloze)))
```

Next, we center the predictor variable, and we scale it to the same standard deviation as the linear cloze probabilities. To implement this scaling, first divide the centered smoothed log cloze probability variable by its standard deviation (effectively creating z-scaled values). As a next step, multiply the z-scaled values by the standard deviation of the non-transformed cloze probability variable. This way, both predictors (log cloze and cloze) have the same standard deviation. We therefore expect them to have a similar impact on the N400. As a result of this transformation, the same priors can be used for both variables (given that we currently have no specific information about the effect of log cloze probability versus linear cloze probability):

```
df_eeg <- df_eeg %>%
  mutate(c_logscloze = scale(c_logscloze) * sd(c_cloze))
```

Then, run a linear mixed-effects model with log cloze probability instead of linear cloze probability, and again carry out bridge sampling.

```
fit_N400_h_log <- brm(n400 ~ c_logscloze +
                        (c_logscloze | subj) + (c_logscloze | item),
                      prior = priors1,
                      warmup = 2000,
                      iter = 20000,
```

```
                  control = list(adapt_delta = 0.9),
                  save_pars = save_pars(all = TRUE),
                  data = df_eeg)
```

```
margLogLik_log <- bridge_sampler(fit_N400_h_log, silent = TRUE)
```

Next, compare the linear and the log model to each other using Bayes factors.

```
(BF_log_lin <- bayes_factor(margLogLik_log, margLogLik_linear))
```

```
## Estimated Bayes factor in favor of x1 over x2: 6.04167
```

The results show a Bayes factor of 6 of the log model over the linear model. This shows some evidence that log cloze probability is a better predictor of N400 amplitudes than linear cloze probability. This analysis demonstrates that model comparisons using Bayes factor are not limited to nested models, but can also be used for non-nested models. Here, a specific benefit of the Bayes factor is that it considers the functional shape or complexity of the models, instead of merely counting the number of parameters (as is done in information criteria like BIC and AIC).

13.3 The influence of the priors on Bayes factors: Beyond the effect of interest

We saw above that the width (or standard deviation) of the prior distribution for the effect of interest had a strong impact on the results from Bayes factor analyses. Thus, one question is whether only the prior for the effect of interest is important, or whether priors for other model parameters can also impact the resulting Bayes factors in an analysis. It turns out that priors for other model parameters can also be important and impact Bayes factors, especially when there are non-linear components in the model, such as in generalized linear mixed effects models. We investigate this issue by using a simulated data set on a variable that has a Bernoulli distribution; in each trial, subjects can perform either successfully (`pDV = 1`) on a task, or not (`pDV = 0`). The simulated data is from a factorial experimental design, with one between-subject factor F with 2 levels ($F1$ and $F2$), and Table 13.2 shows success probabilities for each of the experimental conditions.

TABLE 13.2: Summary statistics per condition for the simulated data.

Factor A	N data	Means
F1	50	0.98
F2	50	0.70

```
data("df_BF")
str(df_BF)
```

```
## tibble [100 x 3] (S3: tbl_df/tbl/data.frame)
##  $ F  : Factor w/ 2 levels "F1","F2": 1 1 1 1 1 1 1 1 1 1 ...
##  $ pDV: int [1:100] 1 1 1 1 1 1 1 1 1 1 ...
##  $ id : int [1:100] 1 2 3 4 5 6 7 8 9 10 ...
```

Our question now is whether there is evidence for a difference in success probabilities between groups $F1$ and $F2$. As contrasts for the factor F, we use scaled sum coding $(-0.5, +0.5)$.

```
contrasts(df_BF$F) <- c(-0.5, +0.5)
```

Next, we proceed to specify our priors. For the difference between groups ($F1$ versus $F2$), define a normally distributed prior with a mean of 0 and a standard deviation of 0.5. Thus, we do not specify a direction of the difference a priori, and assume not too large effect sizes. Now run two logistic `brms` models, one with the group factor F included, and one without the factor F, and compute Bayes factors using bridge sampling to obtain the evidence that the data provide for the alternative hypothesis that a group difference exists between levels $F1$ and $F2$.

So far, we have only specified the prior for the effect size. The question we are asking now is whether priors on other model parameters can impact the Bayes factor computations for testing the group effect. Specifically, can the prior for the intercept influence the Bayes factor for the group difference? The results show that the answer is yes, in some situations. Let's have a look at this in more detail. Let's assume that we compare two different priors for the intercept. We specify each as a normal distribution with a standard deviation of 0.1, thus, specifying relatively high certainty a priori where the intercept of the data will fall. The only difference that we now specify, is that one time, the prior mean (on the latent logistic scale) is set to 0, corresponding to a prior mean probability of 0.5. In the other condition, we specify a prior mean of 2, corresponding to a prior mean probability of 0.88. When we look at the data (see Table 13.2) we see that the prior mean of 0 (i.e., prior probability for the intercept of 0.5) is not very compatible with the data, whereas the prior mean of 2 (i.e., a prior probability for the intercept of 0.88) is quite closely aligned with the actual data.

We now compute Bayes factors for the group difference (***F*1** versus ***F*2**) by using these different priors for the intercept. Thus, we first fit a null (***M*0**) and alternative (***M*1**) model under a prior that is misaligned with the data (a narrow distribution centered in zero), and perform bridge sampling for these models:

```
# set prior
priors_logit1 <- c(prior(normal(0, 0.1), class = Intercept),
                   prior(normal(0, 0.5), class = b))
# Bayesian GLM: M0
fit_pDV_H0 <- brm(pDV ~ 1,
                  data = df_BF,
                  family = bernoulli(link = "logit"),
                  prior = priors_logit1[-2, ],
                  save_pars = save_pars(all = TRUE))
# Bayesian GLM: M1
fit_pDV_H1 <- brm(pDV ~ 1 + F,
                  data = df_BF,
                  family = bernoulli(link = "logit"),
                  prior = priors_logit1,
                  save_pars = save_pars(all = TRUE))
# bridge sampling
mLL_binom_H0 <- bridge_sampler(fit_pDV_H0, silent = TRUE)
mLL_binom_H1 <- bridge_sampler(fit_pDV_H1, silent = TRUE)
```

Next, get ready for computing Bayes factors by again running the null (***M*0**) and the alternative (***M*1**) model, now assuming a more realistic prior for the intercept (prior mean = 2).

```
priors_logit2 <- c(prior(normal(2, 0.1), class = Intercept),
                   prior(normal(0, 0.5), class = b))
# Bayesian GLM: M0
fit_pDV_H0_2 <- brm(pDV ~ 1,
                    data = df_BF,
                    family = bernoulli(link = "logit"),
                    prior = priors_logit2[-2, ],
```

```
                  save_pars = save_pars(all = TRUE))
# Bayesian GLM: M1
fit_pDV_H1_2 <- brm(pDV ~ 1 + F,
                  data = df_BF,
                  family = bernoulli(link = "logit"),
                  prior = priors_logit2,
                  save_pars = save_pars(all = TRUE))
# bridge sampling
mLL_binom_H0_2 <- bridge_sampler(fit_pDV_H0_2, silent = TRUE)
mLL_binom_H1_2 <- bridge_sampler(fit_pDV_H1_2, silent = TRUE)
```

Based on these models and bridge samples, we can now compute the Bayes factors in support for $M1$ (i.e., in support of a group-difference between $F1$ and $F2$). We can do so for the unrealistic prior for the intercept (prior mean of 0) and the more realistic prior for the intercept (prior mean of 2).

```
(BF_binom_H1_H0 <- bayes_factor(mLL_binom_H1, mLL_binom_H0))
```

```
## Estimated Bayes factor in favor of x1 over x2: 7.19574
```

```
(BF_binom_H1_H0_2 <- bayes_factor(mLL_binom_H1_2, mLL_binom_H0_2))
```

```
## Estimated Bayes factor in favor of x1 over x2: 29.58496
```

The results show that with the realistic prior for the intercept (prior mean = 2), the evidence for the $M1$ is quite strong, with a Bayes factor of $BF_{10} = 29.6$. With the unrealistic prior for the intercept (i.e., prior mean = 0), by contrast, the evidence for the $M1$ is much reduced, $BF_{10} = 7.2$, and now only modest.

Thus, when performing Bayes factor analyses, not only can the priors for the effect of interest (here the group difference) impact the results, under certain circumstances priors for other model parameters can too, such as the prior mean for the intercept here. Such an influence will not always be strong, and can sometimes be negligible. There may be many situations, where the exact specification of the intercept does not have much of an effect on the Bayes factor for a group difference. However, such influences can in principle occur, especially in models with non-linear components. Therefore, it is very important to be careful in specifying realistic priors for all model parameters, also including the intercept. A good way to judge whether prior assumptions are realistic and plausible is prior predictive checks, where we simulate data based on the priors and the model and judge whether the simulated data is plausible and realistic.

13.4 The Bayes factor in Stan

The package `bridgesampling` allows for a straightforward calculation of Bayes factor for Stan models as well. All the limitations and caveats of Bayes factor discussed in this chapter apply to Stan code as much as they apply to `brms` code. The sampling notation (`~`) should not be used in the Stan code; see the online section B.2 for an explanation.

An advantage of using Stan in comparison with `brms` is Stan's flexibility. We revisit the model implemented before in section 8.4.2. We want to assess the evidence for a *positive* effect of attentional load on pupil size against a similar model that assumes no effect. To do this, assume the following likelihood:

$$p\_size_n \sim Normal(\alpha + c\_load_n \cdot \beta_1 + c\_trial \cdot \beta_2 + c\_load \cdot c\_trial \cdot \beta_3, \sigma) \quad (13.13)$$

Define priors for all the βs as before, with the difference that β_1 can only have positive values:

$$\begin{aligned} \alpha &\sim Normal(1000, 500) \\ \beta_1 &\sim Normal_+(0, 100) \\ \beta_2 &\sim Normal(0, 100) \\ \beta_3 &\sim Normal(0, 100) \\ \sigma &\sim Normal_+(0, 1000) \end{aligned} \quad (13.14)$$

The following Stan model is the direct translation of the new priors and likelihood.

```
data {
  int<lower = 1> N;
  vector[N] c_load;
  vector[N] c_trial;
  vector[N] p_size;
}
parameters {
  real alpha;
  real<lower = 0> beta1;
  real beta2;
```

```
  real beta3;
  real<lower = 0> sigma;
}
model {
  target += normal_lpdf(alpha | 1000, 500);
  target += normal_lpdf(beta1 | 0, 100) -
    normal_lccdf(0 | 0, 100);
  target += normal_lpdf(beta2 | 0, 100);
  target += normal_lpdf(beta3 | 0, 100);
  target += normal_lpdf(sigma | 0, 1000)
    - normal_lccdf(0 | 0, 1000);
  target += normal_lpdf(p_size | alpha + c_load * beta1 +
                                 c_trial * beta2 +
                                 c_load .* c_trial * beta3, sigma);
}
```

Fit the model with 20,000 iterations to ensure that the Bayes factor is stable, and increase the `adapt_delta` parameter to avoid warnings:

```
data("df_pupil")
df_pupil <- df_pupil %>%
  mutate(c_load = load - mean(load),
         c_trial = trial - mean(trial))
ls_pupil <- list(p_size = df_pupil$p_size,
                 c_load = df_pupil$c_load,
                 c_trial = df_pupil$c_trial,
                 N = nrow(df_pupil))
pupil_pos <- system.file("stan_models",
                         "pupil_pos.stan",
                         package = "bcogsci")
fit_pupil_int_pos <- stan(file = pupil_pos,
                          data = ls_pupil,
                          warmup = 1000,
                          iter = 20000,
                          control = list(adapt_delta = .95))
```

The null model that we defined has $\beta_1 = 0$ and is written in Stan as follows:

```
data {
  int<lower = 1> N;
  vector[N] c_load;
  vector[N] c_trial;
  vector[N] p_size;
```

```
}
parameters {
  real alpha;
  real beta2;
  real beta3;
  real<lower = 0> sigma;
}
model {
  target += normal_lpdf(alpha | 1000, 500);
  target += normal_lpdf(beta2 | 0, 100);
  target += normal_lpdf(beta3 | 0, 100);
  target += normal_lpdf(sigma | 0, 1000)
    - normal_lccdf(0 | 0, 1000);
  target += normal_lpdf(p_size | alpha + c_trial * beta2 +
                                 c_load .* c_trial * beta3, sigma);
}
```

```
pupil_null <- system.file("stan_models",
                          "pupil_null.stan",
                          package = "bcogsci")
fit_pupil_int_null <- stan(file = pupil_null,
                           data = ls_pupil,
                           warmup = 1000,
                           iter = 20000)
```

Compare the models with bridge sampling:

```
lml_pupil <- bridge_sampler(fit_pupil_int_pos, silent = TRUE)
lml_pupil_null <- bridge_sampler(fit_pupil_int_null, silent = TRUE)
BF_att <- bridgesampling::bf(lml_pupil, lml_pupil_null)
```

```
BF_att
```

```
## Estimated Bayes factor in favor of lml_pupil over lml_pupil_null: 25.16042
```

We find that the data is 25.16 more likely under a model that assumes a positive effect of load than under a model that assumes no effect.

13.5 Bayes factors in theory and in practice

13.5.1 Bayes factors in theory: Stability and accuracy

One question that we can ask here is how stable and accurate the estimates of Bayes factors are. The bridge sampling algorithm needs a lot of posterior samples to obtain stable estimates of the Bayes factor. Running bridge sampling based on a too small effective sample size will yield unstable estimates of the Bayes factor, such that repeated computations will yield radically different Bayes factor values. Moreover, even if the Bayes factor is approximated in a stable way, it is unclear whether this approximate Bayes factor is equal to the true Bayes factor, or whether there is bias in the computation such that the approximate Bayes factor has a wrong value. We show this below.

13.5.1.1 Instability due to the effective number of posterior samples

The number of iterations, which in turn affects the total number of posterior samples can have a strong impact on the robustness of the results of the bridge sampling algorithm (i.e., on the resulting Bayes factor) and there are no good theoretical guarantees that bridge sample will yield accurate estimates of Bayes factors. In the analyses presented above, we set the number of iterations to a very large number of $n = 20{,}000$. The sensitivity analysis therefore took a considerable amount of time. It turns out that the results from this analysis were stable, as shown below.

Running the same analysis with less iterations will induce some instability in the Bayes factor estimates based on the bridge sampling, such that running the same analysis twice would yield different results. Furthermore, due to variations in starting values, bridge sampling itself may be unstable and yield inconsistent results for successive runs on the same posterior samples. This is highly concerning because if the number of effective sample sizes is not large enough, the results published in a paper may not be stable. Indeed, the default number of iterations in `brms` is set as `iter = 2000` (and the default number of warmup iterations is `warmup = 1000`). These defaults were not set to support bridge sampling, i.e., they were not defined for computation of densities to support Bayes factors calculations. These defaults are reasonable only for posterior inference on expectations (e.g., posterior means) for models that are not too complex. However, when using these defaults for estimation of densities and the computation of Bayes factors, instabilities can arise.

As an illustration, we perform the same sensitivity analysis again, now using the default number of 2000 iterations in `brms`. The posterior sampling process now runs much quicker. Moreover, we check the stability of the Bayes factors in the sensitivity analyses by repeating both sensitivity analyses (with $n = 20{,}000$ iterations and with the default number of $n = 2000$ iterations) a second time, to see whether the results for Bayes factors are stable.

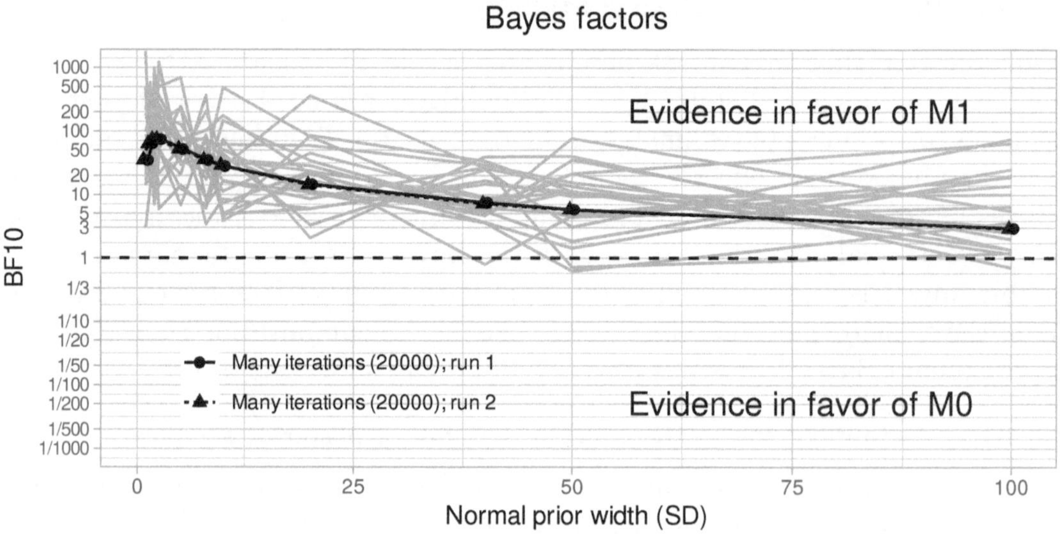

FIGURE 13.4: The effect of the number of samples on a prior sensitivity analysis for the Bayes factor. Black lines show 2 runs with 20,000 iterations. Gray lines show 20 runs with default number of iterations (2000).

The results displayed in Figure 13.4 show that the Bayes factors are highly unstable when the number of iterations is low. They clearly deviate from the Bayes factors estimated with 20,000 iterations, resulting in very unstable estimates. By contrast, the analyses using 20,000 iterations provide nearly the same results in both analyses. The two lines lie virtually directly on top of each other; the points are jittered horizontally for better visibility.

This demonstrates that it is necessary to use a large number of iterations when computing Bayes factors using bridge sampling. In practice, one should compute the sensitivity analysis (or at least one of the models or priors) twice (as we did here) to make sure that the results are stable and sufficiently similar, in order to provide a good basis for reporting results.

By contrast, as mentioned earlier as well, Bayes factors based on the Savage–Dickey method (as implemented in `brms`) can be unstable even when using a large number of posterior samples. This problem can arise especially when the posterior is very far from zero, and thus very large or very small Bayes factors are obtained. (In such cases, the precise value of the Bayes factor may sometimes not matter in practical applications.) However, because of this instability of the Savage-Dickey method in `brms`, it is a good idea to use bridge sampling and to check the stability of the estimates using more than one method.

13.5.1.2 Inaccuracy of Bayes factor estimates: Does the estimate approximate the true Bayes factor well?

An important point about approximate estimates of Bayes factors using bridge sampling is that there are no strong guarantees for their accuracy. That is, even if we can show that the approximated Bayes factor estimate using bridge sampling is stable (i.e., when using sufficient effective samples, see the analyses above), even then it remains unclear whether the Bayes factor estimate actually is close to the true Bayes factor. The estimated Bayes factors based on bridge sampling may be stable, but, in theory, could be biased: they may not be very close to the true (correct) Bayes factor. The technique of (marginal) simulation-based calibration (SBC; Geweke 2004; Talts et al. 2018; Schad, Betancourt, and Vasishth 2020) can be used to investigate this question (SBC is also discussed in section 10.2 in chapter 10). We investigate this question next.[2]

In the SBC approach, the priors are used to simulate data. Then, posterior inference is done on the simulated data and the posterior can be compared to the prior. If the posteriors are equal to the priors, then this supports accurate computations. Applied to Bayes factor analyses, one defines a prior on the hypothesis space, i.e., one defines the prior probabilities for a null and an alternative model, specifying how likely each model is a priori. From these priors, one can randomly draw one hypothesis (model), e.g., $nsim = 500$ times. Thus, in each of 500 draws one randomly chooses one model (either $M0$ or $M1$), with the probabilities given by the model priors. For each draw, one first samples model parameters from their prior distributions, and then uses these sampled model parameters to simulate data. For each simulated data set, one can then compute marginal likelihoods and Bayes factor estimates using posterior samples and bridge sampling, and one can then compute the posterior probabilities for each hypothesis (i.e., how likely each model is a posteriori). As the last, and critical step in marginal SBC, one can then compare the posterior model probabilities to the prior model probabilities. A key result in marginal SBC (also termed data-averaged posterior) is that if the computation of marginal likelihoods and posterior model probabilities is performed accurately (without bias) by the bridge sampling procedure; that is, if the Bayes factor estimate is close to the true Bayes factor, then the posterior model probabilities should be the same as the prior model probabilities.

[2]For details, see Schad, D. J., Nicenboim, B., Bürkner, P. C., Betancourt, M., and Vasishth, S. (2022). Workflow techniques for the robust use of Bayes factors. Psychological Methods.

Here, we perform this marginal SBC (data-averaged posterior) approach. Across the 500 simulations, we systematically vary the prior model probability from zero to one. For each of the 500 simulations we sample a model (hypothesis) from the model prior, then sample parameters from the priors over parameters, use the sampled parameters to simulate data, fit the null and the alternative model on the simulated data, perform bridge sampling for each model, compute the Bayes factor estimate between them, and compute posterior model probabilities. If the bridge sampling works accurately, then the posterior model probabilities should be the same as the prior model probabilities. Given that we varied the prior model probabilities from zero to one, the posterior model probabilities should also vary from zero to one. In Figure 13.5, we plot the posterior model probabilities as a function of the prior model probabilities. If the posterior probabilities are the same as the priors, then the local regression line and all points should lie on the diagonal.

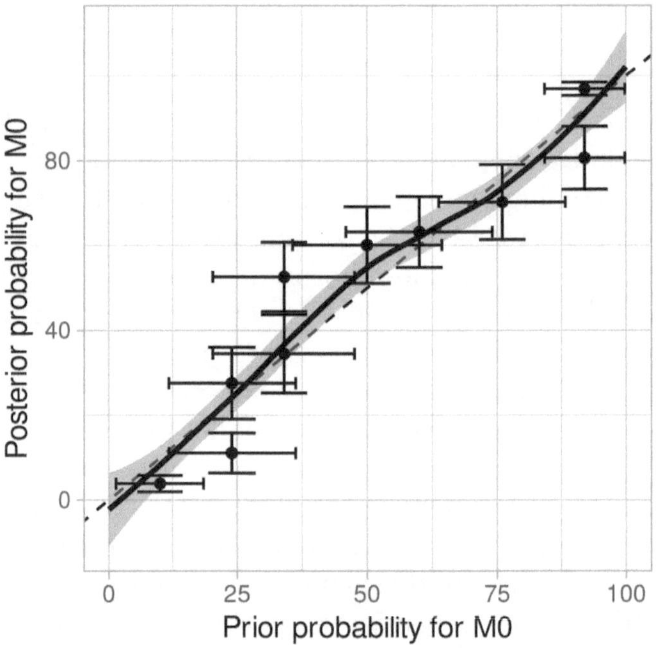

FIGURE 13.5: The posterior probabilities for M0 are plotted as a function of prior probabilities for M0. If the approximation of the Bayes factor using bridge sampling is unbiased, then the data should be aligned along the diagonal (see dashed black line). The thick black line is a prediction from a local regression analysis. The points are average posterior probabilities as a function of a priori selected hypotheses for 50 simulation runs each. Error bars represent 95% confidence intervals.

The results of this analysis in Figure 13.5 show that the local regression line is very close to the diagonal, and that the data points (each summarizing results from 50 simulations, with means and confidence intervals) also lie close to the diagonal. This demonstrates that the estimated posterior model probabilities are close to their a priori values. This result shows that posterior model probabilities, which are based on the Bayes factor estimates from the bridge sampling, are unbiased for a large range of different a priori model probabilities.

This result is very important as it shows one example case where the Bayes factor approximation is accurate. However, this demonstration is valid only for this one specific application case, i.e., with a particular data set, particular models, specific priors for the parameters, and a specific comparison between nested models. Strictly speaking, if one wants to be sure that the Bayes factor estimate is accurate for a particular data analysis, then such a SBC validation analysis would have to be computed for every data analysis. For details, including code, on how to perform such a marginal SBC (data-averaged posterior), see `https://osf.io/y354c/`. However, the fact that the SBC yields such promising results for this first application case also gives some hope that the bridge sampling may be accurate also for other comparable data analysis situations.

Based on these results on the average theoretical performance of Bayes factor estimation, we next turn to a different issue: how Bayes factors depend on and vary with varying data, leading to bad performance in individual cases despite good average performance.

13.5.2 Bayes factors in practice: Variability with the data

13.5.2.1 Variation associated with the data (subjects, items, and residual noise)

A second, and very different, source limiting robustness of Bayes factor estimates derives from the variability that is observed with the data, i.e., among subjects, items, and residual noise. Thus, when repeating an experiment a second time in a replication analysis, using different subjects and items, will lead to different outcomes of the statistical analysis every time a new replication run is conducted. The "dance of p-values" (Cumming 2014) illustrates this well-known limit to robustness in frequentist analyses, where p-values are not consistently significant across studies over repeated replication attempts. Instead, every time a study is repeated, the outcomes produce wildly disparate p-values. This can also be observed when simulating data from some known truth and re-running analyses on simulated data sets.

Additionally, Bayesian analyses ought to incorporate this same kind of variability (also refer to `https://daniellakens.blogspot.com/2016/07/dance-of-bayes-factors.html`). Here we show this type of variability in Bayes factor analyses by looking at a new example data analysis: We look at research on sentence comprehension, and specifically on effects of cue-based retrieval interference (Lewis and Vasishth 2005).

13.5.2.2 An example: The facilitatory interference effect

The experiments investigating the cognitive processes behind a well-researched phenomenon in sentence comprehension will be examined next. The agreement attraction configuration below serves as the example for this discussion. In it, the grammatically incorrect sentence (2) appears more grammatical than the equally grammatically incorrect sentence (1):

(1) The key to the cabinet are in the kitchen.
(2) The key to the cabinets are in the kitchen.

Both sentences are ungrammatical because the subject ("key") does not agree with the verb in number ("are"). When compared to (1), sentences like (2) are frequently found to have shorter reading times at (or immediately after) the verb ("are"); for a meta-analysis, see Jäger, Engelmann, and Vasishth (2017). These shorter reading times are sometimes called "facilitatory interference" (Dillon 2011); facilitatory in this context simply refers to the fact that reading times at the relevant word are shorter in (2) compared to (1), without necessarily implying that processing is easier. One explanation for the shorter reading times is that there is an illusion of grammaticality because the attractor word (cabinets in this case) agrees locally in number with the verb. This is an interesting phenomenon because the plural versus singular feature of the attractor noun ("cabinet/s") is not the subject, and therefore, under the rules of English grammar, is not supposed to agree with the number marking on the verb. That agreement attraction effects are consistently observed indicates that some non-compositional processes are taking place.

Using a computational implementation (formulated in the ACT-R framework, Anderson et al. 2004), an account of agreement attraction effects in language processing explains how retrieval-based working memory mechanisms lead to such agreement attraction effects in ungrammatical sentences (Engelmann, Jäger, and Vasishth 2020; see also Hammerly, Staub, and Dillon 2019; and Yadav et al. 2023). Numerous studies have examined agreement attraction in grammatically incorrect sentences using comparable experimental setups and various dependent measures, including eye tracking and self-paced reading. It is generally believed to be a robust empirical phenomenon, and we choose it for analysis here for that reason.

Here, we look at a self-paced reading study on agreement attraction in Spanish by Lago et al. (2015). For the experimental condition agreement attraction (x; sentence type), we estimate a population-level effect against a null model in which the sentence type population-level effect is not included. For the agreement attraction effect of sentence type, we use sum contrast coding (i.e., -1 and +1). We run a hierarchical model with the following formula in `brms`: `rt ~ x + (x | subj) + (x | item)`, where `rt` is reading time, we have random variation associated with subjects and with items, and we assume that reading times follow a log-normal distribution: `family = lognormal()`.

First, load the data:

```
data("df_lagoE1")
head(df_lagoE1)
```

```
##     subj item  rt  int  x   expt
## 2     S1   I1 588  low -1 lagoE1
## 22    S1  I10 682 high  1 lagoE1
## 77    S1  I13 226  low -1 lagoE1
## 92    S1  I14 580 high  1 lagoE1
## 136   S1  I17 549  low -1 lagoE1
## 153   S1  I18 458 high  1 lagoE1
```

As a next step, determine priors for the analysis of these data.

13.5.2.3 Determine priors using meta-analysis

One good way to obtain priors for Bayesian analyses, and specifically for Bayes factor analyses, is to use results from meta-analyses on the subject. Here, we take the prior for the experimental manipulation of agreement attraction from a published meta-analysis (Jäger, Engelmann, and Vasishth 2017).[3]

The mean effect size (difference in reading time between the two experimental conditions) in the meta-analysis is -22 milliseconds (ms), with 95% $CI = [-36\ -9]$ (Jäger, Engelmann, and Vasishth 2017, Table 4). This means that on average, the target word (i.e., the verb) in sentences such as (2) is on average read 22 milliseconds faster than in sentences such as (1). The size of the effect is measured on the millisecond scale, assuming a normal distribution of effect sizes across studies.

However, individual reading times usually do not follow a normal distribution. Instead, a better assumption about the distribution of reading times is a log-normal distribution. This is what we will assume in the `brms` model. Therefore,

[3]This meta-analysis already includes the data that we want to make inference about; thus, this meta-analysis estimate is not really the right estimate to use, since it involves using the data twice. We ignore this detail here because our goal is simply to illustrate the approach.

to use the prior from the meta-analysis in the Bayesian analysis, we have to transform the prior values from the millisecond scale to log millisecond scale.

We have performed this transformation in the published version of the arXiv paper. Based on these calculations, the prior for the experimental factor of interference effects is set to a normal distribution with mean = −0.03 and standard deviation = 0.009.

One can check that the prior on the slope parameter amounts to assigning a prior on the millisecond scale such that the effect has mean −22 ms, with an approximate 95% CI [−36 − 9] ms:

```
# contrast coding:
high <- 1
low <- -1
# mean:
exp(6 + (- 0.03) * high) -
  exp(6 + (- 0.03) * low)
```

```
## [1] -24.2
```

```
# confidence bounds on log scale:
lower <- - 0.03 - 2 * 0.009
upper <- - 0.03 + 2 * 0.009
# confidence bounds on ms scale:
exp(6 + lower * high) - exp(6 + lower * low)
```

```
## [1] -38.7
```

```
exp(6 + upper * high) - exp(6 + upper * low)
```

```
## [1] -9.68
```

For the other model parameters, we use principled priors.

```
priors <- c(prior(normal(6, 0.5), class = Intercept),
            prior(normal(-0.03, 0.009), class = b),
            prior(normal(0, 0.5), class = sd),
            prior(normal(0, 1), class = sigma),
            prior(lkj(2), class = cor))
```

13.5.2.4 Running a hierarchical Bayesian analysis

Next, run a `brms` model on the data. We use a large number of iterations (`iter = 10000`) with bridge sampling to estimate the Bayes factor of the "full" model, which includes a population-level effect for the experimental condition agreement attraction (`x`; i.e., sentence type). As mentioned above, for the agreement attraction effect of sentence type, we use sum contrast coding (i.e., −1 and +1).

We first show the population-level (or fixed) effects from the posterior analyses:

```
fixef(m1_lagoE1)
```

```
##           Estimate Est.Error  Q2.5 Q97.5
## Intercept     6.01      0.06  5.90  6.13
## x            -0.03      0.01 -0.05 -0.02
```

They show that for the population-level effect `x`, capturing the agreement attraction effect, the 95% credible interval does not overlap with zero. This indicates that there is some hint that the effect may have the expected negative direction, reflecting shorter reading times in the plural condition. As mentioned earlier, this does not provide a direct test of the hypothesis that the effect exists and is not zero. This is not tested here, because we did not specify the null hypothesis of zero effect explicitly. We can, however, draw inferences about this null hypothesis by using the Bayes factor.

Estimate Bayes factors between a full model, where the effect of agreement attraction is included, and a null model, where the effect of agreement attraction is absent, using the command `bayes_factor(lml_m1_lagoE1, lml_m0_lagoE1)`. The Bayes factor BF_{10}, or the strength of the alternative over the null, is calculated by the function.

```
h_lagoE1$bf
```

```
## [1] 6.59
```

With a Bayes factor of 6, the output indicates that the alternative model–which incorporates the population-level effect of agreement attraction–may have some merit. That is, this provides evidence for the alternative hypothesis that there is a difference between the experimental conditions, i.e., a facilitatory effect in the plural condition of the size derived from the meta-analysis.

As discussed earlier, the `bayes_factor` command should be run several times to check the stability of the Bayes factor calculation.

13.5.2.5 Variability of the Bayes factor: Posterior simulations

One way to investigate how variable the outcome of Bayes factor analyses can be (given that the Bayes factor is computed in a stable and accurate way) is to generate some realistic data and compute Bayes factors for these simulated data. A key question then is how to set priors that yield realistic simulated data sets. Here, we use the posteriors from the analysis of a real empirical data set and generate posterior predictive distributions, which serve as our realistic simulated data. Computing the Bayes factor analysis again on the simulated data can provide some insight into how variable the Bayes factor will be.

Here, we perform these simulations. We start with a data set and a model (the Bayesian hierarchical model fitted to the data from Lago et al. 2015), and use the posterior distributions from this model to generate posterior predictive data. In these simulations, one takes posterior samples for the model parameters (i.e., $p(\boldsymbol{\Theta} \mid \boldsymbol{y})$), and for each posterior sample of the model parameters, one can simulate new data $\tilde{\boldsymbol{y}}$ from the model $p(\tilde{\boldsymbol{y}} \mid \boldsymbol{\Theta})$.

```
pred_lagoE1 <- posterior_predict(m1_lagoE1)
```

The question that we are interested in here now is, how much information is contained in this simulated data. That is, we can run Bayesian models on these simulated data sets and compute Bayes factors to test whether in the simulated data there is evidence for agreement attraction effects.

The interesting question here is how variable the outcomes of these Bayes factor analyses will be among various simulated replications of the same study.

Now, using 50 different simulated data sets simulated from the fitted model, we carry out this analysis. For each of these data sets, we can proceed in exactly the same way as we did for the original observed experimental data. That is, we again fit the same `brms` model 50 times, now to the simulated data, and using the same prior as before. For each simulated data set, we use bridge sampling to compute the Bayes factor of the alternative model compared to a null model where the agreement attraction effect (population-level effect predictor of sentence type, `x`) is set to 0. For each simulated predictive data set, we store the resulting Bayes factor. We again use the prior from the meta-analysis.

13.5.2.6 Visualize distribution of Bayes factors

We can now visualize the distribution of Bayes factors (BF_{10}) across the simulated data by plotting a histogram. In this histogram, values greater than one support the alternative model (M1) that there are agreement attraction effects (i.e., the sentence type effect differs from zero), and values of the Bayes factor less than one support the null model (M0), which states that there is no agreement attraction effect (i.e., there is no difference in reading times between experimental conditions).

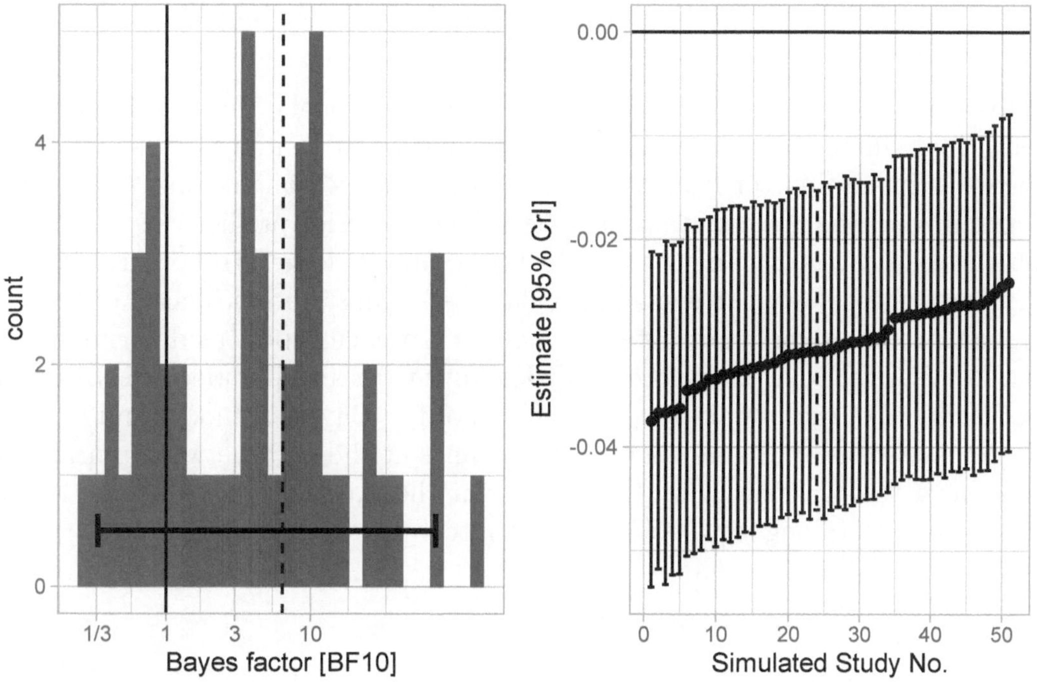

FIGURE 13.6: Estimates of the retrieval interference facilitatory effect and the 95% credible intervals for all simulations (solid lines) and the empirically observed data (dashed line) are shown in the left panel. An illustration of the alternative model's Bayes factors (BF10) over the null model in 50 simulated data sets is shown in the right panel. The horizontal error bar displays 95% of all Bayes factors, the dashed line displays the Bayes factor calculated from the empirical data, and the vertical solid black line displays equal evidence for both hypotheses.

The results (see Figure 13.6) show that the Bayes factors are quite variable. The Bayes factor results differ in that they either provide strong evidence for the alternative model ($BF_{10} > 10$) or moderate evidence for the null model ($BF_{10} < 1/3$), despite the fact that all data sets are simulated from the same distribution. The majority of the simulated data sets support the alternative model with moderate to weak evidence. In other words, this analysis reveals the "dance of the Bayes factors" with simulated repetitions of the same study, similar to the "dance of *p*-values" (Cumming 2014).[4] The variation in these findings demonstrates a very important point that is not widely appreciated: the evidence we get from a particular Bayes factors calculation may not hold up

[4]See `http://daniellakens.blogspot.com/2016/07/dance-of-bayes-factors.html`.

if the same study is replicated. Just obtaining a large Bayes factor alone is not necessarily informative; the variation in the Bayes factor under (hypothetical or actual) repeated sampling needs to be considered as well. For a real-life example, see the replication attempt reported in Schoknecht, Yadav, and Vasishth (2025).

Why are there variations in the Bayes factors amongst the simulated data sets? The difference between the two sentence types' reading times, and thus the experimental effect from which we want to draw conclusions, could vary depending on the noise and uncertainty in the posterior predictive simulations. This is one obvious explanation for why the results could be so different. Plotting the Bayes factors from this simulated data set against the estimated difference in simulated reading times between the two sentence types (as determined by the Bayesian model) therefore provides an interesting perspective. In other words, we take the population-level effects of the Bayesian model and extract the estimated mean difference in reading times at the verb between plural and singular attractor conditions. Then, we plot the Bayes factor as a function of this difference (along with 95% credible intervals).

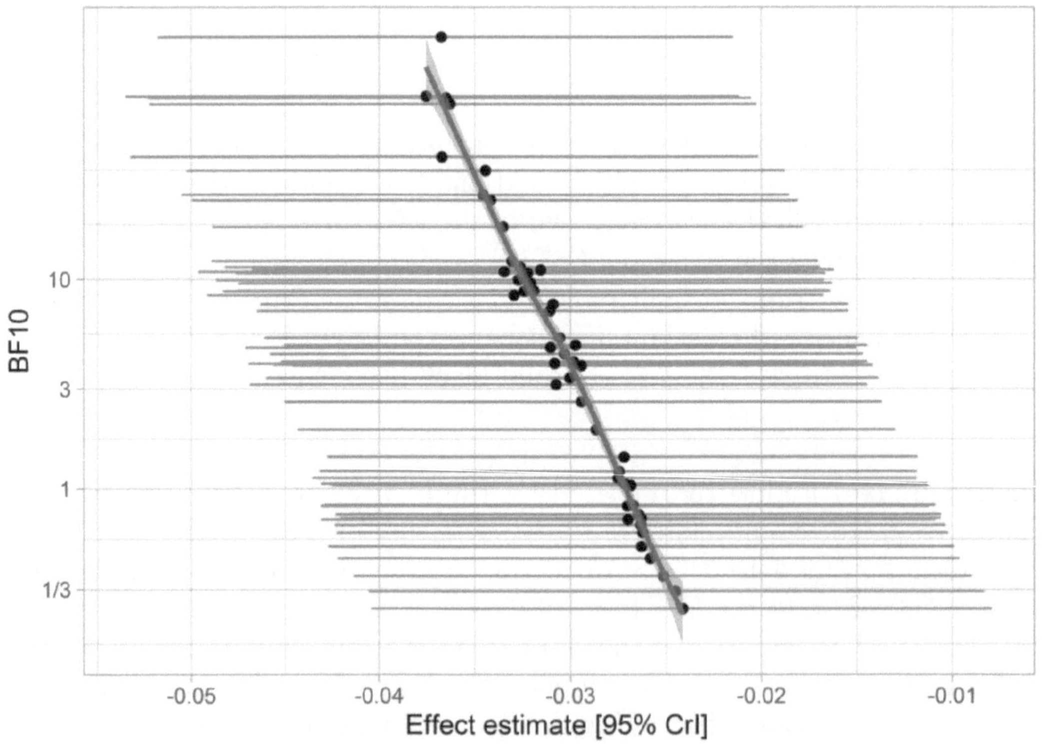

FIGURE 13.7: The Bayes factor (BF10) as a function of the estimate (with 95% credible intervals) of the facilitatory effect of retrieval interference across 50 simulated data sets. The prior is from a meta-analysis.

The findings (illustrated in Figure 13.7) demonstrate that there is quite a lot of variability in the average difference in reading times between experimental conditions amongst the posterior predictive simulations. This suggests that there is little information about the effect of interest in the experimental data and design. Of course, if the data is noisy, Bayes factor analyses based on this simulated data cannot be stable across simulations either. Therefore, as is evident from Figure 13.7, one of the main factors influencing the Bayes factor computations is, in fact, the variation in mean reading times between experimental conditions (other model parameters don't show a close association; see the published version of the arXiv article Schad et al. 2021).

The Bayes factor BF10 in Figure 13.7 increases with the negative difference in reading times, meaning that the more quickly the plural noun condition (in this case, "cabinets" in example sentence 2) is read in comparison to the singular noun condition (i.e., "cabinet"; example sentence 1), the stronger the evidence is in favor of the alternative model. Conversely, when the difference in reading times becomes less negative, that is, when the plural condition (sentence 2) is not read noticeably faster than the singular condition (sentence 1), the Bayes factor BF10 drops to values smaller than 1. Crucially, this behavior arises from the fact that we are utilizing informative priors from the meta-analysis, where the agreement attraction effect's prior mean has a negative value (i.e., a prior mean of −0.03) instead of being centered at a mean of zero. A null model of no effect is therefore more consistent with reading time differences that are less negative or more positive than this prior mean. This also leads to the startling conclusion that, in contrast to the much more variable Bayes factor results, the 95% credible intervals are fairly consistent and do not overlap with zero. This should alarm researchers who exclusively use the 95% credible interval to decide whether an effect is present or not, i.e., to make a discovery claim.

The precise question of whether the data provide more support for the effect size found in the meta-analysis than the absence of any effect is addressed by computing Bayes factors for such a prior with a non-zero mean.

The important lesson to learn from this analysis is that Bayes factors can be quite variable for different data sets assessing the same phenomenon. Individual data sets in the cognitive sciences often do not contain a lot of information about the phenomenon of interest, even when–as is the case here with agreement attraction–the phenomenon is thought to be a relatively robust one. For a more detailed investigation of how Bayes factors can vary with data, in both simulated and real replication studies, we refer the reader to the published version of the arXiv article, and Vasishth, Yadav, et al. (2022).

13.5.3 A cautionary note about Bayes factors

Just like frequentist *p*-values (Wasserstein and Lazar 2016), Bayes factors are easy to misuse and misinterpret, and have the potential to mislead the scientist if used in an automated manner. A recent article (Tendeiro et al. 2023) reviews many of the misuses of Bayes factors analyses in psychology and related areas. As discussed in this chapter, Bayes factors (and Bayesian analysis in general) require a great deal of thought; there is no substitute for sensitivity analyses, and the development of sensible priors. Using default priors and deriving black and white conclusions from Bayes factor analyses is never a good idea.

13.6 Sample size determination using Bayes factors

This section contains text adapted from Vasishth, Yadav, et al. (2022).

When planning a new experiment, it is possible to take what is a fundamentally frequentist approach to work out what sample size one would need in order to cross a certain Bayes factor threshold of evidence.

It may sound surprising to Bayesian modelers that sample size planning is even something to plan for: One of the many advantages of Bayesian modeling is that it is straightforward to plan an experiment without necessarily specifying the sample size in advance (e.g., Spiegelhalter, Abrams, and Myles 2004). Indeed, in our own research, running an experiment until some precision criterion in the posterior distribution is reached (Freedman, Lowe, and Macaskill 1984; Spiegelhalter, Freedman, and Parmar 1994; Kruschke 2014; Kruschke and Liddell 2018) is our method of choice (Jäger et al. 2020; Vasishth et al. 2018; Stone et al. 2023). This approach is possible to implement if one has sufficient financial resources (and time) to keep running an experiment till a particular precision criterion is reached.

However, even when planning a Bayesian analysis, there can be situations where one needs to determine sample size in advance. One important situation where this becomes necessary is when one applies for research funding. In a funding proposal, one obviously has to specify the sample size in advance in order to ask for the necessary funds for conducting the study. Other situations where sample size planning is needed is in the design of clinical trials, the design of replication trials, and when pre-registering experiments and/or preparing registered reports.

There exist good proposals on how to work out sample sizes in advance, specifically in the case of Bayesian analyses. For example, Wang and Gelfand

(2002) aims to ensure that the researcher obtains strong evidence for the effect being estimated.

Here, we unpack the approach taken in Wang and Gelfand (2002). The approach is important because it provides an easy-to-implement workflow for doing sample size calculations using complex hierarchical models of the type we discuss in the present book.

The Wang and Gelfand (2002) approach is as follows. We have adapted the procedure outlined below slightly for our purposes, but the essential ideas are due to these authors.

1. Decide on a distribution for the effect sizes you wish to detect.
2. Choose a criterion that counts as a threshold for a decision. This can be a Bayes factor of, say, 10 (Jeffreys 1939).[5]
3. Then do the following for increasing sample sizes ***n***:
    1. Simulate prior predictive data ***niter*** times (say, ***niter*** = 100) for sample size ***n***; use informative priors (these are referred to as sampling priors in Wang and Gelfand 2002).
    2. Fit the model to the simulated data using regularizing (or principled) priors (these are called fitting priors in Wang and Gelfand 2002), and derive the posterior distribution each time, and compute the Bayes factor using a null model that assumes a zero effect for the parameter of interest.
    3. Display, in one plot, the ***niter*** posterior distributions and the Bayes factors. If the chosen decision criterion is met reasonably well under repeated sampling for a given sample size, choose that sample size.

In Vasishth, Yadav, et al. (2022), we show how this approach can be adapted for the types of models we discuss in the present chapter.

13.7 Summary

Bayes factors are a very important tool in Bayesian data analysis. They allow the researcher to quantify the evidence in favor of certain effects in the data by comparing a full model, which contains a parameter corresponding to the effect of interest, with a null model that does not contain that parameter. We saw that Bayes factor analyses are highly sensitive to priors specified for the

[5]The Bayes factor is just one of many possible performance criteria; see Wang and Gelfand (2002) for some other alternatives.

parameters; this is true both for the parameter corresponding to the effect of interest, but also sometimes for priors relating to other parameters in the model, such as the intercept. It is therefore very important to perform prior predictive checks to select good and plausible priors. Moreover, sensitivity analyses, where Bayes factors are investigated for differing prior assumptions, should be standardly reported in any analysis involving Bayes factors. We studied theoretical aspects of Bayes factors and saw that bridge sampling requires a very large effective sample size in order to obtain stable results for approximate Bayes factors. Therefore, one should always perform a Bayes factor analysis with bridge sampling at least twice to ensure that the results are stable. Bridge sampling comes with no strong guarantees concerning its accuracy, and we saw that marginal simulation-based calibration can be used to evaluate the accuracy of Bayes factor estimates. Finally, we learned that Bayes factors can strongly vary with the data. In the cognitive sciences, the data are–even for relatively robust effects–often not stable due to small effect sizes and limited sample size. Therefore, the resulting Bayes factors can strongly vary across data sets investigating the same phenomenon. As a consequence, only large effect sizes, large sample studies, and/or replication studies can lead to reliable inferences from empirical data in the cognitive sciences.

One topic that was not discussed in detail in this chapter is data aggregation. In repeated measures data, null hypothesis Bayes factor analyses can be performed on the raw data, i.e., without aggregation, by using Bayesian hierarchical models. In an alternative approach, the data are first aggregated by taking the mean per subject and condition, before running null hypothesis Bayes factor analyses on the aggregated data. Inferences (Bayes factors) based on aggregated data can be biased, when either (i) item variability is present in addition to subject variability, or (ii) when the sphericity assumption (inherent in repeated measures ANOVA) is violated (Schad, Nicenboim, and Vasishth 2023). In these cases, aggregated analyses can lead to biased results and should not be used. By contrast, non-aggregated analyses are robust also in these cases and yield accurate Bayes factor estimates.

13.8 Further reading

A detailed explanation on how bridge sampling works can be found in Gronau et al. (2017), and more details about the bridgesampling package can be found in Gronau, Singmann, and Wagenmakers (2020). Wagenmakers et al. (2010) provides a complete tutorial and the mathematical proof of the Savage–Dickey method; also see O'Hagan and Forster (2004) and `https://statproofbook.`

`github.io/P/bf-sddr.html`. Some limitations of the Savage–Dickey approach under certain conditions are discussed in Heck (2019). A comparison between Bayes factor approximation methods is discussed in Oberauer, Musfeld, and Aust (2024). The package `bayestestr` (Makowski, Ben-Shachar, and Lüdecke 2019) can also be used for Bayes factor computations using the Savage–Dickey method. For a Bayes Factor Test calibrated to investigate replication success, see Verhagen and Wagenmakers (2014). A special issue on hierarchical modeling and Bayes factors appears in the journal Computational Brain and Behavior in response to an article by van Doorn et al. (2021). Kruschke and Liddell (2018) discuss alternatives to Bayes factors for hypothesis testing. An argument against null hypothesis testing with Bayes Factors appears in this blog post by Andrew Gelman: `https://statmodeling.stat.columbia.edu/2019/09/10/i-hate-bayes-factors-when-theyre-used-for-null-hypothesis-significance-testing/`. An argument in favor of null hypothesis testing with Bayes Factor as an approximation (but assuming realistic effects) appears in: `https://statmodeling.stat.columbia.edu/2018/03/10/incorporating-bayes-factor-understanding-scientific-information-replication-crisis/`. A visualization of the distinction between Bayes factor and k-fold cross-validation is in a blog post by Fabian Dablander, `https://tinyurl.com/47n5cte4`. Decision theory, which was only mentioned in passing in this chapter, is discussed in Parmigiani and Inoue (2009). Hypothesis testing in its different flavors is discussed in Robert (2022). Alternative ways to test the accuracy of Bayes factors are discussed in Sekulovski, Marsman, and Wagenmakers (2024). When planning studies, Bayes-factor based power calculations can be carried out; an example of such power computations in the context of psycholinguistics, which uses the software tools discussed in the present book, is Vasishth, Yadav, et al. (2022) (also see the references cited there).

14

Cross-validation

A popular way to evaluate and compare models is to investigate their ability to make predictions for "out-of-sample data," that is, to use what we learned from the observed data to predict future or unseen observations. Cross-validation is used to test which of the models under consideration is/are able to learn the most from our data in order to make better predictions. In cognitive science, the goal is only rarely to predict future observations; more often, the researcher wants to compare the performance of different models given some observed data. Nevertheless, there are situations in cognitive science where evaluating the predictive performance of models becomes important.

The objective of cross-validation is to avoid over-optimistic predictions; such over-optimistic predictions would arise if we were to use the data to estimate the parameters of our model, and then use these estimates to predict the same data. That amounts to using the data twice. The basic idea behind cross-validation is that the models are fit with a large subset of the data, the *training set*, and then the fitted models are used to predict a smaller, unseen part of the data, the *held-out set*. In order to treat the entire data set as a held-out set and to evaluate the predictive accuracy using every observation, one successively changes what constitutes the training set and the held-out set by refitting the model and evaluating it in different folds. This ensures that the predictions of the model are tested over the entire data set.

14.1 The expected log predictive density of a model

In order to compare the quality of the posterior predictions of two models, a *utility* function or a *scoring rule* is used (see Gneiting and Raftery 2007 for a review on scoring rules). The logarithmic score rule (Good 1952), shown in Equation (14.1), has been proposed as a reasonable way to assess the posterior predictive distribution of a candidate model $\mathcal{M}_1$ given the data y. This approach is reasonable because it takes into account the uncertainty of the predictions (unlike, for example, using the mean square error). If new observations are

DOI: 10.1201/9780429342646-19

well accounted by the posterior predictive distribution, then the density of the posterior predictive distribution is high and so is its logarithm.

$$u(\mathcal{M}_1, y_{pred}) = \log p(y_{pred}|y, \mathcal{M}_1) \tag{14.1}$$

Unlike the Bayes factor, the prior is absent from Equation (14.1). However, the prior does have a role here: The posterior predictive distribution is based on the posterior distribution $p(\Theta \mid y)$ (where Θ is a vector of all the parameters of the model), which, according to Bayes' rule, depends on both priors and likelihood together. Recall Equation (3.11) in section 3.6, repeated here for convenience:

$$p(y_{pred} \mid y) = \int_{\Theta} p(y_{pred} \mid \Theta) p(\Theta \mid y) \, \mathrm{d}\Theta \tag{14.2}$$

In Equation (14.2), we are implicitly conditioning on the model under consideration:

$$p(y_{pred} \mid y, \mathcal{M}_1) = \int_{\Theta} p(y_{pred} \mid \Theta, \mathcal{M}_1) p(\Theta \mid y, \mathcal{M}_1) \, \mathrm{d}\Theta \tag{14.3}$$

The predicted data, y_{pred}, are unknown to the utility function, so the utility function as presented in Equation (14.1) cannot be evaluated. For this reason, we marginalize over *all possible future data* (calculating $E[\log p(y_{pred}|y, \mathcal{M}_1)]$); this expression is called the expected log predictive density of model $\mathcal{M}_1$:

$$elpd = u(\mathcal{M}_1) = \int_{y_{pred}} p_t(y_{pred}) \log p(y_{pred} \mid y, \mathcal{M}_1) \, \mathrm{d}y_{pred} \tag{14.4}$$

where p_t is the true data generating distribution. If we consider a set of models, the model with the highest $elpd$ is the model with the predictions that are the closest to those of the true data generating process.[1] The intuition behind Equation (14.4) is that we are evaluating the predictive distribution of $\mathcal{M}_1$ over all possible future data weighted by how likely the future data are according to their true distribution. This means that observations that are very likely according to the true model will have a higher weight than unlikely ones.

But we don't know the true data-generating distribution, p_t! If we knew it, we wouldn't be looking for the best model, since we would already know what the best model is.

[1]Maximizing the $elpd$ in Equation (14.4) is also equivalent to minimizing the Kullback–Leibler (KL) divergence from the true data generating distribution $p_t(y_{pred})$ to the *posterior predictive distribution* of the candidate model $\mathcal{M}_1$.

We can use the observed data distribution as a proxy for the true data generating distribution. So instead of weighting the predictive distribution by the true density of all possible future data, we just use the N observations that we have. We can do that because our observations are presumed to be samples from the true distribution of the data: Under this assumption, observations with higher likelihood according to the true distribution of the data will also be obtained more often. This means that instead of integrating, we sum the posterior predictive density of the observations and we give to each observation the same weight; this is valid because observations that are appearing more often in the data should presumably also appear more often in the future (see online section D.3). This quantity is called *log pointwise predictive density* or lpd (Vehtari, Gelman, and Gabry 2017 write this without the $1/N$):

$$lpd = \frac{1}{N} \sum_{n=1}^{N} \log p(y_n | y, \mathcal{M}_1) \tag{14.5}$$

The lpd is an overestimate of $elpd$ for actual future data, because the parameters of the posterior predictive distribution are estimated with the same observations that we are considering out-of-sample. Incidentally, this also explains why posterior predictive checks are generally optimistic, and why good fits cannot be taken too seriously. But they do serve the purpose of identifying very strong model misspecifications.[2]

However, we can obtain a more conservative estimate of the predictive performance of a model using cross-validation (Geisser and Eddy 1979). This is explained next. (As an aside, we mention here that there are also other alternatives to cross-validation; these are presented in Vehtari and Ojanen 2012).

14.2 K-fold and leave-one-out cross-validation

The basic idea of K-fold cross-validation (K-fold-CV) is to split the N observations of our data into K subsets, such that each subset is used as a validation (or held-out) set, D_k, while the remaining set (the training set), D_{-k} is used for estimating the parameters and approximating p_t, the true data distribution. The leave-one-out cross-validation (LOO-CV) method represents a special case of K-fold-CV where, $K = N$ and the training set only excludes one observation. For the general case, K-fold-CV, we estimate $elpd$ as follows.

[2]The double use of the data is also a problem when one relies on information criteria like the Akaike Information Criterion (AIC) or the Bayesian Information Criterion (BIC).

$$\widehat{elpd} = \frac{1}{N} \sum_{n=1}^{N} \log p(y_n | D_{-k}, \mathcal{M}_1) \text{ with } y_n \in D_k \tag{14.6}$$

In Equation (14.6), each observation, y_n, belongs to a certain "validation" fold, D_k, and the predictive accuracy of y_n is evaluated based on a posterior predictive model trained on the set, D_{-k}, which is the complete data set excluding the validation fold that contains the n-th observation. This means that the posterior predictive distribution used to evaluate y_n was derived without having information from that y_n-th observation (in other words, the model was trained without that observation in the subset of the data, D_{-k}).

In K-fold-CV, several observations are held out in the same (validation) fold. This means that the held-out observations are split among K folds, and D_{-k}, the data used to derive the posterior predictive distribution, contain only a proportion of the observations; this proportion is $(1-1/K)$. By contrast, in leave-one-out cross-validation, the held-out data set includes only one observation. That is, D_{-k} contains the entire data set except for one data point, y_n, with $n = 1, \dots, N$, that is D_{-n}. Online section D.4 explains the algorithm in detail.

For the general case of K-fold-CV, Vehtari, Gelman, and Gabry (2017) define the expected log *pointwise* predictive density of the observation y_n as follows:

$$\widehat{elpd}_n = \log p(y_n | D_{-k}, \mathcal{M}_1) \text{ with } y_n \in D_k \tag{14.7}$$

This quantity indicates the predictive accuracy of the model $\mathcal{M}_1$ for a single observation, and it is reported in the package `loo`, which also `brms` relies on. In addition, the `loo` package uses the sum of the expected log pointwise predictive density, $\sum \widehat{elpd}_n$ (Equation (14.6) without $\frac{1}{N}$) as a measure of predictive accuracy (this is referred to as `elpd_loo` or `elpd_kfold` in the packages `loo` and `brms`). For model comparison, the difference between the $\sum \widehat{elpd}_n$ of competing models can be computed, including the standard deviation of the sampling distribution of the difference. It's important to notice that we are calculating an approximation to the expectation $elpd$ that we actually want to compute, and thus we always need to consider its inherent randomness (due to having a limited amount of data), which is quantified by the standard error (Vehtari et al. 2019).

Unlike what is common with information criterion methods (such as the Akaike Information Criterion, AIC, and the Deviance Information Criterion, DIC), a higher $\widehat{elpd}$ means higher predictive accuracy. An alternative to using $\widehat{elpd}$ is to examine $-2 \times \widehat{elpd}$, which is equivalent to deviance, and is called the LOO Information Criterion (LOOIC) (see section 22 of Vehtari 2022).

The approximation to the true data generating distribution is worse when fewer observations are used, and thus ideally we would set $K = N$, and thus compute LOO-CV rather than K-fold-CV. The main advantage of LOO-CV is its robustness, since the training set is as similar as possible to the observed data, and the same observations are never used simultaneously for training and evaluating the predictions. A major disadvantage is the computational burden (Vehtari and Ojanen 2012), since we need to fit a model as many times as the number of observations. The package `loo` provides an approximation to LOO-CV, Pareto smoothed importance sampling leave-one-out (PSIS-LOO; Vehtari et al. 2024; Vehtari, Gelman, and Gabry 2017) which, as we show next, is relatively straightforward to use in `brms` and in Stan models (see `https://mc-stan.org/loo/articles/loo2-with-rstan.html`). However, in some cases, its estimates can be unreliable; this is indicated by the estimated shape parameter $\hat{k}$ of the generalized Pareto distribution (`https://mc-stan.org/loo/reference/pareto-k-diagnostic.html`). The value $\hat{k}$ (which is unrelated to the k in K-fold-CV) estimates how far an approximated individual leave-one-out distribution is from the full distribution. If leaving out an observation substantially alters the posterior, then importance sampling cannot provide a reliable estimate. Very high $\hat{k}$ values often indicate model misspecification, outliers, or mistakes in data processing. The threshold for $\hat{k}$ depends on the sample size (Vehtari et al. 2024) and it is reported by the `loo()` function. In cases where one or several pointwise predictive density have associated large $\hat{k}$, either (i) the problematic predictions can be refitted with exact LOO-CV, (ii) one can try some additional computations using the existing posterior sample based on the moment matching approximation (see `https://mc-stan.org/loo/articles/loo2-moment-matching.html` and Paananen et al. 2021), or (iii) one can abandon PSIS-LOO-CV and use K-fold-CV, with K typically set to 10.

One of the main disadvantages of cross-validation (in comparison with Bayes factor at least) is that the numerical difference in predictive accuracy is hard to interpret. As a rule of thumb, it has been suggested that if the `elpd` difference (`elpd_diff` in the `loo` package) is less than 4, the difference is small, and if it is larger than 4, one should compare that difference to its standard error (`se_diff`) (see section 16 of Vehtari 2022).

14.3 Testing the N400 effect using cross-validation

As in section 13.2 with the Bayes factor, let us revisit section 5.2, where the effect of cloze probability on the N400 average signal was estimated. Consider two models here, a model that includes the effect of cloze probability, such as `fit_N400_sih` from section 5.2.5, and a null model.

Verify the model that was fit; this is a hierarchical model that includes an effect of cloze probability:

```
formula(fit_N400_sih)
```

```
## n400 ~ c_cloze + (c_cloze | subj) + (c_cloze | item)
```

In contrast to the situation with Bayes factor, priors are less critical for cross-validation. Priors are only important in cross-validation to the extent that they affect parameter estimation: As discussed earlier, very narrow priors can bias the posterior; and unrealistically wide priors can lead to convergence problems. The number of samples is also less critical than with Bayes factor; most of the uncertainty in the estimates of the $\widehat{elpd}$ is due to the number of observations. However, a very small number of samples can affect the $\widehat{elpd}$ because the posterior estimation will be affected by the small sample size. We update our previous formula to define a null model as follows:

```
fit_N400_sih_null <- update(fit_N400_sih, ~ . - c_cloze)
```

14.3.1 Cross-validation with PSIS-LOO

Estimating $elpd$ using PSIS-LOO is very straightforward with `brms`, which uses the package `loo` as a back-end. There is no need to refit the model, and `loo` takes care of applying the PSIS approximation to derive estimates and standard errors.

```
(loo_sih <- loo(fit_N400_sih))
```

```
##
## Computed from 4000 by 2863 log-likelihood matrix.
##
##          Estimate   SE
## elpd_loo -11092.5 46.7
## p_loo        81.1  2.7
## looic     22185.0 93.4
## ------
## MCSE of elpd_loo is 0.1.
## MCSE and ESS estimates assume MCMC draws (r_eff in [0.5, 2.1]).
##
```

```
## All Pareto k estimates are good (k < 0.7).
## See help('pareto-k-diagnostic') for details.
```

```
(loo_sih_null <- loo(fit_N400_sih_null))
```

```
##
## Computed from 4000 by 2863 log-likelihood matrix.
##
##           Estimate   SE
## elpd_loo -11095.1 46.5
## p_loo        89.2  3.0
## looic     22190.3 93.0
## ------
## MCSE of elpd_loo is 0.2.
## MCSE and ESS estimates assume MCMC draws (r_eff in [0.4, 1.7]).
##
## All Pareto k estimates are good (k < 0.7).
## See help('pareto-k-diagnostic') for details.
```

The function `loo` reports three quantities with their standard error:

1. `elpd_loo` is the sum of pointwise predictive accuracy (a larger, less negative number indicates better predictions).
2. `p_loo` is an estimate of effective complexity of the model; asymptotically and under certain regularity conditions, `p_loo` can be interpreted as the effective number of parameters. If `p_loo` is larger than the number of data points or parameters, this may indicate a severe model misspecification.
3. `looic` is simply `-2*elpd_loo`, the *elpd* on the deviance scale. This is called the information criterion, and is mainly provided for historical reasons: other information criteria like the AIC (Akaike Information Criterion) and the DIC (Deviance Information Criterion) are commonly used in model selection (Venables and Ripley 2002; Lunn et al. 2012).

It's important to bear in mind that the PSIS-LOO approximation to LOO can only be trusted if there are no large Pareto ($\hat{k}$) estimates. To compare the models, take a look at the difference between `elpd_loo` and the standard error of that difference:

```
loo_compare(loo_sih, loo_sih_null)
```

```
##                   elpd_diff se_diff
## fit_N400_sih       0.0       0.0
## fit_N400_sih_null -2.6       2.5
```

Although the model that includes cloze probability as a predictor has higher predictive accuracy, the difference is smaller than 4 and it's smaller than two SE. This means that from the perspective of LOO-CV, both models are almost indistinguishable! In fact, the same will happen if the model is compared using logarithmic predictability to the linear or null model; see online exercise G.14.1.

It is also possible to check whether the alternative model is making good predictions for some range of values by examining the difference in pointwise predictive accuracy as a function of, for example, cloze probability. In the following plot, we subtract the predictive accuracy of the alternative model from the accuracy of the null model; larger differences can be interpreted as an advantage for the alternative model. However, as far as posterior predictive accuracy goes, both models are quite similar. Figure 14.1 shows that the difference in predictive accuracy is symmetrical with respect to the zero; as we go further from the mean cloze (which is around 0.5), the differences in predictions are larger but they span positive and negative values.

The following code stores the difference in predictive accuracy of the models in a variable and plots it in Figure 14.1.

```
df_eeg <- mutate(df_eeg,
                 diff_elpd = loo_sih$pointwise[, "elpd_loo"] -
                   loo_sih_null$pointwise[, "elpd_loo"])
ggplot(df_eeg, aes(x = cloze, y = diff_elpd)) +
  geom_point(alpha = .4, position = position_jitter(w = .001, h = 0))
```

The expectation was that, similar to the Bayes factor, cross-validation techniques will also show that a model that includes cloze probability as a predictor is superior to a model without it. It is unsettling that the above result does not show this: the effect of cloze probability on the N400 has been replicated in numerous studies.

Before discussing why a large difference was not observed, let us check what K-fold-CV yields.

14.3.2 Cross-validation with K-fold

Estimating *elpd* using k-fold-CV has the advantage of omitting one layer of approximations: the *elpd* based on PSIS-LOO-CV is an approximation of the

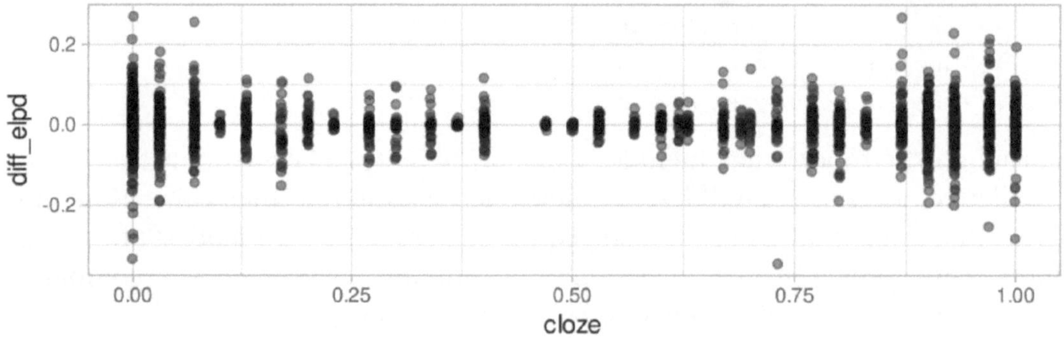

FIGURE 14.1: The difference in predictive accuracy between a model including the effect of cloze and a null model. A larger (more positive) difference indicates an advantage for the model that includes the effect of cloze.

$elpd$ based on *exact* LOO-CV (and we saw how any cross-validation approach gave us an approximation to the true $elpd$). This means that we don't need to worry about $\hat{k}$. However, K-fold also uses a reduced training set in comparison with LOO, worsening the approximation to the true generating process p_t.

Before dividing our data into folds, we need to think about the way the data will be split: The data could be split randomly, but this approach would have the risk that in some of the training sets, observations from a particular subject, for example, could be completely absent. Such an omission will lead to large differences in predictive accuracy between folds. This situation can be avoided by using stratification: split the observations into groups, ensuring that relative category frequencies are approximately preserved in the training and held-out data. This stratification can be achieved by using the `kfold()` function, available in the package `brms`, by setting `folds = "stratified"` and `group = "subj"`, by default `K` is set to 10, but that can be changed. Additionally, since we know the models converge, we use only one chain for each model and parallelize the fitting procedure with the `plan(multisession)` function from the `future` package, returning to the default setting with `plan(sequential)`.

```
library(future)
plan(multisession, workers = 4)
kfold_sih <- kfold(fit_N400_sih,
                   folds = "stratified",
                   group = "subj",
                   chains = 1)
kfold_sih_null <- kfold(fit_N400_sih_null,
                        folds = "stratified",
```

```
                              group = "subj")
plan(sequential)
```

Running K-fold CV takes some time since each model is re-fit K times. Inspect the *elpd* values:

```
kfold_sih
```

```
##
## Based on 10-fold cross-validation.
##
##             Estimate   SE
## elpd_kfold -11093.9 46.4
## p_kfold        82.5  3.3
## kfoldic     22187.8 92.7
```

```
kfold_sih_null
```

```
##
## Based on 10-fold cross-validation.
##
##             Estimate   SE
## elpd_kfold -11100.8 46.5
## p_kfold        94.8  3.8
## kfoldic     22201.6 93.0
```

Compare the two models using `loo_compare` (this function is used for both PSIS-LOO-CV and K-fold-CV):

```
loo_compare(kfold_sih, kfold_sih_null)
```

```
##                   elpd_diff se_diff
## fit_N400_sih        0.0       0.0
## fit_N400_sih_null -6.9       4.2
```

Because the $\widehat{elpd}$ values depend on how the folds were formed, K-fold CV introduces an additional element of randomness compared to (PSIS-)LOO-CV. To verify the robustness of the results, one can rerun the `kfold()` function so that different random configurations of the folds are selected. In this case, however, the results with K-fold-CV and PSIS-LOO-CV are quite similar: The two models can't really be distinguished.

14.3.3 Leave-one-group-out cross-validation

An alternative to splitting the observations randomly using stratification is to treat naturally occurring clusters as folds; this is leave-one-group-out cross-validation (LOGO-CV). The output of LOGO-CV tells us about the capacity of the models for generalizing to unseen clusters. LOGO-CV is implemented next using subjects as the group of interest.

In general, there is some tension between the arguments for K-fold-CV with folds stratified by subject, which ensures data from all subjects in each fold, and for LOGO-CV, which isolates data from different subjects. However, the two approaches have different goals and merits. The first approach answers how well the model generalizes to unseen data from this particular set of subjects. The second approach addresses how well the model generalizes to the behavior of unseen subjects if we were to collect more data.

```
plan(multisession, workers = 4)
logo_sih <- kfold(fit_N400_sih,
                  group = "subj",
                  chains = 1)
logo_sih_null <- kfold(fit_N400_sih_null,
                       group = "subj",
                       chains = 1)
plan(sequential)
```

Running LOGO-CV with subjects takes even longer since each model is refit as many times as there are subjects, in this case 37 times. We can now inspect the *elpd* estimates and evaluate which model generalizes better to unseen subjects.

Compare the models using `loo_compare`.

```
loo_compare(logo_sih, logo_sih_null)
```

```
##                   elpd_diff se_diff
## fit_N400_sih       0.0       0.0
## fit_N400_sih_null -4.0       3.3
```

As before, and as with PSIS-LOO-CV and with K-fold-CV, the two models can't be distinguished. Even though the full model and the null model presented earlier make different predictions, the difference in predictive accuracy is not substantial compared to the variability in the data. In these situations, cross-validation is not very useful for distinguishing very small effect sizes from zero effect sizes; see section 14.5.

14.4 Comparing different likelihoods with cross-validation

One can also compare two models with different likelihoods. Section 3.7.2 in chapter 3 showed how a log-normal distribution was a more appropriate likelihood than a normal distribution for response times data. This was because response times are bounded by zero and right skewed, unlike the symmetrical normal distribution. Let's use PSIS-LOO-CV to compare the predictive accuracy of the Stroop model from section 5.3 in chapter 5, which assumed a log-normal likelihood to fit response times of correct responses with a similar model which assumes a normal likelihood. We can only compare models fit to the same data; we can't compare models fit to different dependent variables (e.g., one raw dependent variable and one log-transformed).

Load the data from `bcogsci`, create a sum-coded predictor (see chapter 6 for more details), and fit the model as in section 5.3.

```
data("df_stroop")
df_stroop <- df_stroop %>%
  mutate(c_cond = if_else(condition == "Incongruent", 1, -1))
```

```
fit_stroop_log <- brm(RT ~ c_cond + (c_cond | subj),
  family = lognormal(),
  prior = c(prior(normal(6, 1.5), class = Intercept),
            prior(normal(0, 1), class = b),
            prior(normal(0, 1), class = sigma),
            prior(normal(0, 1), class = sd),
            prior(lkj(2), class = cor)),
  data = df_stroop)
```

Calculate the $elpd_{loo}$ for the original model with the log-normal likelihood:

```
loo_stroop_log <- loo(fit_stroop_log)
```

```
loo_stroop_log
```

```
##
## Computed from 4000 by 3058 log-likelihood matrix.
##
##           Estimate    SE
## elpd_loo -19859.4  93.6
## p_loo        61.1   4.1
## looic     39718.8 187.2
## ------
## MCSE of elpd_loo is 0.1.
## MCSE and ESS estimates assume MCMC draws (r_eff in [0.6, 2.4]).
##
## All Pareto k estimates are good (k < 0.7).
## See help('pareto-k-diagnostic') for details.
```

The summary shows that $\hat{k}$ estimates are ok.

Now fit a similar model which assumes that the likelihood is a normal distribution. It's important now to change the priors since they are on a different scale (namely, in milliseconds). We choose reasonable but wide priors. A sensitivity analysis can be done if we are unsure about the priors. However, unlike what happened with the Bayes factor in chapter 13, the priors are going to affect cross-validation based model comparison only as far as they have a noticeable effect on the posterior distribution.

```
fit_stroop_normal <- brm(RT ~ c_cond + (c_cond | subj),
  family = gaussian(),
  prior = c(prior(normal(400, 600), class = Intercept),
            prior(normal(0, 100), class = b),
            prior(normal(0, 300), class = sigma),
            prior(normal(0, 300), class = sd),
            prior(lkj(2), class = cor)),
  data = df_stroop)
```

The estimated *elpd* based on PSIS-LOO-CV has several large $\hat{k}$ values.

```
loo_stroop_normal <- loo(fit_stroop_normal)
```

```
loo_stroop_normal
```

```
##
## Computed from 4000 by 3058 log-likelihood matrix.
##
```

```
##           Estimate    SE
## elpd_loo -21596.5 486.6
## p_loo       121.3  70.9
## looic     43192.9 973.1
## ------
## MCSE of elpd_loo is NA.
## MCSE and ESS estimates assume MCMC draws (r_eff in [0.5, 2.5]).
##
## Pareto k diagnostic values:
##                          Count Pct.    Min. ESS
## (-Inf, 0.7]   (good)     3055  99.9%   83
##    (0.7, 1]   (bad)         2   0.1%   <NA>
##    (1, Inf)   (very bad)    1   0.0%   <NA>
## See help('pareto-k-diagnostic') for details.
```

We use exact LOO (rather than its approximation) for the problematic observations. By setting `reloo = TRUE`, the 3 problematic observations with $\hat{k}$ values over 0.7 are re-fit using exact LOO-CV.[3]

```
loo_stroop_normal <- loo(fit_stroop_normal, reloo = TRUE)
```

```
loo_stroop_normal
```

```
##
## Computed from 4000 by 3058 log-likelihood matrix.
##
##           Estimate     SE
## elpd_loo -21700.4  587.3
## p_loo       225.1  174.7
## looic     43400.7 1174.6
## ------
## MCSE of elpd_loo is NaN.
## MCSE and ESS estimates assume MCMC draws (r_eff in [0.5, 2.5]).
##
## All Pareto k estimates are good (k < 0.7).
## See help('pareto-k-diagnostic') for details.
```

Next, compare the models.

[3]An alternative approach is to use the model matching approximation for problematic observations (Paananen et al. 2021), by setting `moment_match = TRUE` in the `loo()` call (and we also need to fit the model with `save_pars = save_pars(all = TRUE)`). In this particular case, this approximation won't solve our problem.

```
loo_compare(loo_stroop_log, loo_stroop_normal)
```

```
##                   elpd_diff se_diff
## fit_stroop_log        0.0       0.0
## fit_stroop_normal -1840.9     531.6
```

Here, cross-validation shows a clear advantage for the model with the log-normal likelihood. Figure 14.2(a) visualizes the pointwise predictive accuracy.

```
df_stroop <- df_stroop %>%
  mutate(diff_elpd = loo_stroop_log$pointwise[, "elpd_loo"] -
           loo_stroop_normal$pointwise[, "elpd_loo"])
ggplot(df_stroop, aes(x = RT, y = diff_elpd)) +
  geom_point(alpha = .4) + xlab("RC (ms)")
```

Figure 14.2(a) shows that at first glance, the advantage of the log-normal likelihood seems to lie in being able to capture extremely slow observations. Figure 14.2(b) zooms in to visualize the pointwise predictive accuracy for observations with response times smaller than two seconds.

```
ggplot(df_stroop, aes(x = RT, y = diff_elpd)) +
  geom_point(alpha = .3) +
  xlab("RC (ms)") +
  geom_hline(yintercept = 0, linetype = "dashed") +
  coord_cartesian(xlim = c(0, 2000), ylim = c(-10, 10))
```

Figure 14.2(b) suggests that the advantage of the log-normal likelihood seems to lie in being able to account for most of the observations in the data set, which occur around the 500 ms.

14.5 Issues with cross-validation

Sivula, Magnusson, and Vehtari (2020) analyzed the behavior of the uncertainty estimate of the ***elpd*** in typical situations. Although they focus on LOO-CV, the consequences are the same for K-fold-CV (and cross-validation in a non-Bayesian context). Sivula, Magnusson, and Vehtari (2020) identified three cases where the uncertainty estimates can perform badly:

1. The models make very similar predictions.

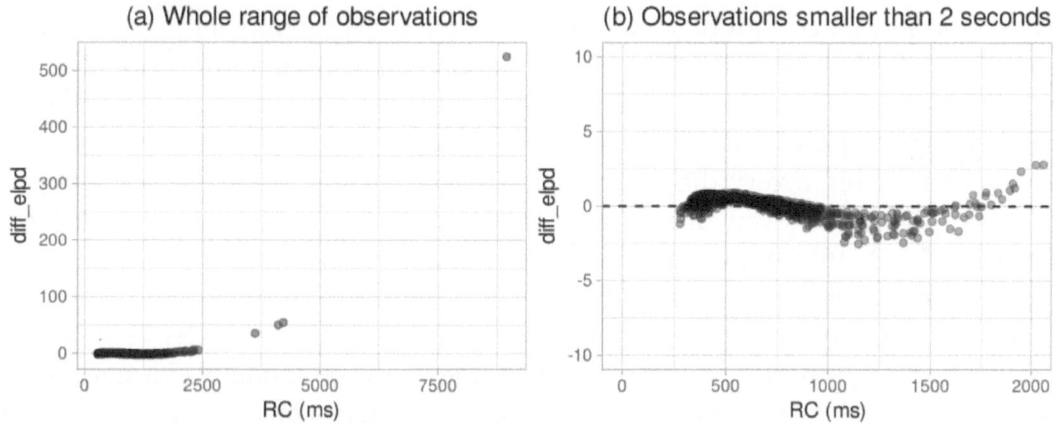

FIGURE 14.2: (a) Difference in predictive accuracy between Stroop models with log-normal and normal likelihoods, where a more positive value favors the log-normal model. (b) Difference in predictive accuracy between the same models, focusing on observations under two seconds, with positive values again favoring the log-normal model.

2. The number of observations is small.
3. The models are misspecified with outliers (influential extreme values) in the data.

When the models make similar predictions (as it is the case with nested models in earlier model comparisons), and when there is not much difference in the predictive performance of the models, the uncertainty estimates will behave badly. In these situations, cross-validation is not very useful for separating very small effect sizes from zero effect sizes. In addition, small differences in the predictive performance cannot reliably be detected by cross-validation if the number of observations is small. However, if the predictions are very similar, Sivula, Magnusson, and Vehtari (2020) show that the same problems persist even with a larger data set.

One of the issues that cross-validation methods face when they are used to compare nested models lies in the way that the exact ***elpd*** is approximated: In cross-validation approximations, out-of-sample observations are used, which are not part of the model that was fit. Every time the predictive accuracy of an observation is evaluated, modeling assumptions are ignored. One of the weaknesses of cross-validation is the high variance in the approximation of the integral over the unknown true data distribution, p_t (section 4 of Vehtari and Ojanen 2012).

Cross-validation methods are sometimes criticized because when a lot of data are available, they will give undue preference to the complex model in comparison to a true simpler model (Gronau and Wagenmakers 2018). This might be true for toy examples using simulated data, where it is possible to obtain essentially unlimited observations, and where a model that is known to be wrong is compared with the true model.[4] However, the problems that occur in practice are often very different: This is because the true model is unknown and very likely not even under consideration in the model comparison (see Navarro 2019). In our experience, in practical settings we are very far from the asymptotic behavior of cross-validation whereby it gives undue preference to a more complex model in comparison to a true simpler model. The main weakness of cross-validation is that, as mentioned above, modeling assumptions are ignored, which prevents it from selecting a more complex model rather than a simple one when there is only a modest gain in predictive fit (Vehtari et al. 2019).

An alternative to the cross-validation approach discussed here for nested models is the projection predictive method (Piironen, Paasiniemi, and Vehtari 2020). However, this approach (which is less general since it is valid only for generalized linear models) has a somewhat different objective. In the projection predictive method, we first build the most complete predictive model, the *reference model*, and then we look for a simpler model that gives similar predictions as the reference model. The idea is that for a given complexity (number of predictors), the model with the smallest predictive discrepancy to the reference model should be selected. See `https://github.com/stan-dev/projpred` for an implementation of this approach. Thus this approach focuses on model simplification rather than on model comparison.

For models that are badly misspecified, the bias in the uncertainty makes their comparison unreliable as well. In this case, posterior predictive checks and possible model refinements are worth considering before carrying out model comparison.

If there is a large number of observations and the models under consideration are different enough from each other, the differences in predictive accuracy will dwarf the variance in the estimate of $elpd$, and cross-validation can be very useful (see also Piironen and Vehtari 2017). An example of this situation appeared in section 14.4. When models are very different, one advantage of cross-validation methods in comparison with the Bayes factor is that the selection of priors is less critical in cross-validation. It is sometimes hard to decide on priors that encode our knowledge for one model, and this difficulty is

[4]If the true model is under consideration among the models being compared, we are under an M_{closed} scenario. However, this is rarely realistic. The most common case is an M_{open} scenario (Bernardo and Smith 2009), where the true model is not included in the set of models being compared.

exacerbated when we want to assign comparable prior information to models with a different number of parameters that might be on a different scale. Given that cross-validation methods are less sensitive to prior specification, different models can be compared on the same footing. Another difficulty with Bayes factors' dependency on priors is that different researchers may disagree on the priors, and selecting an uninformative prior to encompass these different views as a middle ground is not a neutral option for the Bayes factor, while it will barely affect cross-validation. See Nicenboim and Vasishth (2018) and Yadav et al. (2023) for examples from psycholinguistics where K-fold-CV does help in distinguishing between models.

14.6 Cross-validation in Stan

PSIS-LOO-CV and K-fold-CV can also be used with our Stan models, but care has to be taken to store the appropriate log-likelihood in the `generated quantities` block.[5]

14.6.1 PSIS-LOO-CV in Stan

As explained earlier, PSIS-LOO (as implemented in the package `loo`) approximates the likelihood of the held-out data based on the observed data: it's faster (because only one model is fit), and it only requires a minimal modification of the Stan code we need to fit a model. By default, Stan only saves the sum of the log likelihood of each observation (in the parameter `lp__`). The log-likelihood of each observation needs to be stored in the generated quantities block.

We revisit the model implemented in section 8.4.2, which was evaluated using the Bayes factor in section 13.4. Now, we want to compare the predictive performance of a model that assumes an effect of attentional load on pupil size against a similar model that assumes no effect. To do this, we assume the following likelihood:

$$p\_size_n \sim Normal(\alpha + c\_load_n \cdot \beta_1 + c\_trial \cdot \beta_2 + c\_load \cdot c\_trial \cdot \beta_3, \sigma) \quad (14.8)$$

Define priors for all the β's as before:

[5]This means the sampling notation should not be used in this block (see online section B.2).

$$
\begin{aligned}
\alpha &\sim Normal(1000, 500) \\
\beta_{\{1,2,3\}} &\sim Normal(0, 100) \\
\sigma &\sim Normal_{+}(0, 1000)
\end{aligned} \tag{14.9}
$$

Prepare the data as in section 8.4.2:

```
df_pupil <- df_pupil %>%
  mutate(c_load = load - mean(load),
         c_trial = trial - mean(trial))
ls_pupil <- list(c_load = df_pupil$c_load,
                 c_trial= df_pupil$c_trial,
                 p_size = df_pupil$p_size,
                 N = nrow(df_pupil))
```

Add a `generated quantities` block to the model shown below. (It is also possible to run this block in a stand-alone file with the `rstan` function `gqs()`). If the variable name `log_lik` is used in the Stan code, the `loo` package will know where to find the log likelihood of the observations.

Code the effects as `beta1`, `beta2`, `beta3` to more easily compare the model with the one used in the BF chapter, but in this case we could have used a vector or an array instead. This is the model `pupil_cv.stan` shown below:

```
data {
  int<lower = 1> N;
  vector[N] c_load;
  vector[N] c_trial;
  vector[N] p_size;
}
parameters {
  real alpha;
  real beta1;
  real beta2;
  real beta3;
  real<lower = 0> sigma;
}
model {
  // priors including all constants
  target += normal_lpdf(alpha | 1000, 500);
  target += normal_lpdf(beta1 | 0, 100);
  target += normal_lpdf(beta2 | 0, 100);
  target += normal_lpdf(beta3 | 0, 100);
```

```
  target += normal_lpdf(sigma | 0, 1000)
    - normal_lccdf(0 | 0, 1000);
  target += normal_lpdf(p_size | alpha + c_load * beta1 +
                        c_trial * beta2 +
                        c_load .* c_trial * beta3, sigma);

}
generated quantities{
  array[N] real log_lik;
  for (n in 1:N){
    log_lik[n] = normal_lpdf(p_size[n] | alpha + c_load[n] * beta1 +
                             c_trial[n] * beta2 +
                             c_load[n] * c_trial[n] * beta3,
                             sigma);

  }
}
```

For the null model, just omit the term with `beta1` in both the model block and the generated quantities block. This is the model `pupil_null.stan` shown below:

```
data {
  int<lower = 1> N;
  vector[N] c_load;
  vector[N] c_trial;
  vector[N] p_size;
}
parameters {
  real alpha;
  real beta2;
  real beta3;
  real<lower = 0> sigma;
}
model {
  target += normal_lpdf(alpha | 1000, 500);
  target += normal_lpdf(beta2 | 0, 100);
  target += normal_lpdf(beta3 | 0, 100);
  target += normal_lpdf(sigma | 0, 1000)
    - normal_lccdf(0 | 0, 1000);
  target += normal_lpdf(p_size | alpha + c_trial * beta2 +
                                 c_load .* c_trial * beta3, sigma);
}
generated quantities{
```

```
  array[N] real log_lik;
  for (n in 1:N){
    log_lik[n] = normal_lpdf(p_size[n] | alpha + c_trial[n] * beta2 +
                                 c_load[n] * c_trial[n] * beta3, sigma);
  }
}
```

The models can be found in the `bcogsci` package:

```
pupil_cv <- system.file("stan_models",
                        "pupil_cv.stan",
                        package = "bcogsci")
pupil_null <- system.file("stan_models",
                          "pupil_null.stan",
                          package = "bcogsci")
```

Fit the models:

```
fit_pupil_int_ll <- stan(file = pupil_cv,
                         data = ls_pupil)
fit_pupil_int_null_ll <- stan(file = pupil_null,
                              data = ls_pupil)
```

Show summary of predictive accuracy of the models using the function `loo`.

```
(loo_int <- loo(fit_pupil_int_ll))
```

```
##
## Computed from 4000 by 41 log-likelihood matrix.
##
##          Estimate   SE
## elpd_loo   -251.1  5.3
## p_loo         5.3  1.5
## looic       502.2 10.5
## ------
## MCSE of elpd_loo is 0.1.
## MCSE and ESS estimates assume MCMC draws (r_eff in [0.5, 1.4]).
##
## All Pareto k estimates are good (k < 0.7).
## See help('pareto-k-diagnostic') for details.
```

```
(loo_null <- loo(fit_pupil_int_null_ll))
```

```
##
## Computed from 4000 by 41 log-likelihood matrix.
##
##          Estimate  SE
## elpd_loo   -255.3 4.6
## p_loo         4.2 1.1
## looic       510.5 9.1
## ------
## MCSE of elpd_loo is 0.0.
## MCSE and ESS estimates assume MCMC draws (r_eff in [0.7, 1.1]).
##
## All Pareto k estimates are good (k < 0.7).
## See help('pareto-k-diagnostic') for details.
```

```
loo_compare(loo_int, loo_null)
```

```
##        elpd_diff se_diff
## model1  0.0       0.0
## model2 -4.1       3.4
```

As it happened with the cloze probability effect in the previous section, we cannot decide which model has better predictive accuracy according to PSIS-LOO.

14.6.2 K-fold-CV in Stan

To use K-fold-CV (or LOGO-CV) in Stan (as opposed to PSIS-LOO), we need to be careful to store the log-likelihood of the *held-out data*, since we evaluate our model with only this subset of the data. The following example closely follows the vignette `https://cran.r-project.org/web/packages/loo/vignettes/loo2-elpd.html`.

The steps taken are as follows:

1. Split the data in 10 folds.

Since there is only one subject, we don't need to stratify (using `kfold_split_stratified()`), and we use `kfold_split_random()` from the `loo` package.

```
df_pupil$fold <- kfold_split_random(K = 10, N = nrow(df_pupil))
# Show number of obs for each fold:
df_pupil %>%
  group_by(fold) %>%
  count() %>%
  print(n=10)
```

```
## # A tibble: 10 x 2
## # Groups:   fold [10]
##     fold     n
##    <int> <int>
##  1     1     4
##  2     2     4
##  3     3     4
##  4     4     4
##  5     5     4
##  6     6     4
##  7     7     4
##  8     8     4
##  9     9     4
## 10    10     5
```

2. Fit and extract the log pointwise predictive densities for each fold.

Compile the alternative and the null models first with `stan_model()`, and prepare two matrices to store the predictive densities from the held out data. Each matrix has as many rows as post-warmup iterations in the models fit above (1500×4), and as many columns as observations in the data set.

```
pupil_stanmodel <- stan_model(pupil_cv)
pupil_null_stanmodel <- stan_model(pupil_null)
log_pd_kfold <- matrix(nrow = 6000, ncol = nrow(df_pupil))
log_pd_null_kfold <- matrix(nrow = 6000, ncol = nrow(df_pupil))
```

Next, loop over the 10 folds. Each loop carries out the following steps. First, fit each model (i.e., the alternative and null model) to all the observations except the ones belonging to the held-out fold using `sampling()`. As before, since we know the models converge, we use only one chain for each model. The function `sampling()` uses the already-compiled models. Second, compute the log pointwise predictive densities for the held-out fold with `gqs()`. This function produces `generated quantities` based on samples from a posterior (in the `draw` argument) and ignores all the blocks except `generated quantities`.[6] Finally, store the predictive density for the observations of the held-out fold in a matrix by extracting the log likelihood of the held-out data. The output of this loop is a matrix of the log pointwise predictive densities of all the observations.

```
# Loop over the folds
for(k in 1:10){
  # Training set for k
  df_pupil_train <- df_pupil %>%
    filter(fold != k)
  ls_pupil_train <- list(c_load = df_pupil_train$c_load,
                         c_trial= df_pupil_train$c_trial,
                         p_size = df_pupil_train$p_size,
                         N = nrow(df_pupil_train))
  # Held out set for k
   df_pupil_ho <- df_pupil %>%
     filter(fold == k)
  ls_pupil_ho <- list(c_load = df_pupil_ho$c_load,
                      c_trial= df_pupil_ho$c_trial,
                      p_size = df_pupil_ho$p_size,
                      N = nrow(df_pupil_ho))
  # Train the models
  fit_train <- sampling(pupil_stanmodel,
                        iter = 3000,
                        chains = 1,
                        data = ls_pupil_train)
  fit_null_train <- sampling(pupil_null_stanmodel,
                             chains = 1,
                             iter = 3000,
                             data = ls_pupil_train)
  # Generated quantities based on the posterior from the training set
  # and the data from the held out set
  gq_ho <- gqs(pupil_stanmodel,
               draws = as.matrix(fit_train),
               data = ls_pupil_ho)
  gq_null_ho <- gqs(pupil_null_stanmodel,
```

[6]The reader using `cmdstanr` rather than `rstan` might find a cryptic error here. This is because `cmdstanr` expects the parameters not to change. A workaround can be found in https://discourse.mc-stan.org/t/generated-quantities-returns-error-mismatch-between-model-and-fitted-parameters-csv-file/17869/15

```
                    draws = as.matrix(fit_null_train),
                    data = ls_pupil_ho)
  # Extract log likelihood which represents
  # the pointwise predictive density
  log_pd_kfold[, df_pupil$fold == k] <-
    extract_log_lik(gq_ho)
  log_pd_null_kfold[, df_pupil$fold == k] <-
    extract_log_lik(gq_null_ho)
}
```

3. Compute K-fold $\widehat{elpd}$.

Next, evaluate the predictive performance of the model on the 10 folds using `elpd()`.

```
(elpd_pupil_kfold <- elpd(log_pd_kfold))
```

```
##
## Computed from 6000 by 41 log-likelihood matrix using the generic elpd function
##
##      Estimate  SE
## elpd   -242.7 0.7
## ic      485.4 1.3
```

```
(elpd_pupil_null_kfold <- elpd(log_pd_null_kfold))
```

```
##
## Computed from 6000 by 41 log-likelihood matrix using the generic elpd function
##
##      Estimate  SE
## elpd   -248.4 0.7
## ic      496.7 1.3
```

4. Compare the $\widehat{elpd}$ estimates.

```
loo_compare(elpd_pupil_kfold, elpd_pupil_null_kfold)
```

```
##        elpd_diff se_diff
## model1  0.0       0.0
## model2 -5.7       0.6
```

As with PSIS-LOO, we cannot decide which model has better predictive accuracy according to K-fold-CV.

14.7 Summary

In this chapter, we learned how to use K-fold cross-validation and leave-one-out cross-validation, using both built-in functionality in `brms` as well as Stan, in conjunction with the `loo` package. We saw an example of model comparison where cross-validation helped distinguish between the two models (log-normal vs. normal likelihood), and another example where no important differences were found between the models being compared (the N400 data with cloze probability as predictor). In general, cross-validation will be helpful when comparing rather different models (for examples from psycholinguistics, see Nicenboim and Vasishth 2018; and Yadav et al. 2023); when the models are highly similar, it will be difficult to distinguish between them. In particular, for typical psychology and linguistics data sets, it will be difficult to get conclusive results from model comparisons using cross-validation that aim to find evidence for the presence of a population-level (or fixed) effect, if the effect is very small and/or the data are relatively sparse (this is often the case, especially in psycholinguistic data). In such cases, if the aim is to find evidence for a theoretical claim, other model comparison methods like Bayes factors might be more meaningful.

14.8 Further reading

A technical discussion about cross-validation methods can be found in Chapter 7 of Gelman et al. (2014). For a discussion about the advantages and disadvantages of (leave-one-out) cross-validation, see Gronau and Wagenmakers (2018), Vehtari et al. (2019) and Gronau and Wagenmakers (2019). A LOO glossary from the `loo` package can be found in (`https://mc-stan.org/loo/reference/loo-glossary.html`). Cross-validation is still an active area of research, there are multiple websites and blog posts on this topic: Aki Vehtari, the creator of the `loo` package has a comprehensive FAQ about cross-validation in `https://avehtari.github.io/modelselection/CV-FAQ.html`; on Andrew Gelman's blog, Vehtari also discusses the situations where cross-validation can be applied: `https://statmodeling.stat.columbia.edu/2018/08/03/loo-cross-validation-approaches-valid/`.

Part VI

Cognitive modeling with Stan

15

Introduction to cognitive modeling

Until this point in the book, we have been discussing models that specify a generative process for the observed data. This generative process could be as simple as $Y \sim Normal(\mu, \sigma)$ or it could be an elaborate hierarchical model that incorporates multiple variance components. Usually, in these kinds of models, what is of interest is a parameter that represents a so-called effect of interest. Examples that we encountered in the present book are: the effect of word frequency on reading time; effect of relative clause type on reading time; and the effect of attentional load on pupil size.

15.1 What characterizes a computational cognitive model?

One characteristic common to the models seen so far is that no underlying latent cognitive process is specified that elaborates on the generative process that produces the observed dependent variable. For example, in a logistic regression, the correct or incorrect response (say, to a yes/no comprehension question) could be the result of a cascade of alternative steps taken unconsciously (or consciously) by the subject as they generate a response. To make this concrete, a subject could give a yes/no response to a comprehension question after seeing a target sentence by probabilistically processing the sentence deeply or superficially; once the deep/shallow path is taken, the subject might end up giving either a correct or incorrect answer (the latter by misinterpreting the meaning of the sentence). What is observed in the data is a correct or incorrect response, but the reason that that particular response was given could be underlyingly due to deep or superficial processing.

In this book, we use the phrase "computational cognitive modeling" to refer to generative models that are implemented computationally (as a computer program) and that specify latent (unobserved and, usually, unobservable) processes that result in a behavioral or other kind of response. Cognitive modeling as presented in this section goes beyond estimates of "effects" in the sense discussed above; the principal goal is to explain and understand how a particular cognitive process unfolds, often at an individual trial level

DOI: 10.1201/9780429342646-21

and incorporating assumptions about individual-level variation in the way the assumed latent processes are deployed.

Unpacking the latent cognitive process that produces a response has a long history in cognitive science. For example, in sentence processing research, early models like the classic garden-path model (Frazier 1979) seek to spell out the steps that occur when the human sentence processing system (the parser) attempts to build syntactic structure incrementally when faced with a temporarily ambiguous sentence. To make this concrete, consider the sentence: "While Mary bathed the baby happily played in the living room." Compared to the unambiguous baseline sentence "While Mary bathed, the baby happily played in the living room," the garden-path model assumes that in the ambiguous sentence the parser initially connects the noun phrase "the baby" as the grammatical object of the verb "bathed." It is only when the verb "played" is encountered that the parser reassigns "the baby" as the grammatical subject of the verb "played," leading to the correct parse whereby Mary being the one who is bathing, and not that Mary is bathing the baby. This process of reassigning "the baby"'s grammatical role is computationally costly and is called reanalysis in sentence processing. The slowdown observed (e.g., in reading studies) at the verb "played" is often called the garden-path effect.

This kind of paper-pencil model implicitly posits an overly simplistic and deterministic parsing process: Although this is never spelled out in the garden-path model, there is no assumption that the parser could only probabilistically misparse the sentence when it encounters "the baby"; misparses are implicitly assumed to happen every single time such a temporarily ambiguous sentence is encountered. Such a model does not explicitly allow alternative parsing constraints to come into play; by contrast, a computational model allows the empirical consequences of multiple parsing constraints to be considered quantitatively (e.g., Jurafsky 1996; Paape and Vasishth 2022).

15.2 Some advantages of taking the latent-variable modeling approach

Although this kind of simple paper-pencil model is an excellent start toward modeling the latent process of sentence comprehension, just stopping with such a description has several disadvantages. First, no quantitative predictions can be derived; a corollary is that a slowdown at the verb "played" (due to reanalysis) of 10 ms or 500 ms would both be equally consistent with the predictions of the model—the model cannot say anything about how much time the reanalysis process would take. This is a problem for model evaluation, not least because overly large effect sizes observed in data could just be Type M errors and therefore very misleading (Gelman and Carlin 2014; Vasishth et al. 2018). Second, such paper-pencil models encourage an excessive (in fact, exclusive) focus on the average effect size (here, the garden-path effect); the variability among individuals (which would affect the standard error of the estimated effect from data) plays no role in determining whether the model's prediction is consistent with the data. A third problem with such verbally stated models is that it is often not clear what the exact assumptions are. This makes it difficult to establish whether the model's predictions are consistent with observed patterns in the data. As a concrete example of these hidden degrees of freedom, an influential theory in sentence processing is the Dependency Locality Theory or DLT (Gibson 2000). This theory explicitly assumes that interposing new discourse referents intervening between, say, a subject and a verb, will make it more difficult to complete the subject-verb dependency. The status of the intervening discourse referent is crucial for dependency completion cost to increase. A further descriptive elaboration of the model is presented in Warren and Gibson (2002); there, the authors argued that the cost of an intervening discourse referent should be graded according to its discourse accessibility. There was no quantitative implementation of what that implies empirically; for example, do old vs. new discourse referents eliminate dependency completion cost entirely, or is the cost attenuated by some amount (if so, how much?)? Interestingly, in later work, the discourse status of the intervener was ignored; in Gibson and Wu (2013), the intervening discourse referents are all old (previously introduced in an immediately preceding context), but they are assumed to have the same impact on dependency completion as new discourse referents, as described in Gibson (2000). Had the model been computationally implemented, the empirical impact of relaxing the model assumption would have had a quantitative impact on the predictions. This example also illustrates how easy it is to be misled by paper-pencil statements of theories.

Thus, the absence of quantitative predictions, the inability to quantitatively investigate individual-level variation, and hidden degrees of freedom in the model, are three major drawbacks of paper-pencil models. As Roberts and Pashler (2000) have discussed at length, a good fit of a model to data is not merely about the sign of the predicted effect being correct; the model must be able to commit a priori to the uncertainty of the predicted effect, and the estimated effect from the data and its uncertainty need to be compared to the predictions of the model. A good fit ideally requires a tightly constrained quantitative prediction derived from the model that is then validated by comparing the prediction with unseen data; this point has also been eloquently made by the psychologist Meehl (1997). Meehl suggests that the model makes risky numerical predictions (by which he means tightly constrained quantitative predictions), which should then be compared with the observed effect size and its confidence interval—this is essentially the same criterion for a good fit that Roberts and Pashler (2000) also favor.

For example, if a computational implementation of the garden-path model were to exist, one could have derived (through prior specifications on the parameters of the model) prior predictive distributions of the garden-path effect, and compared these predictions to the estimates of the effect from a statistical model fit to new, unseen data (for an example from psycholinguistics, see Vasishth and Engelmann 2022). Using parameteric variation, one could even investigate the implications of the model for individual-level differences (e.g., Yadav et al. 2022); such implications of models are impossible to derive unless the model is implemented computationally. In the absence of an implemented model, one is reduced to classifying subjects into groups (e.g., by working memory capacity measures) and investigating average group-level effects (e.g., Caplan and Waters 1999). This makes the question a binary one: are there individual differences or are there none? The right question to ask about individual differences is a quantitative one (Haaf and Rouder 2019).

15.3 Types of computational cognitive model

There are many different classes of computational cognitive model. For example, Newell (1990) pioneered a cognitive architectures approach, where a model of a particular cognitive process (like sentence processing) occurs within a broader computational framework that defines very general constraints on human information processing. Examples of cognitive architectures are SOAR (Laird 2019), the CAPS family of models (Just, Carpenter, and Varma 1999), and ACT-R (Anderson et al. 2004). Other approaches include connectionist models (e.g., McClelland and Rumelhart 1989) and dynamical systems-based models (e.g., Port and Van Gelder 1995; Tabor and Tanenhaus 1999; Beer 2000; Rabe et al. 2021, 2024; Engbert et al. 2005).

In this book, we focus on Bayesian cognitive models (Lee and Wagenmakers 2014); these are distinct from models that assume that human cognitive processes involve Bayesian inference (e.g., Feldman 2017). The type of model we discuss here has the characteristic that the underlying generative process spells out the latent, probabilistically occurring sub-processes. The latent processes are spelled out by specifying a Bayesian model that allows different events to happen probabilistically in each trial. An example is multinomial processing tree models, which specify a sequence of possible latent sub-processes. Another example is a hierarchical finite mixture process which specifies that, in some proportion of trials, the observed response comes from one distribution, and in another proportion from a different distribution. A third example is the assumption that the observed response (e.g., response time) is the result of an unobserved (latent) race process in the cognitive system. Probabilistic programming languages like Stan allow us to implement such latent process models, allowing for hierarchical structure (individual-level variability). Additionally, we can make prior assumptions about plausible parameter values, this is especially important because the parameters have a meaning in terms of latent processes.

This part of the book introduces these three types of cognitive models using Stan. In many cases, a great deal of cognitive detail is sacrificed for tractability, but this is a characteristic shared by all computational models—by definition, a model is a simplification of the underlying process being modeled (McClelland 2009).

15.4 Summary

The main take-away from this section is that it is possible to specify an underlying generative process for the data that reflects theoretical assumptions in a particular research area. The gain is that: (i) the assumptions of the underlying theory and their consequences become transparent and explicit (Epstein 2008); (ii) one can derive quantitative predictions (along with the uncertainty of those predictions) which, as Roberts and Pashler (2000) point out, are vital for model evaluation; (iii) it becomes possible (at least in principle) to eliminate competing theoretical proposals through quantitative model comparison using benchmark data (Nicenboim and Vasishth 2018; Lissón et al. 2021; Lissón et al. 2023; Yadav et al. 2023); and (iv) the implications of models for average behavior (e.g., Yadav et al. 2023) as well as for individual-level differences (e.g., Yadav et al. 2022) can be investigated.

15.5 Further reading

General textbooks on computational modeling for cognitive science are Busemeyer and Diederich (2010) and Farrell and Lewandowsky (2018). The textbook by Lee and Wagenmakers (2014) focuses on relatively simple computational cognitive models implemented in a Bayesian framework (using the BUGS language). A good free textbook on computational modeling for cognitive science is Blokpoel and Rooij (2021). The entire special issue (Lee 2011a) on hierarchical Bayesian modeling in the *Journal of Mathematical Psychology* is highly recommended (in particular, see the article by Lee 2011b). Wilson and Collins (2019) discuss good practices in the computational modeling of behavioral data using examples from reinforcement learning. Haines et al. (2020) discuss how generative models produce higher test-retest reliability and more theoretically informative parameter estimates than do traditional methods. Bayesian models of individual differences are discussed in Haaf and Rouder (2019). For an overview of the different modeling approaches in cognitive science and the relationships between them, see McClelland (2009). Luce (1991) is a classic book that focuses on modeling response times.

16

Multinomial processing trees

In this chapter, we introduce a widely used class of cognitive models that can be implemented in Stan, the multinomial processing tree. This model is useful in situations where the behavioral response from the subject is one of several possible categorical outcomes. As an example, we will look into a word production task, where we ask individuals with aphasia (a language impairment that is usually due to a cerebrovascular accident or head trauma, Damasio 1992), to name the object shown in a picture, e.g., a picture of a cat. The participant (hereafter, subject) could produce the correct name ("cat"), a semantically and phonologically related but incorrect name ("rat"), a semantically unrelated but phonologically related word ("hat"), or a non-word ("cag"). The researcher may have a theory about how each possible outcome ends up being probabilistically produced. Such a theoretical process model can be expressed as a multinomial processing tree. Before we dive into multinomial processing trees, we discuss the distributions that generalize the binomial and Bernoulli distribution for modeling more than two possible outcomes.

16.1 Modeling multiple categorical responses

One way to model categorical responses is using multinomial or categorical distributions. The categorical responses could be "yes" or "no"; "blue", "red" or "yellow"; "true", "false", or "I don't know"; or more complicated categories. Crucially, the response of each observation can be coded as belonging to only one of several possible categories. The multinomial and the categorical distribution represent two ways of characterizing the underlying generative process for such data.

The *multinomial distribution* is the generalization of the binomial distribution for more than two possible outcomes. Recall that the binomial works like this: in order to randomly generate the number of successes in an observation consisting of 10 trials, with the probability of success of 0.5, one can type:

DOI: 10.1201/9780429342646-22

```
rbinom(1, size = 10, prob = 0.5)
```

```
## [1] 4
```

It is possible to repeatedly generate multiple observations as follows. Suppose five simulated observations are needed, each with 10 trials:

```
rbinom(5, size = 10, prob = 0.5)
```

```
## [1] 6 5 7 7 2
```

Now, suppose that there are $N = 3$ possible answers to a question (yes, no, don't know), and suppose that the probabilities of producing each answer are:

- $P(\text{yes}) = 0.1$
- $P(\text{no}) = 0.1$
- $P(\text{don't know}) = 0.8$

The probabilities must sum to 1, because those are the only three possible outcomes. Given such a situation, it is possible to simulate a single experiment with 10 trials, where each of the three possibilities appears a certain number of times. We do this with `rmultinom()`:

```
(random_sample <- rmultinom(1, size = 10, prob = c(0.1, 0.1, 0.8)))
```

```
##      [,1]
## [1,]    0
## [2,]    2
## [3,]    8
```

The above call returns the result of the random sample: 0 cases of the first answer type, 2 cases of the second; and 8 cases of the third.

Analogously to the binomial function shown above, five observations can be simulated, each having 10 trials:

```
rmultinom(5, size = 10, prob = c(0.1, 0.1, 0.8))
```

```
##      [,1] [,2] [,3] [,4] [,5]
## [1,]    1    3    1    1    3
## [2,]    2    0    2    1    1
## [3,]    7    7    7    8    6
```

The *categorical distribution* is the generalization of the Bernoulli distribution for more than two possible outcomes, and it is the special case of the multinomial distribution when we have only one trial. Recall that the Bernoulli distribution can be used as follows. If we carry out a coin toss (each coin toss counts as a single trial), we will either get a heads or a tails:

```
rbern(5, prob = 0.5)
```

```
## [1] 0 0 1 0 1
```

```
## equivalent rbinom command:
rbinom(5, size = 1, prob = 0.5)
```

```
## [1] 0 0 1 1 1
```

Thus, what the Bernoulli is to the binomial, the categorical is to the multinomial. For example, one can simulate five observations, each of which will give one of the three responses with the given probabilities. We do this with `rcat()` from the `extraDistr` package.

```
rcat(5, prob = c(0.1, 0.1, 0.8), labels = c("yes", "no", "dontknow"))
```

```
## [1] dontknow yes      dontknow dontknow dontknow
## Levels: yes no dontknow
```

The above is analogous to using the multinomial with `size = 1` (a single trial in each experiment). In the output below, the `rmultinom()` function shows which of the three categories is produced.

```
rmultinom(5, size = 1, prob = c(0.1, 0.1, 0.8))
```

```
##      [,1] [,2] [,3] [,4] [,5]
## [1,]    0    0    1    0    0
## [2,]    0    1    0    0    0
## [3,]    1    0    0    1    1
```

With these distributions as background, consider now a simulated situation where multiple responses are possible.

16.1.1 A model for multiple responses using the multinomial likelihood

Impaired picture naming (anomia) is a common form of aphasia. It is assessed as part of most comprehensive aphasia test batteries, since picture naming accuracy is relatively easily obtained and is a reliable test score; in addition, the types of errors that are committed can provide useful information for diagnosis.

In this simulated experiment, the responses are categorized as shown in Table 16.1.

TABLE 16.1: Categorization of responses for the simulated experiment.

Category	Description	Example
Correct	The response matches the target.	cat
Neologism	The response is not a word, but it has a phonological relation to the target.	cag
Formal	The response is a word with only a phonological relation to the target.	hat
Mixed	The response is a word with both a semantic and phonological relation to the target.	rat
NR	All other responses, including omissions, descriptions, non-nouns, etc.	–

First, generate data assuming a multinomial distribution. The outcomes will be determined by a vector $\boldsymbol{\theta}$ (called `true_theta` below in the R code) that indicates the probability of each outcome:

```
(true_theta <- tibble(theta_NR = 0.2,
                      theta_Neologism = 0.1,
                      theta_Formal = 0.2,
                      theta_Mixed = 0.08,
                      theta_Correct =  1 -
  (theta_NR + theta_Neologism + theta_Formal + theta_Mixed)))
```

```
## # A tibble: 1 x 5
##   theta_NR theta_Neologism theta_Formal theta_Mixed theta_Correct
##      <dbl>           <dbl>        <dbl>       <dbl>         <dbl>
## 1      0.2             0.1          0.2        0.08          0.42
```

```
## The probabilities must sum to 1:
sum(true_theta)
```

```
## [1] 1
```

Given this vector of probabilities $\boldsymbol{\theta}$, generate values assuming a multinomial distribution of responses in 100 trials:

```
N_trials <- 100
(ans_mn <- rmultinom(1, N_trials, true_theta))
```

```
##                 [,1]
## theta_NR          23
## theta_Neologism    7
## theta_Formal      18
## theta_Mixed        6
## theta_Correct     46
```

Now, we'll try to recover the probability of each answer with a model with the following likelihood:

$$ans \sim Multinomial(\boldsymbol{\theta}) \tag{16.1}$$

where $\boldsymbol{\theta} = \langle \theta_{nr}, \theta_{neol.}, \theta_{formal}, \theta_{mix}, \theta_{corr} \rangle$.

A common prior for the vector of probability parameters of a multinomial distribution is the Dirichlet distribution, which extends the Beta distribution to cases where more than two categories are available.

$$\boldsymbol{\theta} \sim Dirichlet(\boldsymbol{\alpha}) \tag{16.2}$$

The Dirichlet distribution has a parameter $\boldsymbol{\alpha}$, called the concentration parameter, and it is a vector with the same length as $\boldsymbol{\theta}$. If we set $\boldsymbol{\alpha} = \langle 2, 2, 2, 2, 2 \rangle$, this is analogous to $\sim Beta(2, 2)$. The intuition behind this concentration parameter is that the prior probability distribution of the vector $\boldsymbol{\theta}$ corresponds to having seen each outcome twice.

A Stan model assuming a multinomial likelihood and Dirichlet prior is shown below. Since the elements of $\boldsymbol{\theta}$ should sum to one, we declare this vector, `theta`, is of type `simplex`. The simplex type ensures that its elements sum to one and also constrains them to have non-negative values. In order to generate the vector $\boldsymbol{\alpha}$ that contains five times the value two, we use `rep_vector(2, 5)` (which is similar to `rep(2, 5)` in R).

```
data {
  int<lower = 1> N_trials;
  array[5] int<lower = 0,upper = N_trials> ans;
}
parameters {
```

```
  simplex[5] theta;
}
model {
  target += dirichlet_lpdf(theta | rep_vector(2, 5));
  target += multinomial_lpmf(ans | theta);
}
generated quantities{
  array[5] int pred_ans = multinomial_rng(theta, N_trials);
}
```

Fit the model:

```
# Create a list:
# c(ans_mn) makes a vector out of the matrix ans_mn
data_mn <-  list(N_trials = N_trials,
                 ans = c(ans_mn))
str(data_mn)
```

```
multinom <- system.file("stan_models",
                        "multinom.stan",
                        package = "bcogsci")
fit_mn <- stan(multinom, data = data_mn)
```

Print the posteriors:

```
print(fit_mn, pars = c("theta"))
```

```
##          mean 2.5% 97.5% n_eff Rhat
## theta[1] 0.23 0.15  0.31  4282    1
## theta[2] 0.08 0.04  0.14  4262    1
## theta[3] 0.18 0.12  0.26  4166    1
## theta[4] 0.07 0.03  0.13  3858    1
## theta[5] 0.44 0.35  0.53  4346    1
```

Next, use `mcmc_recover_hist()` in the code below to confirm that the posterior distributions of the elements of $\boldsymbol{\theta}$ are close to the true point values that were set up when simulating the data. See Figure 16.1.

```
as.data.frame(fit_mn) %>%
  select(starts_with("theta")) %>%
```

```
mcmc_recover_hist(true = unlist(true_theta)) +
coord_cartesian(xlim = c(0, 1))
```

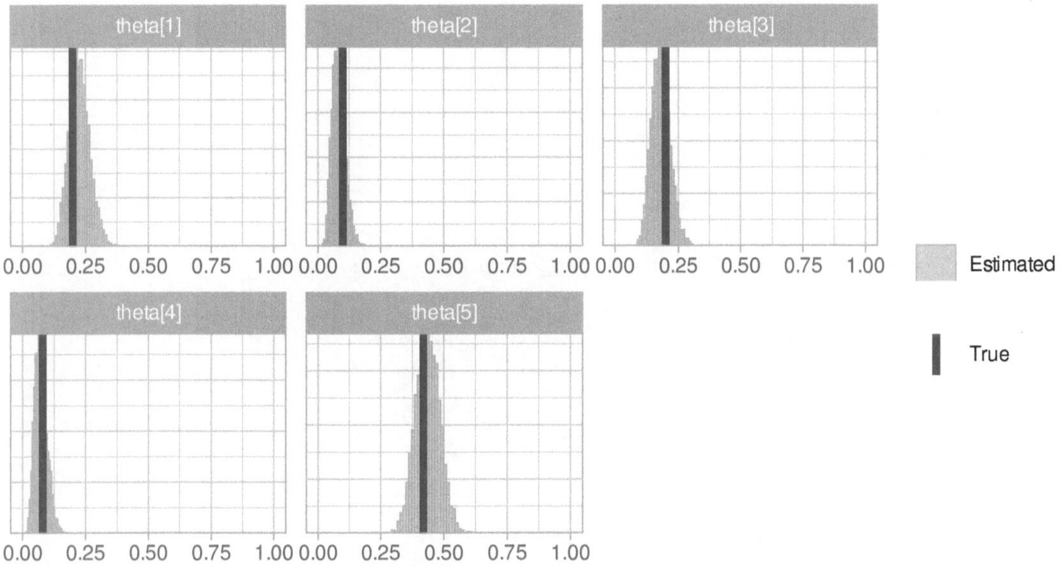

FIGURE 16.1: Posterior distributions and true means of `theta` for the multinomial model defined in `multinom.stan`.

Here, we evaluate whether our model is able to "recover" the true point values of its parameters. By "recover," we mean that the true point values are somewhere inside the posterior distribution of the model.

The frequentist properties of Bayesian models guarantee that if we simulate data several times, 95% of the true values should be inside of the 95% CrI intervals generated by a "well-calibrated" model. Furthermore, if the true values of some parameters are consistently well above or below their posterior distribution, it may mean that there is some problem with the model specification. We follow Cook, Gelman, and Rubin (2006) here, and for now we are going to verify that our model is roughly correct. A more principled (and computationally demanding) approach uses simulation based calibration (SBC) introduced in section 10.2 of chapter 10 (also see Talts et al. 2018; Schad, Betancourt, and Vasishth 2020).

16.1.2 A model for multiple responses using the categorical distribution

Using the same information as above, we can model each response one at a time, instead of aggregating them. Using the categorical distribution gives us more flexibility to define what happens at every trial. However, we are not using the additional flexibility yet. As a result, the next model and the previous one are equivalent.

```
data {
  int<lower = 1> N_obs;
  array[N_obs] int<lower = 1, upper = 5> w_ans;
}
parameters {
  simplex[5] theta;
}
model {
  target += dirichlet_lpdf(theta | rep_vector(2, 5));
  for(n in 1:N_obs)
    target += categorical_lpmf(w_ans[n] | theta);
}
generated quantities{
  array[N_obs] int pred_w_ans;
  for(n in 1:N_obs)
    pred_w_ans[n] = categorical_rng(theta);
}
```

Given the same set of probabilities $\boldsymbol{\theta}$ as above, generate 100 individual observations using `rcat()` from the `extraDistr` package:

```
N_obs <- 100
ans_cat <- rcat(N_obs, prob = as.matrix(true_theta))
```

The above output is how Stan expects to see the data. The data fed into the Stan model is defined as a list as usual:

```
data_cat <-  list(N_obs = N_obs,
                  w_ans = ans_cat)
str(data_cat)
```

```
## List of 2
##  $ N_obs: num 100
##  $ w_ans: num [1:100] 1 3 3 3 1 1 2 3 2 5 ...
```

Fitting the Stan model (`categorical.stan`) should yield approximately the same $\boldsymbol{\theta}$ as with the multinomial likelihood defined in the model `multinom.stan`.

```
categorical <- system.file("stan_models",
                            "categorical.stan",
                            package = "bcogsci")
fit_cat <- stan(categorical, data = data_cat)
```

```
print(fit_cat, pars = c("theta"))
```

```
##          mean 2.5% 97.5% n_eff Rhat
## theta[1] 0.19 0.12  0.27  4359    1
## theta[2] 0.08 0.04  0.14  4838    1
## theta[3] 0.27 0.20  0.36  4185    1
## theta[4] 0.05 0.02  0.10  4919    1
## theta[5] 0.40 0.31  0.49  4160    1
```

The above models estimate the posterior distribution for the probability for each possible response. The slightly different posteriors with the two methods are due to differences in the randomly generated data and the sampling process, not differences in grouping. If we had some experimental manipulation, we could even fit regressions to these parameters. This is called a multinomial logistic regression or categorical regression; see further reading for some examples.

16.2 Modeling picture naming abilities in aphasia with MPT models

Multinomial processing tree (MPT) modeling is a method that estimates latent variables that have a psychological interpretation given categorical data (a review is provided in Batchelder and Riefer 1999). In other words, an MPT model is just one way to model categorical responses following a multinomial or categorical distribution. MPT models assume that the observed response categories result from a sequences of underlying cognitive events which are represented as a binary branching tree. Each binary branching is associated with a parameter that represents the probability of going down either branch. Every successive node is assumed to be independent of the preceding node, allowing us to use the product rule from probability theory to compute the probability of going down a particular path. The leaves of the binary branching tree are the observed response in the data. The goal is to derive posterior distributions of the latent probability parameters specified for the binary branching in the model.

Walker, Hickok, and Fridriksson (2018) created an MPT model that specifies a set of possible internal errors that lead to the various possible response types during a picture naming trial for persons with aphasia. Here we'll explore a simplification of the original model.

The model assumes that when an attempt is made to produce a word, errors in production can arise at the whole word level (lexical level) or the segmental level (phonological level). Semantic errors are assumed to arise from the lexical substitutions, and neologism errors from phonological substitutions. Real-word responses that are phonologically related to the correct target word can arise from substitutions at the lexical or phonological level.

The task for the subject is to view a picture and name the object represented in the picture. Table 16.2 summarizes the psychological interpretation of the parameters of the model. When an attempt is made to retrieve the word from memory, the following possible steps can unfold (this is a simplified version of the original model):

- Either the subject will make some lexical selection, or fail to make a lexical selection, returning a non-response (NR). The probability of making some lexical selection is a, so the probability of a non-response is $1 - a$, as these are only two possibilities at this initial stage of the binary branching tree. Example: the subject sees the picture of a cat, and either produces the response "I don't know," or starts the process of producing a word.
- If a lexical selection is made, the target word is selected with probability t, or some other word is chosen with probability $1 - t$.
- Once a word is selected, either its phonological representation is selected with probability f, or some other (incorrect) phonological representation is selected with probability $1 - f$.
- Once a word is selected, there can be a phonological change that leads to a real, formally related word with probability c, or a neologism with probability $1 - c$. Example: the subject produces either a formally related word "hat," or a neologism like "cag."

The end-result of walking down this tree is that the subject either produces a non-response ("I don't know" or silence), a correct response, a related word, a mixed word, or a neologism. There is more than one way to produce a neologism or a related word, and the posterior probabilities of the various paths will determine the probability of each possible path.

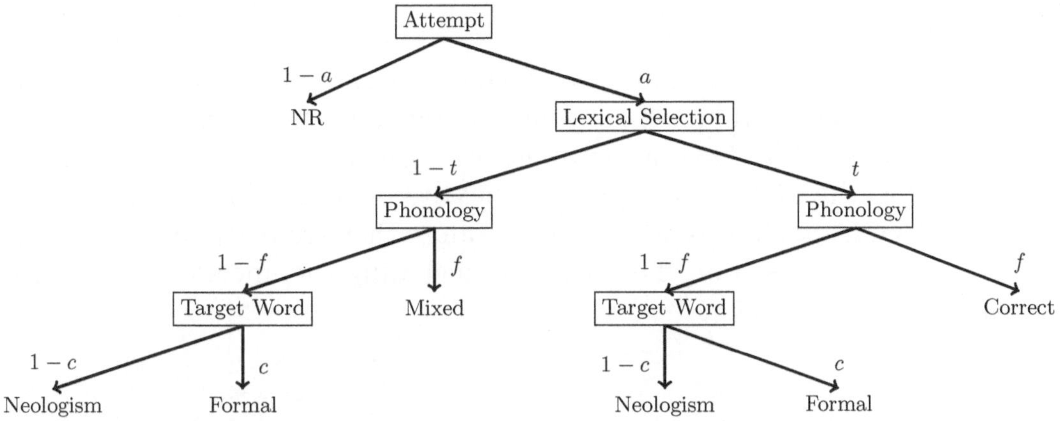

FIGURE 16.2: Representation of a simplification of the MPT used in Walker, Hickok, and Fridriksson (2018).

TABLE 16.2: Psychological interpretation of the parameters of the MPT model.

Param.	Description
a	Probability of initiating an attempt
t	Probability of selecting a target word over competitors
f	Probability of retrieving correct phonemes
c	Probability of a phoneme change in the target creating a real word

16.2.1 Calculation of the probabilities in the MPT branches

By navigating through the branches of the MPT (Figure 16.2), we can calculate the probabilities of the five responses (the categorical outcomes), based on the four underlying parameters assumed in the MPT:

- $P(NR|a,t,f,c) = 1 - a$
- $P(Neologism|a,t,f,c) = a \cdot (1-t) \cdot (1-f) \cdot (1-c) + a \cdot t \cdot (1-f) \cdot (1-c)$
- $P(Formal|a,t,f,c) = a \cdot (1-t) \cdot (1-f) \cdot c + a \cdot t \cdot (1-f) \cdot c$
- $P(Mixed|a,t,f,c) = a \cdot (1-t) \cdot f$
- $P(Correct|a,t,f,c) = a \cdot t \cdot f$

Given that

$$P(NR|a,t,f,c) + P(Neologism|a,t,f,c) + P(Formal|a,t,f,c) + P(Mixed|a,t,f,c) + P(Correct|a,t,f,c) = 1 \quad (16.3)$$

there is no need to characterize every outcome: we can always calculate any one of the remaining responses as one minus the other responses.

16.2.2 A simple MPT model

First, simulate 200 trials assuming no variability between items and subjects. It is convenient to define functions to compute each outcome's probability based on the previous MPT. One needs to assign "true values" to the underlying parameters of the MPT; these point values are only for illustration. Ideally, one should simulate data using parameter values that are realistic; that is, one should use values based on the literature, including the uncertainty of these parameters.

```
# The probabilities of the different answers:
Pr_NR <- function(a, t, f, c)
  1 - a
Pr_Neologism <- function(a, t, f, c)
  a * (1 - t) * (1 - f) * (1 - c) + a * t * (1 - f) * (1 - c)
Pr_Formal <- function(a, t, f, c)
  a * (1 - t) * (1 - f) * c +  a * t * (1 - f) * c
Pr_Mixed <- function(a, t, f, c)
  a * (1 - t) * f
Pr_Correct <- function(a, t, f, c)
  a * t * f
# The true underlying values for simulated data:
a_true <- 0.75
t_true <- 0.9
f_true <- 0.8
c_true <- 0.1
# The probability of the different answers:
Theta <-
  tibble(NR = Pr_NR(a_true, t_true, f_true, c_true),
         Neologism = Pr_Neologism(a_true, t_true, f_true, c_true),
         Formal = Pr_Formal(a_true, t_true, f_true, c_true),
         Mixed = Pr_Mixed(a_true, t_true, f_true, c_true),
         Correct = Pr_Correct(a_true, t_true, f_true, c_true))
N_trials <- 200
(ans <- rmultinom(1, N_trials, c(Theta)))
```

```
##            [,1]
## NR           49
## Neologism    26
## Formal        5
## Mixed        17
## Correct     103
```

The above data can be modeled in Stan as discussed below (see `mpt_mnm.stan`). The probabilities of the different categories go into the `transformed parameters` section because they are derived from the probability parameters in the model. The data are modeled as coming from a multinomial likelihood. If priors are not specified, then a Beta distribution with $a = 1$ and $b = 1$ (a Uniform(0,1) distribution) is assumed for the parameters a, t, f, and c. Unlike $\boldsymbol{\theta}$, the values of these parameters are independent of each other and they do not sum to one. For this reason, we should not use a Dirichlet prior here.

We define the following model:

$$
\begin{aligned}
\theta_{nr} &= 1 - a \\
\theta_{neol.} &= a \cdot (1 - t) \cdot (1 - f) \cdot (1 - c) + a \cdot t \cdot (1 - f) \cdot (1 - c) \\
\theta_{formal} &= a \cdot (1 - t) \cdot (1 - f) \cdot c + a \cdot t \cdot (1 - f) \cdot c \\
\theta_{mix} &= a \cdot (1 - t) \cdot f \\
\theta_{corr} &= a \cdot t \cdot f \\
\boldsymbol{\theta} &= \langle \theta_{nr}, \theta_{neol.}, \theta_{formal}, \theta_{mix}, \theta_{corr} \rangle \\
ans &\sim Multinomial(\boldsymbol{\theta}) \\
a, t, f, c &\sim Beta(2, 2)
\end{aligned}
\tag{16.4}
$$

This translates to the following code:

```
data {
  int<lower = 1> N_trials;
  array[5] int<lower = 0, upper = N_trials> ans;
}
parameters {
  real<lower = 0, upper = 1> a;
  real<lower = 0, upper = 1> t;
  real<lower = 0, upper = 1> f;
  real<lower = 0, upper = 1> c;
}
transformed parameters {
  simplex[5] theta;
  theta[1] = 1 - a; //Pr_NR
  theta[2] = a * (1 - t) * (1 - f) * (1 - c) +
    a * t * (1 - f) * (1 - c); //Pr_Neologism
  theta[3] = a * (1 - t) * (1 - f) * c
    + a * t * (1 - f) * c;  //Pr_Formal
  theta[4] = a * (1 - t) * f; //Pr_Mixed
  theta[5] = a * t * f; //Pr_Correct
}
```

```
model {
  target += beta_lpdf(a | 2, 2);
  target += beta_lpdf(t | 2, 2);
  target += beta_lpdf(f | 2, 2);
  target += beta_lpdf(c | 2, 2);
  target += multinomial_lpmf(ans | theta);
}
generated quantities{
    array[5] int pred_ans;
  pred_ans = multinomial_rng(theta, N_trials);
}
```

Fit the model:

```
data_sMPT <-  list(N_trials = N_trials,
                   ans = c(ans))
```

```
mpt_mnm <- system.file("stan_models",
                       "mpt_mnm.stan",
                       package = "bcogsci")
fit_sMPT <- stan(mpt_mnm, data = data_sMPT)
```

Print out a summary of the posterior of the parameter of interest:

```
print(fit_sMPT, pars = c("a", "t", "f", "c"))
```

```
##   mean 2.5% 97.5% n_eff Rhat
## a 0.75 0.69  0.81  4422    1
## t 0.85 0.78  0.90  4486    1
## f 0.79 0.72  0.85  4050    1
## c 0.20 0.09  0.35  4188    1
```

What the model gives us is posterior distributions of each of the parameters `a`, `t`, `f`, `c`. From these we can derive the probabilities of producing the different observed responses, and the posterior predictive distributions, which could be used for model evaluation.

An important sanity check in modeling is checking whether the model can in principle recover the true parameters that generated the data; see Figure 16.3.

```
as.data.frame(fit_sMPT) %>%
  select(c("a","t","f","c")) %>%
  mcmc_recover_hist(true = c(a_true, t_true, f_true, c_true)) +
  coord_cartesian(xlim = c(0, 1))
```

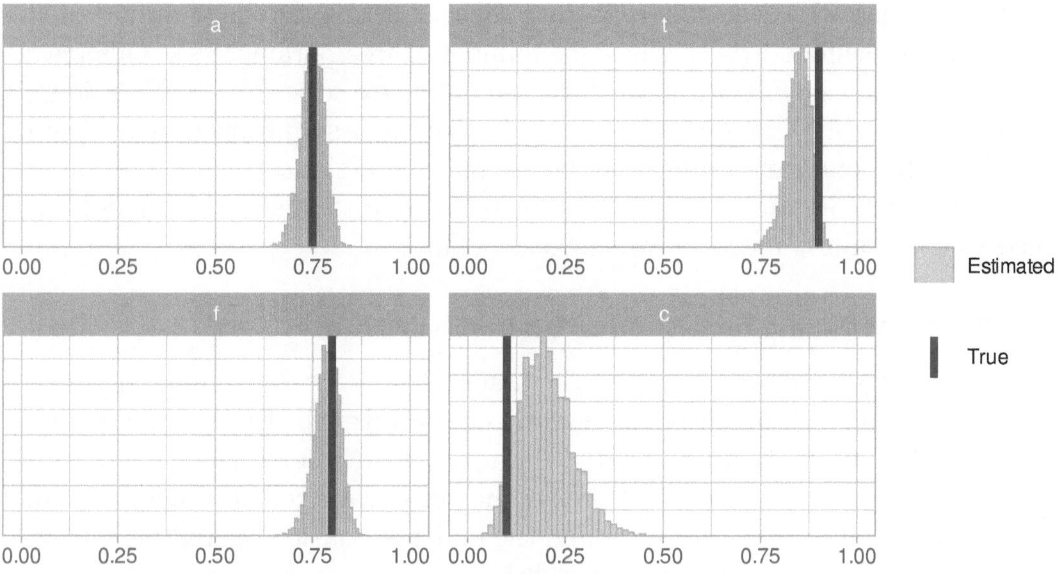

FIGURE 16.3: Posterior distributions and true point values of the parameters of the simple MPT model (`mpt_mnm.stan`).

The above figure shows that the model can indeed recover the true parameters fairly accurately.

The posterior distributions of the $\boldsymbol{\theta}$ parameters can also be summarized:

```
print(fit_sMPT, pars = c("theta"))
```

```
##          mean 2.5% 97.5% n_eff Rhat
## theta[1] 0.25 0.19  0.31  4422    1
## theta[2] 0.13 0.09  0.18  3875    1
## theta[3] 0.03 0.01  0.06  4059    1
## theta[4] 0.09 0.06  0.13  4465    1
## theta[5] 0.50 0.43  0.57  4520    1
```

These posteriors tell us the probability of producing each of the possible responses. This model might be useful for estimating the latent parameters, a, t, f, c, but without further constraints it is unfalsifiable.

Recall that for the multinomial likelihood in section 16.1.1, we had a simplex of size five, which means that we had four free parameters (since the fifth can be deduced based on the others). With five possible answers we can always estimate a vector of probabilities $\boldsymbol{\theta}$ that fits the data, in the same way that with two possible answers (e.g., zeros and ones), we can always estimate a single probability θ that fits the data (using a Bernoulli or binomial likelihood). The MPT that we present here is just reparameterizing the vector $\boldsymbol{\theta}$ of the multinomial likelihood (with the same number of free parameters). This means that it will always achieve a perfect fit; see online exercise G.16.2. This doesn't mean that this MPT model is "useless": Under the assumption that the model is meaningful, one can estimate its latent parameters and this estimation can have theoretical implications. If we want to be able to falsify this model, we'll need to constrain it more, as we suggest below.

16.2.3 An MPT model assuming by-item variability

The use of aggregated data implies the assumption that the estimated parameters do not vary too much between subjects and items. If this assumption is incorrect, the analysis of aggregated data may lead to erroneous conclusions: reliance on aggregated data in the presence of parameter heterogeneity may lead to biased parameter estimates and the underestimation of credible intervals (see also Schad, Nicenboim, and Vasishth 2023).

If it is known that f is affected by the phonological complexity of the individual word (e.g., *cat* is easier to produce than *umbrella*), the previous model does not have a way to include that information.

Simulated data can be generated taking into account the `complexity` of the items. Assume here for simplicity that the complexity of items is given as data and it is scaled and centered; i.e., mean complexity is represented by 0, and the standard deviation is assumed to be 1. We will assume a regression model that determines the parameter, f, as a function of the phonological complexity of each trial.

One important detail is that f is a probability and needs to be bounded between 0 and 1. To make sure that this property is met, the computation of f for each item will be converted to probability space using the logistic function. This is achieved as follows.

Suppose that f' is a linear function of complexity. For example, two parameters α_f and β_f (intercept and slope, respectively) could determine how f' is affected by complexity:

$$f'_j = \alpha_f + complexity_j \cdot \beta_f.$$

The parameters α_f and β_f are defined in an unconstrained log-odds space (they can be any real number). The model that is fit then yields an f'_j value for each item j in log-odds space. The log-odds value f'_j can be converted to a probability value f_{true} by applying the logistic function (or the inverse logit, $logit^{-1}$) to f'. Recall from the generalized linear model discussed earlier that if we have a model in log-odds space:

$$\log\left(\frac{p_j}{1-p_j}\right) = \alpha + \beta \cdot x_j = \mu_j \tag{16.5}$$

Then we can recover the probability p_j by solving for p_j:

$$p_j = \frac{\exp(\mu_j)}{1+\exp(\mu_j)} = \frac{1}{1+\exp(-\mu_j)} \tag{16.6}$$

The above is the logistic or inverse logit function: it takes as input μ_j and returns the corresponding probability p_j. The `plogis()` function in R carries out the calculation shown above.

```
N_obs <- 50
complexity <- rnorm(N_obs) # by default mean = 0, sd = 1
## choose some hypothetical values:
alpha_f <- 0.3
# the negative sign indicates that
# increased complexity will lead to a reduced value of f:
beta_f <- -0.3
# f' as a linear function of complexity:
f_prime <- alpha_f + complexity * beta_f
head(f_prime)
```

```
## [1]  0.468  0.369 -0.168  0.279  0.261 -0.215
```

```
## probabilities f for each item:
f_true <- plogis(f_prime)
head(f_true)
```

```
## [1] 0.615 0.591 0.458 0.569 0.565 0.447
```

This change in our assumptions entails that the probability of each response changes depending on the item associated with each observation. The parameters `theta` now have to be a matrix. This is in R; in Stan, we will code it as an array of simplexes, i.e., an array of non-negative values that sums to 1.

We continue with the functions defined in section 16.2.2, and the same values for `a_true`, `t_true`, and `c_true` as defined in that section. Since most of the equations depend on `f`, and `f` is a vector now, the outcomes are automatically vectors. But this is not the case for `theta_NR_v`, and thus we need to repeat the value.

```
theta_NR_v <- rep(Pr_NR(a_true, t_true, f_true, c_true), N_obs)
theta_Neologism_v <- Pr_Neologism(a_true, t_true, f_true, c_true)
theta_Formal_v <- Pr_Formal(a_true, t_true, f_true, c_true)
theta_Mixed_v <- Pr_Mixed(a_true, t_true, f_true, c_true)
theta_Correct_v <- Pr_Correct(a_true, t_true, f_true, c_true)
theta_item <- matrix(c(theta_NR_v,
                       theta_Neologism_v,
                       theta_Formal_v,
                       theta_Mixed_v,
                       theta_Correct_v),
                     ncol = 5)
dim(theta_item)
```

```
## [1] 50  5
```

```
head(theta_item,n = 3)
```

```
##      [,1]  [,2]   [,3]   [,4]  [,5]
## [1,] 0.25 0.260 0.0289 0.0461 0.415
## [2,] 0.25 0.276 0.0307 0.0443 0.399
## [3,] 0.25 0.366 0.0406 0.0344 0.309
```

Store this in a data frame:

```
sim_data_cx <- tibble(item = 1:N_obs,
                      complexity = complexity,
                      w_ans = c(rcat(N_obs,theta_item)))
sim_data_cx
```

```
## # A tibble: 50 x 3
##    item complexity w_ans
##   <int>      <dbl> <dbl>
## 1     1     -0.560     5
## 2     2     -0.230     2
## 3     3      1.56      2
## # i 47 more rows
```

The following model (saved in `mpt_cat.stan`) is essentially doing the same thing as the previous model but instead of fitting a multinomial to the summary of all the trials, it is fitting a categorical distribution to each individual observation. (This is analogous to the difference between the Bernoulli and binomial distributions).

This is still not an appropriate model for the generative process that we are assuming in this section, because it still ignores the effect of complexity. But it is a good start.

```
data {
  int<lower = 1> N_obs;
  array[N_obs] int<lower = 1, upper = 5> w_ans;
}
parameters {
  real<lower=0,upper=1> a;
  real<lower=0,upper=1> t;
  real<lower=0,upper=1> f;
  real<lower=0,upper=1> c;
}
transformed parameters {
  array[N_obs] simplex[5] theta;
  for(n in 1:N_obs){
    //Pr_NR:
    theta[n, 1] = 1 - a;
    //Pr_Neologism:
    theta[n, 2] = a * (1 - t) * (1 - f) * (1 - c) +
      a * t * (1 - f) * (1 - c);
    //Pr_Formal:
    theta[n, 3] = a * (1 - t) * (1 - f) * c +
      a * t * (1 - f) * c;
    //Pr_Mixed:
    theta[n, 4] = a * (1 - t) * f;
    //Pr_Correct:
    theta[n, 5] = a * t * f;
  }
}
model {
  target += beta_lpdf(a | 2, 2);
  target += beta_lpdf(t | 2, 2);
  target += beta_lpdf(f | 2, 2);
  target += beta_lpdf(c | 2, 2);
  for(n in 1:N_obs)
```

```
    target += categorical_lpmf(w_ans[n] | theta[n]);
}
generated quantities{
    array[N_obs] int pred_w_ans;
  for(n in 1:N_obs)
    pred_w_ans[n] = categorical_rng(theta[n]);
}
```

An important aspect of the previous model is that `theta` is declared as `array[N_obs] simplex[5] theta;`. This means that `theta` is an array of simplexes and thus has now two dimensions: each element of the array (of length `N_obs`) is a simplex and sums to one. That's why we iterate over the `N_obs`. However, one limitation of the previous model is that the latent parameters `a`, `t`, `f`, `c` are declared as `real` and they do not vary in each iteration of the loop. Before moving to the next section, you might want to do online exercise G.16.3, where you are asked to edit the previous chunk of code to incorporate the fact that `f` is now a transformed parameter that depends on the trial information and two new parameters.

16.2.4 A hierarchical MPT

The previous model doesn't take into account that subjects might vary (and neither does the modification to this model that is suggested in online exercise G.16.3). Let's focus on taking into account the differences between subjects.

Different subjects might not be equally motivated to do the task. This can be accounted for by adding hierarchical structure to the parameter `a`, the probability of initiating an attempt. Begin by simulating some data that incorporates by-subject variability.

First, define the number of items and subjects, and the number of observations:

```
N_item <- 20
N_subj <- 30
N_obs <- N_item * N_subj
```

Then, generate a vector for subjects and for items. Assume here that each subject sees each item.

```
subj <- rep(1:N_subj, each = N_item)
item <- rep(1:N_item, time = N_subj)
```

A vector representing complexity is created for the number of items we have, and this vector is repeated as many times as there are subjects:

```
complexity <- rep(rnorm(N_item), times = N_subj)
```

Next, create a data frame with all the above information:

```
(exp_sim <- tibble(subj = subj,
                   item = item,
                   complexity = complexity))
```

```
## # A tibble: 600 x 3
##     subj  item complexity
##    <int> <int>      <dbl>
## 1      1     1     -0.560
## 2      1     2     -0.230
## 3      1     3      1.56
## # i 597 more rows
```

To create subject-level variability in the data, a between-subject standard deviation needs to be defined. This standard deviation represents the deviations of subjects about the grand mean. We are defining this adjustment in log-odds space.

```
# New parameters, in log-odds space:
tau_u_a <- 1.1
## generate subject adjustments in log-odds space:
u_a <- rnorm(N_subj, 0, tau_u_a)
str(u_a)
```

```
##  num [1:30] -1.175 -0.24 -1.129 -0.802 -0.688 ...
```

Given the fixed `a_true` probability value of 0.75, the subject-level values for individual `a_true` can be derived by (a) first converting the overall `a_true` value to log-odds space, (b) adding the by-subject adjustment to this converted overall value, and (c) then converting back to probability space using the logistic or inverse logit (`plogis()`) function. Essentially we generate data assuming the following:

$$
\begin{aligned}
a'_{h,n} &= \alpha_a + u_{a,subj[n]} \\
a_{h,n} &= \text{logit}^{-1}(a'_{h,n})
\end{aligned}
\tag{16.7}
$$

Where $u_{a,subj[n]}$ is a vector with the same length as the total number of observations. The meaning of this notation was explained in section 9.1.

This is done in R as follows:

```
a_true <- 0.75 # as before
## convert the intercept to log-odds space:
alpha_a <- qlogis(a_true)
## a_h' in log-odds space:
a_h_prime <-  alpha_a + u_a[subj]
## convert back to probability space
a_true_h <- plogis(a_h_prime)
str(a_true_h)
```

```
##  num [1:600] 0.481 0.481 0.481 0.481 0.481 ...
```

What this achieves mathematically is adding varying intercepts by subjects to `alpha_a`, and then the values adjusted by subject are saved in probability space.

As before, `f_true` is computed as a function of complexity:

```
alpha_f <- 0.3; beta_f <- -0.3
f_true <- plogis(alpha_f + complexity * beta_f)
```

We continue with the same probability functions and the rest of the true point values remain the same as well.

```
t_true <- 0.9; c_true <- 0.1
Pr_NR <- function(a, t, f, c)
    1 - a
Pr_Neologism <- function(a, t, f, c)
    a * (1 - t) * (1 - f) * (1 - c) + a * t * (1 - f) * (1 - c)
Pr_Formal <- function(a, t, f, c)
    a * (1 - t) * (1 - f) * c +  a * t * (1 - f) * c
Pr_Mixed <- function(a, t, f, c)
    a * (1 - t) * f
Pr_Correct <- function(a, t, f, c)
    a * t * f
```

Now, we can define the probabilities of different outcomes:

```
# Aux. parameters that define the probabilities:
theta_NR_v_h <- Pr_NR(a_true_h, t_true, f_true, c_true)
theta_Neologism_v_h <- Pr_Neologism(a_true_h, t_true, f_true, c_true)
theta_Formal_v_h <- Pr_Formal(a_true_h, t_true, f_true, c_true)
theta_Mixed_v_h <- Pr_Mixed(a_true_h, t_true, f_true, c_true)
theta_Correct_v_h <- Pr_Correct(a_true_h, t_true, f_true, c_true)
theta_h <- matrix(c(theta_NR_v_h,
                    theta_Neologism_v_h,
                    theta_Formal_v_h,
                    theta_Mixed_v_h,
                    theta_Correct_v_h),
                  ncol = 5)
dim(theta_h)
```

```
## [1] 600   5
```

The probability specifications shown above can now generate the simulated data:

```
(sim_data_h <- mutate(exp_sim,
                      w_ans = rcat(N_obs,theta_h)))
```

```
## # A tibble: 600 x 4
##     subj  item complexity w_ans
##    <int> <int>      <dbl> <dbl>
## 1      1     1     -0.560     2
## 2      1     2     -0.230     1
## 3      1     3      1.56      1
## # i 597 more rows
```

Next, define the following model; we omit the steps with f' and a' and directly apply the logistic function to a regression. The parameters t, c do not vary by item or subject and therefore do not have the subscript $_n$. We start by defining relatively weak priors for all the parameters in the following model. (See how we decided on the priors of α and β in the logistic regression example of section 4.3.2).

$$
\begin{aligned}
\alpha_a, \alpha_f &\sim Normal(0, 1.5)\\
\beta_f &\sim Normal(0, 1)\\
t, c &\sim Beta(2, 2)\\
\tau_u &\sim Normal(0, 1)\\
u_a &\sim Normal(0, \tau_{u_a})\\
a_n &= logit^{-1}(\alpha_a + u_{a,subj[n]})\\
f_n &= logit^{-1}(\alpha_f + complexity_n \cdot \beta_f)\\
\theta_{n,nr} &= 1 - a_n\\
\theta_{n,neol.} &= a_n \cdot (1-t) \cdot (1-f_n) \cdot (1-c) + a_n \cdot t \cdot (1-f_n) \cdot (1-c)\\
\theta_{n,formal} &= a_n \cdot (1-t) \cdot (1-f_n) \cdot c + a_n \cdot t \cdot (1-f_n) \cdot c\\
\theta_{n,mix} &= a_n \cdot (1-t) \cdot f_n\\
\theta_{n,corr} &= a_n \cdot t \cdot f_n\\
\theta_n &= \langle \theta_{n,nr}, \theta_{n,neol.}, \theta_{n,formal}, \theta_{n,mix}, \theta_{n,corr} \rangle\\
ans_n &\sim Categorical(\theta_n)
\end{aligned}
\tag{16.8}
$$

The corresponding Stan model `mpt_h.stan` will look like this:

```
data {
  int<lower = 1> N_obs;
  array[N_obs] int<lower = 1, upper = 5> w_ans;
  array[N_obs] real complexity;
  int<lower = 1> N_subj;
  array[N_obs] int<lower = 1, upper = N_subj> subj;
}
parameters {
  real<lower = 0, upper = 1> t;
  real<lower = 0, upper = 1> c;
  real alpha_a;
  real<lower = 0> tau_u_a;
  vector[N_subj] u_a;
  real alpha_f;
  real beta_f;
}
transformed parameters {
  array[N_obs] simplex[5] theta;
  for (n in 1:N_obs){
    real a = inv_logit(alpha_a + u_a[subj[n]]);
    real f = inv_logit(alpha_f + complexity[n] * beta_f);
    //Pr_NR
```

```
    theta[n, 1] = 1 - a;
    //Pr_Neologism
    theta[n, 2] = a * (1 - t) * (1 - f) * (1 - c) +
      a * t * (1 - f) * (1 - c);
    //Pr_Formal
    theta[n, 3] = a * (1 - t) * (1 - f) * c
      + a * t * (1 - f) * c;
    //Pr_Mixed
    theta[n, 4] = a * (1 - t) * f;
    //Pr_Correct
    theta[n, 5] = a * t * f;
  }
}
model {
  target += beta_lpdf(t | 2, 2);
  target += beta_lpdf(c | 2, 2);
  target += normal_lpdf(alpha_a | 0, 1.5);
  target += normal_lpdf(alpha_f | 0, 1.5);
  target += normal_lpdf(beta_f | 0, 1);
  target += normal_lpdf(u_a | 0, tau_u_a);
  target += normal_lpdf(tau_u_a | 0, 1) - normal_lccdf(0 | 0, 1);
  for(n in 1:N_obs)
    target +=  categorical_lpmf(w_ans[n] | theta[n]);
}
generated quantities{
  array[N_obs] int<lower = 1, upper = 5> pred_w_ans;
  for(n in 1:N_obs)
    pred_w_ans[n] = categorical_rng(theta[n]);
}
```

For ease of exposition, we are not using the non-centered parameterization discussed previously in section 9.1.2. We could also apply it here; that will speed up and improve the convergence of the model. See online exercise G.16.4.

It would be a good idea to plot prior predictive distributions for this model, but we skip this step here. Next, fit the model to the simulated data, by first defining the data as a list:

```
sim_list_h <-  list(N_obs = nrow(sim_data_h),
                    w_ans = sim_data_h$w_ans,
                    N_subj = max(sim_data_h$subj),
```

```
                    subj = sim_data_h$subj,
                    complexity = sim_data_h$complexity)
```

```
mpt_h <- system.file("stan_models",
                     "mpt_h.stan",
                     package = "bcogsci")
fit_mpt_h <- stan(mpt_h, data = sim_list_h)
```

Print out a summary of the posterior:

```
print(fit_mpt_h,
      pars = c("t", "c", "tau_u_a", "alpha_a", "alpha_f", "beta_f"))
```

```
##          mean  2.5% 97.5% n_eff Rhat
## t        0.91  0.87  0.94  5039    1
## c        0.11  0.07  0.16  4970    1
## tau_u_a  0.99  0.67  1.42  2163    1
## alpha_a  1.02  0.61  1.43  1452    1
## alpha_f  0.25  0.06  0.45  5384    1
## beta_f  -0.22 -0.41 -0.02  5265    1
```

If we had fit this to real data, we would now conclude that:

(i) given the value of `beta_f`, complexity has an adverse effect on the probability of retrieving the correct phonemes, and

(ii) given the posterior distribution of `tau_u_a`, there is a great deal of variation in the subjects' probability of initiating an attempt at each trial. Furthermore, if we had some expectation about t and c based on previous research, we could conclude that our results are in line (or not) with previous findings.

One could inspect how one unit of complexity is affecting the probability of retrieving the correct phoneme (f). We first derive the value of f for an item of zero complexity (that is $\alpha_f + 0 \times \beta_f$) and then the value of f for an item with a complexity of one ($\alpha_f + 1 \times \beta_f$). We are interested in summarizing the difference between the two:

```
as.data.frame(fit_mpt_h) %>%
  select(alpha_f, beta_f) %>%
```

```
  mutate(f_0 = plogis(alpha_f),
         f_1 = plogis(alpha_f + beta_f),
         diff_f = f_1 - f_0) %>%
  summarize(Estimate = mean(diff_f),
            `2.5%` = quantile(diff_f, 0.025),
            `97.5%` = quantile(diff_f, 0.975))
```

```
##   Estimate   2.5%   97.5%
## 1   -0.054 -0.103 -0.0053
```

It is worth highlighting that MPT models (like other cognitive models) should be experimentally validated before interpreting the parameters as measures of cognitive processes. Particularly powerful are tests of selective influence, where one demonstrates that each of the parameters can be selectively influenced as theoretically predicted, while all other parameters remain unaffected (Schmidt, Erdfelder, and Heck 2023).

One further interesting step could be to develop a competing model that assumes a different latent process, and then comparing the performance of the MPT with this competing model, using Bayes factors or K-fold-CV (or both).

Since we generated the data based on known latent parameters, we also plot the posteriors together with the true point values of the parameters in Figure 16.4. This is something that we can only do with simulated data.

```
as.data.frame(fit_mpt_h) %>%
  select(tau_u_a, alpha_a, t, alpha_f, beta_f, c) %>%
  mcmc_recover_hist(true = c(tau_u_a,
                             qlogis(a_true),
                             t_true, alpha_f,
                             beta_f, c_true))
```

If everything is correctly defined in the model, we should be able to generate posterior predictive data based on our estimates that looks quite similar to the simulated data; see Figure 16.5. The error bars in y_{rep} include 90% of the probability mass of the predictive distribution (this is a default of `ppc_bars()`). In a well-calibrated model, the data (y, here the proportion of answers of each type) should be inside the error bars in 90% of the cases.

```
gen_data <- rstan::extract(fit_mpt_h)$pred_w_ans
ppc_bars(sim_list_h$w_ans, gen_data) +
  ggtitle ("Hierarchical model")
```

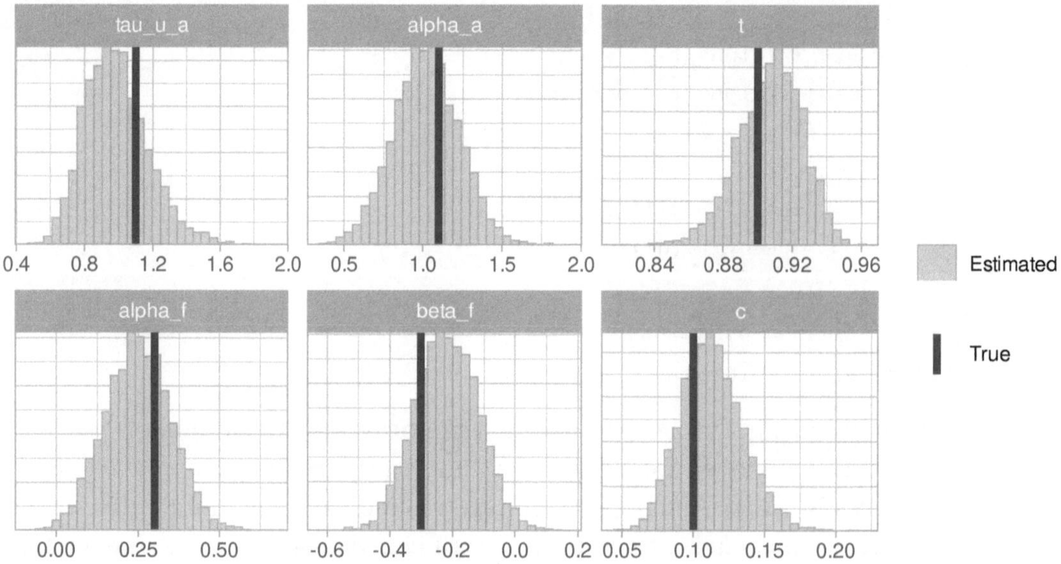

FIGURE 16.4: Posterior of the hierarchical MPT with true point values as vertical lines (model `mpt_h.stan`).

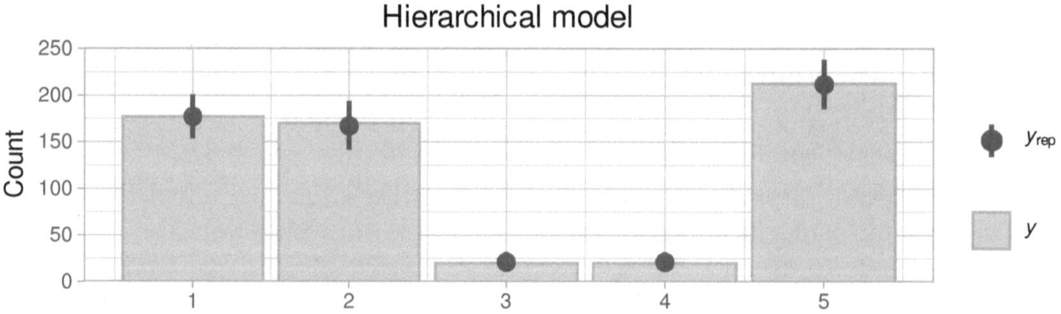

FIGURE 16.5: A posterior predictive check for aggregated data in the hierarchical MPT model.

It is also useful to look at the individual subjects' posteriors; these are shown in Figure 16.6.

```
ppc_bars_grouped(sim_list_h$w_ans,
                 gen_data, group = subj) +
  ggtitle ("By-subject plot for the hierarchical model")
```

But what about the first *non-hierarchical* MPT model (`mpt_cat.stan`)?

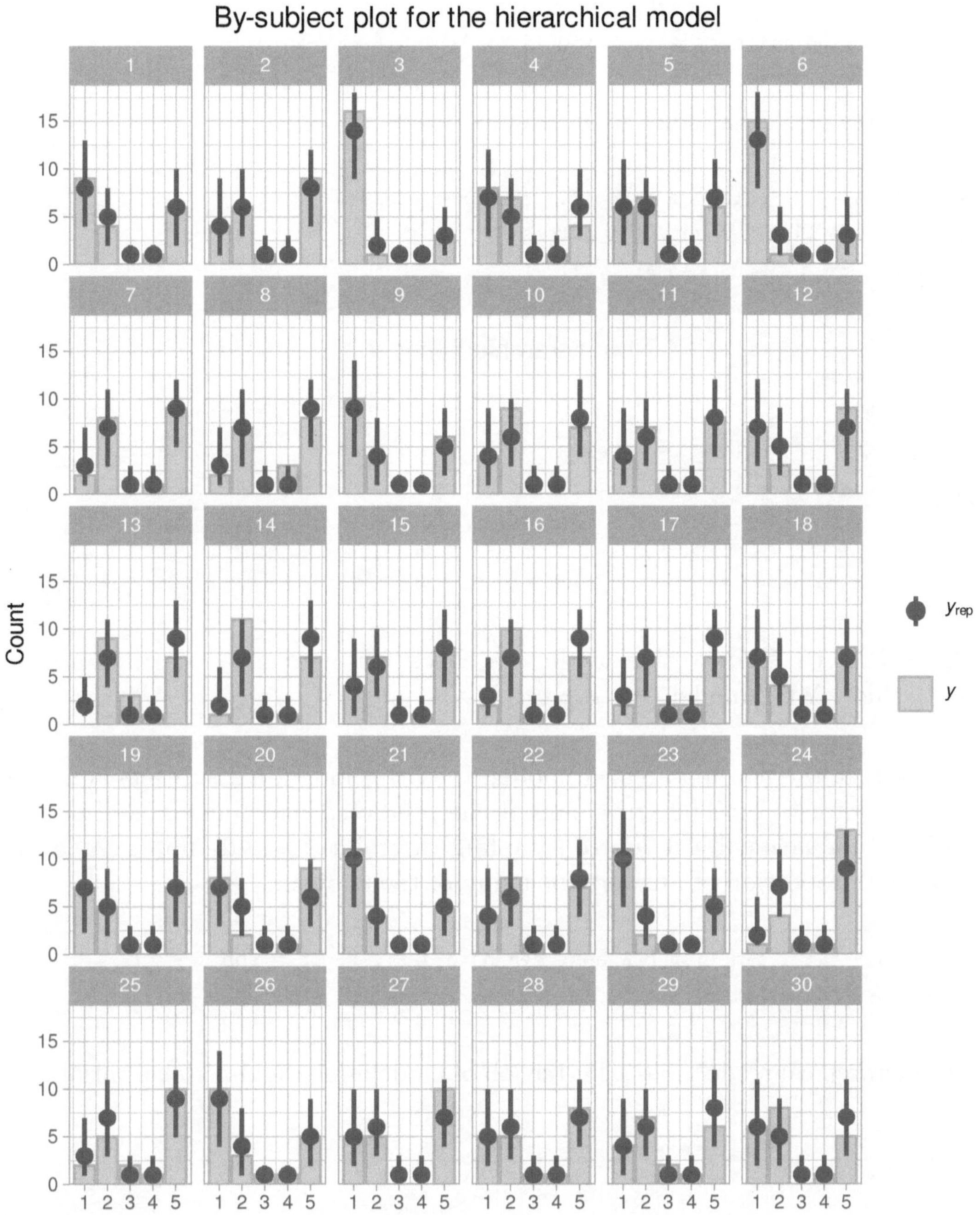

FIGURE 16.6: Individual subjects in the hierarchical MPT model.

```
mpt_cat <- system.file("stan_models",
                       "mpt_cat.stan",
                       package = "bcogsci")
fit_sh <- stan(mpt_cat, data = sim_list_h)
```

The aggregated data looks great (Figure 16.7).

```
gen_data_sMPT <- rstan::extract(fit_sh)$pred_w_ans
ppc_bars(sim_list_h$w_ans, gen_data_sMPT) +
  ggtitle ("Non-hierarchical model")
```

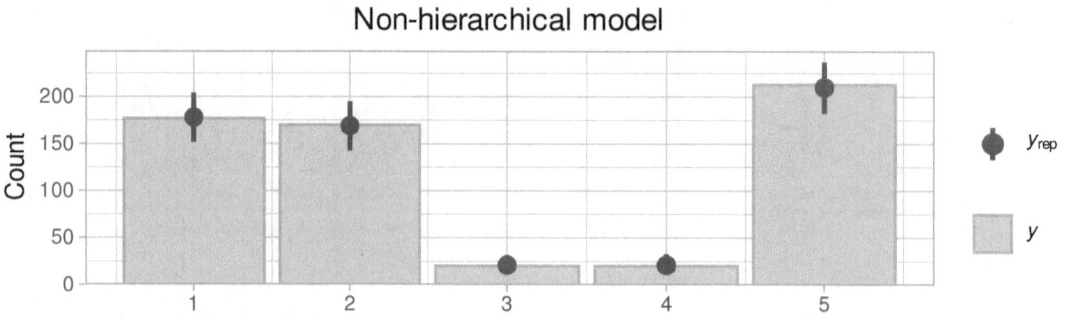

FIGURE 16.7: Posterior predictive check for aggregated data in a non-hierarchical MPT model (`mpt_cat.stan`).

However, in the non-hierarchical model, the fit to individual subjects is not as good (Figure 16.8): The error bars of the predicted distribution is of course identical for each subject (up to sampling variability) and do not include the observed proportion of answers for many of the subjects.

```
ppc_bars_grouped(sim_list_h$w_ans, gen_data_sMPT, group = subj) +
  ggtitle ("By-subject plot for the non-hierarchical model")
```

The hierarchical model does a better job of modeling individual-level variability.

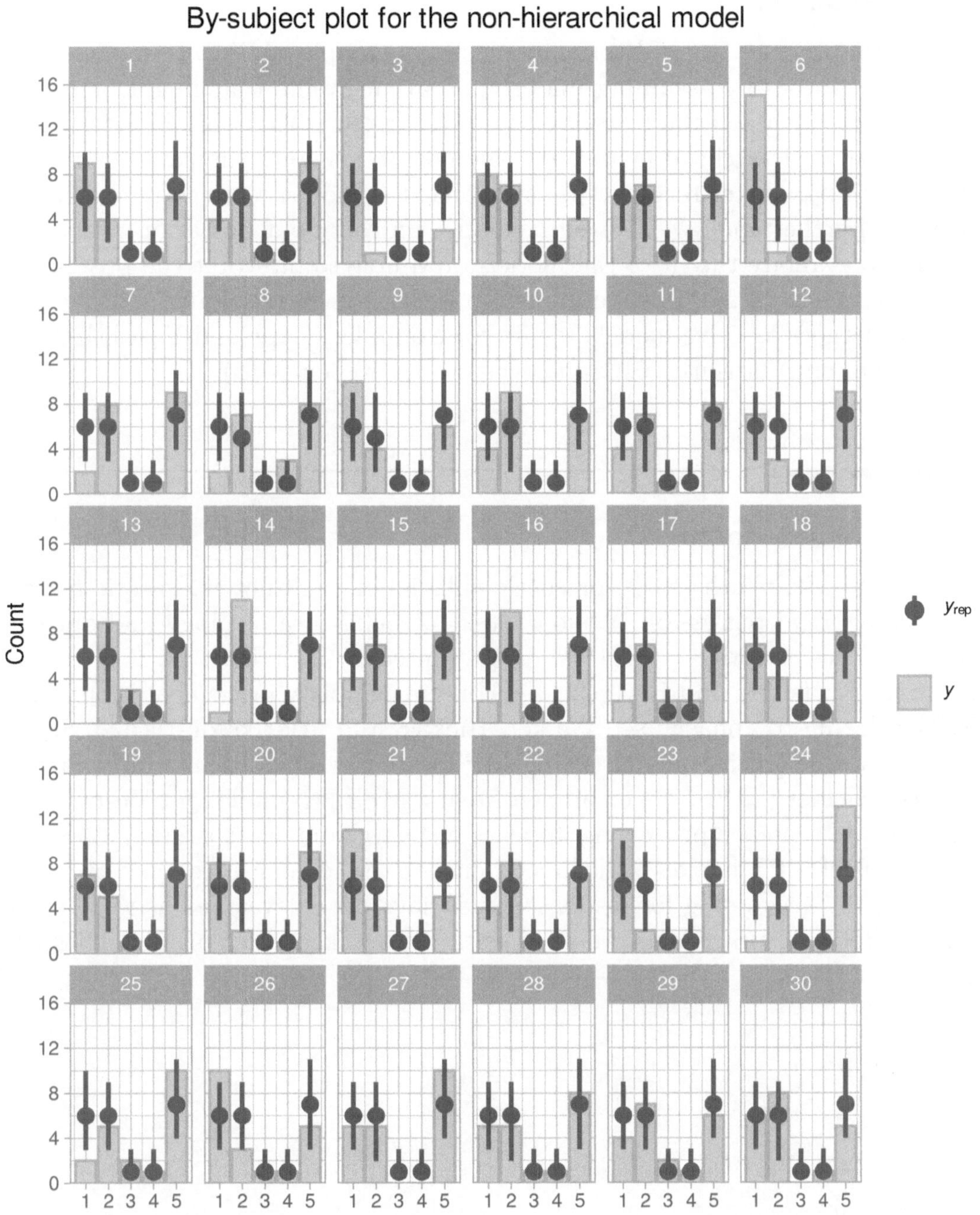

FIGURE 16.8: Individual subjects in the non-hierarchical MPT model (`mpt_cat.stan`).

16.3 Summary

In this chapter, we learned to fit increasingly complex MPTs to model categorical responses by assuming underlying latent events that might or might not happen with a certain probability. We started with a simple model and ended with a hierarchical model. We saw how to generate simulated data and investigate parameter recovery to verify that the models were correctly implemented. Furthermore, we showed how one can carry out posterior predictive checks to evaluate the fit of the models, and how to interpret the posteriors of the parameters. MPTs have a lot of potential as stand-alone models or as parts of larger cognitive models (as in, for example, Paape and Vasishth 2022; Nicenboim and Vasishth 2016; and Klauer and Kellen 2018).

16.4 Further reading

Koster and McElreath (2017) present a tutorial on multinomial logistic regression/categorical regression in the context of behavioral ecology and anthropology. Other tutorials on MPTs are presented by Matzke et al. (2015) and Schmidt, Erdfelder, and Heck (2023). For the complete implementation of an MPT relating to aphasia, see Walker, Hickok, and Fridriksson (2018). Some examples of cognitive models using MPTs are Lee et al. (2020) and Smith and Batchelder (2010). An example of using MPTs to model garden-pathing processes in sentence processing appears in Paape and Vasishth (2022). Some extensions of MPT models that directly encode the temporal sequence of latent processes by accounting for response times have been developed as well (Heck and Erdfelder 2016; Klauer and Kellen 2018; Heck, Erdfelder, and Kieslich 2018).

17

Mixture models

Mixture models integrate multiple data generating processes into a single model. This is especially useful in cases where the data alone don't allow us to fully identify which observations belong to which process. Mixture models are important in cognitive science because many theories of cognition assume that the behavior of subjects in certain tasks is determined by an interplay of different cognitive processes (e.g., response times in schizophrenia in Levy et al. 1993; retrieval from memory in sentence processing in McElree 2000; Nicenboim and Vasishth 2018; fast choices in Ollman 1966; Dutilh et al. 2011; generalized processing tree models in Heck, Erdfelder, and Kieslich 2018). It is important to stress that a mixture distribution of observations is an *assumption* of the latent process developing trial by trial based on a given theory—it doesn't necessarily represent the true generative process. The role of Bayesian modeling is to help us understand the extent to which this assumption is well-founded, by using posterior predictive checks and by comparing different models.

We focus here on the case where we have only two components; each component represents a distinct cognitive process based on the domain knowledge of the researcher. The vector $\mathbf{z}$ serves as a latent *indicator variable* that indicates which of the mixture components an observation y_n belongs to ($n = 1, \dots, N$ is the number of data points). We assume two components, and thus each z_n can be either 0 or 1 (this will allows us to generate z_n with a Bernoulli distribution). We also assume two different generative processes, p_1 and p_2, which generate different distributions of the observations based on a vector of parameters indicated by Θ_1 and Θ_2, respectively. These two processes occur with probability θ and $1 - \theta$, and each observation is generated as follows:

$$
\begin{aligned}
z_n &\sim Bernoulli(\theta) \\
y_n &\sim \begin{cases} p_1(\Theta_1), & \text{if } z_n = 1 \\ p_2(\Theta_2), & \text{if } z_n = 0 \end{cases}
\end{aligned}
\tag{17.1}
$$

We focus on only two components because this type of model is already hard to fit and, as we show in this chapter, it requires plenty of prior information to be able to sample from the posterior in most applied situations. However, the approach presented here can in principle be extended to a larger number of

DOI: 10.1201/9780429342646-23

mixtures by replacing the Bernoulli distribution with a categorical one. This can be done if the number of components in the mixture is finite; the number of components is determined by the researcher.

In order to fit this model, we need to estimate the posterior of each of the parameters contained in the vectors Θ_1 and Θ_2 (intercepts, slopes, group-level effects, etc.), the probability θ, and the indicator variable that corresponds to each observation z_n. One issue that presents itself here is that z_n must be a discrete parameter, and Stan only allows continuous parameters. This is because Stan's algorithm requires the derivatives of the (log) posterior distribution with respect to all parameters, and discrete parameters are not differentiable (since they have "breaks"). In probabilistic programming languages like WinBUGS (Lunn et al. 2012), JAGS (Plummer 2016), PyMC3 (Salvatier, Wiecki, and Fonnesbeck 2016) and Turing (Ge, Xu, and Ghahramani 2018), discrete parameters are possible to use; but not in Stan. In Stan, we can circumvent this issue by marginalizing out the indicator variable z.[1] If p_1 appears in the mixture with probability θ, and p_2 with probability $1 - \theta$, then the joint likelihood is defined as a function of Θ (which concatenates the mixing probability, θ, and the parameters of the p_1 and p_2, Θ_1 and Θ_2), and importantly z_n "disappears":

$$p(y_n|\Theta) = \theta \cdot p_1(y_n|\Theta_1) + (1 - \theta) \cdot p_2(y_n|\Theta_2) \tag{17.2}$$

The intuition behind this formula is that each likelihood function, p_1, p_2 is weighted by its probability of being the relevant generative process. For our purposes, it suffices to say that marginalization works; the reader interested in the mathematics behind marginalization is directed to the further reading section at the end of the chapter.[2]

Even though Stan cannot fit a model with the discrete indicator of the latent class $\mathbf{z}$ that we used in Equation (17.1), this equation will prove very useful when we want to generate synthetic data.

In the following sections, we model a well-known phenomenon (i.e., the speed-accuracy trade-off) assuming an underlying finite mixture process. We start from the verbal description of the model, and then implement the model step by step in Stan.

---

[1]See section 1.6.1.1 in chapter 1 for a review on the concept of marginalization.

[2]As mentioned above, other probabilistic languages that do not rely on Hamiltonian dynamics (exclusively) are able to deal with this. However, even when sampling discrete parameters is possible, marginalization is more efficient (Yackulic et al. 2020): when z_n is omitted we are fitting a model with n fewer parameters.

17.1 A mixture model of the speed-accuracy trade-off: The fast-guess model account

When we are faced with multiple choices that require an immediate decision, we can speed up the decision at the expense of accuracy and become more accurate at the expense of speed; this is called the speed-accuracy trade-off (Wickelgren 1977). The most popular class of models that can incorporate both response times and accuracy, and give an account for the speed-accuracy trade-off is the class of sequential sampling models, which include the drift diffusion model (Ratcliff 1978), the linear ballistic accumulator (Brown and Heathcote 2008), and the log-normal race model (Heathcote and Love 2012; Rouder et al. 2015), which we discuss in chapter 18; for a review see Ratcliff et al. (2016).

However, an alternative model that has been proposed in the past is Ollman's simple fast-guess model (Ollman 1966; Yellott 1967, 1971).[3] Although it has mostly fallen out of favor (but see Dutilh et al. 2011; and Heck and Erdfelder 2020 for more modern variants of this model), it presents a very simple framework using finite mixture modeling that can also account for the speed-accuracy trade-off. In the next sections, we'll use this model to exemplify the use of finite mixtures to represent different cognitive processes.

17.1.1 The global motion detection task

One way to examine the behavior of human and primate subjects when faced with two-alternative forced choices is the detection of the global motion of a random dot kinematogram (Britten et al. 1993). In this task, a subject sees a number of random dots on the screen. A proportion of dots move in a single direction (e.g., right) and the rest move in random directions. The subject's goal is to estimate the overall direction of the movement. One of the reasons for the popularity of this task is that it permits the fine-tuning of the difficulty of trials (Dutilh et al. 2019): The task is harder when the proportion of dots that move coherently (the level of *coherence*) is lower; see Figure 17.1.

Ollman's (1966) fast-guess model assumes that the behavior in this task (and in any other choice task) is governed by two distinct cognitive processes: (i) a guessing mode and (ii) a task-engaged mode. In the guessing mode, responses are fast and accuracy is at chance level. In the task-engaged mode, responses are slower and accuracy approaches 100%. This means that intermediate values

[3] Ollman's original model was meant to be relevant only for means; Yellott (1967, 1971) generalized it to a distributional form.

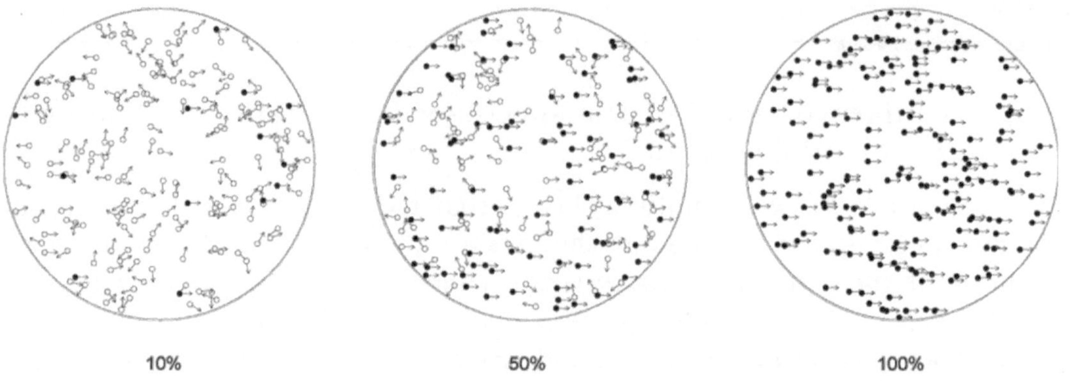

FIGURE 17.1: Three levels of difficulty of the global motion detection task. The figures show a consistent movement to the right with three levels of coherence (10%, 50%, and 100%). The subjects see the dots moving in the direction indicated by the arrows. The subjects do not see the arrows and all the dots look identical in the actual task. Adapted from Han et al. (2018); licensed under CC BY 4.0 (`https://creativecommons.org/licenses/by/4.0/`).

of response times and accuracy can only be achieved by mixing responses from the two modes. Further assumptions of this model are that response times depend on the difficulty of the choice, and that the probability of being on one of the two states depend on the speed incentives during the instructions.

To simplify matters, we ignore the possibility that the accuracy of the choice is also affected by the difficulty of the choice. Also, we ignore the possibility that subjects might be biased to one specific response in the guessing mode, but see exercise G.17.3 in the online materials.

17.1.1.1 The data set

We implement the assumptions behind Ollman's fast-guess model and examine its fit to data of a global motion detection task from Dutilh et al. (2019).

The data set from Dutilh et al. (2019) contains approximately 2800 trials of each of the 20 subjects participating in a global motion detection task and can be found in `df_dots` in the `bcogsci` package. There were two level of coherence, yielding hard and easy trials (`diff`), and the trials where done under instructions that emphasized either accuracy or speed (`emphasis`). More information about the data set can be found by accessing the documentation for the data set (by typing `?df_dots` on the R command line, assuming that the `bcogsci` package is installed).

```
data("df_dots")
df_dots
```

```
## # A tibble: 56,097 x 12
##    subj diff  emphasis    rt   acc fix_dur stim  resp  trial block
##   <int> <chr> <chr>    <dbl> <int>   <dbl> <chr> <chr> <int> <int>
## 1     1 easy  speed      482     1   0.738 R     R         1     6
## 2     1 hard  speed      602     1   0.784 R     R         2     6
## 3     1 hard  speed      381     1   0.651 R     R         3     6
##   block_trial bias
##         <int> <chr>
## 1           1 no
## 2           2 no
## 3           3 no
## # i 56,094 more rows
```

We might think that if the fast-guess model were true, we would see a bimodal distribution, when we plot a histogram of the data. Unfortunately, when two similar distributions are mixed, we won't necessarily see any apparent bimodality; see Figure 17.2.

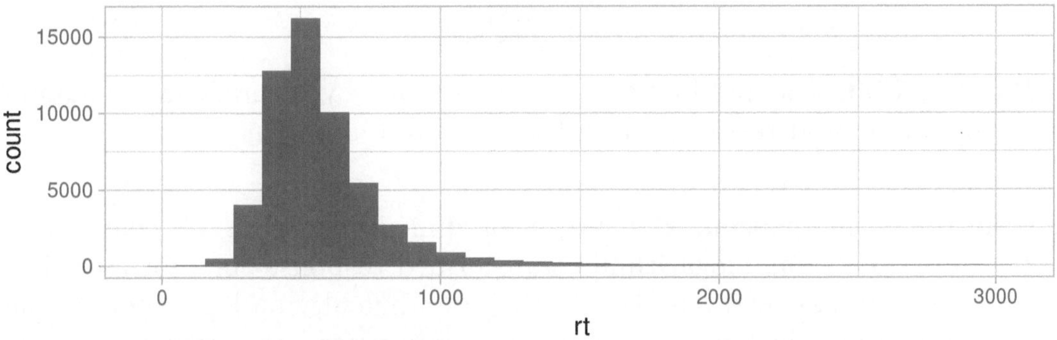

FIGURE 17.2: Distribution of response times in the data of the global motion detection task in Dutilh et al. (2019).

However, Figure 17.3 reveals that incorrect responses were generally faster, and this was especially true when the instructions emphasized accuracy.

17.1.2 A very simple implementation of the fast-guess model

The description of the model makes it clear that an ideal subject who never guesses has a response time that depends only on the difficulty of the trial. As we did in previous chapters, we assume that response times are log-normally distributed, and for simplicity we start by modeling the behavior of a single subject:

$$rt_n \sim LogNormal(\alpha + \beta \cdot x_n, \sigma) \tag{17.3}$$

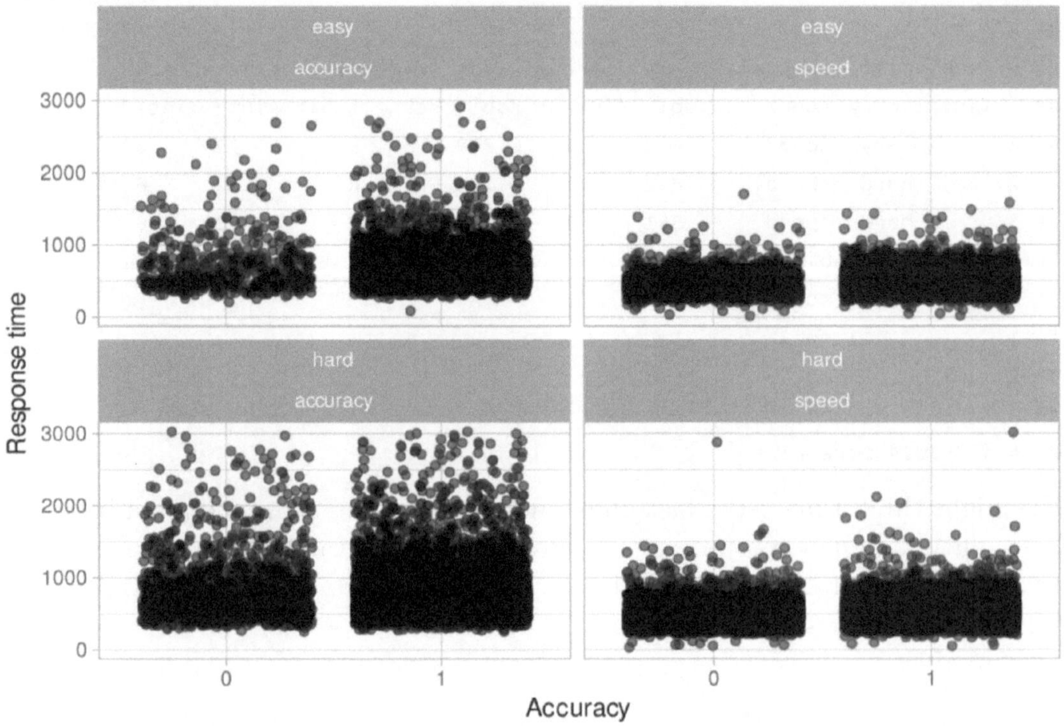

FIGURE 17.3: The distribution of response times by accuracy in the data of the global motion detection task in Dutilh et al. (2019).

In the previous equation, x is larger for difficult trials. If we center x, α represents the average logarithmic transformed response time for a subject engaged in the task, and β is the effect of trial difficulty on log-response time. We assume a non-deterministic process, with a noise parameter σ. Also see the online section A.4 for more information about log-normally distributed response times.

Alternatively, a subject that guesses in every trial would show a response time distribution that is independent of the difficulty of the trial:

$$rt_n \sim LogNormal(\gamma, \sigma_2) \tag{17.4}$$

Here γ represents the average logarithmic transformed response time when a subject only guesses. We assume that responses from the guessing mode might have a different noise component than from the task-engaged mode.

The fast-guess model makes the assumption that during a task, a single subject would behave in these two ways: They would be engaged in the task a proportion of the trials and would guess on the rest of the trials. This means that for a single subject, there is an underlying probability of being engaged in the task,

p_{task}, that determines whether they are actually choosing ($z = 1$) or guessing ($z = 0$):

$$z_n \sim Bernoulli(p_{task}) \tag{17.5}$$

The value of the parameter z in every trial determines the behavior of the subject. This means that the distribution that we observe is a mixture of the two distributions presented before:

$$rt_n \sim \begin{cases} LogNormal(\alpha + \beta \cdot x_n, \sigma), & \text{if } z_n = 1 \\ LogNormal(\gamma, \sigma_2), & \text{if } z_n = 0 \end{cases} \tag{17.6}$$

In order to have a Bayesian implementation, we also need to define some priors. We use priors that encode what we know about response time experiments. These priors are slightly more informative than the ones that we used in section 4.2, but they still can be considered regularizing priors. One can verify this by performing prior predictive checks. As we increase the complexity of our models, it's worth spending some time designing more realistic priors. These will speed up computation and in some cases they will be crucial for solving convergence problems.

$$\begin{aligned} \alpha &\sim Normal(6, 1) \\ \beta &\sim Normal(0, 0.1) \\ \sigma &\sim Normal_+(0.5, 0.2) \end{aligned} \tag{17.7}$$

$$\begin{aligned} \gamma &\sim Normal(6, 1) \\ \sigma_2 &\sim Normal_+(0.5, 0.2) \end{aligned} \tag{17.8}$$

For now, we allow all values for the probability of having an engaged response equal likelihood; we achieve this by setting the following prior to p_{task}:

$$p_{task} \sim Beta(1, 1) \tag{17.9}$$

This represents a flat, uninformative prior over the probability parameter p_{task}.

Before we fit our model to the real data, we generate synthetic data to make sure that our model is working as expected.

We first define the number of observations, predictors, and fixed point values for each of the parameters. We assume 1000 observations and two levels of difficulty, x, coded -0.5 (easy) and 0.5 (hard). The point values chosen for the parameters are relatively realistic (based on our previous experience on response time experiments). Although in the priors we try to encode the range of possible values for the parameters, in this simulation we assume only one instance of this possible range:

```
N <- 1000
# level of difficulty:
x <- c(rep(-0.5, N/2), rep(0.5, N/2))
# Parameters true point values:
alpha <- 5.8
beta <- 0.05
sigma <- 0.4
sigma2 <- 0.5
gamma <- 5.2
p_task <- 0.8
# Median time
c("engaged" = exp(alpha), "guessing" = exp(gamma))
```

```
##  engaged guessing
##      330      181
```

For generating a mixture of response times, we use the indicator of a latent class, `z`.

```
z <- rbern(n = N, prob = p_task)
rt <- if_else(z == 1,
              rlnorm(N,
                     meanlog = alpha + beta * x,
                     sdlog = sigma),
              rlnorm(N,
                     meanlog = gamma,
                     sdlog = sigma2))
df_dots_simdata1 <- tibble(trial = 1:N, x = x, rt = rt)
```

We verify that our simulated data is realistic, that is, it's in the same range as the original data; see Figure 17.4.

```
ggplot(df_dots_simdata1, aes(rt)) +
  geom_histogram()
```

To implement the mixture model defined in Equation (3.14) in Stan, the discrete parameter z needs to be marginalized out:

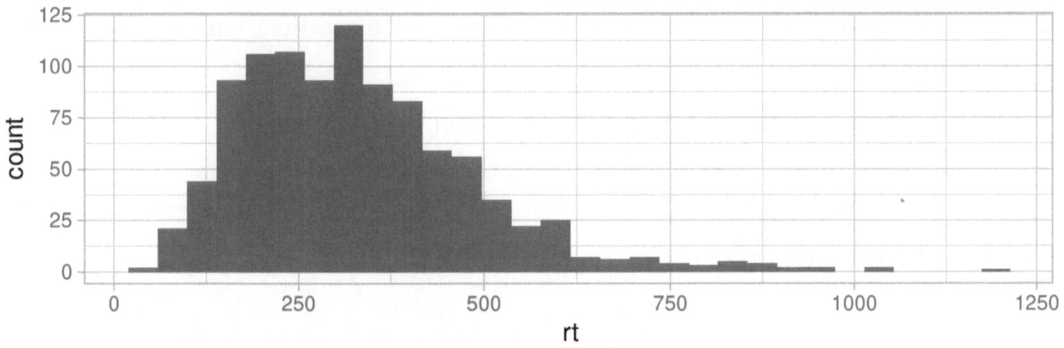

FIGURE 17.4: Response times in the simulated data (`df_dots_simdata1`) that follows the fast-guess model.

$$\begin{aligned} p(rt_n|\Theta) = p_{task} \cdot LogNormal(rt_n|\alpha + \beta \cdot x_n, \sigma) + \\ (1 - p_{task}) \cdot LogNormal(rt_n|\gamma, \sigma_2) \end{aligned} \tag{17.10}$$

In addition, Stan requires the likelihood to be defined in log-space:

$$\begin{aligned} \log(p(rt|\Theta)) = \log(p_{task} \cdot LogNormal(rt_n|\alpha + \beta \cdot x_n, \sigma) + \\ (1 - p_{task}) \cdot LogNormal(rt_n|\gamma, \sigma_2)) \end{aligned} \tag{17.11}$$

A "naive" implementation in Stan would look like the following (recall that `_lpdf` functions provide log-transformed densities):

```
target += log(
        p_task * exp(lognormal_lpdf(rt[n] | alpha + x[n] * beta, sigma)) +
              (1-p_task) * exp(lognormal_lpdf(rt[n] | gamma, sigma2)));
```

However, we need to take into account that $log(A \pm B)$ can be numerically unstable (i.e., prone to underflow/overflow, see Blanchard, Higham, and Higham 2020). Stan provides several functions to deal with different special cases of logarithms of sums and differences. Here we need `log_sum_exp(x, y)` that corresponds to `log(exp(x) + exp(y))` and `log1m(x)` that corresponds to `log(1-x)`.

First, we need to take into account that the first summand of the logarithm, `p_task * exp(lognormal_lpdf(rt[n] | alpha + x[n] * beta, sigma))` corresponds to `exp(x)`, and the second one, `(1-p_task) * exp(lognormal_lpdf(rt[n] | gamma, sigma2))` to `exp(y)` in `log_sum_exp(x, y)`. This means that we need to first apply the logarithm to each of them to use them as arguments of `log_sum_exp(x, y)`:

```
target += log_sum_exp(
              log(p_task) + lognormal_lpdf(rt[n] | alpha +
              x[n] * beta, sigma),
              log(1-p_task) + lognormal_lpdf(rt[n] | gamma, sigma2));
```

Now we can just replace `log(1-p_task)` by the more stable `log1m(p_task)`:

```
target += log_sum_exp(
              log(p_task) + lognormal_lpdf(rt[n] | alpha +
              x[n] * beta, sigma),
              log1m(p_task) + lognormal_lpdf(rt[n] | gamma, sigma2));
```

The complete model (`mixture_rt.stan`) is shown below:

```
data {
  int<lower = 1> N;
  vector[N] x;
  vector[N] rt;
}
parameters {
  real alpha;
  real beta;
  real<lower = 0> sigma;
  real gamma; //guessing
  real<lower = 0> sigma2;
  real<lower = 0, upper = 1> p_task;
}
model {
  // priors for the task component
  target += normal_lpdf(alpha | 6, 1);
  target += normal_lpdf(beta | 0, 0.1);
  target += normal_lpdf(sigma | 0.5, 0.2)
    - normal_lccdf(0 | 0.5, 0.2);
  // priors for the guessing component
  target += normal_lpdf(gamma | 6, 1);
  target += normal_lpdf(sigma2 | 0.5, 0.2)
    - normal_lccdf(0 | 0.5, 0.2);
  target += beta_lpdf(p_task | 1, 1);
  // likelihood
  for(n in 1:N)
    target +=
      log_sum_exp(log(p_task) +
                  lognormal_lpdf(rt[n] | alpha + x[n] * beta, sigma),
                  log1m(p_task) +
                  lognormal_lpdf(rt[n] | gamma, sigma2));
}
```

Call the Stan model `mixture_rt.stan`, and fit it to the simulated data. First, we set up the simulated data as a list structure:

```
ls_dots_simdata <- list(N = N, rt = rt, x = x)
```

Then fit the model:

```
mixture_rt <- system.file("stan_models",
                          "mixture_rt.stan",
                          package = "bcogsci")
fit_mix_rt <- stan(mixture_rt, data = ls_dots_simdata)
```

```
## Warning: The largest R-hat is 1.74, indicating chains have not mixed.
## Running the chains for more iterations may help. See
## https://mc-stan.org/misc/warnings.html#r-hat
##
##  Warning: Bulk Effective Samples Size (ESS) is too low, indicating
## posterior means and medians may be unreliable. Running the chains for
## more iterations may help. See
## https://mc-stan.org/misc/warnings.html#bulk-ess
##
##  Warning: Tail Effective Samples Size (ESS) is too low, indicating
## posterior variances and tail quantiles may be unreliable. Running the
## chains for more iterations may help. See
## https://mc-stan.org/misc/warnings.html#tail-ess
```

There are a lot of warnings, the Rhats are too large, and number of effective samples is too low:

```
print(fit_mix_rt)
```

```
##              mean     2.5%    97.5% n_eff Rhat
## alpha        5.80     5.09     6.02     2 2.47
## beta         0.06    -0.09     0.19     5 1.30
## sigma        0.34     0.18     0.54     3 1.83
## gamma        5.62     5.21     5.97     3 1.81
## sigma2       0.47     0.27     0.56     3 2.15
## p_task       0.37     0.14     0.78    46 1.06
## lp__     -6385.68 -6390.09 -6382.78    15 1.09
```

The traceplots show clearly that the chains aren't mixing; see Figure 17.5.

```
traceplot(fit_mix_rt) +
  scale_x_continuous(breaks = NULL)
```

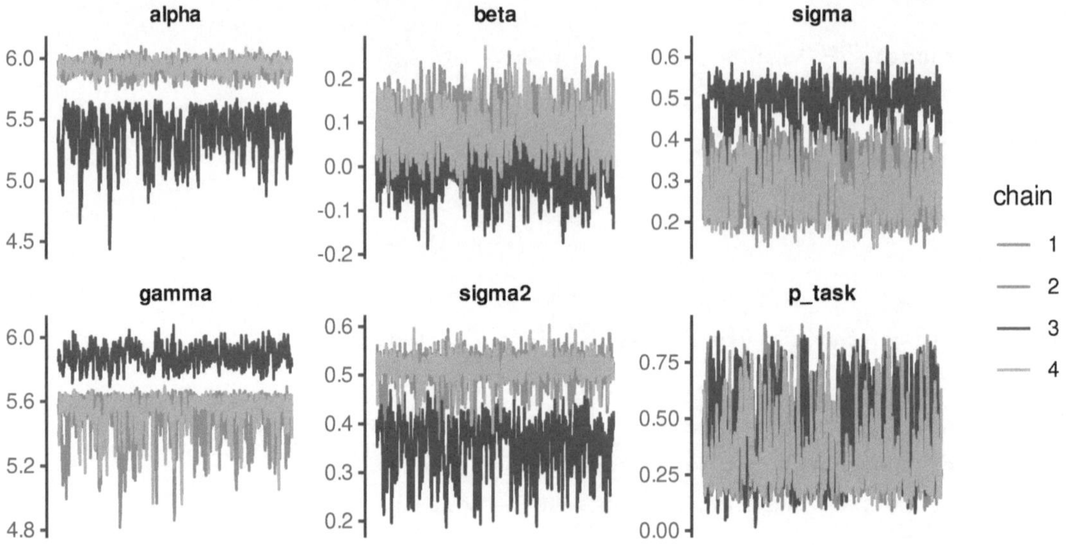

FIGURE 17.5: Traceplots from the model `mixture_rt.stan` fit to simulated data.

The problem with this model is that the mixture components (i.e., the fast-guesses and the engaged mode) are underlyingly exchangeable and thus the posterior is multimodal and the model does not converge. Each chain (and each iteration) doesn't know how each component was identified by the rest of the chains (e.g., in some chains and iterations, $z = 0$ corresponds to fast guesses, whereas in other cases, it corresponds to deliberate responses). However, we do have information that can identify the components: According to the theoretical model, we know that the average response in the engaged mode, represented by α, should be slower than the average response in the guessing mode, γ.

Even though the theoretical model assumes that guesses are faster than engaged responses, this is not explicit in our computational model. That is, our model lacks some of the theoretical information that we have, namely that the distribution of engaged response times should be slower than the distribution of guessing times. This can be encoded with *order constraints* or *inequality constraints*: a strong prior for γ, where we assume that the upper bound of its prior distribution is truncated at the value α:

$$\gamma \sim Normal(6, 1), \text{for } \gamma < \alpha \tag{17.12}$$

This would be enough to make the model converge.

Another softer constraint that we could add to our implementation is the assumption that subjects are generally more likely to be trying to do the task than just guessing. If this assumption is correct, we also improve the accuracy of our estimation of the posterior of the model. (The opposite is also true: If subjects are not trying to do the task, this assumption will be unwarranted and our prior information will lead us further from the "true" values of the parameters). The following prior has the probability density concentrated near 1.

$$p_{task} \sim Beta(8, 2) \tag{17.13}$$

Plotting this prior confirms where most of the probability mass lies; see Figure 17.6.

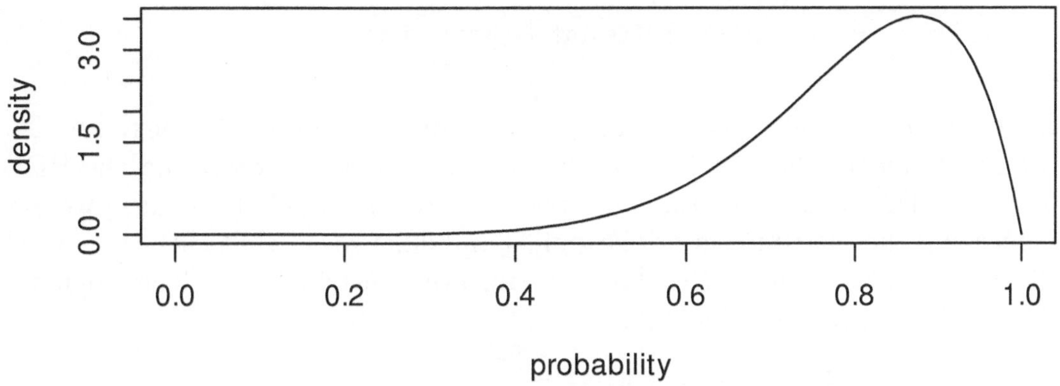

FIGURE 17.6: A density plot for the $Beta(8, 2)$ prior on p_{task}.

The Stan code for this model is shown below as `mixture_rt2.stan`.

```
data {
  int<lower = 1> N;
  vector[N] x;
  vector[N] rt;
}
parameters {
  real alpha;
  real beta;
  real<lower = 0> sigma;
  real<upper = alpha> gamma;
  real<lower = 0> sigma2;
  real<lower = 0, upper = 1> p_task;
}
model {
  target += normal_lpdf(alpha | 6, 1);
  target += normal_lpdf(beta | 0, 0.3);
  target += normal_lpdf(sigma | 0.5, 0.2)
```

```
    - normal_lccdf(0 | 0.5, 0.2);
  target += normal_lpdf(gamma | 6, 1) -
    normal_lcdf(alpha | 6, 1);
  target += normal_lpdf(sigma2 | 0.5, 0.2)
    - normal_lccdf(0 | 0.5, 0.2);
  target += beta_lpdf(p_task | 8, 2);
  for(n in 1:N)
    target +=
      log_sum_exp(log(p_task) +
                  lognormal_lpdf(rt[n] | alpha + x[n] * beta, sigma),
                  log1m(p_task) +
                  lognormal_lpdf(rt[n] | gamma, sigma2)) ;
}
```

Once we change the upper bound of `gamma` in the `parameters` block, we also need to truncate the distribution in the `model` block by correcting the PDF with its CDF. This correction is carried out using the CDF because we are truncating the distribution at the right-hand side; recall that earlier we used the complement of the CDF when we truncate a distribution at the left-hand side); see the online section A.2.

```
  target += normal_lpdf(gamma | 6, 1) -
    normal_lcdf(alpha | 6, 1);
```

Fit this model (call it `mixture_rt2.stan`) to the same simulated data set that we used before:

```
mixture_rt2 <- system.file("stan_models",
                           "mixture_rt2.stan",
                           package = "bcogsci")
fit_mix_rt2 <- stan(mixture_rt2, data = ls_dots_simdata)
```

Now the summaries and traceplots look fine; see Figure 17.7.

```
print(fit_mix_rt2)
```

```
##            mean   2.5%   97.5% n_eff Rhat
## alpha      5.84   5.75    5.94  1283    1
## beta       0.05  -0.03    0.13  2096    1
## sigma      0.39   0.31    0.45  1067    1
## gamma      5.26   4.92    5.56  1126    1
## sigma2     0.47   0.35    0.58  1233    1
## p_task     0.70   0.37    0.90  1026    1
```

```
## lp__    -6387.25 -6391.88 -6384.61  1260     1
```

```
traceplot(fit_mix_rt2) +
  scale_x_continuous(breaks = NULL)
```

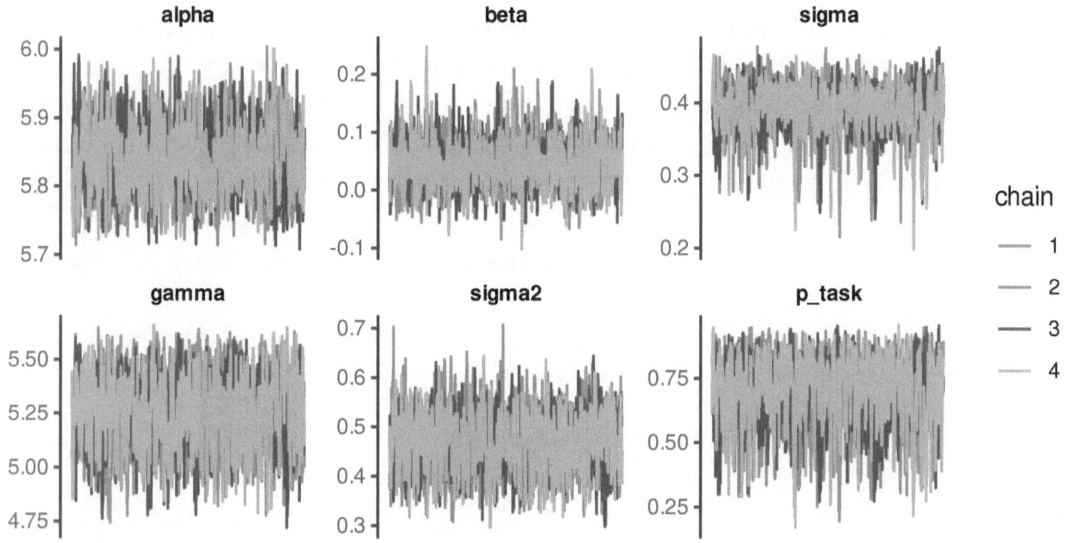

FIGURE 17.7: Traceplots from the model `mixture_rt2.stan` fit to simulated data.

17.1.3 A multivariate implementation of the fast-guess model

A problem with the previous implementation of the fast-guess model is that we ignore the accuracy information in the data. We can implement a version that is closer to the verbal description of the model: In particular, we also want to model the fact that accuracy is at chance level in the fast-guessing mode and that accuracy approaches 100% during the task-engaged mode.

This means that the mixture affects two pairs of distributions:

$$z_n \sim Bernoulli(p_{task}) \tag{17.14}$$

The response time distribution

$$rt_n \sim \begin{cases} LogNormal(\alpha + \beta \cdot x_n, \sigma), & \text{if } z_n = 1 \\ LogNormal(\gamma, \sigma_2), & \text{if } z_n = 0 \end{cases} \tag{17.15}$$

and an accuracy distribution

$$acc_n \sim \begin{cases} Bernoulli(p_{correct}), & \text{if } z_n = 1 \\ Bernoulli(0.5), & \text{if } z_n = 0 \end{cases} \tag{17.16}$$

We have a new parameter $p_{correct}$, which represents the probability of making a correct answer in the engaged mode. The verbal description says that it is closer to 100%, and here we have the freedom to choose whatever prior we believe represents for us values that are close to 100% accuracy. We translate this belief into a prior as follows; our prior choice is relatively informative but does not impose a hard constraint; if a subject consistently shows relatively low (or high) accuracy, $p_{correct}$ will change accordingly:

$$p_{correct} \sim Beta(995, 5) \tag{17.17}$$

In our simulated data, we assume that the global motion detection task is done by a very accurate subject, with an accuracy of 99.9%.

First, simulate response times, as done earlier:

```
N <- 1000
x <- c(rep(-0.5, N/2), rep(0.5, N/2)) # difficulty
alpha <- 5.8
beta <- 0.05
sigma <- 0.4
sigma2 <- 0.5
gamma <- 5.2 # fast guess location
p_task <- 0.8 # prob of being on task
z <- rbern(n = N, prob = p_task)
rt <- if_else(z == 1,
              rlnorm(N,
                     meanlog = alpha + beta * x,
                     sdlog = sigma),
              rlnorm(N,
                     meanlog = gamma,
                     sdlog = sigma2))
```

Simulate accuracy and include both response times and accuracy in the simulated data set:

```
p_correct <- 0.999
acc <- ifelse(z, rbern(n = N, p_correct),
                 rbern(n = N, 0.5))
```

```
df_dots_simdata3 <- tibble(trial = 1:N,
                           x = x,
                           rt = rt,
                           acc = acc) %>%
                    mutate(diff = if_else(x == 0.5, "hard", "easy"))
```

Plot the simulated data in Figure 17.8. This time we can see the effect of task difficulty on the simulated response times and accuracy:

```
ggplot(df_dots_simdata3, aes(x = factor(acc), y = rt)) +
  geom_point(position = position_jitter(width = 0.4, height = 0),
             alpha = 0.5) +
  facet_wrap(diff ~ .) +
  xlab("Accuracy") +
  ylab("Response time")
```

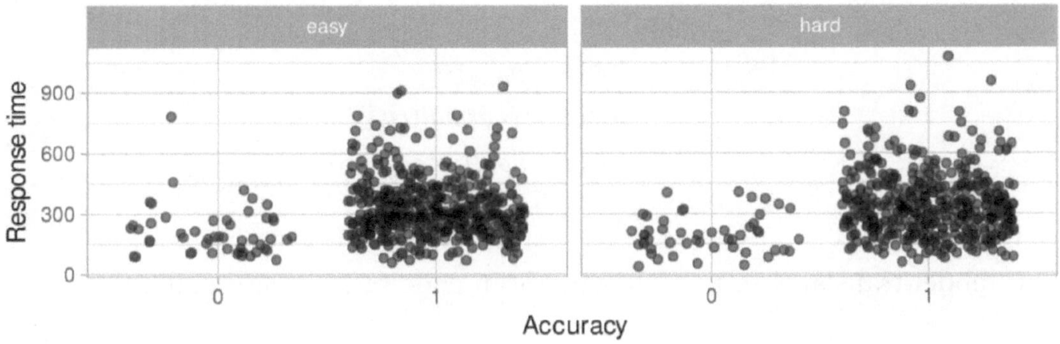

FIGURE 17.8: Response times by accuracy, accounting for task difficulty in the simulated data (`df_dots_simdata3`) that follows the fast-guess model.

Next, we need to marginalize out the discrete parameters from the joint distribution of accuracy and response times. However, we assume that conditional on the latent indicator parameter z, response times and accuracy are independent. For this reason, we can multiply the likelihoods for `rt` and `acc` within each component.

$$
\begin{aligned}
p(rt, acc|\Theta) =& p_{task} \cdot \\
& LogNormal(rt_n|\alpha + \beta \cdot x_n, \sigma) \cdot \\
& Bernoulli(acc_n|p_{correct}) \\
& + \\
& (1 - p_{task}) \cdot \\
& LogNormal(rt_n|\gamma, \sigma_2) \cdot \\
& Bernoulli(acc_n|0.5)
\end{aligned}
\tag{17.18}
$$

In log-space:

$$
\begin{aligned}
\log(p(rt, acc|\Theta)) = \log(\exp(& \\
& \log(p_{task}) + \\
& \log(LogNormal(rt_n|\alpha + \beta \cdot x_n, \sigma)) + \\
& \log(Bernoulli(acc_n|p_{correct}))) \\
+ & \\
\exp(& \\
& \log(1 - p_{task}) + \\
& \log(LogNormal(rt_n|\gamma, \sigma_2)) + \\
& \log(Bernoulli(acc_n|0.5))) \\
) &
\end{aligned}
\tag{17.19}
$$

Our model translates to the following Stan code (`mixture_rtacc.stan`):

```
data {
  int<lower = 1> N;
  vector[N] x;
  vector[N] rt;
  array[N] int acc;
}
parameters {
  real alpha;
  real beta;
  real<lower = 0> sigma;
  real<upper = alpha> gamma;
  real<lower = 0> sigma2;
  real<lower = 0, upper = 1> p_correct;
  real<lower = 0, upper = 1> p_task;
}
model {
```

```
  target += normal_lpdf(alpha | 6, 1);
  target += normal_lpdf(beta | 0, 0.3);
  target += normal_lpdf(sigma | 0.5, 0.2)
    - normal_lccdf(0 | 0.5, 0.2);
  target += normal_lpdf(gamma | 6, 1) -
    normal_lcdf(alpha | 6, 1);
  target += normal_lpdf(sigma2 | 0.5, 0.2)
    - normal_lccdf(0 | 0.5, 0.2);
  target += beta_lpdf(p_correct | 995, 5);
  target += beta_lpdf(p_task | 8, 2);
  for(n in 1:N){
    target +=
      log_sum_exp(log(p_task) +
                  lognormal_lpdf(rt[n] | alpha + x[n] * beta, sigma) +
                  bernoulli_lpmf(acc[n] | p_correct),
                  log1m(p_task) +
                  lognormal_lpdf(rt[n] | gamma, sigma2) +
                  bernoulli_lpmf(acc[n] | 0.5));
  }
}
```

Next, set up the data in list format:

```
ls_dots_simdata <- list(N = N,
                        rt = rt,
                        x = x,
                        acc = acc)
```

Then fit the model:

```
mixture_rtacc <- system.file("stan_models",
                             "mixture_rtacc.stan",
                             package = "bcogsci")
fit_mix_rtacc <- stan(mixture_rtacc, data = ls_dots_simdata)
```

We see that our model can be fit to both response times and accuracy, and its parameters estimates have sensible values (given the fixed parameters we used to generate our simulated data).

```
print(fit_mix_rtacc)
```

```
##                mean     2.5%    97.5% n_eff Rhat
```

```
## alpha          5.80     5.77     5.83  4126    1
## beta           0.08     0.02     0.13  6376    1
## sigma          0.39     0.37     0.42  4586    1
## gamma          5.17     5.09     5.26  5625    1
## sigma2         0.45     0.40     0.51  4795    1
## p_correct      1.00     0.99     1.00  5897    1
## p_task         0.80     0.76     0.84  5440    1
## lp__       -6611.20 -6615.68 -6608.48  1782    1
```

We will evaluate the recovery of the parameters more carefully when we deal with the hierarchical version of the fast-guess model in section 17.1.5. Before we extend this model hierarchically, let us also take into account the instructions given to the subjects.

17.1.4 An implementation of the fast-guess model that takes instructions into account

The actual global motion detection experiment that we started with has another manipulation that can help us to evaluate better the fast-guess model. In some trials, the instructions emphasized accuracy (e.g., "Be as accurate as possible") and in others speed (e.g., "Be as fast as possible"). The fast-guess model also assumes that the probability of being in one of the two states depends on the speed incentives given during the instructions. This entails that now p_{task} depends on the instructions x_2, where we encode a speed incentive with -0.5 and an accuracy incentive with 0.5. Essentially, we need to fit the following regression:

$$\alpha_{task} + x_2 \cdot \beta_{task} \tag{17.20}$$

As we did with MPT models in the previous chapter (in section 16.2.3), we need to bound the previous regression between 0 and 1; we achieve this using the logistic or inverse logit function:

$$p_{task} = logit^{-1}(\alpha_{task} + x_2 \cdot \beta_{task}) \tag{17.21}$$

This means that we need to interpret $\alpha_{task} + x_2 \cdot \beta_{task}$ in log-odds space, which has the range $(-\infty, \infty)$ rather than the probability space $[0, 1]$; also see section 16.2.3.

The likelihood (defined before in section 17.1.3) now depends on the value of x_2 for the specific row:

$$z_n \sim Bernoulli(p_{task_n}) \tag{17.22}$$

A response time distribution is defined:

$$rt_n \sim \begin{cases} LogNormal(\alpha + \beta \cdot x_n, \sigma), & \text{if } z_n = 1 \\ LogNormal(\gamma, \sigma_2), & \text{if } z_n = 0 \end{cases} \tag{17.23}$$

and an accuracy distribution is defined as well:

$$acc_n \sim \begin{cases} Bernoulli(p_{correct}), & \text{if } z_n = 1 \\ Bernoulli(0.5), & \text{if } z_n = 0 \end{cases} \tag{17.24}$$

The only further change in our model is that rather than a prior on p_{task}, we now need priors for α_{task} and β_{task}, which are on the log-odds scale.

For β_{task}, we assume an effect that can be rather large and we won't assume a direction a priori (for now):

$$\beta_{task} \sim Normal(0, 1) \tag{17.25}$$

This means that the subject could be affected by the instructions in the expected way, with an increased probability to be task-engaged, leading to better accuracy when the instructions emphasize accuracy ($\beta_{task} > 0$). Alternatively, the subject might behave in an unexpected way, with a decreased probability to be task-engaged, leading to worse accuracy when the instructions emphasize accuracy ($\beta_{task} < 0$). The latter situation, $\beta_{task} < 0$, could represent the instructions being misunderstood. It's certainly possible to include priors that encode the expected direction of the effect instead: $Normal_+(0, 1)$. Unless there is a compelling reason to constrain the prior in this way, following Cromwell's rule (Box E.2 in the online chapter E), we leave open the possibility of the β parameter having negative values.

How can we choose a prior for α_{task} that encodes the same information that we had in the previous model in p_{task}? One possibility is to create an auxiliary parameter p_{btask}, that represents the baseline probability of being engaged in the task, with the same prior that we use in the previous section, and then transform it to an unconstrained space for our regression with the logit function:

$$\begin{aligned} p_{btask} &\sim Beta(8, 2) \\ \alpha_{task} &= logit(p_{btask}) \end{aligned} \tag{17.26}$$

To verify that our priors make sense, in Figure 17.9 we plot the difference in prior predicted probability of being engaged in the task under the two emphasis conditions:

```
Ns <- 1000 # number of samples for the plot
# Priors
p_btask <- rbeta(n = Ns, shape1 = 8, shape2 = 2)
beta_task <- rnorm(n = Ns, mean = 0, sd = 1)
# Predicted probability of being engaged
p_task_easy <- plogis(qlogis(p_btask) + 0.5 * beta_task)
p_task_hard <- plogis(qlogis(p_btask) + -0.5 * beta_task)
# Predicted difference
diff_p_pred <- tibble(diff = p_task_easy - p_task_hard)
```

```
diff_p_pred %>%
  ggplot(aes(diff)) +
  geom_histogram()
```

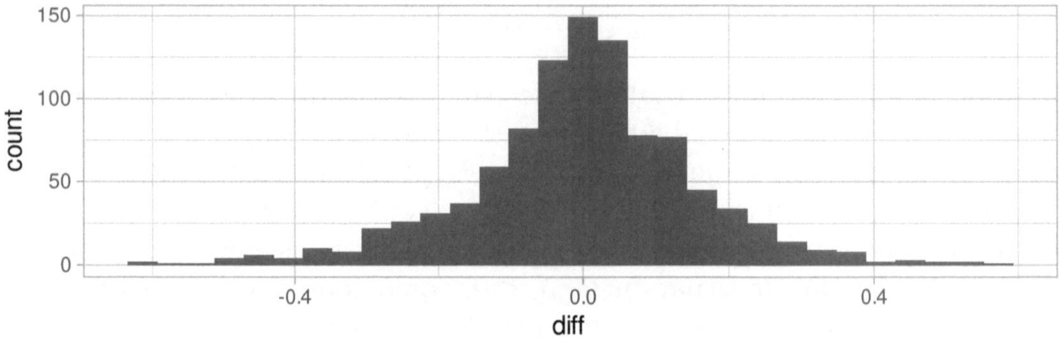

FIGURE 17.9: The difference in prior predicted probability of being engaged in the task under the two emphasis conditions for the simulated data (`diff_p_pred`) that follows the fast-guess model.

Figure 17.9 shows that we are predicting a priori that the difference in p_{task} will tend to be smaller than ± 0.3, which seems to make sense intuitively. If we had more information about the likely range of variation, we could of course have adapted the prior to reflect that belief.

We are ready to generate a new data set, by deciding on some fixed values for β_{task} and p_{btask}:

```
N <- 1000
x <- c(rep(-0.5, N/2), rep(0.5, N/2)) # difficulty
x2 <- rep(c(-0.5, 0.5), N/2) # instructions
# Verify that the predictors are crossed in a 2x2 design :
```

```
predictors <- tibble(x, x2)
xtabs(~ x + x2, predictors)
```

```
##       x2
## x      -0.5 0.5
##   -0.5  250 250
##   0.5   250 250
```

```
alpha <- 5.8
beta <- 0.05
sigma <- 0.4
sigma2 <- 0.5
gamma <- 5.2
p_correct <- 0.999
# New true point values:
p_btask <- 0.85
beta_task <- 0.5
# Generate data:
alpha_task <- qlogis(p_btask)
p_task <- plogis(alpha_task + x2 * beta_task)
z <- rbern(N, prob = p_task)
rt <- ifelse(z,
             rlnorm(N, meanlog = alpha + beta * x, sdlog = sigma),
             rlnorm(N, meanlog = gamma, sdlog = sigma2))
acc <- ifelse(z, rbern(N, p_correct),
                 rbern(N, 0.5))
df_dots_simdata4 <- tibble(trial = 1:N,
                           x = x,
                           rt = rt,
                           acc = acc,
                           x2 = x2) %>%
          mutate(diff = if_else(x == 0.5, "hard", "easy"),
                 emphasis = ifelse(x2 == 0.5, "accuracy", "speed"))
```

We can generate a plot now where both the difficulty of the task and the instructions are manipulated; see Figure 17.10.

```
ggplot(df_dots_simdata4, aes(x = factor(acc), y = rt)) +
  geom_point(position = position_jitter(width = .4, height = 0),
             alpha = 0.5) +
```

```
  facet_wrap(diff ~ emphasis) +
  xlab("Accuracy") +
  ylab("Response time")
```

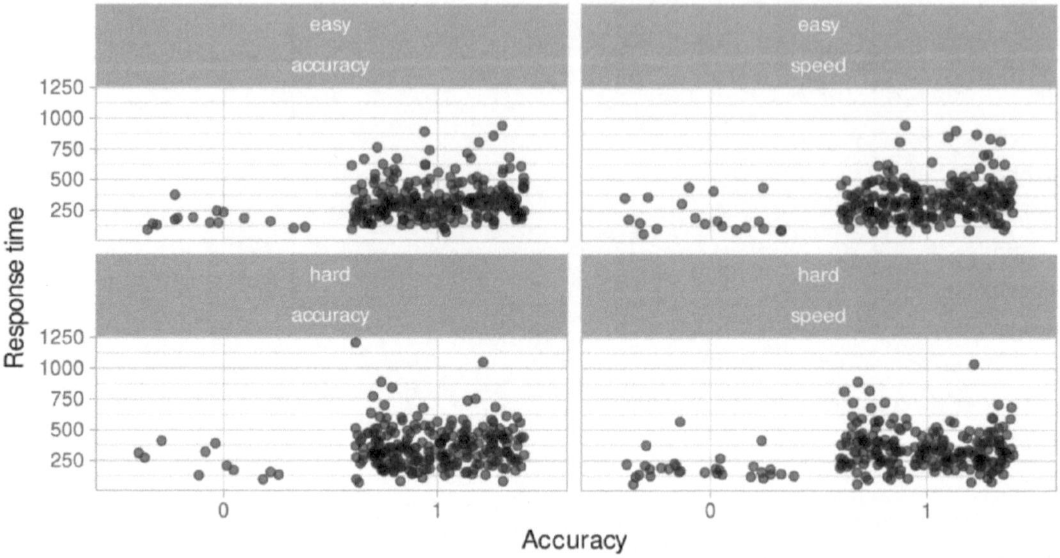

FIGURE 17.10: Response times and accuracy by the difficulty of the task and the instructions type for the simulated data (`df_dots_simdata4`) that follows the fast-guess model.

In the Stan implementation, `log_inv_logit(x)` is applying the logistic (or inverse logit) function to `x` to transform it into a probability and then applying the logarithm; `log1m_inv_logit(x)` is applying the logistic function to `x`, and then applying the logarithm to its complement $(1 - p)$. We do this because rather than having `p_task` in probability space, we have `lodds_task` in log-odds space:

```
real lodds_task = logit(p_btask) + x2[n] * beta_task;
```

The parameter `lodds_task` estimates the mixing probabilities in log-odds:

```
target += log_sum_exp(log_inv_logit(lodds_task) + ...,
                      log1m_inv_logit(lodds_task) + ...)
```

We also add a `generated quantities` block that can be used for further (prior or posterior) predictive checks. In this block, we do use `z` as an indicator of the latent class (task-engaged mode or fast-guessing mode), since we do not estimate `z`, but rather generate it based on the parameter's posteriors.

We use the dummy variable `onlyprior` to indicate whether we use the data or we only sample from the priors. One can always do the predictive checks in

R, transforming the code that we wrote for the simulation into a function, and writing the priors in R. However, it can be simpler to take advantage of Stan output format and rewrite the code in Stan. One downside of this is that the `stanfit` object that stores the model output can become too large for the memory of the computer. Another downside is reduced robustness, as it is more likely that we overlook an error if we only work in Stan rather than re-implementing the code in different programming languages (e.g., R and Stan) (Cooper and Guest 2014).

The code shown below is available in the `bcogsci` package and is called `mixture_rtacc2.stan`.

```
data {
  int<lower = 1> N;
  vector[N] x;
  vector[N] rt;
  array[N] int acc;
  vector[N] x2; //speed or accuracy emphasis
  int<lower = 0, upper = 1> onlyprior;
}
parameters {
  real alpha;
  real beta;
  real<lower = 0> sigma;
  real<upper = alpha> gamma;
  real<lower = 0> sigma2;
  real<lower = 0, upper = 1> p_correct;
  real<lower = 0, upper = 1> p_btask;
  real beta_task;
}
model {
  target += normal_lpdf(alpha | 6, 1);
  target += normal_lpdf(beta | 0, 0.1);
  target += normal_lpdf(sigma | 0.5, 0.2)
    - normal_lccdf(0 | 0.5, 0.2);
  target += normal_lpdf(gamma | 6, 1)-
  normal_lcdf(alpha | 6, 1);
  target += normal_lpdf(sigma2 | 0.5, 0.2)
    - normal_lccdf(0 | 0.5, 0.2);
  target += normal_lpdf(beta_task | 0, 1);
  target += beta_lpdf(p_correct | 995, 5);
  target += beta_lpdf(p_btask | 8, 2);
  if(onlyprior != 1)
```

```
    for(n in 1:N){
      real lodds_task = logit(p_btask) + x2[n] * beta_task;
      target +=
        log_sum_exp(log_inv_logit(lodds_task)+
                 lognormal_lpdf(rt[n] | alpha + x[n] * beta, sigma) +
                 bernoulli_lpmf(acc[n] | p_correct),
                 log1m_inv_logit(lodds_task) +
                 lognormal_lpdf(rt[n] | gamma, sigma2) +
                 bernoulli_lpmf(acc[n] | 0.5));
    }
}
generated quantities {
  array[N] real rt_pred;
  array[N] real acc_pred;
  array[N] int z;
  for(n in 1:N){
    real lodds_task = logit(p_btask) + x2[n] * beta_task;
    z[n] = bernoulli_rng(inv_logit(lodds_task));
    if(z[n] == 1){
      rt_pred[n] = lognormal_rng(alpha + x[n] * beta, sigma);
      acc_pred[n] = bernoulli_rng(p_correct);
    } else{
      rt_pred[n] = lognormal_rng(gamma, sigma2);
      acc_pred[n] = bernoulli_rng(0.5);
    }
  }
}
```

Before fitting the model to the simulated data, we perform prior predictive checks.

17.1.4.1 Prior predictive checks

Generate prior predictive distributions, by setting `onlyprior` to `1`.

```
ls_dots_simdata <- list(N = N,
                        rt = rt,
                        x = x,
                        x2 = x2,
                        acc = acc,
                        onlyprior = 1)
mixture_rtacc2 <- system.file("stan_models",
```

```
                                    "mixture_rtacc2.stan",
                                    package = "bcogsci")
```

```
fit_mix_rtacc2_priors <- stan(mixture_rtacc2,
                              data = ls_dots_simdata)
```

We plot prior predictive distributions of response times as follows in Figure 17.11(a), by setting `y = rt_pred` using `ppd_dens_overlay()`.

```
rt_pred <- extract(fit_mix_rtacc2_priors)$rt_pred
ppd_dens_overlay(rt_pred[1:100,]) +
  coord_cartesian(xlim = c(0, 2000))
```

Some of the predictive data sets contain responses that are too large, and some of them have too much probability mass close to zero, but there is nothing clearly wrong in the prior predictive distributions (considering that the model hasn't "seen" the data yet).

If we want to plot the prior predicted distribution of differences in response time conditioning on task difficulty, we need to define a new function. Then we use the `bayesplot` function `ppc_stat()` that takes as an argument of `stat` any summary function; see Figure 17.11(a).

```
meanrt_diff <- function(rt){
  mean(rt[x == 0.5]) -
    mean(rt[x == -0.5])
}
ppd_stat(rt_pred, stat = meanrt_diff)
```

We find that the range of response times look reasonable. There are, however, always more checks that can be done; examples are plotting other summary statistics or predictions conditioned on other aspects of the data.

17.1.4.2 Fit to simulated data

Fit the model to data, by setting `onlyprior = 0`:

```
ls_dots_simdata <- list(N = N,
                        rt = rt,
                        x = x,
```

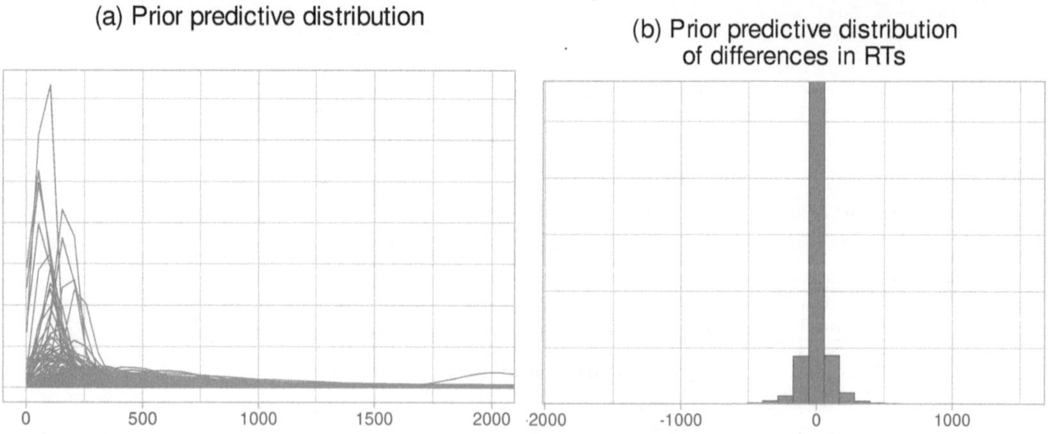

FIGURE 17.11: (a) Prior predictive distributions of response times from the fast-guess model (`mixture_rtacc2.stan`). (b) Prior predictive distribution of response time differences, using the same model and prior settings, conditioned on task difficulty.

```
                        x2 = x2,
                        acc = acc,
                        onlyprior = 0)
```

```
fit_mix_rtacc2 <- stan(mixture_rtacc2, data = ls_dots_simdata)
```

```
print(fit_mix_rtacc2,
      pars = c("alpha", "beta", "sigma", "gamma", "sigma2",
               "p_correct", "p_btask", "beta_task"))
```

```
##           mean 2.5% 97.5% n_eff Rhat
## alpha     5.82 5.78  5.85  4875    1
## beta      0.05 0.00  0.11  4943    1
## sigma     0.40 0.38  0.42  4286    1
## gamma     5.12 5.00  5.24  3055    1
## sigma2    0.46 0.38  0.54  3247    1
## p_correct 0.99 0.99  1.00  3784    1
## p_btask   0.86 0.83  0.89  3803    1
## beta_task 0.62 0.15  1.10  5922    1
```

We see that we fit the model without problems. Before we evaluate the recovery of the parameters more carefully, we implement a hierarchical version of the fast-guess model.

17.1.5 A hierarchical implementation of the fast-guess model

So far we have evaluated the behavior of one simulated subject. We discussed before (in the context of distributional regression models, in section 5.2.6, and in the MPT modeling chapter 16) that, in principle, every parameter in a model can be made hierarchical. However, this doesn't guarantee that we'll learn anything from the data for those parameters, or that our model will converge. A safe approach here is to start simple, using simulated data. If a model converges on simulated data, it does not guarantee convergence on real data. However, if a model fails to converge on simulated data, it is very likely to fail with real data as well.

For our hierarchical version, we assume that both response times and the effect of task difficulty vary by subject and that different subjects have different guessing times. This entails the following change to the response time distribution:

$$rt_n \sim \begin{cases} LogNormal(\alpha + u_{subj[n],1} + x_n \cdot (\beta + u_{subj[n],2}), \sigma), & \text{if } z_n = 1 \\ LogNormal(\gamma + u_{subj[n],3}, \sigma_2), & \text{if } z_n = 0 \end{cases} \tag{17.27}$$

We assume that the three vectors of u (adjustment to the intercept and slope of the task-engaged distribution, and the adjustment to the guessing time distribution) follow a multivariate normal distribution centered on zero. For simplicity and lack of any prior knowledge about this experiment design and method, we assume the same (weakly informative) prior distribution for the three variance components and the same regularizing LKJ prior for the correlation matrix $\mathbf{R_u}$ that contains the three correlations between the adjustments $(\rho_{u_{1,2}}, \rho_{u_{1,3}}, \rho_{u_{2,3}})$:

$$\begin{aligned} \boldsymbol{u} &\sim \mathcal{N}(0, \Sigma_u) \\ \tau_{u_1}, \tau_{u_2}, \tau_{u_3} &\sim Normal_+(0, 0.5) \\ \mathbf{R_u} &\sim LKJcorr(2) \end{aligned} \tag{17.28}$$

Before we fit the model to the real data set, we simulate data again; this time we simulate 20 subjects, each of whom delivers a total of 100 trials (each subject sees 25 trials for each of the four conditions).

```
# Build the simulate data set
N_subj <- 20
N_trials <- 100
# Parameters' true point values
alpha <- 5.8
```

```r
beta <- 0.05
sigma <- 0.4
sigma2 <- 0.5
gamma <- 5.2
beta_task <- 0.1
p_btask <- 0.85
alpha_task <- qlogis(p_btask)
p_correct <- 0.999
tau_u <- c(0.2, 0.005, 0.3)
rho <- 0.3
```

```r
## Build the data set here:
N <- N_subj * N_trials
stimuli <- tibble(x = rep(c(rep(-0.5, N_trials / 2),
                            rep(0.5, N_trials / 2)), N_subj),
                  x2 = rep(rep(c(-0.5, 0.5), N_trials / 2), N_subj),
                  subj = rep(1:N_subj, each = N_trials),
                  trial = rep(1:N_trials, N_subj))
stimuli
```

```
## # A tibble: 2,000 x 4
##        x    x2  subj trial
##    <dbl> <dbl> <int> <int>
## 1   -0.5  -0.5     1     1
## 2   -0.5   0.5     1     2
## 3   -0.5  -0.5     1     3
## # i 1,997 more rows
```

```r
# Build the correlation matrix for the adjustments u:
Cor_u <- matrix(rep(rho, 9), nrow = 3)
diag(Cor_u) <- 1
Cor_u
```

```
##      [,1] [,2] [,3]
## [1,]  1.0  0.3  0.3
## [2,]  0.3  1.0  0.3
## [3,]  0.3  0.3  1.0
```

```r
# Variance covariance matrix for subjects:
Sigma_u <- diag(tau_u, 3, 3) %*% Cor_u %*% diag(tau_u, 3, 3)
```

```
# Create the correlated adjustments:
u <- mvrnorm(n = N_subj, c(0, 0, 0), Sigma_u)
# Check whether they are correctly correlated
# (There will be some random variation,
# but if you increase the number of observations,
# you'll be able to obtain more exact values below):
cor(u)
```

```
##       [,1]  [,2]  [,3]
## [1,] 1.000 0.514 0.314
## [2,] 0.514 1.000 0.245
## [3,] 0.314 0.245 1.000
```

```
# Check the SDs:
sd(u[, 1]); sd(u[, 2]); sd(u[, 3])
```

```
## [1] 0.221
```

```
## [1] 0.00586
```

```
## [1] 0.295
```

```
# Create the simulated data:
df_dots_simdata <- stimuli %>%
  mutate(z = rbern(N, prob = plogis(alpha_task + x2 * beta_task)),
         rt = ifelse(z,
             rlnorm(N, meanlog = alpha + u[subj, 1] +
                              (beta + u[subj, 2]) * x,
                    sdlog = sigma),
             rlnorm(N, meanlog = gamma + u[subj, 3],
                    sdlog = sigma2)),
         acc = ifelse(z, rbern(n = N, p_correct),
              rbern(n = N, 0.5)),
         diff = if_else(x == 0.5, "hard", "easy"),
         emphasis = ifelse(x2 == 0.5, "accuracy", "speed"))
```

Verify that the distribution of the simulated response times conditional on the simulated accuracy and the experimental manipulations make sense; see Figure 17.12.

```
ggplot(df_dots_simdata, aes(x = factor(acc), y = rt)) +
  geom_point(position = position_jitter(width = 0.4,
                                         height = 0), alpha = 0.5) +
  facet_wrap(diff ~ emphasis) +
  xlab("Accuracy") +
  ylab("Response time")
```

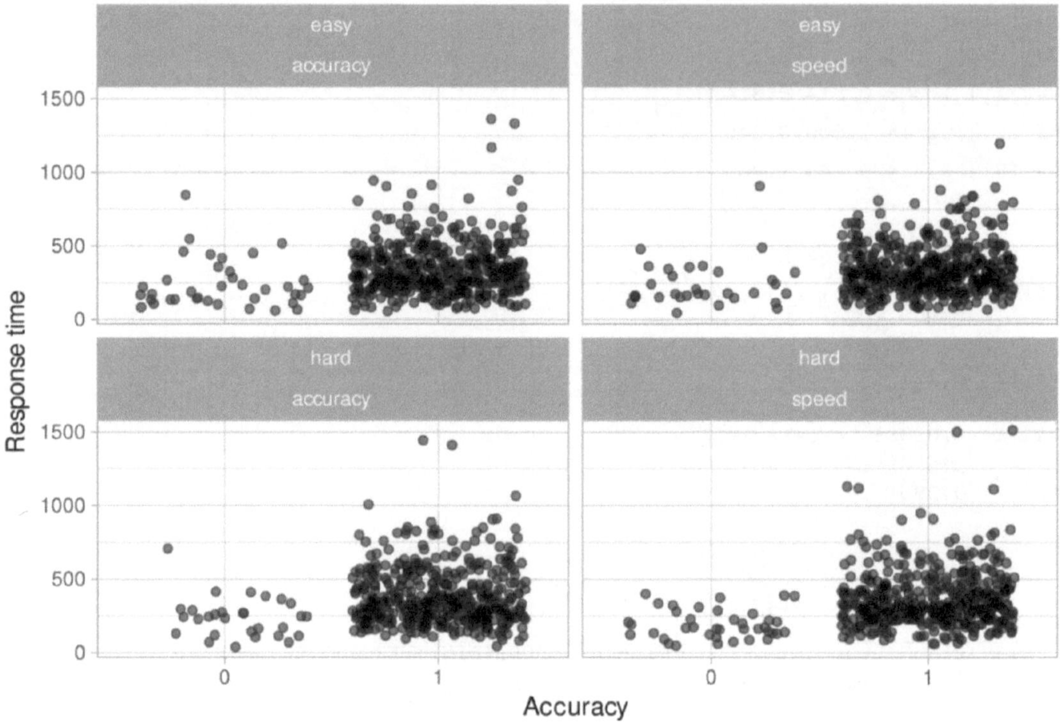

FIGURE 17.12: The distribution of response times conditional on the simulated accuracy and the experimental manipulations for the simulated hierarchical data (`df_dots_simdata`) that follows the fast-guess model.

We implement the model in Stan as follows in `mixture_h.stan`. The hierarchical extension uses the Cholesky factorization for the group-level effects (as in section 9.1.3).

```
data {
  int<lower = 1> N;
  vector[N] x;
  vector[N] rt;
  array[N] int acc;
  vector[N] x2;
  int<lower = 1> N_subj;
```

```
  array[N] int<lower = 1, upper = N_subj> subj;
}
parameters {
  real alpha;
  real beta;
  real<lower = 0> sigma;
  real<upper = alpha> gamma;
  real<lower = 0> sigma2;
  real<lower = 0, upper = 1> p_correct;
  real<lower = 0, upper = 1> p_btask;
  real beta_task;
  vector<lower = 0>[3]  tau_u;
  matrix[3, N_subj] z_u;
  cholesky_factor_corr[3] L_u;
}
transformed parameters {
  matrix[N_subj, 3] u;
  u = (diag_pre_multiply(tau_u, L_u) * z_u)';
}
model {
  target += normal_lpdf(alpha | 6, 1);
  target += normal_lpdf(beta | 0, 0.1);
  target += normal_lpdf(sigma | 0.5, 0.2)
    - normal_lccdf(0 | 0.5, 0.2);
  target += normal_lpdf(gamma | 6, 1) -
    normal_lcdf(alpha | 6, 1);
  target += normal_lpdf(sigma2 | 0.5, 0.2)
    - normal_lccdf(0 | 0.5, 0.2);
  target += normal_lpdf(tau_u | 0, 0.5)
    - 3* normal_lccdf(0 | 0, 0.5);
  target += normal_lpdf(beta_task | 0, 1);
  target += beta_lpdf(p_correct | 995, 5);
  target += beta_lpdf(p_btask | 8, 2);
  target += lkj_corr_cholesky_lpdf(L_u | 2);
  target += std_normal_lpdf(to_vector(z_u));
  for(n in 1:N){
    real lodds_task = logit(p_btask) + x2[n] * beta_task;
    target +=
      log_sum_exp(log_inv_logit(lodds_task) +
          lognormal_lpdf(rt[n] | alpha + u[subj[n], 1] +
                         x[n] * (beta + u[subj[n], 2]), sigma) +
          bernoulli_lpmf(acc[n] | p_correct),
```

```
          log1m_inv_logit(lodds_task) +
          lognormal_lpdf(rt[n] | gamma + u[subj[n], 3], sigma2) +
          bernoulli_lpmf(acc[n] |0.5));
  }
}
generated quantities {
  corr_matrix[3] rho_u = L_u * L_u';
}
```

Save the model code and fit it to the simulated data:

```
ls_dots_simdata <- list(N = N,
                          rt = df_dots_simdata$rt,
                          x = df_dots_simdata$x,
                          x2 = df_dots_simdata$x2,
                          acc = df_dots_simdata$acc,
                          subj = df_dots_simdata$subj,
                          N_subj = N_subj)
mixture_h <- system.file("stan_models",
                          "mixture_h.stan",
                          package = "bcogsci")
fit_mix_h <- stan(file = mixture_h,
                   data = ls_dots_simdata,
                   iter = 3000,
                   control = list(adapt_delta = 0.9))
```

Print the posterior summary:

```
print(fit_mix_h, pars = c("alpha", "beta", "sigma", "gamma",
                           "sigma2", "p_correct", "p_btask",
                           "beta_task", "tau_u",
                           "rho_u[1,2]", "rho_u[1,3]",
                           "rho_u[2,3]"))
```

```
##             mean  2.5% 97.5% n_eff Rhat
## alpha       5.81  5.71  5.91   931    1
## beta        0.05  0.01  0.09  9395    1
## sigma       0.40  0.38  0.41  8989    1
## gamma       5.20  5.02  5.38  3374    1
## sigma2      0.52  0.45  0.59  6808    1
## p_correct   0.99  0.99  1.00  6554    1
## p_btask     0.87  0.85  0.90  7969    1
```

```
## beta_task    0.24 -0.10  0.58 10705    1
## tau_u[1]     0.23  0.16  0.31  1774    1
## tau_u[2]     0.03  0.00  0.09  2552    1
## tau_u[3]     0.32  0.18  0.49  2335    1
## rho_u[1,2]   0.05 -0.68  0.74  8948    1
## rho_u[1,3]   0.37 -0.10  0.72  3874    1
## rho_u[2,3] -0.02 -0.74  0.72  1091    1
```

We see that we can fit the hierarchical extension of our model to simulated data. Next, we'll evaluate whether we can recover the true point values of the parameters.

17.1.5.1 Recovery of the parameters

By "recovering" the true values of the parameters, we mean that the true point values are somewhere inside the bulk of the posterior distribution of the model.

In in Figure 17.13, we use `mcmc_recover_hist()` to compare the posterior distributions of the relevant parameters of the model with their true point values.

```
df_fit_mix_h <- fit_mix_h %>% as.data.frame() %>%
  select(c("alpha", "beta", "sigma", "gamma", "sigma2",
           "p_correct","p_btask", "beta_task", "tau_u[1]",
           "tau_u[2]", "tau_u[3]", "rho_u[1,2]", "rho_u[1,3]",
           "rho_u[2,3]"))
true_values <- c(alpha, beta, sigma, gamma, sigma2,
                 p_correct, p_btask, beta_task, tau_u[1],
                 tau_u[2], tau_u[3], rep(rho,3))
mcmc_recover_hist(df_fit_mix_h, true_values) +
  theme(axis.text.x = element_text(angle = 45, vjust = 1, hjust = 1))
```

The model seems to be underestimating the probability of subjects being correct (`p_correct`) and the amount of noise (`sigma`). However, the numerical differences are relatively small. We can be relatively certain that the model is not seriously misspecified. As mentioned in previous chapters, a more principled (and computationally demanding) approach uses simulation-based calibration introduced in section 10.2 of chapter 10 (also see Talts et al. 2018; Schad, Betancourt, and Vasishth 2020). An intermediate approach would be to re-run the simulation above a few times with a larger number of observations and/or with different true, data-generating parameters to see whether the estimates behave as expected.

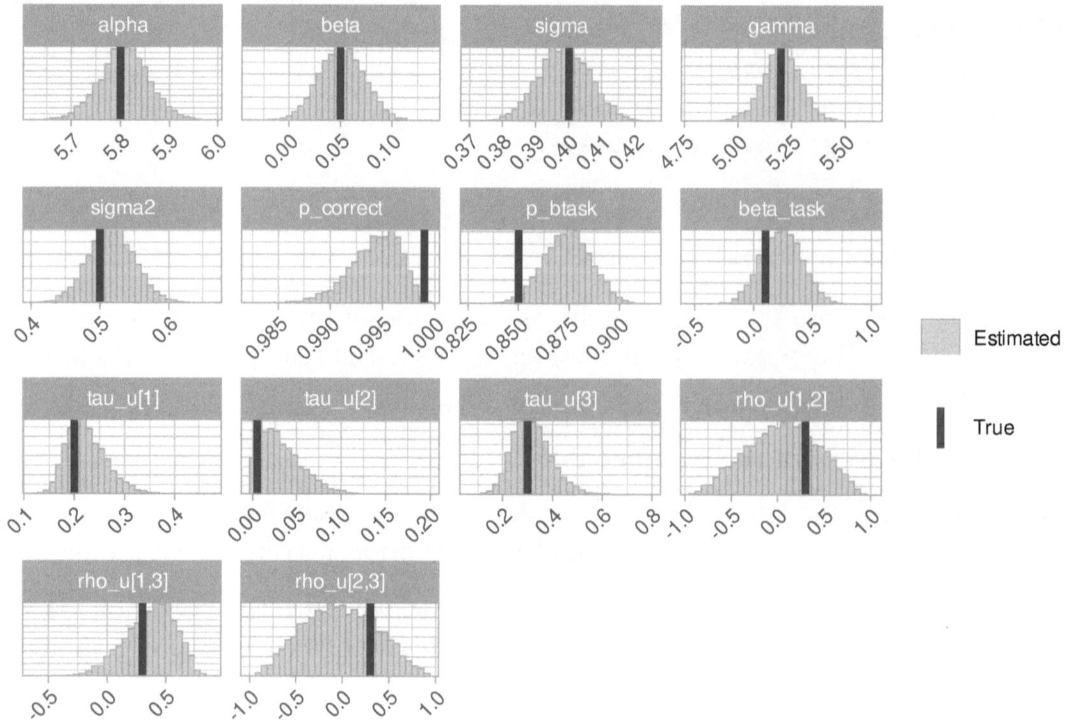

FIGURE 17.13: Posterior distributions of the main parameters of the mixture model `fit_mix_h` together with their true point values.

17.1.5.2 Fitting the model to real data

After verifying that our model works as expected, we are ready to fit it to real data. We code the predictors x and x_2 as we did for the simulated data:

```
df_dots <- df_dots %>%
  mutate(x = if_else(diff == "easy", -0.5, 0.5),
         x2 = if_else(emphasis == "accuracy", 0.5, -0.5))
```

The main obstacle now is that fitting the entire data set takes around 12 hours! We'll sample 600 observations of each subject as follows:

```
df_dots_data_short <- df_dots %>%
  group_by(subj) %>%
  sample_n(600)
```

```
ls_dots_data_short <-
  list(N = nrow(df_dots_data_short),
```

```
       rt = df_dots_data_short$rt,
       x = df_dots_data_short$x,
       x2 = df_dots_data_short$x2,
       acc = df_dots_data_short$acc,
       subj = as.numeric(df_dots_data_short$subj),
       N_subj = length(unique(df_dots_data_short$subj)))
fit_mix_data <- stan(file = mixture_h,
                     data = ls_dots_data_short,
                     control = list(adapt_delta = .9,
                                    max_treedepth = 12))
```

```
## Warning: The largest R-hat is NA, indicating chains have not mixed.
## Running the chains for more iterations may help. See
## https://mc-stan.org/misc/warnings.html#r-hat
##
##  Warning: Bulk Effective Samples Size (ESS) is too low, indicating
## posterior means and medians may be unreliable. Running the chains for
## more iterations may help. See
## https://mc-stan.org/misc/warnings.html#bulk-ess
##
##  Warning: Tail Effective Samples Size (ESS) is too low, indicating
## posterior variances and tail quantiles may be unreliable. Running the
## chains for more iterations may help. See
## https://mc-stan.org/misc/warnings.html#tail-ess
```

The model has not converged at all!

```
print(fit_mix_data,
      pars = c("alpha", "beta", "sigma", "gamma", "sigma2",
               "p_correct","p_btask", "beta_task", "tau_u",
               "rho_u[1,2]", "rho_u[1,3]", "rho_u[2,3]"))
```

```
##             mean  2.5% 97.5% n_eff  Rhat
## alpha       6.40  6.31  6.49     3  1.54
## beta        0.12  0.06  0.16     2  2.70
## sigma       0.27  0.22  0.29     2 11.44
## gamma       6.15  6.04  6.34     2  2.76
## sigma2      0.25  0.19  0.42     2 26.07
## p_correct   0.94  0.92  1.00     2 10.92
## p_btask     0.84  0.69  0.92     2  6.46
## beta_task   3.89  0.72  5.52     2  7.34
```

```
## tau_u[1]    0.15  0.10  0.21    87  1.04
## tau_u[2]    0.04  0.01  0.07     5  1.27
## tau_u[3]    0.16  0.11  0.24     6  1.22
## rho_u[1,2] 0.36 -0.16  0.77  3100  1.00
## rho_u[1,3] 0.23 -0.24  0.60    13  1.10
## rho_u[2,3] 0.31 -0.20  0.72   978  1.01
```

The traceplots in Figure 17.14 show that the chains are not mixing at all. It seems that the posterior is multimodal, and there are at least two combinations of parameters that would fit the data equally well. (Rerunning the model might sometimes reveal three combinations.)

```
traceplot(fit_mix_data) +
  scale_x_continuous(breaks = NULL)
```

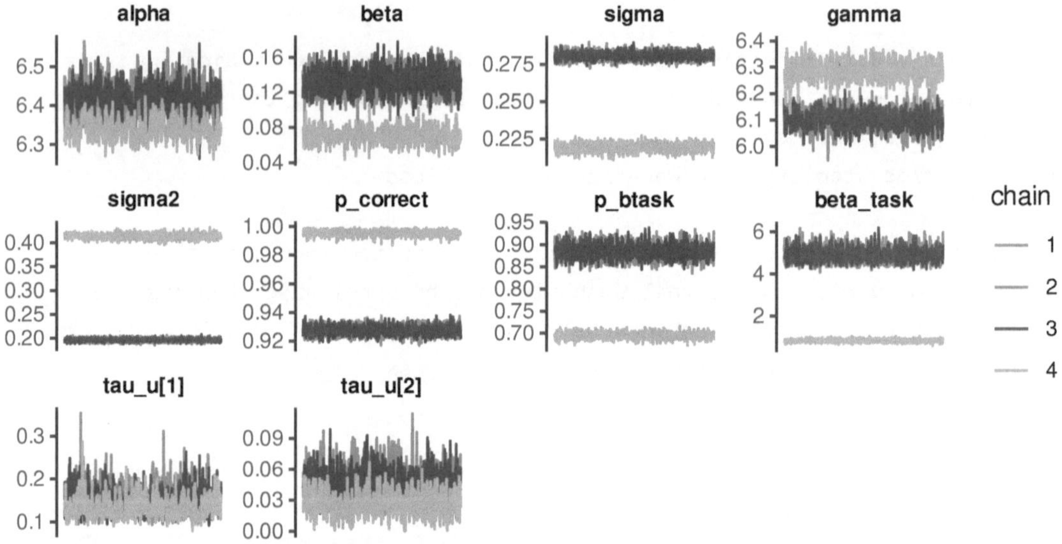

FIGURE 17.14: Traceplots from the hierarchical model (`mixture_h.stan`) fit to (a subset) of the real data. The traceplot shows clearly that the posterior has at least two modes, though rerunning the model might sometimes reveal three modes.

What should we do now? It can be a good idea to back off and simplify the model. Once the simplified model converges, we can think about adding further complexity. The verbal description of our model says that the accuracy in the task-engaged mode should be close to 100%. To simplify the model, we'll assume that it's exactly 100%. This entails the following:

$$p_{correct} = 1 \tag{17.29}$$

We adapt our Stan code in `mixture_h2.stan`, reflecting the assumption that `p_correct` has a fixed value; this parameter is now in a block called `transformed data`. There, we assign to `p_correct` the value of `1`.

```
data {
  int<lower = 1> N;
  vector[N] x;
  vector[N] rt;
  array[N] int acc;
  vector[N] x2;
  int<lower = 1> N_subj;
  array[N] int<lower = 1, upper = N_subj> subj;
}
transformed data {
  real p_correct = 1;
}
parameters {
  real alpha;
  real beta;
  real<lower = 0> sigma;
  real<upper = alpha> gamma; //guessing
  real<lower = 0> sigma2;
  real<lower = 0, upper = 1> p_btask;
  real beta_task;
  vector<lower = 0>[3]  tau_u;
  matrix[3, N_subj] z_u;
  cholesky_factor_corr[3] L_u;
}
transformed parameters {
  matrix[N_subj, 3] u;
  u = (diag_pre_multiply(tau_u, L_u) * z_u)';
}
model {
  target += normal_lpdf(alpha | 6, 1);
  target += normal_lpdf(beta | 0, 0.1);
  target += normal_lpdf(sigma | 0.5, 0.2)
    - normal_lccdf(0 | 0.5, 0.2);
  target += normal_lpdf(gamma | 6, 1) -
    normal_lcdf(alpha | 6, 1);
  target += normal_lpdf(sigma2 | 0.5, 0.2)
    - normal_lccdf(0 | 0.5, 0.2);
  target += normal_lpdf(tau_u | 0, 0.5)
```

```
    - 3* normal_lccdf(0 | 0, 0.5);
  target += normal_lpdf(beta_task | 0, 1);
  target += beta_lpdf(p_btask | 8, 2);
  target += lkj_corr_cholesky_lpdf(L_u | 2);
  target += std_normal_lpdf(to_vector(z_u));
  for(n in 1:N){
    real lodds_task = logit(p_btask) + x2[n] * beta_task;
    target +=
      log_sum_exp(log_inv_logit(lodds_task) +
          lognormal_lpdf(rt[n] | alpha + u[subj[n], 1] +
                         x[n] * (beta + u[subj[n], 2]), sigma) +
          bernoulli_lpmf(acc[n] | p_correct),
          log1m_inv_logit(lodds_task) +
          lognormal_lpdf(rt[n] | gamma + u[subj[n], 3], sigma2) +
          bernoulli_lpmf(acc[n] |0.5));
  }
}
generated quantities {
  corr_matrix[3] rho_u = L_u * L_u';
}
```

Fit the model again to the same data:

```
fit_mixs_data <- stan(file = mixture_h2,
                      data = ls_dots_data_short,
                      control = list(adapt_delta = .99,
                                     max_treedepth = 12))
```

The model has now converged:

```
print(fit_mixs_data,
      pars = c("alpha", "beta", "sigma", "gamma", "sigma2",
               "p_btask", "beta_task", "tau_u",
               "rho_u[1,2]", "rho_u[1,3]", "rho_u[2,3]"))
```

```
##            mean  2.5% 97.5% n_eff Rhat
## alpha      6.34  6.29  6.39   642 1.00
## beta       0.07  0.05  0.09  2868 1.00
## sigma      0.22  0.21  0.22  5309 1.00
## gamma      6.29  6.21  6.35  1548 1.00
## sigma2     0.41  0.40  0.42  5154 1.00
## p_btask    0.69  0.67  0.70  5946 1.00
```

```
## beta_task   0.79  0.65  0.93  5046 1.00
## tau_u[1]    0.13  0.10  0.19  1049 1.00
## tau_u[2]    0.03  0.01  0.05  1100 1.00
## tau_u[3]    0.19  0.13  0.27  1828 1.00
## rho_u[1,2] 0.40 -0.20  0.82  4323 1.00
## rho_u[1,3] 0.10 -0.31  0.49  2065 1.00
## rho_u[2,3] 0.27 -0.31  0.76   620 1.01
```

The traceplots in Figure 17.15 show that this times the chains are mixing well.

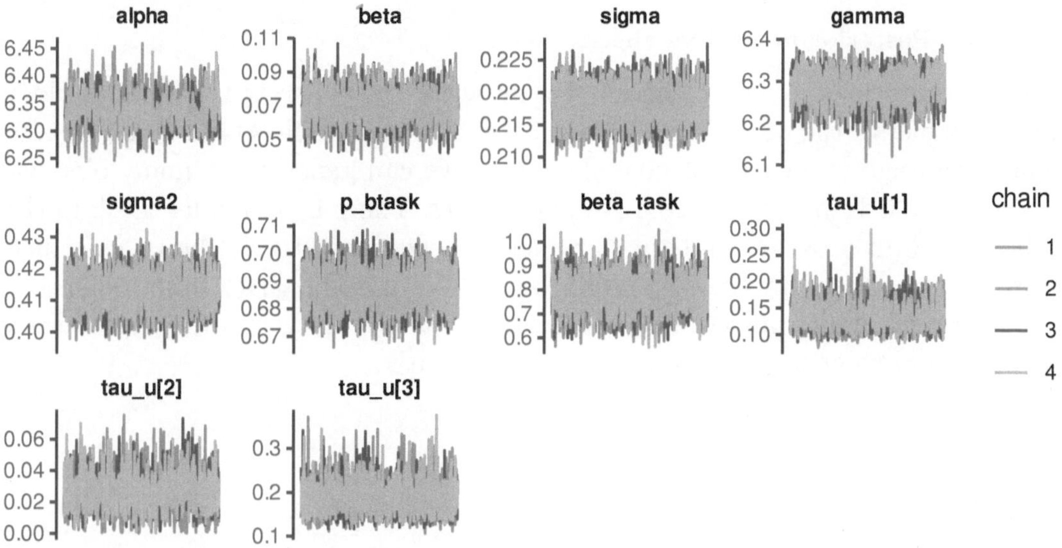

FIGURE 17.15: Traceplots from the simplified hierarchical model (`mixture_h2.stan`, assuming that `p_correct = 1`) fit to (a subset) of the real data. The traceplot shows that chains are mixing well.

What can we say about the fit of the model now?

Under the assumptions that we have made (e.g., that there are two processing modes, response times are affected by the difficulty of the task in the task-engaged mode, accuracy is not affected by the difficulty of the task and is perfect at the task-engaged mode, etc.), we can look at the parameters and conclude the following:

- The instructions seemed to have a strong effect on the processing mode of the subjects (`beta_task` is relatively high), and in the expected direction (emphasis in accuracy led to a higher probability to be in the task engaged mode).
- The guessing mode seemed to be much noisier than the task-engaged mode (compare `sigma` with `sigma2`).

If we want to know whether our model achieves descriptive adequacy, we need to look at the posterior predictive distributions of the model. However, by using posterior predictive checks, we won't be able to conclude that our model is not overfitting. Our success in fitting the fast-guess model to real data does not entail that the model is a good account of the data. It just means that it's flexible enough to fit the data. One further step would be to test whether each parameter can be selectively influenced by specific experimental manipulations as theoretically predicted. Another step could be to develop a competing model and then compare the performance of the models using Bayes factors or cross-validation.

17.1.5.3 Posterior predictive checks

For the posterior predictive checks, we can write the generated quantities block in a new file; in the `bcogsci` package, this code is in the file `mixture_h2_gen.stan`. The advantage of creating a new file is that we can generate as many observations as needed *after estimating the parameters*. There is no model block in the following Stan program and the parameters together with all the transformed parameters appear in the `parameters` block. We use the `gqs()` function in the `rstan` library, which allows us to use the posterior draws from a previously fitted model to generate posterior predicted data.

```
data {
  int<lower = 1> N;
  vector[N] x;
  vector[N] rt;
  array[N] int acc;
  vector[N] x2;
  int<lower = 1> N_subj;
  array[N] int<lower = 1, upper = N_subj> subj;
}
transformed data{
  real p_correct = 1;
}
parameters {
  real alpha;
  real beta;
  real<lower = 0> sigma;
  real<upper = alpha> gamma; //guessing
  real<lower = 0> sigma2;
  real<lower = 0, upper = 1> p_btask;
  real beta_task;
  vector<lower = 0>[3]  tau_u;
  matrix[3, N_subj] z_u;
```

```
  cholesky_factor_corr[3] L_u;
  matrix[N_subj, 3] u;
}
generated quantities {
  array[N] real rt_pred;
  array[N] real acc_pred;
  array[N] int z;
  for(n in 1:N){
    real lodds_task = logit(p_btask) + x2[n] * beta_task;
    z[n] = bernoulli_rng(inv_logit(lodds_task));
    if(z[n]==1){
      rt_pred[n] = lognormal_rng(alpha + u[subj[n],1] +
                                 x[n] * (beta + u[subj[n], 2]), sigma);
      acc_pred[n] = bernoulli_rng(p_correct);
    } else{
      rt_pred[n] = lognormal_rng(gamma + u[subj[n], 3], sigma2);
      acc_pred[n] = bernoulli_rng(0.5);
    }
  }
}
```

Generate responses from 500 simulated experiments as follows:

```
mixture_h2_gen <- system.file("stan_models",
                              "mixture_h2_gen.stan",
                              package = "bcogsci")
gen_model <- stan_model(mixture_h2_gen)
# The argument of the matrix `drop` needs to be set to FALSE,
# otherwise R will simplify the matrix into a vector.
# The two commas in the line below are not a mistake!
draws_par <- as.matrix(fit_mixs_data)[1:500, , drop = FALSE]
gen_mix_data <- gqs(gen_model,
                    data = ls_dots_data_short,
                    draws = draws_par)
```

First, take a look at the general distribution of response times generated by the posterior predictive model and by our real data in Figure 17.16(a).

```
rt_pred <- extract(gen_mix_data)$rt_pred
ppc_dens_overlay(y = ls_dots_data_short$rt, yrep = rt_pred[1:100,]) +
  coord_cartesian(xlim = c(0, 2000))
```

We see that the distribution of the observed response times is narrower than the predictive distribution. We are generating response times that are more spread out than the real data.

Next, examine the effect of the experimental manipulation in Figure 17.16(b): The posterior predictive check reveals that the model underestimates the observed effect of the experimental manipulation: the observed difference between response times is well outside the bulk of the predictive distribution.

```
meanrt_diff <- function(rt){
  mean(rt[ls_dots_data_short$x == 0.5]) -
    mean(rt[ls_dots_data_short$x == -0.5])
}
ppc_stat(ls_dots_data_short$rt,
         yrep = rt_pred,
         stat = meanrt_diff)
```

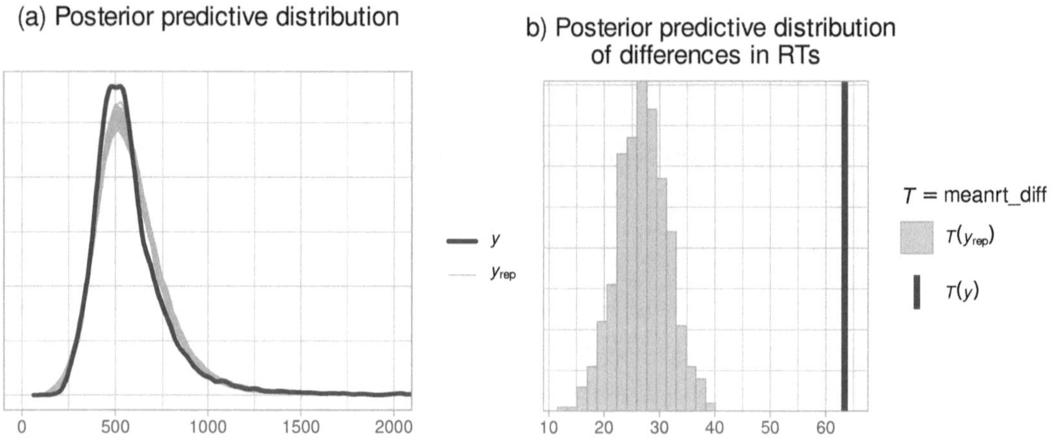

FIGURE 17.16: (a) Posterior predictive distribution of the hierarchical fast-guess model (`mixture_h2_gen.stan`) compared to observed response times. (b) Posterior predictive distribution of the response time difference due to experimental manipulation, using the same model; the vertical bar indicates the observed difference in the data.

Another important posterior predictive check includes comparing the fit of the model using a quantile probability plot, which is presented in the next chapter.

We also look at some instances of the predictive distribution. Figure 17.17 shows a simulated data set in black overlaid onto the real observations in gray. As we noticed in Figure 17.16(a), the model is predicting less variability than what we find in the data, especially when the emphasis is on accuracy.

```
acc_pred <- extract(gen_mix_data)$acc_pred
df_dots_pred <-
  tibble(rt = ls_dots_data_short$rt,
         acc = ls_dots_data_short$acc,
         difficulty = ifelse(ls_dots_data_short$x == 0.5,
                             "hard", "easy"),
         emphasis = ifelse(ls_dots_data_short$x2 == -0.5,
                           "speed", "accuracy"),
         acc_pred1 = acc_pred[1,],
         rt_pred1 = rt_pred[1,])
```

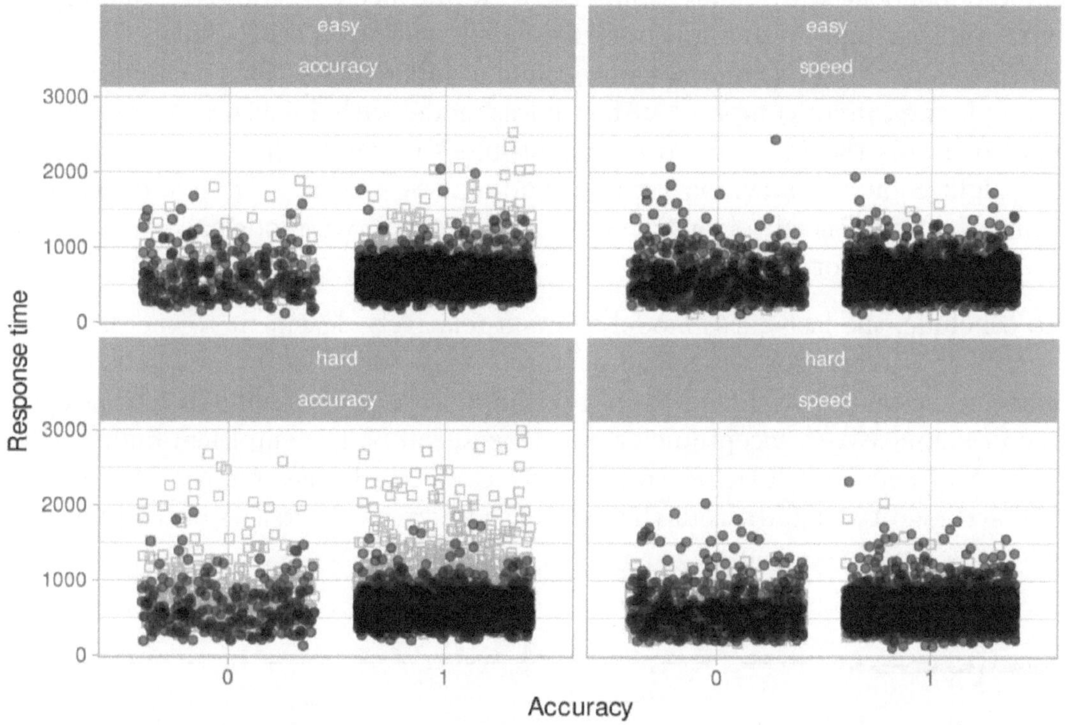

FIGURE 17.17: A simulated (posterior predictive) data set in black overlaid onto the observations in gray (based on `mixture_h2_gen.stan`).

If we would like to compare this model with a competing one using cross-validation, we would need to calculate the point-wise log-likelihood in the generated block:

```
generated quantities {
  array[N] real log_lik;
  for(n in 1:N){
    real lodds_task = logit(p_btask) + x2[n] * beta_task;
```

```
      log_lik[n] += log_sum_exp(log_inv_logit(lodds_task) +
                                 lognormal_lpdf(rt[n] | alpha + u[subj[n], 1] +
                                   x[n] * (beta + u[subj[n], 2]), sigma) +
                                 bernoulli_lpmf(acc[n] | p_correct),
                               log1m_inv_logit(lodds_task) +
                                 lognormal_lpdf(rt[n] | gamma +
                                 u[subj[n], 3], sigma2) +
                                 bernoulli_lpmf(acc[n] | 0.5));
  }
}
```

It is important to bear in mind that we can only compare models on the same dependent variable(s). That is, we would need to compare this model with another one fit to the same dependent variables and also in the same scale: accuracy (`0` or `1`) and response time in milliseconds. This means that, for example, we cannot compare our fast-guess model with an accuracy-only model. It also means that to compare our fast-guess model with a model based on left/right choices (known as stimulus coding, see section 18.1.1) and response times, we would need to reparameterize one of the two models; see the online exercise G.18.3 for chapter 18.

To conclude, the fast-guess model shows a relatively decent fit to the data and is able to account for the speed-accuracy trade-off. The model shows some inaccuracies that could lead to its revision and improvement. To what extent the inaccuracies are acceptable or not depends on (i) the empirical finding that we want to account for (for example, we can already assume that the model will struggle to fit data sets that show slow errors); and (ii) its comparison with a competing account.

17.2 Summary

In this chapter, we learned how to fit increasingly complex two-component mixture models using Stan, starting with a simple model and ending with a fully hierarchical model. We saw how to ensure model convergence via prior constraints, and how to evaluate model fit using the usual prior and posterior predictive checks, and to investigate parameter recovery. Such mixture models are notoriously difficult to fit, but they have a lot of potential in cognitive science applications, especially in developing computational models of different kinds of cognitive processes.

17.3 Further reading

The reader interested in a deeper understanding of marginalization is referred to Pullin, Gurrin, and Vukcevic (2021). Betancourt discusses problems of identification in Bayesian mixture models in a case study (https://mc-stan.org/users/documentation/case-studies/identifying_mixture_models.html). An in-depth treatment of finite mixture modeling can be found in Frühwirth-Schnatter (2006). An in-depth treatment of the fast-guess model and other mixture models of response times is provided in Chapter 7 of Luce (1991).

18

A simple accumulator model to account for choice response time

As mentioned in chapter 17, the most popular class of cognitive-process models that can incorporate both response times and accuracy are sequential sampling models (for a review, see Ratcliff et al. 2016). This class of model includes, among others, the drift diffusion model (Ratcliff 1978), the linear ballistic accumulator (Brown and Heathcote 2008), and the log-normal race model (Heathcote and Love 2012; Rouder et al. 2015). We discuss the log-normal race model in the current chapter. Sequential sampling or evidence-accumulation models are based on the idea that decisions are made by gathering evidence from the environment (e.g., the computer screen in many experiments) until sufficient evidence is gathered and a threshold of evidence is reached. The log-normal race model seems to be the simplest sequential sampling model that can account for the joint distribution of response times and response choice or accuracy (Heathcote and Love 2012; Rouder et al. 2015).

This model belongs to the subclass of *race models*, where the evidence for each response grows gradually in time in separate racing accumulators, until a threshold is reached. A response is made when one of these accumulators first reaches the threshold and wins the race against the other accumulators. This model is sometimes referred as deterministic (or non-stochastic, and ballistic), since the noise only affects the rate of accumulation of evidence before each race starts, but once the accumulator starts accumulating evidence, the rate is fixed. This means that a given accumulator can be faster or slower in different trials (or between choices) but its rate of accumulation will be fixed during a trial (or within choices). Brown and Heathcote (2005) claim that even though it is clear that a range of factors might cause within-choice noise, the behavioral effects might sometimes be small enough to ignore (this is in contrast to models such as the drift diffusion model, where both types of noise are present).

The two main advantages of the log-normal race model in comparison with other sequential sampling models are that: (i) the log-normal race model is very simple, making it easy to extend hierarchically; (ii) it is relatively easy to avoid convergence issues; and (iii) it is straightforward to model more than two choices. This specific model is presented next for pedagogical purposes because it is relatively easy to derive its likelihood given some reasonable assumptions.

DOI: 10.1201/9780429342646-24

However, even though the log-normal race is a "legitimate" cognitive model (see Further Reading for examples), the majority of the literature fits choice response times with the linear ballistic accumulator and/or the drift diffusion model, which provide more flexibility to the modeler.

The next section explains how the log-normal race model is implemented, using data from a lexical decision task.

18.1 Modeling a lexical decision task

In a lexical decision task, a subject is presented with a string of letters on the screen and they need to decide whether the string is a word or a non-word; see Figure 18.1. In the example developed below, a subset of 600 words and 600 non-words from 20 subjects ($600 \times 2 \times 20$ data points) are used from the data of the British Lexicon project (Keuleers et al. 2012). The data are stored as the object `df_blp` in the package `bcogsci`. In this data set, the lexicality of the string (word or non-word) is indicated in the column `lex`. The goal is to investigate how word frequency, shown in the column `freq` (frequency is counted per million words using the British National Corpus), affects the lexical decision task as quantified by accuracy and response time. For more details about the data set, type `?df_blp` on the R command line after loading the library `bcogsci`.

```
data("df_blp")
df_blp
```

```
## # A tibble: 24,000 x 8
##    subj block lex       trial string    acc    rt  freq
##   <dbl> <dbl> <chr>     <dbl> <fct>   <dbl> <dbl> <dbl>
## 1     1    57 non-word 28263 paybods      1   591     0
## 2     1    53 non-word 26414 lunned       1   621     0
## 3     1    49 non-word 24333 pertrax      1   575     0
## # i 23,997 more rows
```

The following code chunk adds 0.01 to the frequency column. A frequency of 0.01 corresponds to a word that appears only once in the corpus; 0.01 is added in order to avoid word frequencies of zero. The frequencies are then log-transformed to compress their range of values (see Brysbaert, Mandera, and Keuleers 2018 for a more in-depth treatment of word frequencies) and centered. It also creates a new variable that sum-codes the lexicality of the each given string (either a word, 0.5, or a non-word, −0.5).

FIGURE 18.1: Two trials in a lexical decision task. For the first trial, `rurble`, the correct answer would be to press the key on a keyboard or a response console that is mapped to the "non-word" response, for the second trial, `monkey`, the correct answer would be to press the key that is mapped to the "word" response.

```
df_blp <- df_blp %>%
  mutate(lfreq = log(freq + 0.01),
         c_lfreq = lfreq - mean(lfreq),
         c_lex = ifelse(lex == "word", 0.5, -0.5))
```

If one wants to study the effect of frequency on words, the "traditional" way to analyze these data would be to fit response times and choice data in two separate models on words, ignoring non-words. One model would be fit on the response times of correct responses, and a second model on the accuracy. These two models are fit below.

To fit the response times model, subset the correct responses given to strings that are words:

```
df_blp_word_c <- df_blp %>%
  filter(acc == 1, lex == "word")
```

Fit a hierarchical model with a log-normal likelihood and log-transformed frequency as a predictor (using `brms` here) and regularizing priors.

```
fit_rt_word_c <- brm(rt ~ c_lfreq + (c_lfreq | subj),
                     data = df_blp_word_c,
                     family = lognormal,
                     prior = c(prior(normal(6, 1.5), class = Intercept),
                               prior(normal(0, 1), class = b),
                               prior(normal(0, 1), class = sigma),
                               prior(normal(0, 1), class = sd),
                               prior(lkj(2), class = cor)),
                     iter = 3000)
```

Show the estimate of the effect of log-frequency on the log-ms scale.

```
posterior_summary(fit_rt_word_c, variable = "b_c_lfreq")
```

```
##           Estimate Est.Error    Q2.5   Q97.5
## b_c_lfreq  -0.0381   0.00261 -0.0434 -0.0329
```

To fit the accuracy model, subset the responses given to strings that are words:

```
df_blp_word <- df_blp %>%
  filter(lex == "word")
```

Fit a hierarchical model with a Bernoulli likelihood (and logit link) using log-transformed frequency as a predictor (using `brms`) and relatively weak priors:

```
fit_acc_word <- brm(acc ~ c_lfreq + (c_lfreq | subj),
                    data = df_blp_word,
                    family = bernoulli(link = logit),
                    prior = c(prior(normal(0, 1.5), class = Intercept),
                      prior(normal(0, 1), class = b),
                      prior(normal(0, 1), class = sd),
                      prior(lkj(2), class = cor)),
                    iter = 3000)
```

Show the estimate of the effect of log-frequency on the log-odds scale:

```
posterior_summary(fit_acc_word, variable = "b_c_lfreq")
```

```
##           Estimate Est.Error  Q2.5 Q97.5
```

```
## b_c_lfreq    0.572    0.0261 0.522 0.625
```

For this specific data set, it does not matter whether response times or accuracy are chosen as the dependent variable, since both yield results with a similar interpretation: More frequent words are identified more easily, that is, with shorter reading times (this is evident from the negative sign on the estimate of the effect), and with higher accuracy (positive sign on the estimate). However, it might be the case that some data set shows divergent directions in response times and accuracy. For example, more frequent words might take longer to identify, leading to a slowdown in response time as frequency increases, but might still be identified more accurately.

Furthermore, two models are fit above, treating response times and accuracy as independent. In reality, there is plenty of evidence that they are related (e.g., the speed-accuracy trade-off). Even in these data, as frequency increases, correct answers are given faster, and most errors are for low-frequency words (see Figure 18.2).

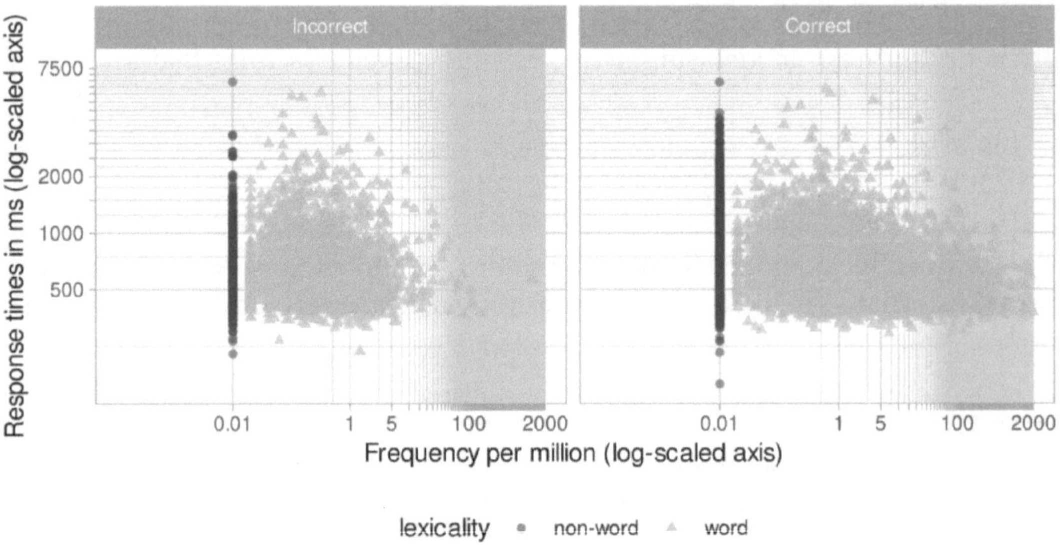

FIGURE 18.2: The distribution of response times for words and non-words, and correct and incorrect answers.

A powerful way to convey the relationship between response times and accuracy is using *quantile probability plots* (Ratcliff and Tuerlinckx 2002; these are closely related to the latency probability plots of Audley and Pike 1965).[1]

A quantile probability plot shows quantiles of the response time distribution

[1]Other useful tools to investigate errors and response time patterns are plots of *conditional accuracy functions* (Maanen, Katsimpokis, and Campen 2018) and plots of *defective cumulative distributions of response times* (Brown and Heathcote 2008).

(typically 0.1, 0.3, 0.5, 0.7, and 0.9) for correct and incorrect responses on the y-axis against probabilities of correct and incorrect responses for experimental conditions on the x-axis. The plot is built by first aggregating the data.

To display a quantile probability plot, create a custom function `qpf()` that takes as arguments a data set grouped by an experimental condition (e.g., words vs non-words, here by `lex`), and the quantiles that need to be displayed (by default, 0.1, 0.3, 0.5, 0.7, 0.9). The function works as follows: First, calculate the desired quantiles of the response times for incorrect and correct responses by condition (these are stored in `rt_q`). Second, calculate the proportion of incorrect and correct responses by condition (these are stored in `p`); because this information is needed for each quantile, repeat it for the number of quantiles chosen (here, five times). Last, record the quantile that each response time and response probability corresponds to (this is recorded in `q`), and whether it corresponds to an incorrect or a correct response (this information is stored in `response`).

```
qpf <- function(df_grouped,
                quantiles = c(0.1, 0.3, 0.5, 0.7, 0.9)) {
  df_grouped %>% summarize(
    rt_q = list(c(quantile(rt[acc == 0], quantiles),
                  quantile(rt[acc == 1], quantiles))),
    p = list(c(rep(mean(acc == 0), length(quantiles)),
               rep(mean(acc == 1), length(quantiles)))),
    q = list(rep(quantiles, 2)),
    response = list(c(rep("incorrect", length(quantiles)),
                      rep("correct", length(quantiles))))) %>%
    # Since the summary contains a list in each column,
    # we unnest it to have the following number of rows:
    # number of quantiles x groups x 2 (incorrect, correct)
    unnest(cols = c(rt_q, p, q, response))
}
df_blp_lex_q <- df_blp %>%
  group_by(lex) %>%
  qpf()
```

The aggregated data look like this:

```
df_blp_lex_q %>% print(n = 10)
```

```
## # A tibble: 20 x 5
##    lex       rt_q     p     q response
##    <chr>    <dbl> <dbl> <dbl> <chr>
```

```
##  1 non-word  433. 0.0521    0.1 incorrect
##  2 non-word  521. 0.0521    0.3 incorrect
##  3 non-word  613  0.0521    0.5 incorrect
##  4 non-word  779. 0.0521    0.7 incorrect
##  5 non-word 1110  0.0521    0.9 incorrect
##  6 non-word  448  0.948     0.1 correct
##  7 non-word  513  0.948     0.3 correct
##  8 non-word  575  0.948     0.5 correct
##  9 non-word  666  0.948     0.7 correct
## 10 non-word  905  0.948     0.9 correct
## # i 10 more rows
```

Plot the data by joining the points that belong to the same quantiles with lines. Given that incorrect responses in most tasks occur in less than 50% of the trials and correct responses occur in a complementary distribution (i.e., in more than 50% of the trials), incorrect responses usually appear in the left half of the plot, and correct ones in the right half. The code that appears below produces Figure 18.3.

```r
ggplot(df_blp_lex_q, aes(x = p, y = rt_q)) +
  geom_vline(xintercept = 0.5, linetype = "dashed") +
  geom_point(aes(shape = lex)) +
  geom_line(aes(group = interaction(q, response))) +
  ylab("RT quantiles (ms)") +
  scale_x_continuous("Response proportion", breaks = seq(0, 1, .2)) +
  scale_shape_discrete("Lexicality") +
  annotate("text", x = 0.40, y = 500, label = "incorrect") +
  annotate("text", x = 0.60, y = 500, label = "correct")
```

The vertical spread among the lines shows the shape of the response time distribution. The lower quantile lines correspond to the left part of the response time distribution, and the higher quantiles to the right part of the distribution. Since the response time distribution is long tailed and right skewed, the higher quantiles are more spread apart than the lower quantiles.

A quantile probability plot can also be used to corroborate the observation that high-frequency words are easier to recognize. To do that, subset the data to only words, and group the strings according to their "frequency group" (that is, according to the quantile of frequency that the strings belong to). Whereas we previously aggregated over all the observations, ignoring subjects, we can also aggregate by subjects first, and then average the results. This will prevent some idiosyncratic responses from subjects from dominating in the plot. (We can also plot individual quantile probability plots by subject). Apart from the

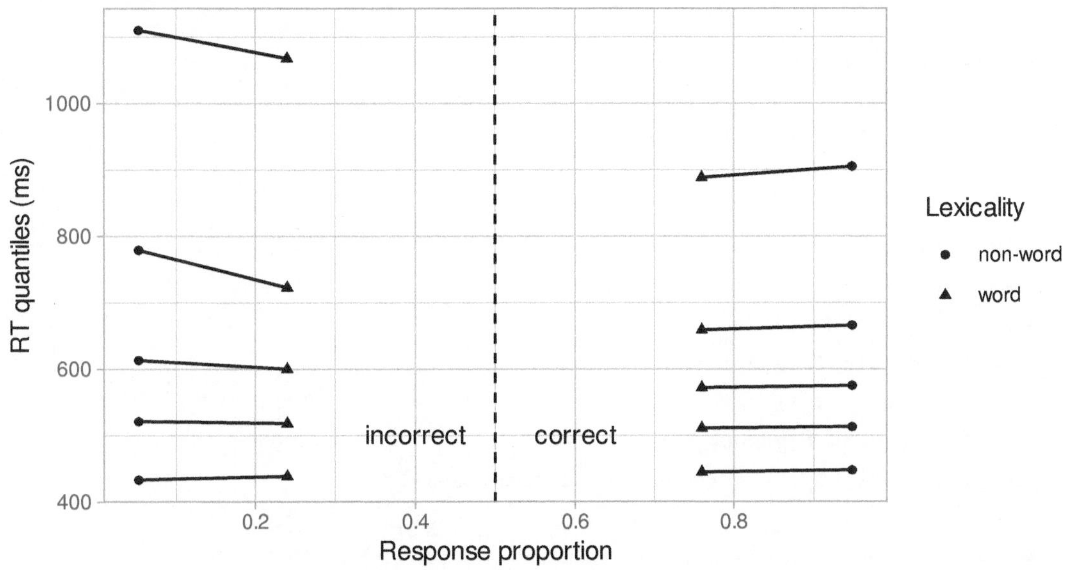

FIGURE 18.3: Quantile probability plots showing 0.1, 0.3, 0.5, 0.7, and 0.9-th response time quantiles plotted against proportion of incorrect responses (left) and proportion of correct responses (right) for strings that are words and non-words.

fact that the aggregation is by subjects, the code below follows the same steps as before, and the result is shown in Figure 18.4. The plot shows that for more frequent words, accuracy improves and responses are faster.

```
df_blp_freq_q <- df_blp %>%
  # Subset only words:
  filter(lex == "word") %>%
  # Create 5 word frequencies group
  mutate(freq_group =
           cut(lfreq,
               quantile(lfreq, c(0,0.2,0.4, 0.6, 0.8, 1)),
               include.lowest = TRUE,
               labels =
                 c("0-0.2", "0.2-0.4",
                   "0.4-0.6", "0.6-0.8", ".8-1"))
         ) %>%
  # Group by condition  and subject:
  group_by(freq_group, subj) %>%
  # Apply the quantile probability function:
  qpf() %>%
  # Group again removing subject:
```

```
  group_by(freq_group, q, response) %>%
  # Get averages of all the quantities:
  summarize(rt_q = mean(rt_q),
            p = mean(p))
# Plot
ggplot(df_blp_freq_q, aes(x = p, y = rt_q)) +
  geom_point(shape = 4) +
  geom_text(
    data = df_blp_freq_q %>%
      filter(q == 0.1),
    aes(label = freq_group), nudge_y = 12) +
  geom_line(aes(group = interaction(q, response))) +
  ylab("RT quantiles (ms)") +
  scale_x_continuous("Response proportion", breaks = seq(0, 1, 0.2)) +
  annotate("text", x = 0.40, y = 900, label = "incorrect") +
  annotate("text", x = 0.60, y = 900, label = "correct")
```

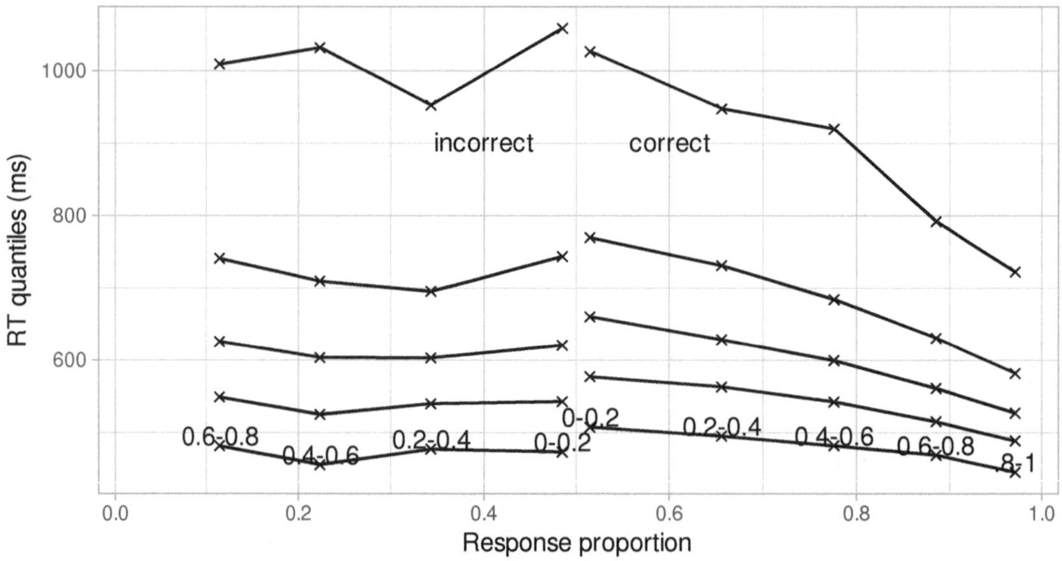

FIGURE 18.4: Quantile probability plot showing 0.1, 0.3, 0.5, 0.7, and 0.9-th response times quantiles plotted against proportion of incorrect responses (left) and proportion of correct responses (right) for words of different frequency. Word frequency is grouped according to quantiles: The first group is words with frequencies smaller than the 0.2-th quantile, the second group is words with frequencies smaller than the 0.4-th quantile and larger than the 0.2-th quantile, and so forth.

So far, several ways were shown to describe the data by representing them graphically. Next, we turn to modeling the data.

18.1.1 Modeling the lexical decision task with the log-normal race model

The log-normal race model is used here to examine the effect of word frequency in both response times and choice (word vs. non-word) in the lexical decision task presented earlier. In this example, the log-normal race model is limited to fitting two choices; as mentioned earlier, this model can in principle fit more than two choices. When modeling a task with two choices, there are two ways to account for the data: either fit the response times and the accuracy (i.e., accuracy coding: correct vs. incorrect), or fit the response times and actual responses (i.e., stimulus coding: in this case word vs. non-word). In this example, we will use the stimulus-coding approach.

The following code chunk adds a new column that incorporates the actual choice made (as `word` vs. `non-word` in `choice` and as `1` vs. `2` in `nchoice`):

```
df_blp <- df_blp %>%
  mutate(choice = ifelse((lex == "word" &
                          acc == 1) |
                         (lex == "non-word" &
                          acc == 0), "word", "non-word"),
         nchoice = ifelse(choice == "word", 1, 2))
```

To start modeling the data, think about the behavior of one synthetic subject. This subject simultaneously accumulates evidence for the response, "word" in one accumulator, and for "non-word" in another independent accumulator. Unlike other sequential sampling models, an increase in evidence for one choice doesn't necessarily reduce the evidence for the other choices. Rouder et al. (2015) points out that it might seem odd to assume that we accumulate evidence for a non-word in the same manner as we accumulate evidence for a word, since non-words may be conceptualized as the absence of a word. However, they stress that this approach is closely related to novelty detection, where the salience of never-before experienced stimuli seems to indicate that novelty is psychologically represented as more than the absence of familiarity. Nevertheless, notions of words and non-word evidence accumulation are indeed controversial (see Dufau, Grainger, and Ziegler 2012). The alternative approach of fitting accuracy rather than stimuli discussed before doesn't really circumvent the problem. This is because when the correct answer is `word`, we assume that the "correct" accumulator accumulates evidence for `word`, and the incorrect one for `non-word`, and the other way around when the correct answer is `non-word`.

18.1.2 A generative model for a race between accumulators

To build a generative model of the task based on the log-normal race model, start by spelling out the assumptions. In a race of accumulators model, the assumption is that the decision time T taken for each accumulator of evidence to reach the threshold at distance to the decision threshold D is simply defined by

$$T = D/V \tag{18.1}$$

where the denominator V is the rate (velocity, sometimes also called drift rate) of evidence accumulation.

The log-normal race model assumes that the rate V in each trial is sampled from a log-normal distribution:

$$V \sim LogNormal(\mu_v, \sigma_v) \tag{18.2}$$

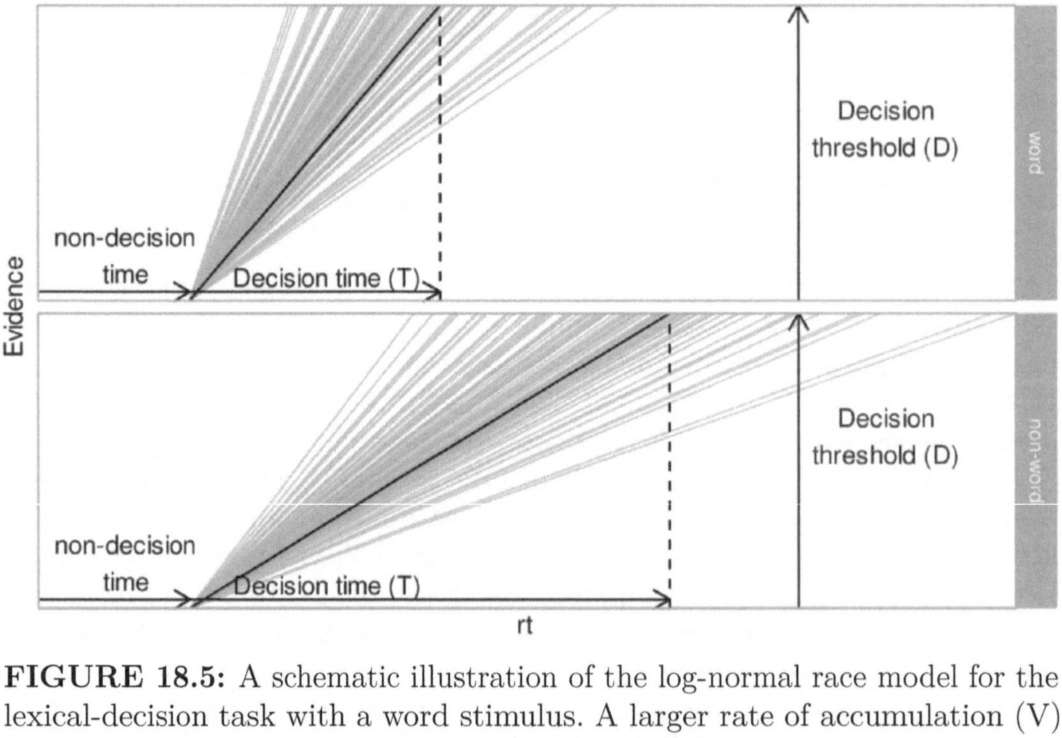

FIGURE 18.5: A schematic illustration of the log-normal race model for the lexical-decision task with a word stimulus. A larger rate of accumulation (V) leads to a larger slope. Here, the choice of word is selected.

A log-normal distribution is partly justified by the work by Ulrich and Miller (1993) (also see online section A.4), and as discussed later, it is very convenient mathematically.

For simplicity, assume that the distance D to the threshold is kept constant. This might not be a good assumption if the experiment is designed so that subjects change their threshold depending on speed or accuracy incentives (that was not the case in this experiment), or if the subject gets fatigued as the experiment progresses, or if there is reason to believe that there might be random fluctuations in this threshold. Later in this chapter, we will discuss what happens if this assumption is relaxed.

In each trial n, a string is presented, and both the word and non-word accumulators of evidence need to be modeled. For the model to capture the behavior of the subjects, evidence needs to be accumulated faster for the word accumulator than for the non-word accumulator when the string presented is a word, and vice versa when the string is a non-word.

Assume that, for trial n, the rate of accumulation of evidence for words, $V_{w,n}$, and the rate for non-words, $V_{nw,n}$, are generated as follows:

$$\begin{aligned} V_{w,n} &\sim LogNormal(\mu_{v_{w,n}}, \sigma) \\ V_{nw,n} &\sim LogNormal(\mu_{v_{nw,n}}, \sigma) \end{aligned} \tag{18.3}$$

The location μ_{v_w} of the distribution of rates of accumulation of evidence for a word w is a function of the lexicality of the string presented (only a word will increase this rate of accumulation and not a non-word), frequency (i.e., high-frequency words might be easier to identify, leading to a faster rate of accumulation than with low-frequency words), and bias (i.e., a subject might have a tendency to answer that a string is a word rather than non-word or vice-versa, regardless of the stimuli).

This assumption can be modeled with a linear regression over μ_{v_w}, with parameters that represent the bias to categorize a string as a word, α_w, the effect of lexicality, β_{lex_w}, and the effect of log-frequency β_{lfreq_w}.

$$mu_{v_{w,n}} = \alpha_w + lex_n \cdot \beta_{lex_w} + lfreq_n \cdot \beta_{lfreq_w} \tag{18.4}$$

For the non-word accumulator, the location for the rate of accumulation of evidence is defined similarly:

$$mu_{v_{nw,n}} = \alpha_{nw} + lex_n \cdot \beta_{lex_{nw}} + lfreq_n \cdot \beta_{lfreq_{nw}} \tag{18.5}$$

The accumulators reach the threshold at time $T_{w,n}$ (for words) and $T_{nw,n}$ (for non-words)

$$\begin{aligned} T_{w,n} &= D_w / V_{w,n} \\ T_{nw,n} &= D_{nw} / V_{nw,n} \end{aligned} \tag{18.6}$$

The choice for trial n corresponds to the accumulator with the shortest time for that trial,

$$choice_n = \begin{cases} word, & \text{if } T_{w,n} < T_{nw,n} \\ non\text{-}word, & \text{otherwise} \end{cases} \tag{18.7}$$

and T_n, the time taken to make a decision for trial n is the minimum of the time $T_{w,n}$ and $T_{nw,n}$:

$$T_n = min(T_{w,n}, T_{nw,n}) \tag{18.8}$$

We also need to take into account that not all the time spent in the task involves making the decision: Time is spent fixating the gaze on the screen, pressing a button, etc. We'll add a shift to the distribution, representing the minimum amount of time that a subject needs for all the peripheral processes that happened before and after the decision (also see Rouder 2005). We represent this non-decision time with T_0. Although some variation in the non-decision time is highly likely, we use a constant as an approximation. This simplification is reasonable if the variation in non-decision time is small relative to the variation in decision time (Heathcote and Love 2012).

$$rt_n = T_0 + T_n \tag{18.9}$$

The following chunk of code generates synthetic data for one subject, by setting true point values to the parameters and translating the previous equations to R. The true values are relatively arbitrary and were decided by trial and error until a relatively realistic distribution of response times was obtained. Considering that this is only one subject (unlike what was shown in previous figures), Figure 18.6 looks relatively fine. (One can also inspect the quantile probability plots of individual subjects in the real data set and compare it to the synthetic data).

First, set a seed to always generate the same pseudo-random values, take a subset of the data set to keep the same structure of the data frame for our simulated subject, and set true point values. The distance to the decision threshold, D, is set to the arbitrary constant value of 1800 so that, with the other parameters having magnitudes similar to those typically found in response time studies, the simulated reaction times have a realistic magnitude.

```
set.seed(123)
df_blp_1subj <- df_blp %>%
  filter(subj == 1)
# Set the same threshold to both accumulators
D <- 1800
alpha_w <- 0.8
beta_wlex <- 0.5
beta_wlfreq <- 0.2
alpha_nw <- 1
beta_nwlex <- -0.5
beta_nwlfreq <- -0.05
sigma <- 0.8
T_0 <- 150
```

Second, generate the locations of both accumulators, `mu_v_w` and `mu_v_nw`, for every trial. This means that both variables are vectors of length, `N`, the number of trials:

```
mu_v_w <- alpha_w + df_blp_1subj$c_lfreq * beta_wlfreq +
  df_blp_1subj$c_lex * beta_wlex
mu_v_nw <- alpha_nw + df_blp_1subj$c_lfreq * beta_nwlfreq +
  df_blp_1subj$c_lex * beta_nwlex
N <- nrow(df_blp_1subj)
```

Third, generate values for the rates of accumulation, `V_w` and `V_nw`, for every trial. Use those rates to calculate `T_w` and `T_nw`, how long it will take for each accumulator to reach its threshold (assumed to be the same for both accumulators) for every trial:

```
V_w <- rlnorm(N, mu_v_w, sigma)
V_nw <- rlnorm(N, mu_v_nw, sigma)
T_w <-  D / V_w
T_nw <- D / V_nw
```

Fourth, calculate the time it takes to reach a decision in every trial, `T_n` as the by-trial minimum between `T_w` and `T_nw`. Similarly, store the winner accumulator for each trial in `accumulator_winner`:

```
T_n <- pmin(T_w, T_nw)
accumulator_winner <- ifelse(T_w == pmin(T_w, T_nw),
```

```
                    "word",
                    "non-word")
```

Finally, add this information to the data frame that now indicates choice, time, and accuracy for each trial:

```
df_blp1_sim <- df_blp_1subj %>%
  mutate(rt = T_0 + T_n,
         choice = accumulator_winner,
         nchoice = ifelse(choice == "word", 1, 2)) %>%
  mutate(acc = ifelse(lex == choice, 1, 0))
```

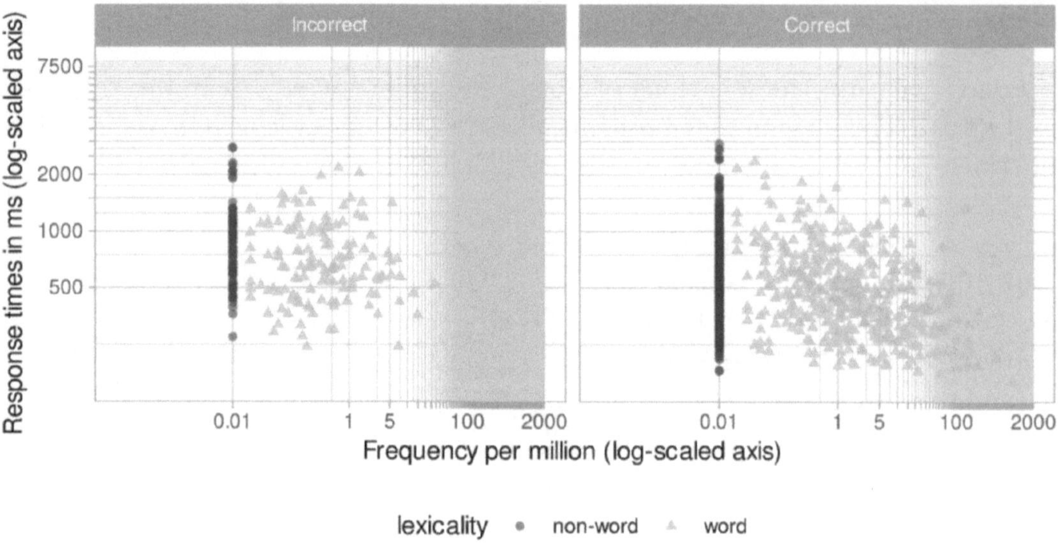

FIGURE 18.6: The distribution of response times for words and non-words, and correct and incorrect answers for the synthetic data of one subject.

18.1.3 Fitting the log-normal race model

A first issue that that arises when attempting to fit the log-normal race model is that we need to fit its likelihood to the decision time T, which is the ratio of D and V; there are two situations that are compatible with our assumptions and are mathematically simple:

(1) If we assume that D is a constant k and only V is a random variable, then $T = k/V$, and

$$\log(T) = \log(k/V) = \log(k) - \log(V) \tag{18.10}$$

Since V is log-normally distributed, $\log(V) \sim Normal(\mu_v, \sigma_v)$, and $\log(k)$ is a constant:

$$\begin{aligned} \log(T) &\sim Normal(\log(k) - \mu_v, \sigma_v) \\ T &\sim LogNormal(\log(k) - \mu_v, \sigma_v) \end{aligned} \tag{18.11}$$

(2) If both D and V are random variables we need a ratio or quotient distribution function. While for arbitrary distributions this requires solving (sometimes extremely complex) integrals (see, for example, Nelson 1981), a log-normally distributed time is not uniquely predicted by assuming that distance is a constant. It also follows if distance is also a log-normally distributed variable and both distributions are independent: If we assume that $D \sim LogNormal(\mu_d, \sigma_d)$, then T is the ratio of two random variables D/V, and

$$\log(T) = \log(D/V) = \log(D) - \log(V) \tag{18.12}$$

We have a difference of independent, normally distributed random variables. It follows from random variable theory (see, e.g., Ross 2002) that:

$$\begin{aligned} \log(T) &\sim Normal(\mu_d - \mu_v, \sqrt{\sigma_d^2 + \sigma_v^2}) \\ T &\sim LogNormal(\mu_d - \mu_v, \sqrt{\sigma_d^2 + \sigma_v^2}) \end{aligned} \tag{18.13}$$

From Equations (18.11) and (18.13), it should be clear that the threshold and accumulation rate cannot be disentangled: a manipulation that affects the rate or the decision threshold will affect the location of the distribution in the same way (also see Rouder et al. 2015). Another important observation is that T won't have a log-normal distribution when D has any other distributional form.

Following Rouder et al. (2015), we assume that the noise parameter is the same for each accumulator, since this means that contrasts between finishing time distributions are captured completely by contrasts of the locations of the log-normal distributions. We discuss at the end of the chapter why one would need to relax this assumption (also see online exercise G.18.2).

In each trial n, with an accumulator for words, indicated with the subscript w, and one for non-words, indicated with nw, we can model the time it takes for each accumulator to get to the threshold D in the following way. For the word accumulator,

$$
\begin{aligned}
\mu_{w,n} &= \mu_{d_w} - \mu_{v_{w,n}} \\
\mu_{w,n} &= \mu_{d_w} - (\alpha_w + lex_n \cdot \beta_{lex_w} + lfreq_n \cdot \beta_{lfreq_w}) \\
\mu_{w,n} &= (\mu_{d_w} - \alpha_w) - lex_n \cdot \beta_{lex_w} - lfreq_n \cdot \beta_{lfreq_w} \\
\mu_{w,n} &= \alpha'_w - lex_n \cdot \beta_{lex_w} - lfreq_n \cdot \beta_{lfreq_w} \\
T_{w,n} &\sim LogNormal(\mu_{w,n}, \sigma)
\end{aligned}
\tag{18.14}
$$

The parameter α'_w absorbs the location of the threshold distribution (μ_{d_w}, which we assumed is independent of the accumulator; that is, it is equal to μ_d) minus the intercept of the rate distribution ($\mu_{v_{w,n}}$), and represents a bias. As α'_w gets smaller, the accumulator will be more likely to reach the threshold first, all things being equal, biasing the responses to `word`.

Similarly, for the non-word accumulator,

$$
\begin{aligned}
\mu_{nw,n} &= \alpha'_{nw} - lex_n \cdot \beta_{lex_{nw}} - lfreq_n \cdot \beta_{lfreq_{nw}} \\
T_{nw,n} &\sim LogNormal(\mu_{nw,n}, \sigma)
\end{aligned}
\tag{18.15}
$$

The only observed time is the one associated with the winner accumulator, the response selected s, which corresponds to the faster accumulator:

$$
T_{accum=s,n} \sim LogNormal(\mu_{accum=s,n}, \sigma) \tag{18.16}
$$

If we only fit the observed finishing times of the accumulators, we're always ignoring that in a given trial the accumulator that lost was slower than the accumulator for which we have the latency; this means that we underestimate the time it takes to reach the threshold and we overestimate the rate of accumulation of both accumulators. This can be treated as a problem of *censored data*, where for each trial the slower observations are not known.

Since the potential decision time for the accumulator that wasn't selected is definitely longer than the one of the winner accumulator, we obtain the likelihood for each unobserved time by integrating out all the possible decision times that the accumulator could have; that is, from the time it took for the winner accumulator to reach the threshold to infinitely large decision times. In other words, we are calculating how likely is each possible value of the latent, unknown values of a random variable $T_{accum \neq s,n}$, knowing that they are always more than or equal to the observed (single) value of $T_{accum=s,n}$.

$$
P(T_{accum \neq s,n} \geq T_{accum=s,n}) = \int_{T_{accum=s,n}}^{\infty} LogNormal(T | \mu_{a \neq s,n}, \sigma) \, dT \tag{18.17}
$$

This integral is the complement of the CDF of the log-normal distribution evaluated at $T_{accum=s,n}$.

$$P(T_{accum\neq s,n}) = 1 - LogNormal\_CDF(T_{accum=s,n}|\mu_{accum\neq s,n}, \sigma) \tag{18.18}$$

where $LogNormal\_CDF(T_{accum=s,n}|\mu_{accum\neq s,n}, \sigma)$ is a convenient shorthand for the CDF of the log-normal distribution with parameters $\mu_{accum\neq s,n}, \sigma$ evaluated at $T_{accum=s,n}$.

So far we have been fitting the decision time T, but our dependent variable is response times, rt, the sum of the decision time T and the non-decision time T_0. This requires a change of variables in our model, $T_n = rt_n - T_0$, since rt but not T is available as data. Here, the Jacobian is 1, since $\left|d\frac{rt_n - T0}{drt_n}\right| = 1$. In a change of variables, we must multiply the likelihood by the Jacobian (for details, see section 10.1). Multiplying the likelihood by one, or equivalently, adding $\log(1) = 0$ to the log-likelihood, does not alter the likelihood (or the log-likelihood). Therefore, we do not need to code the Jacobian in the Stan code.[2]

To sum up, our model can be stated as follows:

$$\begin{aligned}
T_n &= rt_n - T_0 \\
\mu_{w,n} &= \alpha'_w - lex_n \cdot \beta_{lex_w} - lfreq_n \cdot \beta_{lfreq_w} \\
\mu_{nw,n} &= \alpha'_{nw} - lex_n \cdot \beta_{lex_{nw}} - lfreq_n \cdot \beta_{lfreq_{nw}} \\
T_n &\sim \begin{cases} LogNormal(\mu_{w,n}, \sigma) \text{ if } choice = \text{word} \\ LogNormal(\mu_{nw,n}, \sigma) \text{ otherwise} \end{cases}
\end{aligned} \tag{18.19}$$

Rather than trying to estimate all the censored observations, we integrate them out:[3]

$$P(T_{censored}) = \begin{cases} 1 - LogNormal\_CDF(T_n|\mu_{nw,n}, \sigma), \text{ if } choice = \text{word} \\ 1 - LogNormal\_CDF(T_n|\mu_{w,n}, \sigma), \text{ otherwise} \end{cases} \tag{18.21}$$

[2]Adding the Jacobian adjustment would be necessary if one assumes that the non-decision time has its own distribution rather than being constant.

[3]One could estimate the censored times by fitting log-normal distributions that are truncated at T_n since this is the minimum possible time for each censored observation:

$$T_{censored,n} \sim \begin{cases} LogNormal(\mu_{nw,n}, \sigma) \text{ with } T_{censored,n} > T_n, \text{ if } choice = \text{word} \\ LogNormal(\mu_{w,n}, \sigma) \text{ with } T_{censored,n} > T_n, \text{ otherwise} \end{cases} \tag{18.20}$$

We need priors for all the parameters. An added complication here is the prior for the non-decision time, T_0: we need to make sure that it's strictly positive and also that it's smaller than the shortest observed response time. This is because the decision time for each observation, T_n should also be strictly positive:

$$\begin{aligned} T_n = rt_n - T_0 &> 0 \\ rt_n &> T_0 \\ min(\mathbf{rt}) &> T_0 \end{aligned} \tag{18.22}$$

We thus truncate the prior of T_0 so that the values lie between zero and $min(\mathbf{rt})$, the minimum value of the vector of response times. Although the prior of T_0 is informed by the data, the generative model underlying this approach is not. When generating data–before the simulated response times are available–one first draws a value of T_0 from a normal distribution truncated at zero. Then, decision times for each trial n, T_n, are generated, and finally, the simulated (or predicted) response time for each trial, rt_n, is the sum of T_0 and T_n. In other words, there is no circularity in this process. Given the time it takes to fixate the gaze on the screen and a minimal motor response time, centering the prior of T_0 in 150 ms seems reasonable. The rest of the priors are on the log scale. One should use prior predictive checks to verify that the order of magnitude of all the priors is appropriate. We skip this step here and present the priors below:

$$\begin{aligned} T_0 &\sim Normal(150, 100) \text{ with } 0 < T_0 < min(rt_n) \\ \boldsymbol{\alpha} &\sim Normal(6, 1) \\ \boldsymbol{\beta} &\sim Normal(0, 0.5) \\ \sigma &\sim Normal_+(0.5, 0.2) \end{aligned} \tag{18.23}$$

where $\boldsymbol{\alpha}$ is a vector $\langle \alpha'_n, \alpha'_{nw} \rangle$, and $\boldsymbol{\beta}$ is a vector of all the β used in the likelihoods.

To translate the model into Stan, we need a normal distribution truncated so that the values lie between zero and $min(\mathbf{rt})$ for the prior of `T_0`. This means dividing the original distribution by the difference of the CDFs evaluated at these two points; see online section A.2. In log-space, this is a difference between the log-transformed original distribution and the logarithm of the difference of the CDFs. The function `log_diff_exp()` is a more stable version of this last operation (i.e., less prone to underflow/overflow, see Blanchard, Higham, and Higham 2020). What `log_diff_exp()` does is to take the log of the difference of the exponent of two functions. In this case the functions are two log-CDFs.

```
target += normal_lpdf(T_0 | 150, 100)
```

```
        - log_diff_exp(normal_lcdf(min(rt) | 150, 100),
                       normal_lcdf(0 | 150, 100));
```

We implement the likelihood of each joint observation of response time and choice with an if-else clause that calls the likelihood of the accumulator that corresponds to the choice selected in the trial `n`, and the complement CDF for the accumulator that was not selected:

```
    if(nchoice[n] == 1)
      target += lognormal_lpdf(T[n] | alpha[1] -
                                c_lex[n] * beta[1] -
                                c_lfreq[n] * beta[2] , sigma)  +
        lognormal_lccdf(T[n] | alpha[2] -
                         c_lex[n] * beta[3] -
                         c_lfreq[n] * beta[4], sigma);
    else
       target += lognormal_lpdf(T[n] | alpha[2] -
                                 c_lex[n] * beta[3] -
                                 c_lfreq[n] * beta[4], sigma) +
        lognormal_lccdf(T[n] | alpha[1] -
                         c_lex[n] * beta[1] -
                         c_lfreq[n] * beta[2], sigma);
  }
```

The complete Stan code for this model is shown below as `lnrace.stan`:

```
data {
  int<lower = 1> N;
  vector[N] c_lfreq;
  vector[N] c_lex;
  vector[N] rt;
  array[N] int nchoice;
}
parameters {
  array[2] real alpha;
  array[4] real beta;
  real<lower = 0> sigma;
  real<lower = 0, upper = min(rt)> T_0;
}
model {
  vector[N] T = rt - T_0;
  target += normal_lpdf(alpha | 6, 1);
  target += normal_lpdf(beta | 0, 0.5);
  target += normal_lpdf(sigma | 0.5, 0.2)
```

```
    - normal_lccdf(0 | 0.5, 0.2);
  target += normal_lpdf(T_0 | 150, 100)
    - log_diff_exp(normal_lcdf(min(rt) | 150, 100),
                   normal_lcdf(0 | 150, 100));
  for(n in 1:N){
    if(nchoice[n] == 1)
      target += lognormal_lpdf(T[n] | alpha[1] -
                               c_lex[n] * beta[1] -
                               c_lfreq[n] * beta[2], sigma)  +
        lognormal_lccdf(T[n] | alpha[2] -
                        c_lex[n] * beta[3] -
                        c_lfreq[n] * beta[4], sigma);
    else
       target += lognormal_lpdf(T[n] | alpha[2] -
                                c_lex[n] * beta[3] -
                                c_lfreq[n] * beta[4], sigma) +
        lognormal_lccdf(T[n] | alpha[1] -
                        c_lex[n] * beta[1] -
                        c_lfreq[n] * beta[2], sigma);
  }
}
```

Store the data in a list and fit the model. Some warnings might appear during the warm-up, but these warnings can be ignored since they no longer appear afterward, and all the convergence checks look fine (omitted here):

```
lnrace <- system.file("stan_models",
                      "lnrace.stan",
                      package = "bcogsci")
ls_blp1_sim <- list(N = nrow(df_blp1_sim),
                    rt = df_blp1_sim$rt,
                    nchoice = df_blp1_sim$nchoice,
                    c_lex = df_blp1_sim$c_lex,
                    c_lfreq = df_blp1_sim$c_lfreq)
fit_blp1_sim <- stan(lnrace, data = ls_blp1_sim)
```

Print the parameters values:

```
print(fit_blp1_sim, pars = c("alpha", "beta", "T_0", "sigma"))
```

```
##             mean   2.5%  97.5% n_eff Rhat
## alpha[1]    6.67   6.60   6.73  4725    1
```

```
## alpha[2]     6.53    6.46    6.59  3903     1
## beta[1]      0.36    0.17    0.56  3407     1
## beta[2]      0.21    0.18    0.24  3323     1
## beta[3]     -0.46   -0.68   -0.23  2962     1
## beta[4]     -0.07   -0.12   -0.02  3069     1
## T_0        143.24  130.39  152.14  3362     1
## sigma        0.78    0.74    0.82  3569     1
```

As in previous chapters, `mcmc_recover_hist()` can be used to compare the posterior distributions of the relevant parameters of the model with their true point values (Figure 18.7). Indeed, Figure 18.7 shows that the model recovers the parameters reasonably well. First, however, we need to reparameterize the true values, since D cannot be known, and we don't fit V, but rather D/V, with V log-normally distributed. Then, obtain an estimate of α', rather than α, such that $\alpha' = log(mu_d) - \alpha$.

```
true_values <- c(alphapw = log(D) - alpha_w,
                 alphapnw = log(D) - alpha_nw,
                 beta_wlex = beta_wlex,
                 beta_wlfreq = beta_wlfreq,
                 beta_nwlex = beta_nwlex,
                 beta_nwlfreq = beta_nwlfreq,
                 sigma = sigma,
                 T_0 = T_0)
estimates <- as.data.frame(fit_blp1_sim) %>%
  select(- lp__)
mcmc_recover_hist(estimates, true_values) +
   theme(axis.text.x = element_text(angle = 45, vjust = 1, hjust=1))
```

Before moving on to a more complex version of this model, it's worth spending some time making the code more modular. Encapsulate the likelihood of the log-normal race model by writing it as a function. The function has four arguments, the decision time `T`, the choice `nchoice` (this will only work with two

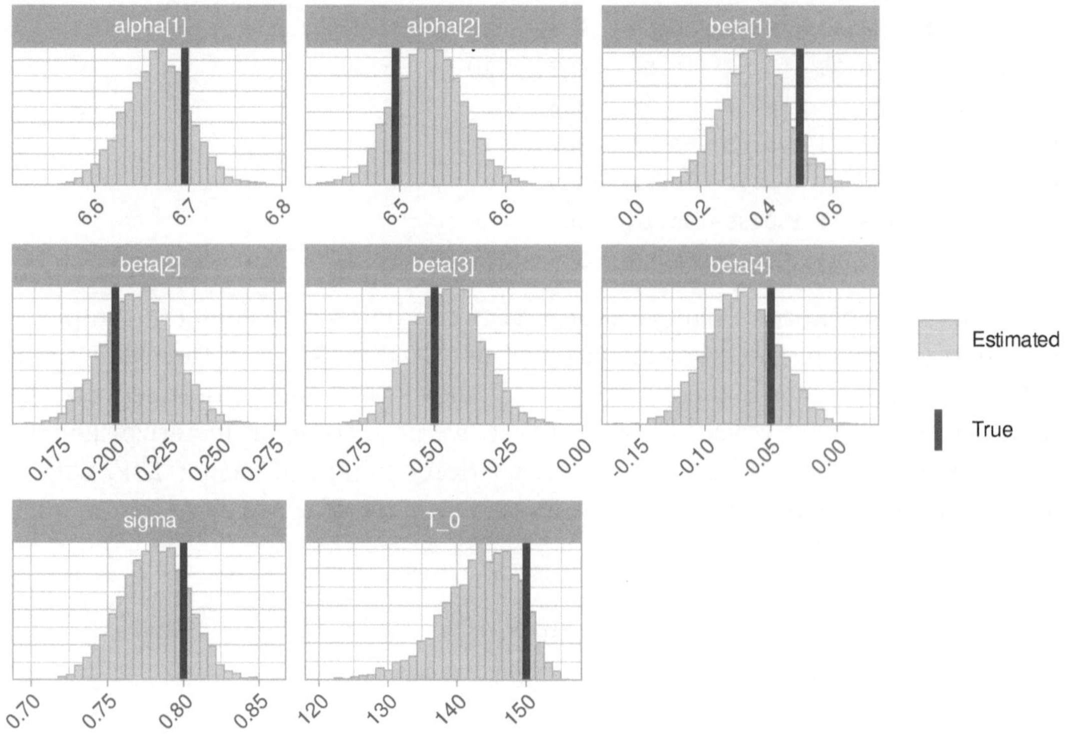

FIGURE 18.7: Posterior distributions of the main parameters of the log-normal race model `fit_blp1_sim` together with their true point values.

choices, `1` and `2`), an array of locations `mu` (which again we implicitly assume to have two elements), and a common scale `sigma`.

```
functions {
  real lognormal_race2_lpdf(real T, int nchoice,
                            array[] real mu, real sigma){
    real lpdf;
    if(nchoice == 1)
        lpdf = lognormal_lpdf(T | mu[1] , sigma)  +
          lognormal_lccdf(T | mu[2], sigma);
      else
        lpdf = lognormal_lpdf(T | mu[2], sigma) +
          lognormal_lccdf(T | mu[1], sigma);
    return lpdf;
  }
}
```

Next, for each iteration `n` of the original for loop, generate an auxiliary variable `T` which contains the decision time for the current trial, and `mu` as an array of size two that contains all the parameters that affect the location at each trial. This will allow us to use our new function as follows in the `model` block:[4]

```
array[N] real log_lik;
for(n in 1:N){
  real T = rt[n] - T_0;
  array[2] real mu = {alpha[1]  -
                      c_lex[n] * beta[1]  -
                      c_lfreq[n] * beta[2],
                      alpha[2]  -
                      c_lex[n] * beta[3] -
                      c_lfreq[n] * beta[4]};
  log_lik[n] = lognormal_race2_lpdf(T | nchoice[n], mu, sigma);
}
```

The variable `log_lik` contains the log-likelihood for each trial. We must not forget to add the total log-likelihood to the `target` variable. This is done simply by `target += sum(log_lik)`.[5]

The complete Stan code for this model can be found in the `bcogsci` package as `lnrace_mod.stan`, it is left for the reader to verify that the results are the same as from the non-modular model `lnrace.stan` fit earlier.

18.1.4 A hierarchical implementation of the log-normal race model

A simple hierarchical version of the previous model assumes that all the parameters $\boldsymbol{\alpha}$ and $\boldsymbol{\beta}$ have by-subject adjustments:

$$
\begin{aligned}
\mu_{w,n} &= \alpha'_w + u_{subj[n],1} - lex_n \cdot (\beta_{lex_w} + u_{subj[n],2}) \\
&\quad - lfreq_n \cdot (\beta_{lfreq_w} + u_{subj[n],3}) \\
\mu_{nw,n} &= \alpha'_{nw} + u_{subj[n],4} - lex_n \cdot (\beta_{lex_{nw}} + u_{subj[n],5}) \\
&\quad - lfreq_n \cdot (\beta_{lfreq_{nw}} + u_{subj[n],6})
\end{aligned}
\tag{18.24}
$$

Similarly to the hierarchical implementation of the fast-guess model in section 17.1.5, assume that $\boldsymbol{u}$ is a matrix with as many rows as subjects and six columns. Also assume that $\boldsymbol{u}$ follows a multivariate normal distribution centered at zero.

[4]This for-loop can also be implemented in the `transformed parameter` block; the advantage of doing this is that the log-likelihood of each observation can be used, for example, for cross-validation; the disadvantage is that the R object might be very large, because it will store the log-likelihood during the warm-up period as well.

[5]There are two advantages of iterating first and then adding the total log likelihood to `target`: (i) we can use the variable `log_lik` for model comparison with cross-validation (see chapter 14) without the need to repeating code in the generated quantities and (ii) using `sum` and adding to `target` once is slighter faster than adding to `target` at each iteration.

For lack of more information, we assume the same (weakly informative) prior distribution for the six variance components $\tau_{u_1}, \tau_{u_2}, \dots, \tau_{u_6}$ with a somewhat smaller effect than we assumed for the prior of σ. As with previous hierarchical models, we assign a regularizing LKJ prior for the correlations between the adjustments:[6]

$$\begin{aligned} \boldsymbol{u} &\sim \mathcal{N}(0, \Sigma_u) \\ \tau_{u_1}, \tau_{u_2}, \dots, \tau_{u_6} &\sim Normal_+(0.1, 0.1) \\ \mathbf{R}_u &\sim LKJcorr(2) \end{aligned} \tag{18.25}$$

Before fitting the model to the real data, verify that it works with simulated data. To create synthetic data of several subjects, repeat the same generative process used before, and add the by-subject adjustments `u` in the same way as in section 17.1.5. This version of the log-normal race model assumes that all the parameters α and β have by-subject adjustments; that is, six adjustments. To simplify the model, ignore the possibility of an adjustment for the non-decision time T_0, but see Nicenboim and Vasishth (2018) for an implementation of the log-normal race model with a hierarchical non-decision time. For simplicity, all the adjustments `u` are normally distributed with the same standard deviation of 0.2, and they have a 0.3 correlation between pairs of `u`'s; see `tau_u` and `rho` below.

First, set a seed, take a subset of the data set to keep the same structure, set true point values, and auxiliary variables that indicate the number of observations, subjects, etc.

```
set.seed(42)
df_blp_sim <- df_blp %>%
  group_by(subj)  %>%
  slice_sample(n = 100) %>%
  ungroup()
D <- 1800
alpha_w <- 0.8
beta_wlex <- 0.5
beta_wlfreq <- 0.2
alpha_nw <- 1
beta_nwlex <- -0.5
beta_nwlfreq <- -0.05
```

[6]There are 15 correlations since there are 15 ways to choose 2 variables out of 6 for specifying the pairwise correlations, where order doesn't matter. This is calculated with $\binom{6}{2}$ which is `choose(6, 2)` in R.

```
sigma <- 0.8
T_0 <- 150
N <- nrow(df_blp_sim)
N_subj <- max(df_blp_sim$subj)
N_adj <- 6
tau_u <- rep(0.2, N_adj)
rho <- 0.3
R_u <- matrix(rep(rho, N_adj * N_adj), nrow = N_adj)
diag(R_u) <- 1
Sigma_u <- diag(tau_u, N_adj, N_adj) %*%
  R_u %*% diag(tau_u, N_adj, N_adj)
u <- mvrnorm(n = N_subj, rep(0, N_adj), Sigma_u)
subj <- df_blp_sim$subj
```

Second, generate the locations of both accumulators, `mu_v_w` and `mu_v_nw`, for every trial:

```
mu_v_w <- alpha_w + u[subj, 1] +
  df_blp_sim$c_lfreq * (beta_wlfreq + u[subj, 2]) +
  df_blp_sim$c_lex * (beta_wlex + u[subj, 3])
mu_v_nw <- alpha_nw + u[subj, 4] +
  df_blp_sim$c_lfreq * (beta_nwlfreq + u[subj, 5]) +
  df_blp_sim$c_lex * (beta_nwlex + u[subj, 6])
```

Third, generate values for the rates of accumulation and use those rates to calculate `T_w` and `T_nw`.

```
V_w <- rlnorm(N, mu_v_w, sigma)
V_nw <- rlnorm(N, mu_v_nw, sigma)
T_w <-  D / V_w
T_nw <- D / V_nw
```

Fourth, calculate the time it takes to reach to a decision and the winner accumulator for each trial.

```
T_n<- pmin(T_w, T_nw)
accumulator_winner <- ifelse(T_w == pmin(T_w, T_nw),
                             "word",
                             "non-word")
```

Finally, add this information to the data frame.

```
df_blp_sim <- df_blp_sim %>%
  mutate(rt = T_0 + T_n,
         choice = accumulator_winner,
         nchoice = ifelse(choice == "word", 1, 2),
         acc = ifelse(lex == choice, 1, 0))
```

The Stan code for this model implements the non-centered parameterization for correlated adjustments (see section 9.1.3 for more details). The model is shown below as `lnrace_h.stan`:

```
functions {
  real lognormal_race2_lpdf(real T, int nchoice,
                            array[] real mu, real sigma){
    real lpdf;
    if(nchoice == 1)
        lpdf = lognormal_lpdf(T | mu[1] , sigma)  +
          lognormal_lccdf(T | mu[2], sigma);
      else
        lpdf = lognormal_lpdf(T | mu[2], sigma) +
          lognormal_lccdf(T | mu[1], sigma);
    return lpdf;
  }
}
data {
  int<lower = 1> N;
  int<lower = 1> N_subj;
  vector[N] c_lfreq;
  vector[N] c_lex;
  vector[N] rt;
  array[N] int nchoice;
  array[N] int subj;
}
transformed data{
  real min_rt = min(rt);
  real max_rt = max(rt);
  int N_re = 6;
}
parameters {
  array[2] real alpha;
  array[4] real beta;
```

```
  real<lower = 0> sigma;
  real<lower = 0, upper = min(rt)> T_0;
  vector<lower = 0>[N_re] tau_u;
  matrix[N_re, N_subj] z_u;
  cholesky_factor_corr[N_re] L_u;
}
transformed parameters {
  matrix[N_subj, N_re] u;
  u = (diag_pre_multiply(tau_u, L_u) * z_u)';
  }
model {
  array[N] real log_lik;
  target += normal_lpdf(alpha | 6, 1);
  target += normal_lpdf(beta | 0, 0.5);
  target += normal_lpdf(sigma | 0.5, 0.2)
    - normal_lccdf(0 | 0.5, 0.2);
  target += normal_lpdf(T_0 | 150, 100)
    - log_diff_exp(normal_lcdf(min(rt) | 150, 100),
                   normal_lcdf(0 | 150, 100));
  target += normal_lpdf(tau_u | 0.1, 0.1)
    - N_re * normal_lccdf(0 | 0.1, 0.1);
  target += lkj_corr_cholesky_lpdf(L_u | 2);
  target += std_normal_lpdf(to_vector(z_u));
  for(n in 1:N){
    real T = rt[n] - T_0;
    array[2] real mu = {alpha[1] + u[subj[n], 1] -
                  c_lex[n] * (beta[1] + u[subj[n], 2]) -
                  c_lfreq[n] * (beta[2] + u[subj[n], 3]),
                  alpha[2] + u[subj[n], 4] -
                  c_lex[n] * (beta[3] + u[subj[n], 5]) -
                  c_lfreq[n] * (beta[4] + u[subj[n], 6])};
    log_lik[n] = lognormal_race2_lpdf(T | nchoice[n], mu, sigma);
  }
  target += sum(log_lik);
}
generated quantities {
  corr_matrix[N_re] R_u = L_u * L_u';
}
```

Store the simulated data in a list and fit it.

```
lnrace_h <- system.file("stan_models",
                        "lnrace_h.stan",
                        package = "bcogsci")
ls_blp_h_sim <- list(N = nrow(df_blp_sim),
                     N_subj = max(df_blp_sim$subj),
                     subj = df_blp_sim$subj,
                     rt = df_blp_sim$rt,
                     nchoice = df_blp_sim$nchoice,
                     c_lex = df_blp_sim$c_lex,
                     c_lfreq = df_blp_sim$c_lfreq)
fit_blp_h_sim <- stan(lnrace_h, data = ls_blp_h_sim,
                      control = list(adapt_delta = 0.99,
                                     max_treedepth = 14))
```

The code below compares the posterior distributions of the relevant parameters of the model with their true values, and plots them in Figure 18.8. The true value for all the correlations was 0.3, but we need to correct the sign depending on whether there was a minus sign in front of the adjustment when we built `mu_v_w` and `mu_v_wn` or not: For example, there is no minus before `u[subj, 1]`, but there is one before `u[subj, 2]`, thus the true correlation between `u[subj, 1]` and `u[subj, 2]` we generated should be negative (plus times minus is minus); and there is a minus before `u[subj, 3]` and thus the correlation between `u[subj, 2]` and `u[subj, 3]` should be positive (minus times minus is positive).

```
R_us <- c(paste0("R_u[1,", 2:6 , "]"),
          paste0("R_u[2,", 3:6 , "]"),
          paste0("R_u[3,", 4:6 , "]"),
          paste0("R_u[4,", 5:6 , "]"),
          "R_u[5,6]")
corrs <- rho * c(-1, -1, 1, -1, -1, 1, -1, 1,
                 1, -1, 1, 1, -1, -1, 1)
true_values <- c(log(D) - alpha_w,
                 log(D) - alpha_nw,
                 beta_wlex,
                 beta_wlfreq,
                 beta_nwlex,
                 beta_nwlfreq,
                 T_0,
                 sigma,
                 tau_u,
```

```
                  corrs)
par_names = c("alpha",
              "beta",
              "T_0",
              "sigma",
              "tau_u",
              R_us)
estimates <- as.data.frame(fit_blp_h_sim) %>%
  select(contains(par_names))
mcmc_recover_hist(estimates, true_values) +
  theme(axis.text.x = element_text(angle = 45, vjust = 1, hjust=1),
        legend.position = "bottom")
```

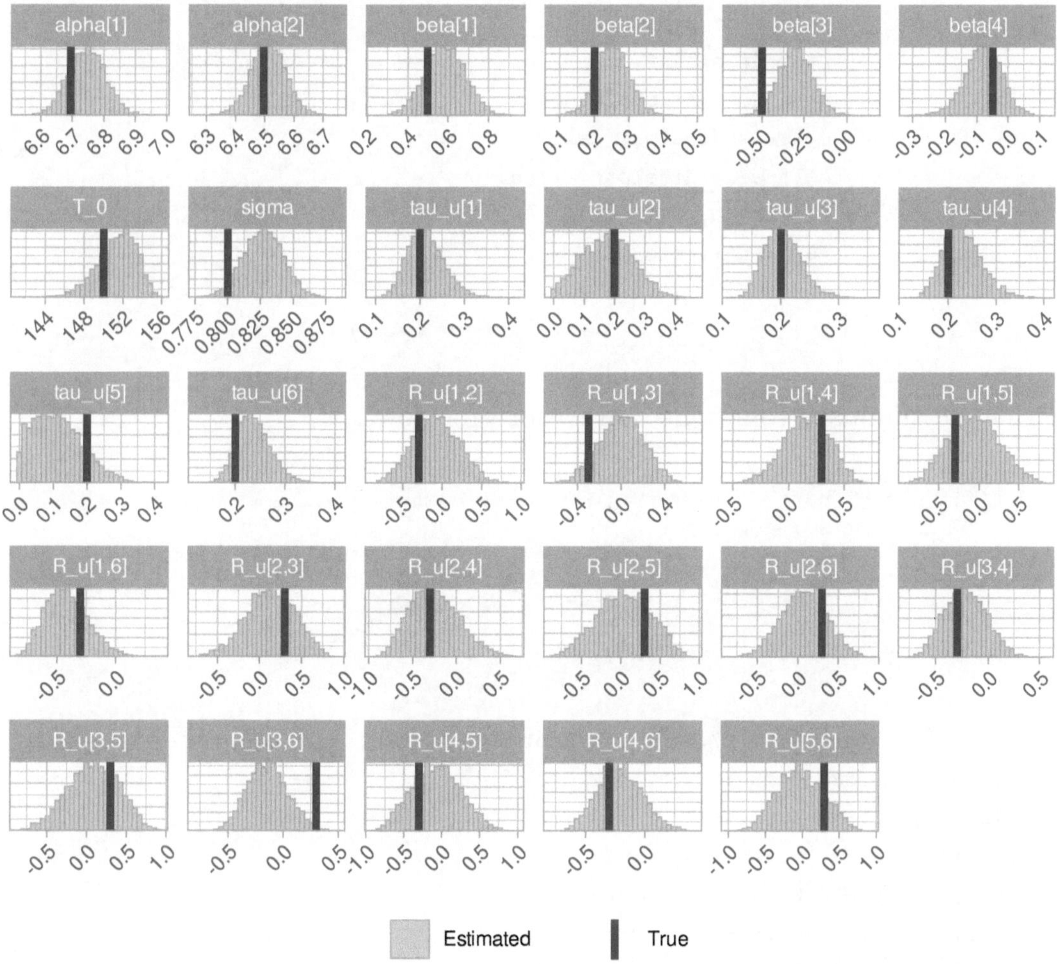

FIGURE 18.8: Posterior distributions of the main parameters of the log-normal race model `fit_blp_h_sim` together with their true values.

Figure 18.8 shows that we can recover the true values quite well, even though there is a great deal of uncertainty over the posteriors of the correlations. As mentioned in previous chapters, a more principled (and computationally demanding) approach uses simulation-based calibration; this was introduced in section 10.2 (also see Talts et al. 2018; Schad, Betancourt, and Vasishth 2020).

We are now ready to fit the model to the observed data.

Create a list with the real data and fit the same Stan model:

```
lnrace_h <- system.file("stan_models",
                        "lnrace_h.stan",
                        package = "bcogsci")
ls_blp_h <- list(N = nrow(df_blp),
                 N_subj = max(df_blp$subj),
                 subj = df_blp$subj,
                 rt = df_blp$rt,
                 nchoice = df_blp$nchoice,
                 c_lex = df_blp$c_lex,
                 c_lfreq = df_blp$c_lfreq)
fit_blp_h <- stan(lnrace_h,
                  data = ls_blp_h,
                  control = list(adapt_delta = 0.99,
                                 max_treedepth = 14))
```

Print the summary (omit the correlations for now).

```
print(fit_blp_h, pars = c("alpha",
                          "beta",
                          "T_0",
                          "sigma",
                          "tau_u"))
```

```
##            mean  2.5% 97.5% n_eff Rhat
## alpha[1]   6.83  6.75  6.90   487 1.01
## alpha[2]   6.64  6.57  6.70   663 1.01
## beta[1]    0.37  0.31  0.42  1987 1.00
## beta[2]    0.06  0.06  0.07  1068 1.00
## beta[3]   -0.14 -0.18 -0.10  2369 1.00
## beta[4]   -0.07 -0.07 -0.06  3658 1.00
## T_0       11.29  9.95 12.46  8304 1.00
## sigma      0.34  0.34  0.35  7296 1.00
```

```
## tau_u[1]  0.17  0.13  0.23  1136 1.00
## tau_u[2]  0.10  0.07  0.15  2320 1.00
## tau_u[3]  0.02  0.01  0.02  2019 1.00
## tau_u[4]  0.14  0.10  0.18  1486 1.00
## tau_u[5]  0.08  0.05  0.12  2282 1.00
## tau_u[6]  0.01  0.01  0.02  1251 1.00
```

Even though the model converged, the posterior summary shows a clear problem with the model: The estimate for the non-decision time, T_0 is less than 12 milliseconds! This is just not possible; physiological research (Clark, Fan, and Hillyard 1994) shows that the eye-to-brain lag, the time it takes for the visual features on the screen until they are propagated from the retina to the brain, is at least 50 milliseconds. Besides identifying the stimuli, the subjects also need to initiate a motor response which takes at least 100 milliseconds. Then how is it possible that we obtained this extremely fast non-decision time? The reason is that the parameter `T_0` is only constrained to be between zero and the shortest response time.

Verify what is the shortest response time in the data set and how many observations are below 150 milliseconds.

```
min(df_blp$rt)
```

```
## [1] 16
```

```
sum(df_blp$rt < 150)
```

```
## [1] 2
```

This shows that some responses must have been initiated even before the stimulus was presented! Next, we deal with *contaminant* observations (Ratcliff and Tuerlinckx 2002).

18.1.5 Dealing with contaminant responses

So far we have assumed that all the observations were coming from responses done after a decision was made. But what happens if there are anticipations or lapses of attention where the subject responds either before the stimuli is presented or after the stimulus was presented, but without attending to the stimulus? We are in a situation analogous to what we described before in chapter 17 with the fast-guess model (Ollman 1966). There, we assumed that the behavior of a subject would be the mixture of two distributions, one corresponding to a guessing mode of responses and another one to a task-engaged mode.

However, there are two major differences from the fast-guess model. First, we assume here that guesses can be fast (e.g., anticipations) as well as slow. Second, in the present case guesses occur in a minority of the cases, and choice and response times are mostly explained by the log-normal race model. The distribution that corresponds to these guesses is sometimes called a *contaminant distribution* (Ratcliff and Tuerlinckx 2002). When the contaminant response times are outside the usual range of response times (either shorter or longer), they can cause major problems in data analysis, distorting estimates. As we saw before, extremely short response times caused by anticipating the response can make it virtually impossible to estimate the non-decision time.

One way to address this problem is to assume that the responses come from a mixture distribution. If one is reasonably confident about which data points are contaminants, a more straightforward approach is to simply remove those points. However, a mixture distribution allows us to account for uncertainty regarding data points that are not obviously extreme. Following Ratcliff and Tuerlinckx (2002), we model the data as a mixture of the sequential sampling model (in this case, the log-normal race model) and a uniform distribution bounded by the minimum and maximum observed response times.[7]

The new likelihood function will look as follows:

$$
\begin{aligned}
p(rt_n, choice_n) =& \theta_c \cdot Uniform(rt_n | min(rt), max(rt)) \cdot \\
& Bernoulli(choice_n | \theta_{bias}) + \\
& (1 - \theta_c) \cdot p_{lnrace}(rt_n, choice_n | \mu, \sigma)
\end{aligned}
\tag{18.26}
$$

The first term of the sum represents the contaminant component that occurs with probability θ_c and has a likelihood that depends on the response time, represented with the uniform PDF, and on the response given, represented with a Bernoulli PMF. When a subject is guessing, the likelihood of each choice depends on θ_{bias}.

[7]The main argument of Ratcliff and Tuerlinckx (2002) is that this method allows for good recovery of the model's parameters as well as the proportion of contaminant responses. However, the assumption that the contaminant responses are uniformly distributed is not well justified. Furthermore, using the data to define the model parameters is usually problematic and might lead to overfitting. A more thorough approach would require us to have some assumptions about the contaminant responses, which might be a combination of fast presses and slow lapses of attention due to mind wandering. We could define a mixture model with a more appropriate distribution than the uniform distribution that have used here, with informative priors for the mixing proportion.

The second term of the likelihood represents the log-normal race model that occurs with probability $1 - \theta_c$. We use $p_{lnrace}(rt_n, choice_n)$ as a shorthand for the following function (which we have already used in the models before):

$$p_{lnrace}(rt_n, choice_n) = \begin{cases} LogNormal(T_n|\mu_{w,n}, \sigma)\cdot \\ (1 - LogNormal\_CDF(T_n|\mu_{nw,n}, \sigma)), \text{ if } choice = \text{word} \\ \\ LogNormalT_n|\mu_{nw,n}, \sigma)\cdot \\ (1 - LogNormal\_CDF(T_n|\mu_{w,n}, \sigma)), \text{ otherwise} \end{cases} \tag{18.27}$$

To simplify the model, we assume that contaminant responses are completely random (i.e., there is no bias to word or non-word). This assumption is encoded in the model by setting $\theta_{bias} = 0.5$.[8] This makes $Bernoulli(choice_n|\theta_{bias}) = 0.5$.

For this model to converge, we need to assume that θ_c is much smaller than 1. This is a sensible assumption for this particular model, since the contaminant distribution is assumed to only happen in a minority of the cases. We set the following prior on θ_c:

$$\theta_c \sim Beta(0.9, 70) \tag{18.28}$$

By setting the first parameter of the beta distribution to a number smaller than 1, we get a distribution of possible probabilities with a "horn" on the left; see Figure 18.9. Our prior belief for θ_c has mean 0.007, and its 95% CrI is $[0, 0.048]$.[9]

We also want to "push" the non-decision time further from zero to get more realistic values. For this reason we increase the informativity of the prior of T_0. A log-normal prior discourages values too close to zero, even with a similar location (on log-scale) than the truncated normal prior. We settle on the following prior (see Figure 18.10):

$$T_0 \sim LogNormal(log(150), 0.6) \tag{18.29}$$

Finally, we want to be able to account for response times that are actually faster than the non-decision time. If the observed response time, `rt`, is smaller

[8]This is, of course, just an assumption that could be verified. But we'll see that the model is already quite complex and achieving convergence is not trivial.

[9]The average is calculated as follows: $0.9/(0.9 + 70)$; this is because the mean of a beta distribution with parameters a, b is $a/(a + b)$. The 95% quantile can be calculated in R with `qbeta(c(0.025, 0.975), 0.9, 70)`.

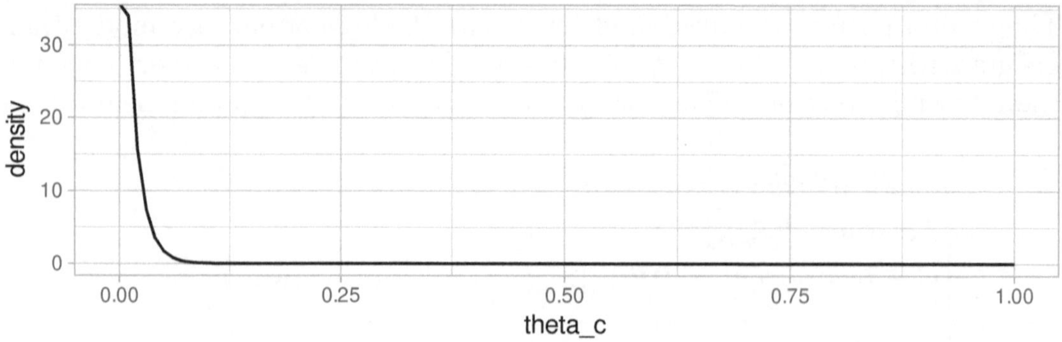

FIGURE 18.9: Prior distribution for the probability of contaminated responses θ_c. Most of the probability mass is close to 0.

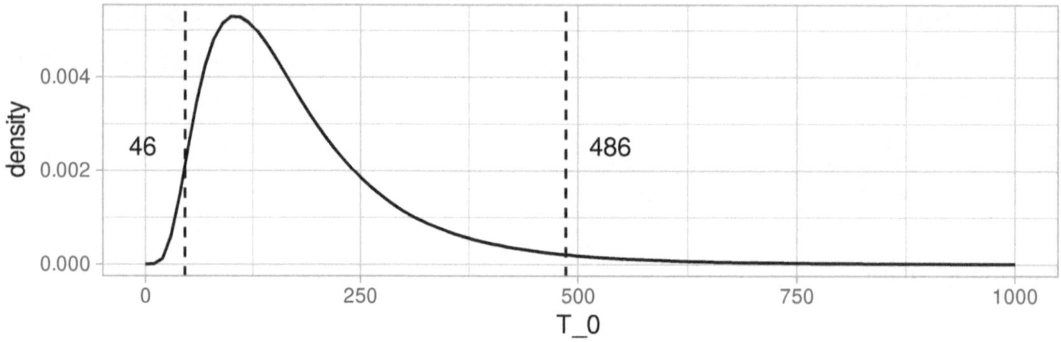

FIGURE 18.10: Prior distribution for T_0. The dashed lines show the 95% credible interval.

than the non-decision time, `T_0`, we can be sure that the observation belongs to the contaminant distribution, because otherwise the decision time `T` should be negative. This means that in this case, the log-normal race likelihood is 0 (and its logarithm is negative infinity). When $T_n < 0$, that is $rt_n < T_0$, the log-likelihood of our model is $log(\theta_c \cdot Uniform(rt_n|min, max) \cdot 0.5 + (1-\theta_c) \cdot p_{lnrace})$ with $p_{lnrace} = 0$, and we only use the following code:

```
target += log(theta_c) + uniform_lpdf(rt[n] | min_rt, max_rt) +
             log(0.5);
```

This also means that we need to relax the constraints on `T_0`; it doesn't need to be smaller than the smallest observed response time, since some of the observations are responses from the contaminant distribution:

```
real<lower = 0> T_0;
```

When $T_n > 0$, that is $rt_n > T_0$, the likelihood is a mixture of the contaminant distribution and the log-normal race model as defined in Equation (18.26). We use `log_sum_exp()` exactly as we did in sections 17.1.2 and 17.1.3 for the fast-guess model. We fit a mixture distribution between a contaminant distribution and the log-normal race model.

```
target += log_sum_exp(
                      log(theta_c) +
                      uniform_lpdf(rt[n] | min_rt, max_rt)
                      + log(0.5),
                      log1m(theta_c) +
                      lognormal_race2_lpdf(T[n] | nchoice[n],
                                           mu, sigma));
```

Finally, the complete Stan code for this model is shown below as `lnrace_cont.stan`:

```
functions {
  real lognormal_race2_lpdf(real T, int nchoice,
                            array[] real mu, real sigma){
    real lpdf;
    if(nchoice == 1)
        lpdf = lognormal_lpdf(T | mu[1] , sigma)  +
          lognormal_lccdf(T | mu[2], sigma);
      else
        lpdf = lognormal_lpdf(T | mu[2], sigma) +
          lognormal_lccdf(T | mu[1], sigma);
    return lpdf;
  }
}
data {
  int<lower = 1> N;
  int<lower = 1> N_subj;
  vector[N] c_lfreq;
  vector[N] c_lex;
  vector[N] rt;
  array[N] int nchoice;
  array[N] int subj;
}
transformed data{
  real min_rt = min(rt);
  real max_rt = max(rt);
  int N_re = 6;
}
parameters {
  array[2] real alpha;
```

```
  array[4] real beta;
  real<lower = 0> sigma;
  real<lower = 0> T_0;
  real<lower = 0, upper = 1> theta_c;
  vector<lower = 0>[N_re] tau_u;
  matrix[N_re, N_subj] z_u;
  cholesky_factor_corr[N_re] L_u;
}
transformed parameters {
  matrix[N_subj, N_re] u;
  u = (diag_pre_multiply(tau_u, L_u) * z_u)';
}
model {
  array[N] real log_lik;
  target += normal_lpdf(alpha | 6, 1);
  target += normal_lpdf(beta | 0, 0.5);
  target += normal_lpdf(sigma | 0.5, 0.2)
    - normal_lccdf(0 | 0.5, 0.2);
  target += lognormal_lpdf(T_0 | log(150), 0.6);
  target += beta_lpdf(theta_c | 0.9, 70);
  target += normal_lpdf(tau_u | 0.1, 0.1)
    - N_re * normal_lccdf(0 | 0.1, 0.1);
  target += lkj_corr_cholesky_lpdf(L_u | 2);
  target += std_normal_lpdf(to_vector(z_u));
    for(n in 1:N){
    real T = rt[n] - T_0;
    if(T > 0){
    array[2] real mu = {alpha[1] + u[subj[n], 1] -
                  c_lex[n] * (beta[1] + u[subj[n], 2]) -
                  c_lfreq[n] * (beta[2] + u[subj[n], 3]),
                  alpha[2] + u[subj[n], 4] -
                  c_lex[n] * (beta[3] + u[subj[n], 5]) -
                  c_lfreq[n] * (beta[4] + u[subj[n], 6])};
    log_lik[n] = log_sum_exp(log(theta_c) +
                           uniform_lpdf(rt[n] | min_rt, max_rt) +
                           log(.5),
                           log1m(theta_c) +
                           lognormal_race2_lpdf(T | nchoice[n],
                                                 mu, sigma));
    } else {
      // T < 0, observed time is smaller than the non-decision time
      log_lik[n] = log(theta_c) +
```

```
        uniform_lpdf(rt[n] | min_rt, max_rt) +
        log(.5);
    }
  }
  target += sum(log_lik);
}
```

In practice, we should verify that this model can recover the true values of its parameters by first simulating data and fitting the model to simulated data, and using simulation-based calibration. We skip these steps here.

Store the real data in a list and fit the model:

```
lnrace_h_cont <- system.file("stan_models",
                             "lnrace_h_cont.stan",
                             package = "bcogsci")
ls_blp_h <- list(N = nrow(df_blp),
                 N_subj = max(df_blp$subj),
                 subj = df_blp$subj,
                 rt = df_blp$rt,
                 nchoice = df_blp$nchoice,
                 c_lex = df_blp$c_lex,
                 c_lfreq = df_blp$c_lfreq)
fit_blp_h_cont <- stan(lnrace_h_cont,
                       data = ls_blp_h,
                       control = list(adapt_delta = 0.99,
                                      max_treedepth = 14))
```

This model takes more than a day to finish in a relatively powerful computer and, disappointingly, it doesn't converge; this non-convergence is apparent from the traceplots in Figure 18.11. We'll see later that even if the converging model finishes faster, it still takes a considerable amount of time. If one has a powerful computer available (with, for example, multiple cores), it is possible to parallelize the sampling further than what we did so far. This is possible with special functions that allow for *multithreading*, which is discussed in Stan Development Team (2024) (Chapter 27 of the User's guide).

The traceplots in Figure 18.11 show that the chains get stuck and don't mix well. It seems that there is not enough information to constrain the model. If we look at the parameter `theta_c`, which represents the mixing proportion between the contaminant and the log-normal distribution, we see that its chains are getting stuck at very unlikely values that are over 0.25. Although in general is not recommended to cut off values from a prior just because they're

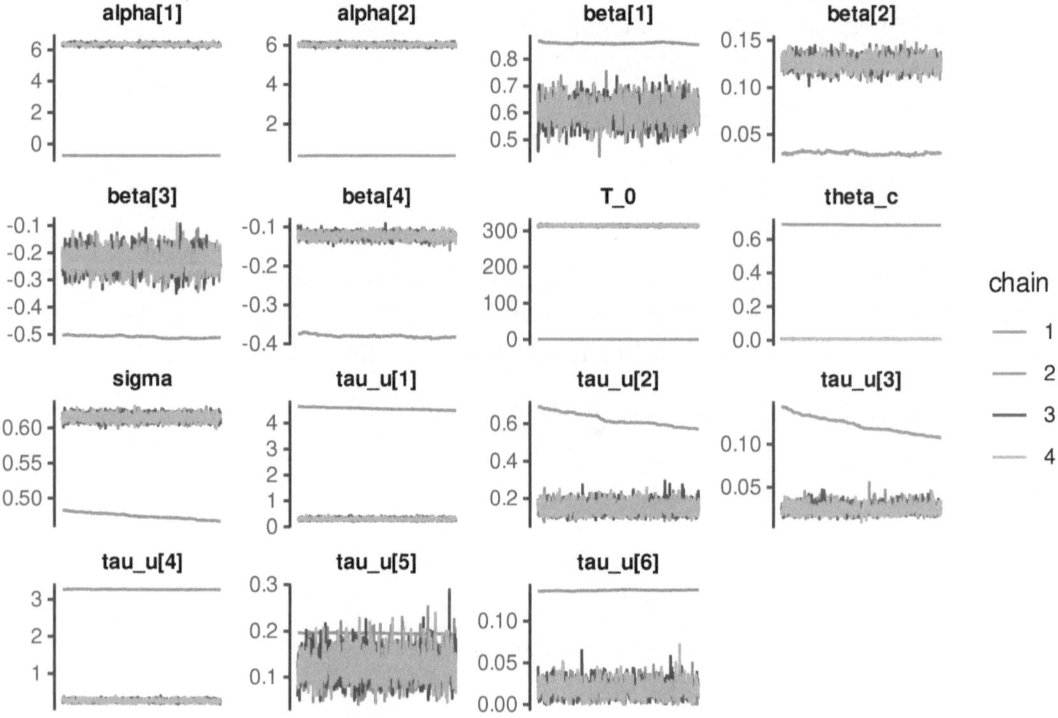

FIGURE 18.11: The traceplots of the `fit_blp_h` shows that the chains are clearly not mixing well for the real data set.

unlikely, in this case restricting the parameter `theta_c` to be smaller than 0.1 helps solving convergence problems. To truncate the prior for θ_c in Stan, we declare the parameter to have an upper bound of 0.1:

```
real<lower = 0, upper = 0.1> theta_c;
```

Change its prior distribution in the `model` block to the following:

```
target += beta_lpdf(theta_c | 0.9, 70) -
  beta_lcdf(0.1 | 0.9, 70);
```

If we were to fit this new model, we would see that some chains of `T_0` mix in values around 300 ms and some other chains (sometimes) get stuck in values very close to zero. This indicates that the model needs more information regarding the non-decision time. Another issue that slows down convergence is that this parameter `T_0` is on a different scale (with a value above 100) than the rest of the parameters (with values below 10). Rather than sampling from `T_0`, we sample from `lT_0` in a new model, so that `T_0 = exp(lT_0)`. We assign the following prior to `lT_0`:

$$lT_0 \sim Normal(log(200), 0.3) \tag{18.30}$$

This is mathematically equivalent to assigning the following prior to T_0:

$$T_0 \sim LogNormal(log(200), 0.3) \tag{18.31}$$

The new version of the model is omitted here, but can be found as `lnrace_h_contb.stan` in the `bcogsci` package. Fit the new modified model to the same data.

```
lnrace_h_contb <- system.file("stan_models",
                              "lnrace_h_contb.stan",
                              package = "bcogsci")
fit_blp_h_contb <- stan(lnrace_h_contb,
                        data = ls_blp_h,
                        control = list(adapt_delta = 0.99,
                                       max_treedepth = 14))
```

This time the model takes considerably less time, but it still takes nine hours. However, the model does converge and the posterior distribution does make sense now.

Print the summary of the main parameters:

```
print(fit_blp_h_contb, pars = c("alpha",
                                "beta",
                                "T_0",
                                "sigma",
                                "theta_c",
                                "tau_u"))
```

```
##              mean   2.5%  97.5% n_eff Rhat
## alpha[1]     6.36   6.23   6.50   355    1
## alpha[2]     6.02   5.91   6.14   643    1
## beta[1]      0.61   0.53   0.69  1863    1
## beta[2]      0.13   0.11   0.14  1705    1
## beta[3]     -0.23  -0.30  -0.16  2365    1
## beta[4]     -0.12  -0.14  -0.11  3579    1
## T_0        315.69 312.51 318.56  4788    1
## sigma        0.62   0.61   0.62  5167    1
## theta_c      0.01   0.00   0.01  7875    1
## tau_u[1]     0.29   0.23   0.37  1425    1
## tau_u[2]     0.15   0.10   0.21  2368    1
## tau_u[3]     0.02   0.02   0.04  1733    1
```

```
## tau_u[4]    0.26    0.20    0.33  1885     1
## tau_u[5]    0.12    0.07    0.18  2705     1
## tau_u[6]    0.02    0.00    0.04  1374     1
```

Print the summary of the correlations:

```
print(fit_blp_h_contb, pars = R_us)
```

```
##             mean  2.5% 97.5% n_eff Rhat
## R_u[1,2] -0.06 -0.41  0.30  2741    1
## R_u[1,3]  0.20 -0.16  0.53  2134    1
## R_u[1,4]  0.61  0.33  0.81  1847    1
## R_u[1,5] -0.01 -0.39  0.36  3043    1
## R_u[1,6] -0.07 -0.53  0.40  4379    1
## R_u[2,3] -0.38 -0.72  0.05  1826    1
## R_u[2,4]  0.06 -0.31  0.41  1696    1
## R_u[2,5] -0.51 -0.85 -0.04  2696    1
## R_u[2,6] -0.35 -0.78  0.24  3178    1
## R_u[3,4]  0.04 -0.30  0.40  1822    1
## R_u[3,5]  0.35 -0.11  0.72  3229    1
## R_u[3,6]  0.01 -0.53  0.53  4507    1
## R_u[4,5] -0.33 -0.65  0.04  4307    1
## R_u[4,6] -0.06 -0.52  0.43  4717    1
## R_u[5,6]  0.05 -0.51  0.62  3152    1
```

What can we say about the fit of the model now?

Under the assumptions that we have made, we can look at the parameters and conclude the following:

- All other things being equal, there is an overall bias to respond `non-word` rather than `word`. We can deduce this because the parameters α represent the boundary separation of each accumulator minus their rate of accumulation (see Equation (18.14)). A smaller `alpha` indicates a closer boundary of evidence and/or a faster rate for a given accumulator. In this case `alpha[2]` is smaller than `alpha[1]`. However, the fact that `tau_u[1]` and `tau_u[4]` are relatively large suggests large individual differences. Furthermore, the fact that the correlation `R_u[1,4]` is large and positive indicates that subjects who are cautious in the accumulation of evidence for words are also cautious for non-words.
- The task seems to have been well-understood given that, when a `word` appears, the rate of accumulation of the word accumulator increases (`beta[1]` > 0), and the rate of the non-word accumulator decreases (`beta[3]` < 0).

- As expected, and replicating previous findings in the literature, words with higher frequency are easier to identify correctly as words, compared to lower frequency words (`beta[2]` > 0 and `beta[4]` < 0).
- The non-decision time (`T_0`) is relatively long, considering that the normal reading of words in a sentence takes around 200–400 ms (Rayner 1998).
- The proportion of contaminant responses (`theta_c`) is quite small (1%), but without taking them into account, the non-decision time would not be possible to estimate.

Our assumptions include both the likelihood we have chosen and the priors. It's clear from the difficulties fitting the data that the model is very sensitive to the choice of priors. Since there seem to be not enough information in the data, we need to provide information through the prior distributions.

18.2 Posterior predictive check with the quantile probability plots

As in the previous chapter, in a new file, we can write the generated quantities block for posterior predictive checks. The advantage is that we can generate as many observations as needed after estimating the parameters. There is no model block in the following Stan program. The `gqs()` function needs a Stan model without transformed parameters. For this reason, this model includes `u` in the `parameters` block rather than in the `transformed parameters` block. The complete Stan code for this model is shown below as `lnrace_h_contb_gen.stan`.

```
data {
  int<lower = 1> N;
  int<lower = 1> N_subj;
  vector[N] c_lfreq;
  vector[N] c_lex;
  vector[N] rt;
  array[N] int nchoice;
  array[N] int subj;
}
transformed data{
  real min_rt = min(rt);
  real max_rt = max(rt);
  int N_re = 6;
}
parameters {
  array[2] real alpha;
  array[4] real beta;
```

```
  real<lower = 0> sigma;
  real<lower = 0> T_0;
  real<lower = 0, upper = 0.1> theta_c;
  vector<lower = 0>[N_re] tau_u;
  matrix[N_subj, N_re] u;
}
model {
}
generated quantities {
  array[N] real rt_pred;
  array[N] real nchoice_pred;
  for(n in 1:N){
    real T = rt[n] - T_0;
    array[2] real mu = {alpha[1] + u[subj[n], 1] -
                        c_lex[n] * (beta[1] + u[subj[n], 2]) -
                        c_lfreq[n] * (beta[2] + u[subj[n], 3]),
                        alpha[2] + u[subj[n], 4] -
                        c_lex[n] * (beta[3] + u[subj[n], 5]) -
                        c_lfreq[n] * (beta[4] + u[subj[n], 6])};
    real cont = bernoulli_rng(theta_c);
    if(cont == 1){
      rt_pred[n] = uniform_rng(min_rt, max_rt);
      nchoice_pred[n] = bernoulli_rng(0.5) + 1;
    } else {
      real accum1 = lognormal_rng(mu[1], sigma);
      real accum2 = lognormal_rng(mu[2], sigma);
      rt_pred[n] = fmin(accum1, accum2) + T_0;
      nchoice_pred[n] = (accum1 > accum2) + 1;
    }
  }
}
```

Compile the model `lnrace_h_contb_gen.stan` and subset 200 samples of each parameter appearing in the `parameters` block using the previous model:

```
lnrace_h_gen <- system.file("stan_models",
                            "lnrace_h_contb_gen.stan",
                            package = "bcogsci")
gen_model <- stan_model(lnrace_h_gen)
draws_par <- as.matrix(fit_blp_h_contb,
                       pars = c("alpha",
                                "beta",
```

```
                          "sigma",
                          "T_0",
                          "theta_c",
                          "tau_u",
                          "u"))[1:200, , drop = FALSE]
```

Use the function `gqs()` (this function draws samples of generated quantities from the Stan model) to generate responses from 200 simulated experiments:

```
gen_race_data <- gqs(gen_model,
                     data = ls_blp_h,
                     draws = draws_par)
```

One can examine the general distribution of response times generated by the posterior predictive model, or the effect of the experimental manipulations on response times, as we did in section 17.1.5.3.

However, it is more informative to look at the quantile probability plot of the posterior predictive distribution. In order to create this plot, first extract the predicted response times and choice, and match each one to their corresponding observation of the corresponding simulation. Before we can use `qpf()`, we need a data frame identical to the one with the observed data, `df_blp`, but one that includes response time and choice for each observation of each simulation (indicated with the column `sim`). We do this in the code below. The code yields a data frame called `df_blp_pred_qpf`, which has the quantile probability information of the observed data and each one of the 200 simulations.

First, extract the array of predicted reading times from `gen_race_data` and transform it to a long format:

```
df_rt <- rstan::extract(gen_race_data)$rt_pred %>%
         # Convert a matrix of 200 x 24000 (iter x N obs)
         # into a data.frame where each column is
         # V1,...,V24000:
         as.data.frame() %>%
         # Add a column which identifies each iter as a simulation:
         mutate(sim = 1:n()) %>%
         # Pivot the data frame so that it has length 200 * 24000.
         # Each row indicates:
         # - sim: from which simulation the observation is coming
         # - obs_id: identifies the 24000 observations
```

```
        # - rt_pred: simulated RT
        # Since each observation is in a column starting with V
        # `names_prefix` removes the "V"
        pivot_longer(cols = -sim,
                     names_to = "obs_id",
                     names_prefix = "V",
                     values_to = "rt_pred") %>%
        # Make sure that obs_id is a number (and not a
        # number represented as a character):
        mutate(obs_id = as.numeric(obs_id))
df_rt
```

```
## # A tibble: 4,800,000 x 3
##     sim obs_id rt_pred
##   <int>  <dbl>   <dbl>
## 1     1      1    742.
## 2     1      2   1011.
## 3     1      3    549.
## # i 4,799,997 more rows
```

Second, extract the array of predicted choice (1 for words and 2 for non-words) and transform it into a long format:

```
df_nchoice <- rstan::extract(gen_race_data)$nchoice_pred %>%
                        as.data.frame() %>%
                        mutate(sim = 1:n()) %>%
                        pivot_longer(cols = -sim,
                                     names_to = "obs_id",
                                     names_prefix = "V",
                                     values_to = "nchoice_pred") %>%
                        mutate(obs_id = as.numeric(obs_id))
df_nchoice
```

```
## # A tibble: 4,800,000 x 3
##     sim obs_id nchoice_pred
##   <int>  <dbl>        <dbl>
## 1     1      1            2
## 2     1      2            2
## 3     1      3            2
## # i 4,799,997 more rows
```

Third, create a new data frame with the characteristics of the stimuli the predictions and observations. The predictions come from 200 simulated data set, each simulation is indexed with `sim` (whereas for the empirical observations, `sim` is set to `NA`):

```
df_blp_main <- df_blp %>%
  select(subj, lex, lfreq, rt, nchoice) %>%
  mutate(obs_id = 1:n())
df_blp_pred <- left_join(df_rt, df_nchoice) %>%
  left_join(select(df_blp_main, -rt, -nchoice)) %>%
  rename(rt = rt_pred, nchoice = nchoice_pred) %>%
  bind_rows(df_blp_main) %>%
  mutate(acc = ifelse((nchoice == 1 & lex == "word") |
                      (nchoice == 2 & lex == "non-word"), 1, 0))
df_blp_pred
```

```
## # A tibble: 4,824,000 x 8
##      sim obs_id    rt nchoice  subj lex      lfreq   acc
##    <int>  <dbl> <dbl>   <dbl> <dbl> <chr>    <dbl> <dbl>
## 1      1      1  742.       2     1 non-word -4.61     1
## 2      1      2 1011.       2     1 non-word -4.61     1
## 3      1      3  549.       2     1 non-word -4.61     1
## # i 4,823,997 more rows
```

Finally, create a data frame with the results of the quantile probability function applied by simulation and by frequency:

```
df_blp_pred_qpf <- df_blp_pred  %>%
  # Subset only words
  filter(lex == "word") %>%
  # Create 5 word frequencies group
  mutate(freq_group =
           cut(lfreq,
               quantile(lfreq, c(0, 0.2, 0.4, 0.6, 0.8, 1)),
               include.lowest = TRUE,
               labels =
                 c("0-0.2", "0.2-0.4", "0.4-0.6", "0.6-0.8", "0.8-1"))) %>%
  # Group by condition  and subject
  group_by(freq_group, sim, subj) %>%
  # Apply the quantile probability function
  qpf() %>%
  # Group again removing subj
```

```
  group_by(freq_group, sim, q, response) %>%
  # Get averages of all the quantities
  summarize(rt_q = mean(rt_q),
            p = mean(p))
```

Now, plot the results with the code shown below (Figure 18.12).

```
ggplot(df_blp_pred_qpf %>% filter(sim < 200), aes(x = p, y = rt_q)) +
  geom_point(shape = 4, alpha = 0.1) +
  geom_line(aes(group = interaction(q, response, sim)),
            alpha = 0.1,
            color = "grey") +
  ylab("log-transformed RT quantiles (ms) ") +
  xlab("Response proportion")  +
  geom_point(data = df_blp_pred_qpf %>% filter(is.na(sim)),
             shape = 4) +
  geom_line(data = df_blp_pred_qpf %>% filter(is.na(sim)),
            aes(group = interaction(q, response))) +
  coord_trans(y = "log")
```

Figure 18.12 shows the quantile probability plots of the observed and simulated data. We would expect the observed quantile probability summaries (in black) to fall within those of the posterior predictive distribution (in gray).

The fit is clearly bad. One major issue is the misfit at the left side of the plot, this is because the model is unable to capture fast errors: Low probability responses must be slower, because having a slow rate and "losing" the race often is what makes their probability low. Other sequential sampling models such as the linear ballistic accumulator and the drift diffusion model can account for fast errors, under the assumption that they happen in the trials where there is a strong initial bias toward the wrong response; this bias occurs due to random variation in the starting points of the accumulators. This characterization of fast errors is not possible for the log-normal race model, because bias and rate effects combine to determine distribution location (Heathcote and Love 2012). Heathcote and Love (2012) point out that the log-normal race model can still produce fast errors if the scale of the accumulator that corresponds to the incorrect choice is larger than the scale of the accumulator of the correct choice. We leave it as an exercise for the reader to verify that a log-normal race model with a scale that depends on the stimuli improves the fit (see online exercise G.18.2).

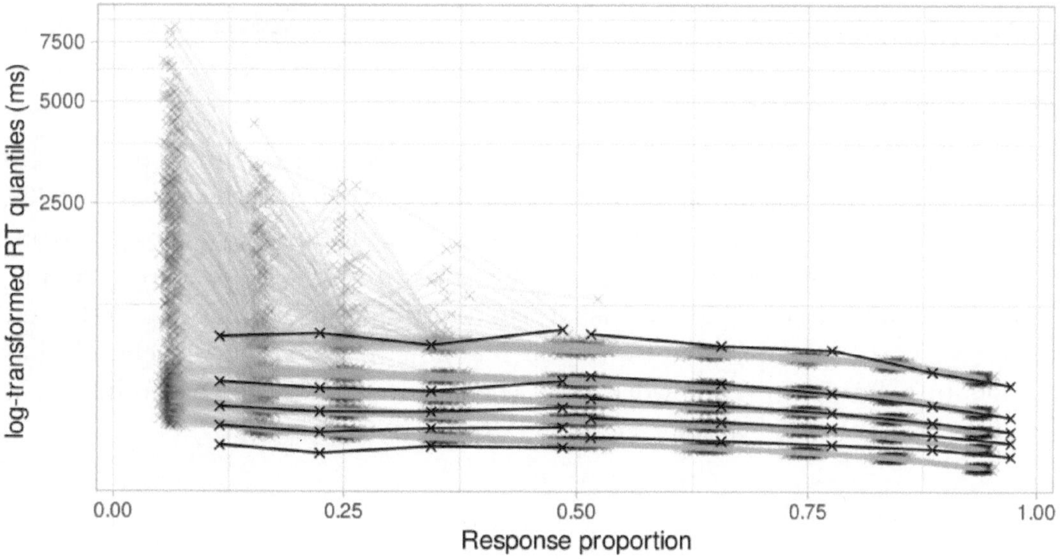

FIGURE 18.12: Quantile probability plots showing 0.1, 0.3, 0.5, 0.7, and 0.9 response times quantiles plotted against proportion of incorrect responses (left) and proportion of correct responses (right) for words of different frequency. Word frequency are grouped according to quantiles: The first group is words with frequencies smaller than the 0.2-th quantile, the second group is words with frequencies smaller than the 0.4-th quantile and larger than the 0.2-th quantile, and so forth. The summary of the observed data is plotted in black and the summaries of the synthetic data sets are plotted in gray.

18.3 Summary

In this chapter, we learned to fit what seems to be the simplest sequential sampling model, a log-normal race model with equal scale (or variances) starting with one subject, continuing with a fully hierarchical model, and finally incorporating a contaminant distribution by using a mixture model. We saw how to evaluate model fit of response time and choice using quantile probability plots. We saw that the log-normal race model with equal scale is unable to account for fast errors and a more complex model is needed. Crucially, many of the techniques explained in this chapter (e.g., including a shift in the distribution, mixture distributions for dealing with contaminated responses, and quantile probability plots) can be used with virtually any type of model that fits response times and choice. One downside of this type of model is that they take a long time to fit.

18.4 Further reading

The log-normal race model was first introduced in Heathcote and Love (2012), and its first Bayesian implementation is described in Rouder et al. (2015). The log-normal race model is closely connected to the retrieval process from memory that is assumed in the cognitive architecture ACT-R; see, for example, Nicenboim and Vasishth (2018); Fisher, Houpt, and Gunzelmann (2022); Lissón et al. (2021). Heathcote et al. (2019) outline how to fit evidence-accumulation models in a Bayesian framework with a custom R package that relies on a Differential-Evolution sampler (DE-MCMC; Turner et al. 2013).

19

In closing

If you have made it this far into the book, congratulations! This book attempted to provide an introduction to Bayesian data analysis and Bayesian computational cognitive modeling specifically aimed at researchers working in cognitive science, construed broadly. Hopefully, this book has provided a flavor and some hands-on examples of how useful, important, and tractable the Bayesian approach is for typical research problems encountered when analyzing data from planned (behavioral) experiments.

A key idea in Bayesian modeling is uncertainty quantification: we are not just interested in the estimates of our parameters, but also in how unsure we are about the parameter values. This shift in focus toward uncertainty quantification, and away from binary conclusions like "the effect is significant" or "the effect is not significant," is useful because it furnishes a more realistic picture of what we can conclude from the model. Such a focus on uncertainty quantification also teaches us to embrace uncertainty as a fact of life. Journal articles from many different fields reporting the results of experiments often foster an illusion of robustness in their conclusions; the recent replication crisis (e.g., Nosek et al. 2022) has underlined just how illusory this certainty is.

There do arise situations where uncertainty quantification is not the primary question of interest. When the research question really has the form, is the effect present or is there evidence against it, the Bayesian methodology does provide a tool: Bayes factors analyses. However, as we have discussed in this book, the Bayes factor is a multiple-edged sword: the researcher has to spend some energy and time conducting a sensitivity analysis under a range of priors, and the conclusions from a Bayes factor analysis are often going to be much more nuanced than those based on frequentist p-values. Thus, even here, there is a different kind of uncertainty about what one learns from the data—a lot can depend on which priors one adopts for the target parameter being estimated. Moreover, even if the Bayes factor decisively shows evidence for or against an effect, how robust this conclusion would be under replication depends to a large extent on the design properties of the study that the conclusion is based on. Specifically, if the false-discovery rate is high (in frequentist statistics, this relates to the statistical power of the design), even a large Bayes factor in favor of an effect may not tell us much. A key take-away from this is to temper one's

DOI: 10.1201/9780429342646-25

excitement when one finds strong evidence one way or another, to study the design properties of one's experiment and to wait to see whether the results will replicate.

Perhaps the greatest advantage of the Bayesian approach is that one can flexibly adapt one's model to the research problem at hand, and to use the model to specify the hypothesized generative process that is assumed to have produced the data. This is in stark contrast to the standard approach in psychology and linguistics, where canned statistical models are used to answer questions like "is the effect present or absent?," without specifying the latent processes that are assumed to have produced the data. Of course, if one specifies a detailed process model, this increases the complexity of the model. Such an increased complexity naturally comes with a cost. One has to think about how to formulate the model, and this can involve complications like reparameterization, and can cost a lot of time and effort. The computations can also take time. However, the bottom line here is that one can either get fast answers to the wrong question, or slow (albeit uncertain!) answers to the real question one wants to answer. There is no free lunch.

Bayesian modeling comes with another cost: one has to spend some time checking that the software used works as intended. This means not ignoring the warnings that the software issues, but it also means that we should carry out gold-standard checks to make sure that the model recovers the parameters as it is supposed to. Simulation of data is an important tool here. Although this model validation process can be time-consuming and computationally expensive, it is important to make sure that the model behaves as it should when we know what the ground truth is. Model validation is a process that even frequentist modelers could benefit from; for example, it is rare in psychology and linguistics for researchers to check whether the linear mixed models they use for inference could even in principle recover the model parameters accurately, or produce realistic data.

In closing, we hope that the reader finds this book useful in their research, and we look forward to seeing more and more work in cognitive science that uses this powerful inferential tool.

References

Adams, Robert A., and Christopher Essex. 2018. *Calculus: A Complete Course*. Pearson.

Allaire, J. J., Yihui Xie, Christophe Dervieux, Jonathan McPherson, Javier Luraschi, Kevin Ushey, Aron Atkins, et al. 2024. *rmarkdown: Dynamic Documents for R*. https://github.com/rstudio/rmarkdown.

Anderson, John R., Dan Bothell, Michael D Byrne, Scott Douglass, Christian Lebiere, and Yulin Qin. 2004. "An Integrated Theory of the Mind." *Psychological Review* 111 (4): 1036–60.

Ashby, F. Gregory. 1982. "Testing the Assumptions of Exponential, Additive Reaction Time Models." *Memory & Cognition* 10 (2): 125–34.

Ashby, F. Gregory, and James T. Townsend. 1980. "Decomposing the Reaction Time Distribution: Pure Insertion and Selective Influence Revisited." *Journal of Mathematical Psychology* 21 (2): 93–123.

Audley, R. J., and A. R. Pike. 1965. "Some Alternative Stochastic Models of Choice 1." *British Journal of Mathematical and Statistical Psychology* 18 (2): 207–25.

Auguie, Baptiste. 2017. *GridExtra: Miscellaneous Functions for "Grid" Graphics*. https://CRAN.R-project.org/package=gridExtra.

Aust, Frederik. 2019. *citr: RStudio Add-in to Insert Markdown Citations*. https://CRAN.R-project.org/package=citr.

Aust, Frederik, and Marius Barth. 2020. *papaja: Create APA Manuscripts with R Markdown*. https://github.com/crsh/papaja.

Baayen, R. Harald, Douglas J. Davidson, and Douglas M. Bates. 2008. "Mixed-Effects Modeling with Crossed Random Effects for Subjects and Items." *Journal of Memory and Language* 59 (4): 390–412.

Baguley, Thomas. 2012. *Serious Stats: A Guide to Advanced Statistics for the Behavioral Sciences*. Macmillan International Higher Education.

Barr, Dale J., Roger P. Levy, Christoph Scheepers, and Harry J. Tily. 2013. "Random Effects Structure for Confirmatory Hypothesis Testing: Keep It Maximal." *Journal of Memory and Language* 68 (3): 255–78.

Barth, Marius. 2023. *tinylabels: Lightweight Variable Labels*. https://github.com/mariusbarth/tinylabels.

Batchelder, William H., and David M. Riefer. 1999. "Theoretical and Empirical Review of Multinomial Process Tree Modeling." *Psychonomic Bulletin & Review* 6 (1): 57–86.

Bates, Douglas M., Martin Mächler, Ben Bolker, and Steve Walker. 2015. "Fitting Linear Mixed-Effects Models Using lme4." *Journal of Statistical Software* 67 (1): 1–48. https://doi.org/10.18637/jss.v067.i01.

Bates, Douglas M., Martin Maechler, and Mikael Jagan. 2023. *Matrix: Sparse and Dense Matrix Classes and Methods.* https://Matrix.R-forge.R-project.org.

Beer, Randall D. 2000. "Dynamical Approaches to Cognitive Science." *Trends in Cognitive Sciences* 4 (3): 91–99.

Bengtsson, Henrik. 2021. "A Unifying Framework for Parallel and Distributed Processing in R Using Futures." *The R Journal* 13 (2): 208–27. https://doi.org/10.32614/RJ-2021-048.

Bennett, Charles H. 1976. "Efficient Estimation of Free Energy Differences from Monte Carlo Data." *Journal of Computational Physics* 22 (2): 245–68. https://doi.org/10.1016/0021-9991(76)90078-4.

Bernardo, José M., and Adrian F. M. Smith. 2009. *Bayesian Theory.* Vol. 405. John Wiley & Sons.

Betancourt, Michael J. 2016. "Identifying the Optimal Integration Time in Hamiltonian Monte Carlo." *arXiv Preprint arXiv:1601.00225.* http://arxiv.org/abs/1601.00225.

———. 2017. "A Conceptual Introduction to Hamiltonian Monte Carlo." *arXiv Preprin: arXiv:1701.02434.* http://arxiv.org/abs/1701.02434.

Betancourt, Michael J., and Mark Girolami. 2015. "Hamiltonian Monte Carlo for Hierarchical Models." *Current Trends in Bayesian Methodology with Applications* 79 (30): 2–4.

Bever, Thomas G. 1970. "The Cognitive Basis for Linguistic Structures." *Cognition and the Development of Language.*

Bishop, Christopher M. 2006. *Pattern Recognition and Machine Learning.* Springer.

Blanchard, Pierre, Desmond J. Higham, and Nicholas J. Higham. 2020. "Accurately computing the log-sum-exp and softmax functions." *IMA Journal of Numerical Analysis* 41 (4): 2311–30. https://doi.org/10.1093/imanum/draa038.

Blitzstein, Joseph K., and Jessica Hwang. 2014. *Introduction to Probability.* Chapman & Hall/CRC Press.

Blokpoel, Mark, and Iris van Rooij. 2021. *Theoretical Modeling for Cognitive Science and Psychology.* https://computationalcognitivescience.github.io/lovelace/.

Blumberg, Eric J., Matthew S. Peterson, and Raja Parasuraman. 2015. "Enhancing Multiple Object Tracking Performance with Noninvasive Brain Stimulation: A Causal Role for the Anterior Intraparietal Sulcus." *Frontiers in Systems Neuroscience* 9: 3. https://doi.org/10.3389/fnsys.2015.00003.

Bolker, Ben. 2018. "Contrasts." https://computationalcognitivescience.github.io/lovelace/.

Box, George E. P., and David R. Cox. 1964. "An Analysis of Transformations." *Journal of the Royal Statistical Society. Series B (Methodological)*, 211–52.

Britten, Kenneth H., Michael N. Shadlen, William T. Newsome, and J. Anthony Movshon. 1993. "Responses of Neurons in Macaque MT to Stochastic Motion Signals." *Visual Neuroscience* 10 (6): 1157–69. https://doi.org/10.1017/S0952523800010269.

Brown, Scott D., and Andrew Heathcote. 2005. "A Ballistic Model of Choice Response Time." *Psychological Review* 112 (1): 117.

———. 2008. "The Simplest Complete Model of Choice Response Time: Linear Ballistic Accumulation." *Cognitive Psychology* 57 (3): 153–78. https://doi.org/10.1016/j.cogpsych.2007.12.002.

Brysbaert, Marc, Paweł Mandera, and Emmanuel Keuleers. 2018. "The Word Frequency Effect in Word Processing: An Updated Review." *Current Directions in Psychological Science* 27 (1): 45–50. https://doi.org/10.1177/0963721417727521.

Burger, Edward B., and Michael Starbird. 2012. *The 5 Elements of Effective Thinking*. Princeton University Press.

Bürki, Audrey, Francois-Xavier Alario, and Shravan Vasishth. 2023. "When Words Collide: Bayesian Meta-Analyses of Distractor and Target Properties in the Picture-Word Interference Paradigm." *Quarterly Journal of Experimental Psychology* 76 (6): 1410–30. https://doi.org/https://doi.org/10.1177/17470218221114644.

Bürki, Audrey, Shereen Elbuy, Sylvain Madec, and Shravan Vasishth. 2020. "What Did We Learn from Forty Years of Research on Semantic Interference? A Bayesian Meta-Analysis." *Journal of Memory and Language* 114. https://doi.org/10.1016/j.jml.2020.104125.

Bürkner, Paul-Christian. 2017. "brms: An R Package for Bayesian Multilevel Models Using Stan." *Journal of Statistical Software* 80 (1): 1–28. https://doi.org/10.18637/jss.v080.i01.

Bürkner, Paul-Christian. 2024. *brms: Bayesian Regression Models Using "Stan"*. https://github.com/paul-buerkner/brms.

Bürkner, Paul-Christian, and Emmanuel Charpentier. 2020. "Modelling Monotonic Effects of Ordinal Predictors in Bayesian Regression Models." *British Journal of Mathematical and Statistical Psychology*. https://doi.org/https://doi.org/10.1111/bmsp.12195.

Bürkner, Paul-Christian, and Matti Vuorre. 2019. "Ordinal Regression Models in Psychology: A Tutorial." *Advances in Methods and Practices in Psychological Science* 2 (1): 77–101. https://doi.org/https://doi.org/10.1177/2515245918823199.

Busemeyer, Jerome R, and Adele Diederich. 2010. *Cognitive Modeling*. Sage.

Caplan, David, and G. S. Waters. 1999. "Verbal Working Memory and Sentence Comprehension." *Behavioral and Brain Science* 22: 77–94. https://doi.org/https://doi.org/10.1017/S0140525X99001788.

Carlin, Bradley P., and Thomas A Louis. 2008. *Bayesian Methods for Data Analysis*. CRC Press.

Carpenter, Bob, Andrew Gelman, Matthew D. Hoffman, Daniel Lee, Ben Goodrich, Michael J. Betancourt, Marcus Brubaker, Jiqiang Guo, Peter Li, and Allen Riddell. 2017. "Stan: A Probabilistic Programming Language." *Journal of Statistical Software* 76 (1):1–32.

Chang, Winston. 2018. *webshot: Take Screenshots of Web Pages*. https://CRAN.R-project.org/package=webshot.

Chen, Stanley F., and Joshua Goodman. 1999. "An Empirical Study of Smoothing Techniques for Language Modeling." *Computer Speech & Language* 13 (4): 359–94. https://doi.org/https://doi.org/10.1006/csla.1999.0128.

Cheng, Joe. 2018. *miniUI: Shiny UI Widgets for Small Screens*. https://CRAN.R-project.org/package=miniUI.

Christensen, Ronald, Wesley Johnson, Adam Branscum, and Timothy Hanson. 2011. "Bayesian Ideas and Data Analysis." CRC Press.

Clark, Vincent P., Silu Fan, and Steven A. Hillyard. 1994. "Identification of Early Visual Evoked Potential Generators by Retinotopic and Topographic Analyses." *Human Brain Mapping* 2 (3): 170–87.

Conway, Andrew R. A., Michael J. Kane, Michael F. Bunting, D. Zach Hambrick, Oliver Wilhelm, and Randall W. Engle. 2005. "Working Memory Span Tasks: A Methodological Review and User's Guide." *Psychonomic Bulletin & Review* 12 (5): 769–86.

Cook, Samantha R., Andrew Gelman, and Donald B. Rubin. 2006. "Validation of Software for Bayesian Models Using Posterior Quantiles." *Journal of Computational and Graphical Statistics* 15 (3): 675–92. https://doi.org/10.1198/106186006X136976.

Cooper, Richard P., and Olivia Guest. 2014. "Implementations Are Not Specifications: Specification, Replication and Experimentation in Computational Cognitive Modeling." *Cognitive Systems Research* 27: 42–49.

Cox, Christopher Martin Mikkelsen, Tamar Keren-Portnoy, Andreas Roepstorff, and Riccardo Fusaroli. 2022. "A Bayesian Meta-Analysis of Infants' Ability to Perceive Audio–Visual Congruence for Speech." *Infancy* 27 (1): 67–96.

Cumming, Geoff. 2014. "The New Statistics: Why and How." *Psychological Science* 25 (1): 7–29.

Dablander, Fabian, Karoline Huth, Quentin F. Gronau, Alexander Etz, and Eric-Jan Wagenmakers. 2022. "A Puzzle of Proportions: Two Popular Bayesian Tests Can Yield Dramatically Different Conclusions." *Statistics in Medicine* 41 (8): 1319–33.

Damasio, Antonio R. 1992. "Aphasia." *New England Journal of Medicine* 326 (8): 531–39. https://doi.org/10.1056/NEJM199202203260806.

de Finetti, Bruno. 1931. "Funcione Caratteristica Di Un Fenomeno Aleatorio." *Atti Dela Reale Accademia Nazionale Dei Lincei, Serie 6. Memorie, Classe Di Scienze Fisiche, Mathematice E Naturale* 4: 251–99.

DeLong, Katherine A., Thomas P. Urbach, and Marta Kutas. 2005. "Probabilistic Word Pre-Activation During Language Comprehension Inferred from Electrical Brain Activity." *Nature Neuroscience* 8 (8): 1117–21. https://doi.org/10.1038/nn1504.

Denckla, Martha Bridge, and Rita G. Rudel. 1976. "Rapid 'Automatized' Naming (R.A.N.): Dyslexia Differentiated from Other Learning Disabilities." *Neuropsychologia* 14 (4): 471–79. https://doi.org/https://doi.org/10.1016/0028-3932(76)90075-0.

Dickey, James M., and B. P. Lientz. 1970. "The Weighted Likelihood Ratio, Sharp Hypotheses About Chances, the Order of a Markov Chain." *The Annals of Mathematical Statistics* 41 (1): 214–26. https://www.jstor.org/stable/2239734.

Dillon, Brian W. 2011. "Structured Access in Sentence Comprehension." PhD thesis, University of Maryland.

Dobson, Annette J., and Adrian Barnett. 2011. *An Introduction to Generalized Linear Models.* CRC Press.

Drton, Mathias. 2013. *SIN: A SINful Approach to Selection of Gaussian Graphical Markov Models.* https://CRAN.R-project.org/package=SIN.

Duane, Simon, A. D. Kennedy, Brian J. Pendleton, and Duncan Roweth. 1987. "Hybrid Monte Carlo." *Physics Letters B* 195 (2): 216–22. https://doi.org/10.1016/0370-2693(87)91197-X.

Dufau, Stéphane, Jonathan Grainger, and Johannes C. Ziegler. 2012. "How to Say 'No' to a Nonword: A Leaky Competing Accumulator Model of Lexical Decision." *Journal of Experimental Psychology: Learning, Memory, and Cognition* 38 (4): 1117. https://doi.org/https://doi.org/10.1037/a0026948.

Dutilh, Gilles, Jeffrey Annis, Scott D. Brown, Peter Cassey, Nathan J. Evans, Raoul P. P. P. Grasman, Guy E. Hawkins, et al. 2019. "The Quality of Response Time Data Inference: A Blinded, Collaborative Assessment of the Validity of Cognitive Models." *Psychonomic Bulletin & Review* 26 (4): 1051–69. https://doi.org/https://doi.org/10.3758/s13423-017-1417-2.

Dutilh, Gilles, Eric-Jan Wagenmakers, Ingmar Visser, and Han L. J. van der Maas. 2011. "A Phase Transition Model for the Speed-Accuracy Trade-Off in Response Time Experiments." *Cognitive Science* 35 (2): 211–50. https://doi.org/10.1111/j.1551-6709.2010.01147.x.

Ebersole, Charles R., Olivia E. Atherton, Aimee L. Belanger, Hayley M. Skulborstad, Jill M. Allen, Jonathan B. Banks, Erica Baranski, et al. 2016. "Many Labs 3: Evaluating Participant Pool Quality Across the Academic Semester via Replication." *Journal of Experimental Social Psychology* 67: 68–82. https://doi.org/https://doi.org/10.1016/j.jesp.2015.10.012.

Eddelbuettel, Dirk, Romain Francois, J. J. Allaire, Kevin Ushey, Qiang Kou, Nathan Russell, Inaki Ucar, Douglas M. Bates, and John Chambers. 2024. *Rcpp: Seamless R and C++ Integration.* https://www.rcpp.org.

Engbert, Ralf, Antje Nuthmann, Eike M. Richter, and Reinhold Kliegl. 2005. "SWIFT: A Dynamical Model of Saccade Generation During Reading." *Psychological Review* 112: 777–813. https://doi.org/https://doi.org/10.1037/0033-295X.112.4.777.

Engelmann, Felix, Lena A. Jäger, and Shravan Vasishth. 2020. "The Effect of Prominence and Cue Association in Retrieval Processes: A Computational Account." *Cognitive Science* 43 (12): e12800. https://doi.org/10.1111/cogs.12800.

Epstein, Joshua M. 2008. "Why Model?" *Journal of Artificial Societies and Social Simulation* 11 (4): 12. https://www.jasss.org/11/4/12.html.

Faraway, Julian James. 2002. *Practical Regression and ANOVA using R.* Vol. 168. Citeseer.

———. 2016. *Extending the Linear Model with R: Generalized Linear, Mixed Effects and Nonparametric Regression Models.* Chapman & Hall/CRC Press.

Farrell, Simon, and Stephan Lewandowsky. 2018. *Computational Modeling of Cognition and Behavior.* Cambridge University Press.

Feldman, Jacob. 2017. "What Are the 'True' Statistics of the Environment?" *Cognitive Science* 41 (7): 1871–1903. https://doi.org/https://doi.org/10.1111/cogs.12444.

Fieller, Nick. 2016. *Basics of Matrix Algebra for Statistics with R.* Boca Raton, FL: CRC Press.

Fisher, Christopher R., Joseph W. Houpt, and Glenn Gunzelmann. 2022. "Fundamental Tools for Developing Likelihood Functions Within ACT-R." *Journal of Mathematical Psychology* 107: 102636. https://doi.org/https://doi.org/10.1016/j.jmp.2021.102636.

Fox, John. 2009. *A Mathematical Primer for Social Statistics.* Vol. 159. Sage.

———. 2015. *Applied Regression Analysis and Generalized Linear Models.* Sage Publications.

Francois, Romain. 2017. *Bibtex: Bibtex Parser.* https://CRAN.R-project.org/package=bibtex.

Frank, Stefan L., Leun J. Otten, Giulia Galli, and Gabriella Vigliocco. 2015. "The ERP Response to the Amount of Information Conveyed by Words in Sentences." *Brain and Language* 140: 1–11. https://doi.org/10.1016/j.bandl.2014.10.006.

Frazier, Lyn. 1979. "On Comprehending Sentences: Syntactic Parsing Strategies." PhD thesis, Amherst: University of Massachusetts.

Freedman, Laurence S., D. Lowe, and P. Macaskill. 1984. "Stopping Rules for Clinical Trials Incorporating Clinical Opinion." *Biometrics* 40 (3): 575–86.

Friendly, Michael, John Fox, and Phil Chalmers. 2020. *Matlib: Matrix Functions for Teaching and Learning Linear Algebra and Multivariate Statistics.* https://CRAN.R-project.org/package=matlib.

Frühwirth-Schnatter, Sylvia. 2006. *Finite Mixture and Markov Switching Models.* Springer Series in Statistics. Springer. https://doi.org/10.1007/978-0-387-35768-3.

Gabry, Jonah, Rok Češnovar, Andrew Johnson, and Steve Bronder. 2024. *cmdstanr: R Interface to "CmdStan".* https://mc-stan.org/cmdstanr/.

Gabry, Jonah, and Tristan Mahr. 2024. *bayesplot: Plotting for Bayesian Models.* https://mc-stan.org/bayesplot/.

Gamerman, Dani, and Hedibert F. Lopes. 2006. *Markov chain Monte Carlo: Stochastic simulation for Bayesian inference.* CRC Press.

Ge, Hong, Kai Xu, and Zoubin Ghahramani. 2018. "Turing: A Language for Flexible Probabilistic Inference." In *Proceedings of Machine Learning Research*, edited by Amos Storkey and Fernando Perez-Cruz, 84:1682–90. Playa Blanca, Lanzarote, Canary Islands: PMLR. http://proceedings.mlr.press/v84/ge18b.html.

Geisser, Seymour, and William F. Eddy. 1979. "A Predictive Approach to Model Selection." *Journal of the American Statistical Association* 74 (365): 153–60.

Gelman, Andrew, and John B. Carlin. 2014. "Beyond Power Calculations: Assessing Type S (Sign) and Type M (Magnitude) Errors." *Perspectives on Psychological Science* 9 (6): 641–51. https://doi.org/https://doi.org/10.1177/1745691614551642.

Gelman, Andrew, John B. Carlin, Hal S. Stern, David B. Dunson, Aki Vehtari, and Donald B. Rubin. 2014. *Bayesian Data Analysis.* Third Edition. Chapman & Hall/CRC Press.

Gelman, Andrew, and Jennifer Hill. 2007. *Data Analysis Using Regression and Multilevel/Hierarchical Models.* Cambridge University Press.

Gelman, Andrew, Daniel P. Simpson, and Michael J. Betancourt. 2017. "The Prior Can Often Only Be Understood in the Context of the Likelihood." *Entropy* 19 (10): 555. https://doi.org/10.3390/e19100555.

Gelman, Andrew, Aki Vehtari, Daniel P. Simpson, Charles C. Margossian, Bob Carpenter, Yuling Yao, Lauren Kennedy, Jonah Gabry, Paul-Christian Bürkner, and Martin Modrák. 2020. "Bayesian Workflow." *arXiv Preprint arXiv:2011.01808.*

Geweke, John. 2004. "Getting It Right: Joint Distribution Tests of Posterior Simulators." *Journal of the American Statistical Association* 99 (467): 799–804.

Gibson, Edward. 2000. "Dependency Locality Theory: A Distance-Based Theory of Linguistic Complexity." In *Image, Language, Brain: Papers from the First Mind Articulation Project Symposium*, edited by Alec Marantz, Yasushi Miyashita, and Wayne O'Neil. MIT Press.

Gibson, Edward, and H.-H. Iris Wu. 2013. "Processing Chinese Relative Clauses in Context." *Language and Cognitive Processes* 28 (1-2): 125–55. https://doi.org/https://doi.org/10.1080/01690965.2010.536656.

Gill, Jeff. 2006. *Essential Mathematics for Political and Social Research.* Cambridge University Press.

Gneiting, Tilmann, and Adrian E. Raftery. 2007. "Strictly Proper Scoring Rules, Prediction, and Estimation." *Journal of the American Statistical Association* 102 (477): 359–78. https://doi.org/10.1198/016214506000001437.

Gohel, David, Hadley Wickham, Lionel Henry, and Jeroen Ooms. 2019. *gdtools: Utilities for Graphical Rendering.* https://CRAN.R-project.org/package=gdtools.

Good, I. J. 1952. "Rational Decisions." *Journal of the Royal Statistical Society. Series B (Methodological)* 14 (1): 107–14. http://www.jstor.org/stable/2984087.

Goodrich, Ben, Jonah Gabry, Imad Ali, and Sam Brilleman. 2018. "Rstanarm: Bayesian Applied Regression Modeling via Stan." http://mc-stan.org/.

Goodrich, Ben, Andrew Gelman, Bob Carpenter, Matthew D. Hoffman, Daniel Lee, Michael J. Betancourt, Marcus Brubaker, et al. 2024. *StanHeaders: C++ Header Files for Stan.* https://mc-stan.org/.

Grant, Robert, and Gian Luca Di Tanna. 2025. *Bayesian Meta-Analysis: A Practical Introduction.* Cambridge University Press.

Green, Peter J. 1998. "Penalized Likelihood." In *Encyclopedia of Statistical Sciences*, 2:578–86. Wiley.

Gronau, Quentin F., Alexandra Sarafoglou, Dora Matzke, Alexander Ly, Udo Boehm, Maarten Marsman, David S. Leslie, Jonathan J. Forster, Eric-Jan Wagenmakers, and Helen Steingroever. 2017. "A Tutorial on Bridge Sampling." *Journal of Mathematical Psychology* 81: 80–97. https://doi.org/10.1016/j.jmp.2017.09.005.

Gronau, Quentin F., Henrik Singmann, and Eric-Jan Wagenmakers. 2020. "bridgesampling: An R Package for Estimating Normalizing Constants." *Journal of Statistical Software* 92 (10): 1–29. https://doi.org/10.18637/jss.v092.i10.

Gronau, Quentin F., and Eric-Jan Wagenmakers. 2018. "Limitations of Bayesian Leave-One-Out Cross-Validation for Model Selection." *Computational Brain & Behavior*. https://doi.org/10.1007/s42113-018-0011-7.

———. 2019. "Rejoinder: More Limitations of Bayesian Leave-One-Out Cross-Validation." *Computational Brain & Behavior* 2 (1): 35–47. https://doi.org/%20https://doi.org/10.1007/s42113-018-0022-4.

Guo, Jiqiang, Jonah Gabry, Ben Goodrich, Andrew Johnson, Sebastian Weber, and Hamada S. Badr. 2024. *rstan: R Interface to Stan*. https://mc-stan.org/rstan/.

Haaf, Julia M., and Jeffrey N Rouder. 2019. "Some Do and Some Don't? Accounting for Variability of Individual Difference Structures." *Psychonomic Bulletin & Review* 26 (3): 772–89.

Haines, Nathaniel, Peter D. Kvam, Louis H. Irving, Colin Smith, Theodore P. Beauchaine, Mark A. Pitt, Woo-Young Ahn, and Brandon M Turner. 2020. "Learning from the Reliability Paradox: How Theoretically Informed Generative Models Can Advance the Social, Behavioral, and Brain Sciences." https://ccs-lab.github.io/pdfs/papers/haines2020_reliability.pdf.

Hammerly, Christopher, Adrian Staub, and Brian W. Dillon. 2019. "The Grammaticality Asymmetry in Agreement Attraction Reflects Response Bias: Experimental and Modeling Evidence." *Cognitive Psychology* 110: 70–104. https://doi.org/https://doi.org/10.1016/j.cogpsych.2019.01.001.

Han, Ding, Jana Wegrzyn, Hua Bi, Ruihua Wei, Bin Zhang, and Xiaorong Li. 2018. "Practice Makes the Deficiency of Global Motion Detection in People with Pattern-Related Visual Stress More Apparent." *PLOS ONE* 13 (2): 1–13. https://doi.org/10.1371/journal.pone.0193215.

Harrell Jr., Frank E. 2015. *Regression Modeling Strategies: With Applications to Linear Models, Logistic and Ordinal Regression, and Survival Analysis*. Springer.

Harris, Christopher M., and Jonathan Waddington. 2012. "On the Convergence of Time Interval Moments: Caveat Sciscitator." *Journal of Neuroscience Methods* 205 (2): 345–56. https://doi.org/https://doi.org/10.1016/j.jneumeth.2012.01.017.

Harris, Christopher M., Jonathan Waddington, Valerio Biscione, and Sean Manzi. 2014. "Manual Choice Reaction Times in the Rate-Domain." *Frontiers in Human Neuroscience* 8: 418. https://doi.org/10.3389/fnhum.2014.00418.

Hartmann, Raphael, Lea Johannsen, and Karl Christoph Klauer. 2020. "rtmpt: An R Package for Fitting Response-Time Extended Multinomial Processing Tree Models." *Behavior Research Methods* 52 (3): 1313–38.

Hayes, Taylor R., and Alexander A. Petrov. 2016. "Mapping and Correcting the Influence of Gaze Position on Pupil Size Measurements." *Behavior Research Methods* 48 (2): 510–27. https://doi.org/10.3758/s13428-015-0588-x.

Heathcote, Andrew, Yi-Shin Lin, Angus Reynolds, Luke Strickland, Matthew Gretton, and Dora Matzke. 2019. "Dynamic Models of Choice." *Behavior Research Methods* 51 (2): 961–85. https://doi.org/https://doi.org/10.3758/s13428-018-1067-y.

Heathcote, Andrew, and Jonathon Love. 2012. "Linear Deterministic Accumulator Models of Simple Choice." *Frontiers in Psychology* 3: 292. https://doi.org/10.3389/fpsyg.2012.00292.

Heck, Daniel W. 2019. "A Caveat on the Savage–Dickey Density Ratio: The Case of Computing Bayes Factors for Regression Parameters." *British Journal of Mathematical and Statistical Psychology* 72 (2): 316–33.

Heck, Daniel W, and Edgar Erdfelder. 2016. "Extending Multinomial Processing Tree Models to Measure the Relative Speed of Cognitive Processes." *Psychonomic Bulletin &Amp; Review* 23 (5): 1440–65. https://doi.org/10.3758/s13423-016-1025-6.

———. 2020. "Benefits of Response Time-Extended Multinomial Processing Tree Models: A Reply to Starns (2018)." *Psychonomic Bulletin &Amp; Review* 27 (3): 571–80. https://doi.org/10.3758/s13423-019-01663-0.

Heck, Daniel W, Edgar Erdfelder, and Pascal J. Kieslich. 2018. "Generalized Processing Tree Models: Jointly Modeling Discrete and Continuous Variables." *Psychometrika* 83 (4): 893–918. https://doi.org/10.1007/s11336-018-9622-0.

Heck, Daniel W, and Eric-Jan Wagenmakers. 2016. "Adjusted Priors for Bayes Factors Involving Reparameterized Order Constraints." *Journal of Mathematical Psychology* 73: 110–16. https://doi.org/https://doi.org/10.1016/j.jmp.2016.05.004.

Heister, Julian, Kay-Michael Würzner, and Reinhold Kliegl. 2012. "Analysing Large Datasets of Eye Movements During Reading." *Visual Word Recognition* 2: 102–30.

Hem, Ingeborg Gullikstad, Geir-Arne Fuglstad, and Andrea Riebler. 2022. "makemyprior: Intuitive Construction of Joint Priors for Variance Parameters in R." https://arxiv.org/abs/2105.09712.

Henrich, Joseph, Steven J. Heine, and Ara Norenzayan. 2010. "The Weirdest People in the World?" *Behavioral and Brain Sciences* 33 (2-3): 61–83. https://doi.org/10.1017/S0140525X0999152X.

Hester, Jim, Gábor Csárdi, Hadley Wickham, Winston Chang, Martin Morgan, and Dan Tenenbaum. 2021. *Remotes: R Package Installation from Remote Repositories, Including 'Github'*. https://CRAN.R-project.org/package=remotes.

Higgins, Julian, and Sally Green. 2008. *Cochrane Handbook for Systematics Reviews of Interventions*. Wiley-Blackwell.

Hoffman, Matthew D., and Andrew Gelman. 2014. "The No-U-Turn Sampler: Adaptively Setting Path Lengths in Hamiltonian Monte Carlo." *Journal of Machine Learning Research* 15 (1): 1593–1623. http://dl.acm.org/citation.cfm?id=2627435.2638586.

Hubel, Kerry A., Bruce Reed, E. William Yund, Timothy J. Herron, and David L. Woods. 2013. "Computerized Measures of Finger Tapping: Effects of Hand Dominance, Age, and Sex." *Perceptual and Motor Skills* 116 (3): 929–52. https://doi.org/https://doi.org/10.2466/25.29.PMS.116.3.929-952.

Izrailev, Sergei. 2024. *Tictoc: Functions for Timing R Scripts, as Well as Implementations of "Stack" and "Stacklist" Structures*. https://github.com/jabiru/tictoc.

Jäger, Lena A., Felix Engelmann, and Shravan Vasishth. 2017. "Similarity-Based Interference in Sentence Comprehension: Literature review and Bayesian meta-analysis." *Journal of Memory and Language* 94: 316–39. https://doi.org/https://doi.org/10.1016/j.jml.2017.01.004.

Jäger, Lena A., Daniela Mertzen, Julie A. Van Dyke, and Shravan Vasishth. 2020. "Interference Patterns in Subject-Verb Agreement and Reflexives Revisited: A Large-Sample Study." *Journal of Memory and Language* 111. https://doi.org/https://doi.org/10.1016/j.jml.2019.104063.

JASP Team. 2019. "JASP (Version 0.11.1)[Computer software]." https://jasp-stats.org/.

Jaynes, Edwin T. 2003. *Probability Theory: The Logic of Science*. Cambridge University Press.

Jeffreys, Harold. 1939. *Theory of Probability*. Clarendon Press.

Johnson, Norman L., Samuel Kotz, and Narayanaswamy Balakrishnan. 1995. *Continuous Univariate Distributions, Volume 2*. Vol. 289.

Jordan, Dominic, and Peter Smith. 2008. *Mathematical Techniques: An Introduction for the Engineering, Physical, and Mathematical Sciences*. Oxford University Press.

Jurafsky, Daniel. 1996. "A Probabilistic Model of Lexical and Syntactic Access and Disambiguation." *Cognition* 20: 137–94. https://doi.org/https://doi.org/10.1207/s15516709cog2002_1.

Just, Marcel Adam, Patricia A. Carpenter, and S. Varma. 1999. "Computational Modeling of High-Level Cognition and Brain Function." *Human Brain Mapping* 8: 128–36. https://doi.org/https://doi.org/10.1002/(SICI)1097-0193(1999)8:2/3%3C128::AID-HBM10%3E3.0.CO;2-G.

Kass, Robert E., and Adrian E. Raftery. 1995. "Bayes Factors." *Journal of the American Statistical Association* 90 (430): 773–95. https://doi.org/10.1080/01621459.1995.10476572.

Kerns, G. J. 2014. *Introduction to Probability and Statistics Using R*. Second Edition.

Keuleers, Emmanuel, Paula Lacey, Kathleen Rastle, and Marc Brysbaert. 2012. "The British Lexicon Project: Lexical Decision Data for 28,730 Monosyllabic and Disyllabic English Words." *Behavior Research Methods* 44 (1): 287–304. https://doi.org/https://doi.org/10.3758/s13428-011-0118-4.

Kim, Shinyoung, Hyunji Moon, Martin Modrák, and Teemu Säilynoja. 2024. *SBC: Simulation Based Calibration for Rstan/Cmdstanr Models*. https://hyunjimoon.github.io/SBC/.

Klauer, Karl Christoph, and David Kellen. 2018. "RT-MPTs: Process Models for Response-Time Distributions Based on Multinomial Processing Trees with Applications to Recognition Memory." *Journal of Mathematical Psychology* 82: 111–30. https://doi.org/https://doi.org/10.1016/j.jmp.2017.12.003.

Kolmogorov, Andreĭ Nikolaevich. 1933. *Foundations of the Theory of Probability: Second English Edition*. Courier Dover Publications.

Koster, Jeremy, and Richard McElreath. 2017. "Multinomial Analysis of Behavior: Statistical Methods." *Behavioral Ecology and Sociobiology* 71 (9): 138. https://doi.org/10.1007/s00265-017-2363-8.

Kruschke, John K. 2014. *Doing Bayesian Data Analysis: A tutorial with R, JAGS, and Stan*. Academic Press.

Kruschke, John K., and Torrin M. Liddell. 2018. "The Bayesian New Statistics: Hypothesis Testing, Estimation, Meta-Analysis, and Power Analysis from a Bayesian Perspective."

Psychonomic Bulletin & Review 25 (1): 178–206. https://doi.org/https://doi.org/10.3758/s13423-016-1221-4.

Kutas, Marta, and Kara D. Federmeier. 2011. "Thirty Years and Counting: Finding Meaning in the N400 Component of the Event-Related Brain Potential (ERP)." *Annual Review of Psychology* 62 (1): 621–47. https://doi.org/10.1146/annurev.psych.093008.131123.

Kutas, Marta, and Steven A. Hillyard. 1980. "Reading Senseless Sentences: Brain Potentials Reflect Semantic Incongruity." *Science* 207 (4427): 203–5. https://doi.org/10.1126/science.7350657.

———. 1984. "Brain Potentials During Reading Reflect Word Expectancy and Semantic Association." *Nature* 307 (5947): 161–63. https://doi.org/10.1038/307161a0.

Lago, Sol, Diego Shalom, Mariano Sigman, Ellen F. Lau, and Colin Phillips. 2015. "Agreement Processes in Spanish Comprehension." *Journal of Memory and Language* 82: 133–49. https://doi.org/https://doi.org/10.1016/j.jml.2015.02.002.

Laird, John E. 2019. *The Soar Cognitive Architecture*. MIT Press.

Laird, Nan M., and James H. Ware. 1982. "Random-Effects Models for Longitudinal Data." *Biometrics*, 963–74. https://doi.org/https://doi.org/10.2307/2529876.

Lambert, Ben. 2018. *A Student's Guide to Bayesian Statistics*. Sage.

Laurinavichyute, Anna. 2020. "Similarity-Based Interference and Faulty Encoding Accounts of Sentence Processing." Dissertation, University of Potsdam. https://publishup.uni-potsdam.de/frontdoor/index/index/docId/50966.

Lee, Michael D, ed. 2011a. "Special Issue on Hierarchical Bayesian Models." *Journal of Mathematical Psychology* 55 (1). https://www.sciencedirect.com/journal/journal-of-mathematical-psychology/vol/55/issue/1.

———. 2011b. "How Cognitive Modeling Can Benefit from Hierarchical Bayesian Models." *Journal of Mathematical Psychology* 55 (1): 1–7. https://doi.org/10.1016/j.jmp.2010.08.013.

Lee, Michael D, Jason R. Bock, Isaiah Cushman, and William R. Shankle. 2020. "An Application of Multinomial Processing Tree Models and Bayesian Methods to Understanding Memory Impairment." *Journal of Mathematical Psychology* 95: 102328. https://doi.org/https://doi.org/10.1016/j.jmp.2020.102328.

Lee, Michael D, and Eric-Jan Wagenmakers. 2014. *Bayesian Cognitive Modeling: A Practical Course*. Cambridge University Press.

Lee, Peter M. 2012. *Bayesian Statistics: An Introduction*.

Levy, Dan. 2021. *Maxims for Thinking Analytically: The Wisdom of Legendary Harvard Professor Richard Zeckhauser*. Dan Levy.

Levy, Deborah L., Philip S. Holzman, Steven Matthysse, and Nancy R. Mendell. 1993. "Eye Tracking Dysfunction and Schizophrenia: A Critical Perspective." *Schizophrenia Bulletin* 19 (3): 461–536. https://doi.org/https://doi.org/10.1093/schbul/19.3.461.

Lewis, Richard L., and Shravan Vasishth. 2005. "An Activation-Based Model of Sentence Processing as Skilled Memory Retrieval." *Cognitive Science* 29: 1–45. https://doi.org/10.1207/s15516709cog0000_25.

Lidstone, George James. 1920. "Note on the General Case of the Bayes-Laplace Formula for Inductive or a Posteriori Probabilities." *Transactions of the Faculty of Actuaries* 8 (182-192): 13.

Lindgren, Finn, and Håvard Rue. 2015. "Bayesian Spatial Modelling with R-INLA." *Journal of Statistical Software* 63 (1): 1–25. https://doi.org/10.18637/jss.v063.i19.

Lissón, Paula, Dario Paape, Dorothea Pregla, Frank Burchert, Nicole Stadie, and Shravan Vasishth. 2023. "Similarity-Based Interference in Sentence Comprehension in Aphasia: A Computational Evaluation of Two Models of Cue-Based Retrieval." *Computational Brain and Behavior*. https://doi.org/10.1007/s42113-023-00168-3.

Lissón, Paula, Dorothea Pregla, Bruno Nicenboim, Dario Paape, Mick van het Nederend, Frank Burchert, Nicole Stadie, David Caplan, and Shravan Vasishth. 2021. "A Computational Evaluation of Two Models of Retrieval Processes in Sentence Processing in Aphasia." *Cognitive Science* 45 (4): e12956. https://onlinelibrary.wiley.com/doi/full/10.1111/cogs.12956.

Loftus, Geoffrey R. 1978. "On Interpretation of Interactions." *Memory & Cognition* 6 (3): 312–19.

Logačev, Pavel, and Noyan Dokudan. 2021. "A Multinomial Processing Tree Model of RC Attachment." In *Proceedings of the Workshop on Cognitive Modeling and Computational Linguistics*, 39–47. Online: Association for Computational Linguistics. https://www.aclweb.org/anthology/2021.cmcl-1.4.

Logačev, Pavel, and Shravan Vasishth. 2016. "A Multiple-Channel Model of Task-Dependent Ambiguity Resolution in Sentence Comprehension." *Cognitive Science* 40 (2): 266–98. https://doi.org/10.1111/cogs.12228.

Lucas, Antoine, Dirk Eddelbuettel with contributions by, Jarek Tuszynski, Henrik Bengtsson, Simon Urbanek, Mario Frasca, Bryan Lewis, Murray Stokely, et al. 2021. *Digest: Create Compact Hash Digests of R Objects*. https://CRAN.R-project.org/package=digest.

Luce, R. Duncan. 1991. *Response Times: Their Role in Inferring Elementary Mental Organization*. Oxford University Press.

Lunn, David J., Chris Jackson, David J. Spiegelhalter, Nichola G. Best, and Andrew Thomas. 2012. *The BUGS Book: A Practical Introduction to Bayesian Analysis*. Vol. 98. CRC Press.

Lunn, David J., Andrew Thomas, Nichola G. Best, and David J. Spiegelhalter. 2000. "WinBUGS-A Bayesian Modelling Framework: Concepts, Structure, and Extensibility." *Statistics and Computing* 10 (4): 325–37.

Lynch, Scott Michael. 2007. *Introduction to Applied Bayesian Statistics and Estimation for Social Scientists*. Springer.

Maanen, Leendert van, Dimitris Katsimpokis, and A. Dilene van Campen. 2018. "Fast and Slow Errors: Logistic Regression to Identify Patterns in Accuracy–Response Time Relationships." *Behavior Research Methods* 51 (5): 2378–89. https://doi.org/10.3758/s13428-018-1110-z.

MacKay, David J. C. 2003. *Information Theory, Inference and Learning Algorithms*. Cambridge University Press.

MacLeod, Colin M. 1991. "Half a Century of Research on the Stroop Effect: An Integrative Review." *Psychological Bulletin* 109 (2): 163. https://doi.org/https://doi.org/10.1037/0033-2909.109.2.163.

Mahowald, Kyle, Ariel James, Richard Futrell, and Edward Gibson. 2016. "A Meta-Analysis of Syntactic Priming in Language Production." *Journal of Memory and Language* 91: 5–27. https://doi.org/https://doi.org/10.1016/j.jml.2016.03.009.

Makowski, Dominique, Mattan S. Ben-Shachar, and Daniel Lüdecke. 2019. "BayestestR: Describing Effects and Their Uncertainty, Existence and Significance Within the Bayesian Framework." *Journal of Open Source Software* 4 (40): 1541. https://doi.org/https://doi.org/10.21105/joss.01541.

Mathot, Sebastiaan. 2018. "Pupillometry: Psychology, Physiology, and Function." *Journal of Cognition* 1 (1): 16. https://doi.org/10.5334/joc.18.

Matzke, Dora, Conor V. Dolan, William H. Batchelder, and Eric-Jan Wagenmakers. 2015. "Bayesian Estimation of Multinomial Processing Tree Models with Heterogeneity in Participants and Items." *Psychometrika* 80 (1): 205–35. https://doi.org/10.1007/s11336-013-9374-9.

Maxwell, Scott E, Harold D Delaney, and Ken Kelley. 2017. *Designing Experiments and Analyzing Data: A Model Comparison Perspective.* New York, NY: Routledge.

McClelland, James L. 2009. "The Place of Modeling in Cognitive Science." *Topics in Cognitive Science* 1 (1): 11–38. https://doi.org/10.1111/j.1756-8765.2008.01003.x.

McClelland, James L., and David E. Rumelhart. 1989. *Explorations in Parallel Distributed Processing: A Handbook of Models, Programs, and Exercises.* MIT Press.

McCullagh, Peter, and J. A. Nelder. 2019. *Generalized Linear Models.* Second Edition. Chapman & Hall/CRC Press.

McElreath, Richard. 2020. *Statistical Rethinking: A Bayesian Course with Examples in R and Stan.* Chapman & Hall/CRC Press.

———. 2021. *Rethinking: Statistical Rethinking Book Package.*

McElree, Brian. 2000. "Sentence Comprehension Is Mediated by Content-Addressable Memory Structures." *Journal of Psycholinguistic Research* 29 (2): 111–23. https://doi.org/https://doi.org/10.1023/A:1005184709695.

McLean, Mathew William. 2017. "RefManageR: Import and Manage Bibtex and Biblatex References in R." *The Journal of Open Source Software.* https://doi.org/10.21105/joss.00338.

Meehl, Paul E. 1997. "The Problem Is Epistemology, Not Statistics: Replace Significance Tests by Confidence Intervals and Quantify Accuracy of Risky Numerical Predictions." In *What If There Were No Significance Tests?*, edited by L. L. Harlow, S. A. Mulaik, and J. H. Steiger. Erlbaum.

Meng, Xiao-li, and Wing Hung Wong. 1996. "Simulating Ratios of Normalizing Constants via a Simple Identity: A Theoretical Exploration." *Statistica Sinica*, 831–60. https://www.jstor.org/stable/24306045.

Miller, I., and M. Miller. 2004. *John E. Freund's Mathematical Statistics with Applications.* Prentice Hall.

Modrák, Martin, Angie H. Moon, Shinyoung Kim, Paul-Christian Bürkner, Niko Huurre, Kateřina Faltejsková, Andrew Gelman, and Aki Vehtari. 2023. "Simulation-Based Calibration Checking for Bayesian Computation: The Choice of Test Quantities Shapes Sensitivity." *Bayesian Analysis*, 1–28. https://doi.org/10.1214/23-BA1404.

Monnahan, Cole C., James T. Thorson, and Trevor A. Branch. 2017. "Faster Estimation of Bayesian Models in Ecology Using Hamiltonian Monte Carlo." Edited by Robert B. O'Hara. *Methods in Ecology and Evolution* 8 (3): 339–48. https://doi.org/10.1111/2041-210X.12681.

Montgomery, D. C., E. A. Peck, and G. G. Vining. 2012. *An Introduction to Linear Regression Analysis.* 5th ed. Wiley.

Morin, David J. 2016. *Probability: For the Enthusiastic Beginner.* Createspace Independent Publishing Platform.

Müller, Kirill, and Hadley Wickham. 2020. *Tibble: Simple Data Frames.* https://CRAN.R-project.org/package=tibble.

Navarro, Danielle J. 2015. *Learning Statistics with R.* https://learningstatisticswithr.com.

———. 2019. "Between the Devil and the Deep Blue Sea: Tensions Between Scientific Judgement and Statistical Model Selection." *Computational Brain & Behavior* 2 (1): 28–34. https://doi.org/10.1007/s42113-018-0019-z.

Neal, Radford M. 2011. "MCMC Using Hamiltonian Dynamics." In *Handbook of Markov Chain Monte Carlo*, edited by Steve Brooks, Andrew Gelman, Galin Jones, and Xiao-Li Meng. Taylor & Francis. https://doi.org/10.1201/b10905-10.

———. 2003. "Slice Sampling." *Annals of Statistics* 31 (3): 705–67. https://doi.org/10.1214/aos/1056562461.

Nelson, Peter R. 1981. "The Algebra of Random Variables." *Technometrics* 23 (2): 197–98. https://doi.org/10.1080/00401706.1981.10486266.

Newell, Allen. 1990. *Unified Theories of Cognition.* Harvard University Press.

Nicenboim, Bruno. 2018. "The Implementation of a Model of Choice: The (Truncated) Linear Ballistic Accumulator." In *StanCon.* Aalto University, Helsinki, Finland. https://doi.org/10.5281/zenodo.1465990.

Nicenboim, Bruno, Pavel Logačev, Carolina Gattei, and Shravan Vasishth. 2016. "When High-Capacity Readers Slow down and Low-Capacity Readers Speed up: Working Memory and Locality Effects." *Frontiers in Psychology* 7 (280). https://doi.org/10.3389/fpsyg.2016.00280.

Nicenboim, Bruno, Timo B. Roettger, and Shravan Vasishth. 2018. "Using Meta-Analysis for Evidence Synthesis: The case of incomplete neutralization in German." *Journal of Phonetics* 70: 39–55. https://doi.org/https://doi.org/10.1016/j.wocn.2018.06.001.

Nicenboim, Bruno, Daniel J. Schad, and Shravan Vasishth (2025). *bcogsci: Datasets and Models for the Book "Introduction to Bayesian Data Analysis for Cognitive Science."* doi:10.5281/zenodo.15231704, R package version 1.0.0.

Nicenboim, Bruno, and Shravan Vasishth. 2016. "Statistical methods for linguistic research: Foundational Ideas - Part II." *Language and Linguistics Compass* 10 (11): 591–613. https://doi.org/10.1111/lnc3.12207.

———. 2018. "Models of Retrieval in Sentence Comprehension: A Computational Evaluation Using Bayesian Hierarchical Modeling." *Journal of Memory and Language* 99: 1–34. https://doi.org/10.1016/j.jml.2017.08.004.

Nicenboim, Bruno, Shravan Vasishth, Felix Engelmann, and Katja Suckow. 2018. "Exploratory and Confirmatory Analyses in Sentence Processing: A case study of number interference in German." *Cognitive Science* 42 (S4). https://doi.org/10.1111/cogs.12589.

Nicenboim, Bruno, Shravan Vasishth, and Frank Rösler. 2020. "Are Words Pre-Activated Probabilistically During Sentence Comprehension? Evidence from New Data and a Bayesian Random-Effects Meta-Analysis Using Publicly Available Data." *Neuropsychologia* 142. https://doi.org/10.1016/j.neuropsychologia.2020.107427.

Nieuwenhuis, Sander, Birte U Forstmann, and Eric-Jan Wagenmakers. 2011. "Erroneous Analyses of Interactions in Neuroscience: A Problem of Significance." *Nature Neuroscience* 14 (9): 1105–7.

Nieuwland, Mante S., Stephen Politzer-Ahles, Evelien Heyselaar, Katrien Segaert, Emily Darley, Nina Kazanina, Sarah Von Grebmer Zu Wolfsthurn, et al. 2018. "Large-Scale Replication Study Reveals a Limit on Probabilistic Prediction in Language Comprehension." *eLife* 7. https://doi.org/10.7554/eLife.33468.

Nosek, Brian A., Tom E Hardwicke, Hannah Moshontz, Aurélien Allard, Katherine S Corker, Anna Dreber, Fiona Fidler, et al. 2022. "Replicability, Robustness, and Reproducibility in Psychological Science." *Annual Review of Psychology* 73: 719–48. https://doi.org/https://doi.org/10.1146/annurev-psych-020821-114157.

Oakley, Jeremy E. 2024. *SHELF: Tools to Support the Sheffield Elicitation Framework.* https://github.com/OakleyJ/SHELF.

Oberauer, Klaus. 2019. "Working Memory Capacity Limits Memory for Bindings." *Journal of Cognition* 2 (1): 40. https://doi.org/10.5334/joc.86.

Oberauer, Klaus, and Reinhold Kliegl. 2001. "Beyond Resources: Formal Models of Complexity Effects and Age Differences in Working Memory." *European Journal of Cognitive Psychology* 13 (1-2): 187–215. https://doi.org/10.1080/09541440042000278.

Oberauer, Klaus, Philipp Musfeld, and Frederik Aust. 2024. "The Stability of Bayes-Factor Estimates for Testing Effects in Generalized Linear Models: Bridge Sampling and Savage–Dickey Ratio."

O'Hagan, Anthony, Caitlin E. Buck, Alireza Daneshkhah, J. Richard Eiser, Paul H. Garthwaite, David J. Jenkinson, Jeremy E. Oakley, and Tim Rakow. 2006. *Uncertain Judgements: Eliciting Experts' Probabilities.* John Wiley & Sons.

O'Hagan, Anthony, and Jonathan J. Forster. 2004. "Kendall's Advanced Theory of Statistics, Vol. 2B: Bayesian Inference." Wiley.

Ollman, Robert. 1966. "Fast Guesses in Choice Reaction Time." *Psychonomic Science* 6 (4): 155–56. https://doi.org/https://doi.org/10.3758/BF03328004.

Ooms, Jeroen. 2023. *pdftools: Text Extraction, Rendering and Converting of Pdf Documents.* https://docs.ropensci.org/pdftools/.

Paananen, Topi, Juho Piironen, Paul-Christian Bürkner, and Aki Vehtari. 2021. "Implicitly Adaptive Importance Sampling." *Statistics and Computing* 31 (2). https://doi.org/10.1007/s11222-020-09982-2.

Paape, Dario, Serine Avetisyan, Sol Lago, and Shravan Vasishth. 2021. "Modeling Misretrieval and Feature Substitution in Agreement Attraction: A Computational Evaluation." *Cognitive Science* 45 (8). https://doi.org/https://doi.org/10.1111/cogs.13019.

Paape, Dario, Garrett Smith, and Shravan Vasishth. 2024. "Do Local Coherence Effects Exist in English Reduced Relative Clauses?" https://doi.org/https://doi.org/10.31219/osf.io/wpke4.

Paape, Dario, and Shravan Vasishth. 2022. "Estimating the True Cost of Garden-Pathing: A Computational Model of Latent Cognitive Processes." *Cognitive Science* 46 (8): e13186. https://doi.org/https://doi.org/10.1111/cogs.13186.

Papaspiliopoulos, Omiros, Gareth O. Roberts, and Martin Sköld. 2007. "A General Framework for the Parametrization of Hierarchical Models." *Statistical Science* 22 (1): 59–73. https://doi.org/10.1214/088342307000000014.

Parmigiani, Giovanni, and Lurdes Inoue. 2009. *Decision Theory: Principles and Approaches.* John Wiley & Sons.

Picton, T. W., S. Bentin, P. Berg, E. Donchin, Steven A. Hillyard, R. Johnson J. R., G. A. Miller, et al. 2000. "Guidelines for Using Human Event-Related Potentials to Study Cognition: Recording Standards and Publication Criteria." *Psychophysiology* 37 (2): 127–52. https://doi.org/10.1111/1469-8986.3720127.

Piironen, Juho, Markus Paasiniemi, and Aki Vehtari. 2020. "Projective inference in high-dimensional problems: Prediction and feature selection." *Electronic Journal of Statistics* 14 (1): 2155–97. https://doi.org/10.1214/20-EJS1711.

Piironen, Juho, and Aki Vehtari. 2017. "Comparison of Bayesian Predictive Methods for Model Selection." *Statistics and Computing* 27 (3): 711–35. https://doi.org/10.1007/s11222-016-9649-y.

Pinheiro, José C, and Douglas M. Bates. 2000. *Mixed-Effects Models in S and S-PLUS.* Springer-Verlag.

Pitt, Mark A., and In Jae Myung. 2002. "When a Good Fit Can Be Bad." *Trends in Cognitive Sciences* 6 (10): 421–25. https://doi.org/10.1016/S1364-6613(02)01964-2.

Plummer, Martin. 2016. "JAGS Version 4.2.0 User Manual."

———. 2022. "Simulation-Based Bayesian Analysis." *Annual Review of Statistics and Its Application.*

Port, Robert F., and Timothy Van Gelder. 1995. *Mind as Motion: Explorations in the Dynamics of Cognition.* MIT Press.

Pullin, Jeffrey, Lyle Gurrin, and Damjan Vukcevic. 2021. "Statistical Models of Repeated Categorical Ratings: The R Package Rater." http://arxiv.org/abs/2010.09335.

Pylyshyn, Zenon W., and Ron W. Storm. 1988. "Tracking Multiple Independent Targets: Evidence for a Parallel Tracking Mechanism." *Spatial Vision* 3 (3): 179–97. https://doi.org/10.1163/156856888X00122.

Rabe, Maximilian M., Johan Chandra, André Krügel, Stefan A. Seelig, Shravan Vasishth, and Ralf Engbert. 2021. "A Bayesian Approach to Dynamical Modeling of Eye-Movement Control in Reading of Normal, Mirrored, and Scrambled Texts." *Psychological Review*. https://doi.org/10.1037/rev0000268.

Rabe, Maximilian M., Dario Paape, Daniela Metzen, Shravan Vasishth, and Ralf Engbert. 2024. "SEAM: An Integrated Activation-Coupled Model of Sentence Processing and Eye Movements in Reading." *Journal of Memory and Language*. https://doi.org/https://doi.org/10.1016/j.jml.2023.104496.

Rabe, Maximilian M., Shravan Vasishth, Sven Hohenstein, Reinhold Kliegl, and Daniel J. Schad. 2020. "hypr: An R Package for Hypothesis-Driven Contrast Coding." *Journal of Open Source Software* 5 (48): 2134.

Ratcliff, Roger. 1978. "A Theory of Memory Retrieval." *Psychological Review* 85 (2): 59. https://doi.org/10.1037/0033-295X.85.2.59.

Ratcliff, Roger, Philip L. Smith, Scott D. Brown, and Gail McKoon. 2016. "Diffusion Decision Model: Current Issues and History." *Trends in Cognitive Sciences* 20 (4): 260–81. https://doi.org/https://doi.org/10.1016/j.tics.2016.01.007.

Ratcliff, Roger, and Francis Tuerlinckx. 2002. "Estimating Parameters of the Diffusion Model: Approaches to Dealing with Contaminant Reaction Times and Parameter Variability." *Psychonomic Bulletin & Review* 9 (3): 438–81. https://doi.org/https://doi.org/10.3758/BF03196302.

Rayner, K. 1998. "Eye movements in reading and information processing: 20 years of research." *Psychological Bulletin* 124 (3): 372–422. https://doi.org/https://doi.org/10.1037/0033-2909.124.3.372.

R Core Team. 2023. *R: A Language and Environment for Statistical Computing*. Vienna, Austria: R Foundation for Statistical Computing. https://www.R-project.org/.

Resnick, Sidney. 2019. *A Probability Path*. Springer.

Ripley, Brian D. 2023. *MASS: Support Functions and Datasets for Venables and Ripley's MASS*. http://www.stats.ox.ac.uk/pub/MASS4/.

Robert, Christian P. 2022. "50 Shades of Bayesian Testing of Hypotheses." In *Advancements in Bayesian Methods and Implementation*, edited by Arni S. R. Srinivasa Rao, G. Alastair Young, and C. R. Rao, 47:103–20. Handbook of Statistics. Elsevier. https://doi.org/https://doi.org/10.1016/bs.host.2022.06.003.

Roberts, Seth, and Harold Pashler. 2000. "How Persuasive Is a Good Fit? A Comment on Theory Testing." *Psychological Review* 107 (2): 358–67. https://doi.org/https://doi.org/10.1037/0033-295X.107.2.358.

Rodriguez, Josue E., Donald R. Williams, and Philippe Rast. 2022. "Who Is and Is Not 'Average'? Random Effects Selection with Spike-and-Slab Priors." *Psychological Methods*. https://doi.org/https://doi.org/10.1037/met0000535.

Rosenthal, Robert, Ralph L. Rosnow, and Donald B. Rubin. 2000. *Contrasts and Effect Sizes in Behavioral Research: A Correlational Approach.* Cambridge University Press.

Ross, Sheldon. 2002. *A First Course in Probability.* Pearson Education.

Rouder, Jeffrey N. 2005. "Are Unshifted Distributional Models Appropriate for Response Time?" *Psychometrika* 70 (2): 377–81. https://doi.org/10.1007/s11336-005-1297-7.

Rouder, Jeffrey N, Julia M. Haaf, and Joachim Vandekerckhove. 2018. "Bayesian Inference for Psychology, Part IV: Parameter Estimation and Bayes Factors." *Psychonomic Bulletin & Review* 25 (1): 102–13. https://doi.org/https://doi.org/10.3758/s13423-017-1420-7.

Rouder, Jeffrey N, Richard D Morey, Paul L Speckman, and Jordan M Province. 2012. "Default Bayes Factors for Anova Designs." *Journal of Mathematical Psychology* 56 (5): 356–74.

Rouder, Jeffrey N, Jordan M Province, Richard D Morey, Pablo Gomez, and Andrew Heathcote. 2015. "The Lognormal Race: A Cognitive-Process Model of Choice and Latency with Desirable Psychometric Properties." *Psychometrika* 80 (2): 491–513. https://doi.org/10.1007/s11336-013-9396-3.

Rouder, Jeffrey N, Paul L Speckman, Dongchu Sun, Richard D Morey, and Geoffrey Iverson. 2009. "Bayesian t Tests for Accepting and Rejecting the Null Hypothesis." *Psychonomic Bulletin & Review* 16 (2): 225–37. https://doi.org/https://doi.org/10.3758/PBR.16.2.225.

Royall, Richard. 1997. *Statistical Evidence: A Likelihood Paradigm.* Chapman & Hall/CRC Press.

Säilynoja, Teemu, Paul-Christian Bürkner, and Aki Vehtari. 2022. "Graphical Test for Discrete Uniformity and Its Applications in Goodness-of-Fit Evaluation and Multiple Sample Comparison." *Statistics and Computing* 32 (2): 1–21. https://doi.org/https://doi.org/10.1007/s11222-022-10090-6.

Salas, Saturnino L., Garret J. Etgen, and Einar Hille. 2003. *Calculus: One and Several Variables.* Ninth. John Wiley & Sons.

Salvatier, John, Thomas V. Wiecki, and Christopher Fonnesbeck. 2016. "Probabilistic Programming in Python Using PyMC3." *PeerJ Computer Science* 2 (April): e55. https://doi.org/10.7717/peerj-cs.55.

Schad, Daniel J., Michael J. Betancourt, and Shravan Vasishth. 2020. "Toward a Principled Bayesian Workflow in Cognitive Science." *Psychological Methods* 26 (1): 103–26. https://doi.org/https://doi.org/10.1037/met0000275.

Schad, Daniel J., Bruno Nicenboim, Paul-Christian Bürkner, Michael J. Betancourt, and Shravan Vasishth. 2021. "Workflow Techniques for the Robust Use of Bayes Factors." *arXiv Preprint arXiv:2103.08744*.

Schad, Daniel J., Bruno Nicenboim, and Shravan Vasishth. 2023. "Data Aggregation Can Lead to Biased Inferences in Bayesian Linear Mixed Models and Bayesian Analysis of Variance." *Psychological Methods.* https://doi.org/https://doi.org/10.1037/met0000621.

Schad, Daniel J., Shravan Vasishth, Sven Hohenstein, and Reinhold Kliegl. 2020. "How to Capitalize on a Priori Contrasts in Linear (Mixed) Models: A Tutorial." *Journal of Memory and Language* 110 (February): 104038. https://doi.org/10/gf9tjp.

Scheipl, Fabian, Thomas Kneib, and Ludwig Fahrmeir. 2013. "Penalized Likelihood and Bayesian Function Selection in Regression Models." *AStA Advances in Statistical Analysis* 97 (4): 349–85. https://doi.org/10.1007/s10182-013-0211-3.

Schmidt, Oliver, Edgar Erdfelder, and Daniel W Heck. 2023. "How to Develop, Test, and Extend Multinomial Processing Tree Models: A Tutorial." *Psychological Methods*, July. https://doi.org/10.1037/met0000561.

Schoknecht, Pia, Himanshu Yadav, and Shravan Vasishth. 2025. "Do Syntactic and Semantic Similarity Lead to Interference Effects? Evidence from Self-Paced Reading and Event-Related Potentials Using German." *Journal of Memory and Language.*

Schönbrodt, Felix D., and Eric-Jan Wagenmakers. 2018. "Bayes Factor Design Analysis: Planning for Compelling Evidence." *Psychonomic Bulletin & Review* 25 (1): 128–42. https://doi.org/https://doi.org/10.3758/s13423-017-1230-y.

Seber, George A. F., and Allen J. Lee. 2003. *Linear Regression Analysis.* 2nd Edition. John Wiley & Sons.

Sekulovski, Nikola, Maarten Marsman, and Eric-Jan Wagenmakers. 2024. "A Good Check on the Bayes Factor." *PsyArXiv*, March. https://doi.org/10.31234/osf.io/59gj8.

Shiffrin, Richard M, Michael D Lee, Woojae Kim, and Eric-Jan Wagenmakers. 2008. "A Survey of Model Evaluation Approaches with a Tutorial on Hierarchical Bayesian Methods." *Cognitive Science: A Multidisciplinary Journal* 32 (8): 1248–84. https://doi.org/10.1080/03640210802414826.

Sidi, Jonathan, and Daniel Polhamus. 2020. *TexPreview: Compile and Preview Snippets of "Latex"*. https://CRAN.R-project.org/package=texPreview.

Singmann, Henrik, Ben Bolker, Jake Westfall, Frederik Aust, and Mattan S. Ben-Shachar. 2020. *Afex: Analysis of Factorial Experiments.* https://CRAN.R-project.org/package=afex.

Sivula, Tuomas, Måns Magnusson, and Aki Vehtari. 2020. "Uncertainty in Bayesian Leave-One-Out Cross-Validation Based Model Comparison." *arXiv Preprint arXiv:2008.10296.*

Smith, Jared B., and William H. Batchelder. 2010. "Beta-MPT: Multinomial Processing Tree Models for Addressing Individual Differences." *Journal of Mathematical Psychology* 54 (1): 167–83.

Soetaert, Karline, and Peter M. J. Herman. 2009. *A Practical Guide to Ecological Modelling. Using R as a Simulation Platform.* Springer.

Sorensen, Tanner, Sven Hohenstein, and Shravan Vasishth. 2016. "Bayesian Linear Mixed Models Using Stan: A Tutorial for Psychologists, Linguists, and Cognitive Scientists." *Quantitative Methods for Psychology* 12 (3): 175–200.

Spector, Robert H. 1990. "The Pupils." In *Clinical Methods: The History, Physical, and Laboratory Examinations*, edited by H. Kenneth Walker, W. Dallas Hall, and J. Willis Hurst, 3rd ed. Butterworths.

Spector, Tim D., and Simon G. Thompson. 1991. "The Potential and Limitations of Meta-Analysis." *Journal of Epidemiology and Community Health* 45 (2): 89.

Spiegelhalter, David J. 2024. "Why Probability Probably Doesn't Exist (but It Is Useful to Act Like It Does)." *Nature* 636 (8043): 560–63.

Spiegelhalter, David J., Keith R. Abrams, and Jonathan P. Myles. 2004. *Bayesian Approaches to Clinical Trials and Health-Care Evaluation.* Vol. 13. John Wiley & Sons.

Spiegelhalter, David J., Laurence S. Freedman, and Mahesh K. B. Parmar. 1994. "Bayesian Approaches to Randomized Trials." *Journal of the Royal Statistical Society. Series A (Statistics in Society)* 157 (3): 357–416.

Spurdle, Abby. 2020. *Barsurf: Heatmap-Related Plots and Smooth Multiband Color Interpolation.* `https://CRAN.R-project.org/package=barsurf`.

———. 2021. *Bivariate: Bivariate Probability Distributions.* `https://sites.google.com/site/spurdlea/r`.

Spurdle, Abby, and Emil Bode. 2020. *Intoo: Minimal Language-Like Extensions.* `https://CRAN.R-project.org/package=intoo`.

Stan Development Team. 2024. "Stan Modeling Language Users Guide and Reference Manual, Version 2.32." `https://mc-stan.org/docs/2_35/`.

Steyer, Rolf, and Werner Nagel. 2017. *Probability and Conditional Expectation: Fundamentals for the Empirical Sciences.* Vol. 5. John Wiley & Sons.

Stone, Kate, Bruno Nicenboim, Shravan Vasishth, and Frank Roesler. 2023. "Understanding the Effects of Constraint and Predictability in ERP." *Neurobiology of Language.* `https://doi.org/10.1162/nol_a_00094`.

Stroop, J. Ridley. 1935. "Studies of Interference in Serial Verbal Reactions." *Journal of Experimental Psychology* 18 (6): 643.

Štrumbelj, Erik, Alexandre Bouchard-Côté, Jukka Corander, Andrew Gelman, Håvard Rue, Lawrence Murray, Henri Pesonen, Martin Plummer, and Aki Vehtari. 2024. "Past, Present and Future of Software for Bayesian Inference." *Statistical Science* 39 (1): 46–61.

Sutton, Alexander J., Nicky J. Welton, Nicola Cooper, Keith R. Abrams, and A. E. Ades. 2012. *Evidence Synthesis for Decision Making in Healthcare.* Vol. 132. John Wiley & Sons.

Tabor, Whitney, Bruno Galantucci, and Daniel Richardson. 2004. "Effects of Merely Local Syntactic Coherence on Sentence Processing." *Journal of Memory and Language* 50: 355–70.

Tabor, Whitney, and Michael K. Tanenhaus. 1999. "Dynamical Models of Sentence Processing." *Cognitive Science* 23 (4): 491–515.

Talts, Sean, Michael J. Betancourt, Daniel P. Simpson, Aki Vehtari, and Andrew Gelman. 2018. "Validating Bayesian Inference Algorithms with Simulation-Based Calibration." *arXiv Preprint arXiv:1804.06788.*

Tendeiro, Jorge N., Henk A. L. Kiers, Rink Hoekstra, Tsz Keung Wong, and Richard D Morey. 2023. "Diagnosing the Misuse of the Bayes Factor in Applied Research." *Advances in Methods and Practices in Psychological Science.*

Turner, Brandon M, Per B. Sederberg, Scott D. Brown, and Mark Steyvers. 2013. "A Method for Efficiently Sampling from Distributions with Correlated Dimensions." *Psychological Methods* 18 (3): 368.

Turner, R. M., David J. Spiegelhalter, G. Smith, and Simon G. Thompson. 2008. "Bias Modelling in Evidence Synthesis." *Journal of the Royal Statistical Society: Series A (Statistics in Society)* 172 (1): 21–47. https://doi.org/https://doi.org/10.1111/j.1467-985X.2008.00547.x.

Ulrich, Rolf, and Jeff Miller. 1993. "Information Processing Models Generating Lognormally Distributed Reaction Times." *Journal of Mathematical Psychology* 37 (4): 513–25. https://doi.org/10.1006/jmps.1993.1032.

Vaidyanathan, Ramnath, Yihui Xie, J. J. Allaire, Joe Cheng, and Kenton Russell. 2023. *htmlwidgets: HTML Widgets for R.* https://github.com/ramnathv/htmlwidgets.

van Doorn, Johnny, Frederik Aust, Julia M. Haaf, Angelika Stefan, and Eric-Jan Wagenmakers. 2021. "Bayes Factors for Mixed Models." *Computational Brain and Behavior.* https://doi.org/https://doi.org/10.1007/s42113-021-00113-2.

Vasishth, Shravan, Zhong Chen, Qiang Li, and Gueilan Guo. 2013. "Processing Chinese Relative Clauses: Evidence for the Subject-Relative Advantage." *PLoS ONE* 8 (10): 1–14. https://doi.org/https://doi.org/10.1371/journal.pone.0077006.

Vasishth, Shravan, Nicolas Chopin, Robin Ryder, and Bruno Nicenboim. 2017. "Modelling Dependency Completion in Sentence Comprehension as a Bayesian Hierarchical Mixture Process: A Case Study Involving Chinese Relative Clauses." In *Proceedings of Cognitive Science Conference.* London, UK. https://arxiv.org/abs/1702.00564v2.

Vasishth, Shravan, and Felix Engelmann. 2022. *Sentence Comprehension as a Cognitive Process: A Computational Approach.* Cambridge University Press. https://books.google.de/books?id=6KZKzgEACAAJ.

Vasishth, Shravan, and Andrew Gelman. 2021. "How to Embrace Variation and Accept Uncertainty in Linguistic and Psycholinguistic Data Analysis." *Linguistics* 59 (5): 1311–42. https://doi.org/10.1515/ling-2019-0051.

Vasishth, Shravan, Lena A. Jäger, and Bruno Nicenboim. 2017. "Feature overwriting as a finite mixture process: Evidence from comprehension data." In *Proceedings of MathPsych/ICCM Conference.* Warwick, UK. https://arxiv.org/abs/1703.04081.

Vasishth, Shravan, Daniela Mertzen, Lena A. Jäger, and Andrew Gelman. 2018. "The Statistical Significance Filter Leads to Overoptimistic Expectations of Replicability." *Journal of Memory and Language* 103: 151–75. https://doi.org/https://doi.org/10.1016/j.jml.2018.07.004.

Vasishth, Shravan, Daniel J. Schad, Audrey Bürki, and Reinhold Kliegl. 2021. "Linear Mixed Models for Linguistics and Psychology: A Comprehensive Introduction." https://vasishth.github.io/Freq_CogSci/.

Vasishth, Shravan, Himanshu Yadav, Daniel J. Schad, and Bruno Nicenboim. 2022. "Sample Size Determination for Bayesian Hierarchical Models Commonly Used in Psycholinguistics." *Computational Brain and Behavior.* https://doi.org/https://doi.org/10.1007/s42113-021-00125-y.

Vehtari, Aki. 2022. "Cross-validation FAQ." https://web.archive.org/web/20221219223947/https://avehtari.github.io/modelselection/CV-FAQ.html.

Vehtari, Aki, Andrew Gelman, and Jonah Gabry. 2017. "Practical Bayesian Model Evaluation

Using Leave-One-Out Cross-Validation and WAIC." *Statistics and Computing* 27 (5): 1413–32. https://doi.org/10.1007/s11222-016-9696-4.

Vehtari, Aki, Andrew Gelman, Daniel P. Simpson, Bob Carpenter, and Paul-Christian Bürkner. 2021. "Rank-Normalization, Folding, and Localization: An Improved $\widehat{R}$ for Assessing Convergence of MCMC." *Bayesian Analysis* 16 (2): 667–718. https://doi.org/10.1214/20-BA1221.

Vehtari, Aki, and Jouko Lampinen. 2002. "Bayesian Model Assessment and Comparison Using Cross-Validation Predictive Densities." *Neural Computation* 14 (10): 2439–68. https://doi.org/10.1162/08997660260293292.

Vehtari, Aki, and Janne Ojanen. 2012. "A Survey of Bayesian Predictive Methods for Model Assessment, Selection and Comparison." *Statistical Surveys* 6 (0): 142–228. https://doi.org/10.1214/12-ss102.

Vehtari, Aki, Daniel Simpson, Andrew Gelman, Yuling Yao, and Jonah Gabry. 2024. "Pareto Smoothed Importance Sampling." *Journal of Machine Learning Research* 25: 1–57.

Vehtari, Aki, Daniel P. Simpson, Yuling Yao, and Andrew Gelman. 2019. "Limitations of 'Limitations of Bayesian Leave-One-Out Cross-Validation for Model Selection'." *Computational Brain & Behavior* 2 (1): 22–27. https://doi.org/10.1007/s42113-018-0020-6.

Venables, William N., and Brian D. Ripley. 2002. *Modern Applied Statistics with S-PLUS*. Springer.

Verhagen, Josine, and Eric-Jan Wagenmakers. 2014. "Bayesian Tests to Quantify the Result of a Replication Attempt." *Journal of Experimental Psychology: General* 143 (4): 1457–75. https://doi.org/10.1037/a0036731.

Wagenmakers, Eric-Jan, Angelos-Miltiadis Krypotos, Amy H Criss, and Geoffrey Iverson. 2012. "On the Interpretation of Removable Interactions: A Survey of the Field 33 Years After Loftus." *Memory & Cognition* 40: 145–60.

Wagenmakers, Eric-Jan, Michael D Lee, Jeffrey N Rouder, and Richard D Morey. 2020. "The Principle of Predictive Irrelevance or Why Intervals Should Not Be Used for Model Comparison Featuring a Point Null Hypothesis." In *The Theory of Statistics in Psychology: Applications, Use, and Misunderstandings*, edited by Craig W. Gruber, 111–29. Springer International Publishing. https://doi.org/10.31234/osf.io/rqnu5.

Wagenmakers, Eric-Jan, Tom Lodewyckx, Himanshu Kuriyal, and Raoul P. P. P. Grasman. 2010. "Bayesian Hypothesis Testing for Psychologists: A Tutorial on the Savage–Dickey Method." *Cognitive Psychology* 60 (3): 158–89. https://doi.org/https://doi.org/10.1016/j.cogpsych.2009.12.001.

Wahn, Basil, Daniel P. Ferris, W. David Hairston, and Peter König. 2016. "Pupil Sizes Scale with Attentional Load and Task Experience in a Multiple Object Tracking Task." *PLOS ONE* 11 (12): e0168087. https://doi.org/10.1371/journal.pone.0168087.

Walker, Grant M, Gregory Hickok, and Julius Fridriksson. 2018. "A Cognitive Psychometric Model for Assessment of Picture Naming Abilities in Aphasia." *Psychological Assessment* 6: 809–26. https://doi.org/10.1037/pas0000529.

Wang, Fei, and Alan E. Gelfand. 2002. "A Simulation-Based Approach to Bayesian Sample Size Determination for Performance Under a Given Model and for Separating Models." *Statistical Science*, 193–208. http://www.jstor.org/stable/3182824.

Wang, Wei, and Andrew Gelman. 2014. "Difficulty of Selecting Among Multilevel Models Using Predictive Accuracy." *Statistics at Its Interface* 7: 1–8. https://dx.doi.org/10.4310/SII.2015.v8.n2.a3.

Warren, Tessa, and Edward Gibson. 2002. "The Influence of Referential Processing on Sentence Complexity." *Cognition* 85: 79–112. https://doi.org/10.1016/S0010-0277(02)00087-2.

Wasserstein, Ronald L., and Nicole A. Lazar. 2016. "The ASA's Statement on p-Values: Context, Process, and Purpose." *The American Statistician* 70 (2): 129–33. https://doi.org/https://doi.org/10.1080/00031305.2016.1154108.

Wickelgren, Wayne A. 1977. "Speed-Accuracy Tradeoff and Information Processing Dynamics." *Acta Psychologica* 41 (1): 67–85. https://doi.org/10.1016/0001-6918(77)90012-9.

Wickham, Hadley. 2019a. *forcats: Tools for Working with Categorical Variables (Factors).* https://CRAN.R-project.org/package=forcats.

———. 2019b. *stringr: Simple, Consistent Wrappers for Common String Operations.* https://CRAN.R-project.org/package=stringr.

Wickham, Hadley, Mara Averick, Jennifer Bryan, Winston Chang, Lucy D'Agostino McGowan, Romain François, Garrett Grolemund, et al. 2019. "Welcome to the tidyverse." *Journal of Open Source Software* 4 (43): 1686. https://doi.org/10.21105/joss.01686.

Wickham, Hadley, Winston Chang, Lionel Henry, Thomas Lin Pedersen, Kohske Takahashi, Claus O. Wilke, Kara Woo, Hiroaki Yutani, and Teun van den Brand. 2024. *ggplot2: Create Elegant Data Visualisations Using the Grammar of Graphics.* https://ggplot2.tidyverse.org.

Wickham, Hadley, Romain François, Lionel Henry, Kirill Müller, and Davis Vaughan. 2023. *dplyr: A Grammar of Data Manipulation.* https://dplyr.tidyverse.org.

Wickham, Hadley, and Lionel Henry. 2023. *purrr: Functional Programming Tools.* https://purrr.tidyverse.org/.

Wickham, Hadley, Jim Hester, and Romain Francois. 2018. *readr: Read Rectangular Text Data.* https://CRAN.R-project.org/package=readr.

Wickham, Hadley, Davis Vaughan, and Maximilian Girlich. 2024. *Tidyr: Tidy Messy Data.* https://tidyr.tidyverse.org.

Wilke, Claus O. 2024. *cowplot: Streamlined Plot Theme and Plot Annotations for 'Ggplot2'.* https://wilkelab.org/cowplot/.

Wilson, Robert C., and Anne G. E. Collins. 2019. "Ten Simple Rules for the Computational Modeling of Behavioral Data." Edited by Timothy E Behrens. *eLife* 8 (November): e49547. https://doi.org/10.7554/eLife.49547.

Wolodzko, Tymoteusz. 2023. *extraDistr: Additional Univariate and Multivariate Distributions.* https://github.com/twolodzko/extraDistr.

Wu, Fuyun, Elsi Kaiser, and Shravan Vasishth. 2017. "Effects of Early Cues on the Processing of Chinese Relative Clauses: Evidence for Experience-Based Theories." *Cognitive Science* 42: 1101–33.

Xie, Yihui. 2019. *servr: A Simple Http Server to Serve Static Files or Dynamic Documents.* `https://CRAN.R-project.org/package=servr`.

———. 2024a. *bookdown: Authoring Books and Technical Documents with R Markdown.* `https://github.com/rstudio/bookdown`.

———. 2024b. *knitr: A General-Purpose Package for Dynamic Report Generation in R.* `https://yihui.org/knitr/`.

Xie, Yihui, Joe Cheng, and Xianying Tan. 2019. *DT: A Wrapper of the Javascript Library 'Datatables'.* `https://CRAN.R-project.org/package=DT`.

Yackulic, Charles B., Michael Dodrill, Maria Dzul, Jamie S. Sanderlin, and Janice A. Reid. 2020. "A Need for Speed in Bayesian Population Models: A Practical Guide to Marginalizing and Recovering Discrete Latent States." *Ecological Applications* 30 (5): e02112. `https://doi.org/https://doi.org/10.1002/eap.2112`.

Yadav, Himanshu, Dario Paape, Garrett Smith, Brian W. Dillon, and Shravan Vasishth. 2022. "Individual Differences in Cue Weighting in Sentence Comprehension: An evaluation using Approximate Bayesian Computation." *Open Mind.* `https://doi.org/https://doi.org/10.1162/opmi_a_00052`.

Yadav, Himanshu, Garrett Smith, Sebastian Reich, and Shravan Vasishth. 2023. "Number Feature Distortion Modulates Cue-Based Retrieval in Reading." *Journal of Memory and Language* 129. `https://doi.org/10.1016/j.jml.2022.104400`.

Yao, Yuling, Aki Vehtari, Daniel P. Simpson, and Andrew Gelman. 2017. "Using Stacking to Average Bayesian Predictive Distributions." *Bayesian Analysis.* `https://doi.org/10.1214/17-BA1091`.

———. 2018. "Yes, but Did It Work?: Evaluating Variational Inference." In *International Conference on Machine Learning*, 5581–90. PMLR.

Yarkoni, Tal. 2020. "The Generalizability Crisis." *Behavioral and Brain Sciences*, 1–37. `https://doi.org/10.1017/S0140525X20001685`.

Yellott, John I. 1967. "Correction for Guessing in Choice Reaction Time." *Psychonomic Science* 8 (8): 321–22. `https://doi.org/10.3758/BF03331682`.

———. 1971. "Correction for Fast Guessing and the Speed-Accuracy Tradeoff in Choice Reaction Time." *Journal of Mathematical Psychology* 8 (2): 159–99. `https://doi.org/10.1016/0022-2496(71)90011-3`.

Zhu, Hao. 2024. *KableExtra: Construct Complex Table with 'Kable' and Pipe Syntax.* `http://haozhu233.github.io/kableExtra/`.